OXIDATIVE STRESS AND CHRONIC DEGENERATIVE DISEASES - A ROLE FOR ANTIOXIDANTS

Edited by **José A. Morales-González**

Oxidative Stress and Chronic Degenerative Diseases - A Role for Antioxidants
http://dx.doi.org/10.5772/45722
Edited by José A. Morales-González

Contributors

Maria De Lourdes Reis Giada, Nalini Rajamannan, Jose Antonio Morales-Gonzalez, Jose Luis Silencio-Barrita, Maria Del Socorro Santiago Sanchez, María Del Carmen Valadez Vega, Luis Delgado Olivares, Maria Teresa Sumaya-Martínez, Manuel Sánchez Gutierrez, Clara Zúñiga Pérez, José Roberto Villagomez Ibarra, Ernesto Alanís García, Zuli Calderon Ramos, Esther Ramírez Moreno, Mirandeli Bautista, Borut Poljsak, Irina Milisav, Antonio Cilla, Eva María Molina-Trinidad, Alejandro Chehue Romero, Elena Guadalupe Olvera Hernández, Telma Flores Cerón, Angelina Álvarez Chávez, Tomas Alejandro Fregoso Aguilar, Jorge Alberto Mendoza Perez, Manuel Soriano, Claudia Calzada, Maria-Luisa Lazo-De-La-Vega-Monroy, E. Osiris Madrigal-Santillan, Sandra Cruz, Karla Guadalupe Pérez-Avila, Mario Nava-Villalba, Maribel Liñan-Fernández, Germán González-Pérez, Marco Torres-Carmona, Cesar Esquivel-Chirino, Jaime Esquivel, Luis Hernández Mora, Delina Montes Sánchez, Alejandro Zentella-Dehesa, Jose Luis Ventura-Gallegos, María De Lourdes Segura-Valdez, Luis F. Jiménez-García, Lourdes Teresa Agredano-Moreno, Tomás Nepomuceno-Mejía, Rogelio Jaime Fragoso-Soriano, Georgina Álvarez-Fernández, Alma Zamora-Cura, Reyna Lara-Martínez, Glenda Gobe, David M Small

Published by InTech

Janeza Trdine 9, 51000 Rijeka, Croatia

Publishing Process Manager Sandra Bakic
Technical Editor InTech DTP team
Cover InTech Design team

Additional hard copies can be obtained from orders@intechopen.com

Oxidative Stress and Chronic Degenerative Diseases - A Role for Antioxidants, Edited by José A. Morales-González
p. cm.
ISBN 978-953-51-1123-8

For EU product safety concerns:
IN TECH d.o.o., Prolaz Marije Krucifikse Kozulić 3, 51000 Rijeka, Croatia,
info@intechopen.com or visit our website at intechopen.com.

Meet the editor

José Antonio Morales-González carried out his undergraduate rotating internship year at the Hospital de Jesús Public Welfare Institution (IAP), graduating as Surgeon Physician in 1997 from the Faculty of Higher Education (FES)-Iztacala National Autonomous University of Mexico (UNAM). He engaged in Doctoral studies in Biological Sciences (with a specialty in Cellular Biology) at the UNAM (1997-2001). Dr. Morales-Gonzalez has been awarded diverse recognitions: Hospital de Jesus IAP awards (1995 and 1996); Gustavo Baz Prada medal by the UNAM (1997); the Alfonso Caso medal for academic merit by the UNAM (2004); distinguished by the National System of Researchers (SNI) as National Researcher level 1 (2010-2016). He is the author of 42 internationally and nationally published articles. He is also editor and coordinator of 20 specialized books. Dr. Morales-González is a Full-time Titular Professor-Researcher C at the Area Académica de Farmacia, ICSa-UAEH, Mexico.

Contents

Preface

Aerobic organisms, such as humans, possess several metabolic pathways in the process of obtaining energy, utilizing oxygen as final electron acceptor. Thus, this is indispensable for the condition of life to exist. Paradoxically, under certain conditions in which reduction intermediaries, mainly of oxygen, accumulate, Free radicals (FR) are produced, which are capable of producing damage at a molecular level, which manifests in diverse chronic pathologies.

Oxidative stress constitutes an alteration produced by disequilibrium between generation of FR and the antioxidant system, which can lead to a damage state, in particular of the biomolecules. Due to the wide variety of reactions produced by FR, these are related with the development and evolution of diverse illnesses such as the chronic degenerative diseases, for example, atherosclerotic disease, high blood pressure, renal disorders, and obesity, in which FR plays a primordial role in the development of long-term complications of these illnesses.

Diet and nutrition are very important in the promotion and maintenance of health throughout life. For some time, the function has been recognized of diet and nutrition as determining factors of non-transmittable diseases; thus, a healthy diet is one of the pillars of health that has become consolidated in our lifestyle over the past several years. A search has been conducted for all of the properties of foods that are beneficial in increasing or maintaining our state of health. Likewise, there are natural and synthetic molecules that are capable of inactivating FR; these FR scavengers are found classified in groups or families of compounds that are in general denominated "antioxidants". The main objective of mechanisms of defense is that of transforming FR into less-damaging products or to neutralize their malignancy completely, performing this by means of a mechanism called redox potential, through oxido-reduction reactions in which a reducer agent participates that donates electrons or through an oxidating agent that removes electrons from FR.

There are elements in the diet, in addition to their nutritional characteristics, in which the property is recognized for being antioxidant agents. Among the many studied, we find vitamin C, vitamin E, vitamin D, vitamin A, some amino acids, the flavonoids, and some oligoelements. All of these antioxidant elements represent an alternative for the treatment and prevention of chronic degenerative diseases, which represent a very high morbid-mortality rate, worldwide.

The work presented here responds to the need of finding, in a sole document, the effect of oxidative stress at different levels, as well as treatment with antioxidants to revert and diminish damage. On the other hand, it is noteworthy that the work will be published for health professionals and researchers who are expert in the theme, and that it contains current, scientifically based information. Thus, I am convinced that the work carried out will be of utmost usefulness for the active health professional as well as for the health professional in training, with the purpose of creating a novel panorama on study of the theme as well as showing an alternative in the treatment of the chronic diseases that are affecting our population to such a great extent.

Therefore, the objective of this book is to understand the mechanism by which free radicals contribute to the development of complications in chronic degenerative diseases. On knowing how FR are generated, the interaction of FR with cells, and the reactions in which they are involved, it will be possible to establish novel treatment alternatives, which could comprise antioxidants, providing the scientific community with the opportunity to make their researches known and enabling the readers to deepen or broaden their knowledge in this field.

Finally, I dedicate this book to Julia González-González, with all my love for your teachings and example.

Prof. Dr. José Antonio Morales-González
Área Académica de Farmacia,
Instituto de Ciencias de la Salud,
Universidad Autónoma del Estado de Hidalgo, México

Section 1

Cell Biology, Chemical Free Radicals and Antioxidant Defenses

Chapter 1

Cell Nanobiology

María de Lourdes Segura-Valdez,
Lourdes T. Agredano-Moreno,
Tomás Nepomuceno-Mejía,
Rogelio Fragoso-Soriano,
Georgina Álvarez-Fernández, Alma Zamora-Cura,
Reyna Lara-Martínez and Luis F. Jiménez-García

Additional information is available at the end of the chapter

http://dx.doi.org/10.5772/52003

1. Introduction

1.1. Cell nanobiology

We define cell nanobiology as an emergent scientific area trying to approach the study of the *in situ* cell processes ocurring at the nanoscale. Therefore, it is part of cell biology but mainly deals with an interphase between analytical methods such as X-ray crystallography producing models at atomic or molecular resolution, and direct nanoscale imaging with high resolution microscopes such as scanning probe microscopes, electron microscopes and super-resolution microscopes. Several cell structures are involved in nanoscale processes (Figure 1).

1.1.1. An overview of cell structure under a genomic approach of gene expression

The main flow of genetic information represented as the so-called central dogma of molecular biology *in situ*, illustrates the major secretory pathway in the cell (Figure 2).

During this pathway, nanoscale particles represent substrates of different moments. During transcription, nuclear particles are involved in transcription and processing of RNA, both, pre-mRNA and pre-rRNA. pre-mRNA is transcribed and processed in the nucleoplasm while pre-rRNA is transcribed and processed within the nucleolus, the major known ribonucleoproteins structure where ribosome biogenesis and other functions of eukaryotic cell take place. Once in the cytoplasm, translation takes place in the ribosome, also a major ribonucleoprotein particle of 10-15 nm in diameter. When the synthesized protein contains a signal peptide, it is translocated into the rough endoplasmic reticulum, helped by the signal recognition particle or SRP, another major and conserved ribonucleoprotein. The transport to Golgi apparatus by the intermediated zone and the TGN producing the three derivatives from the Golgi apparatus are mediated by vesicles [see 1].

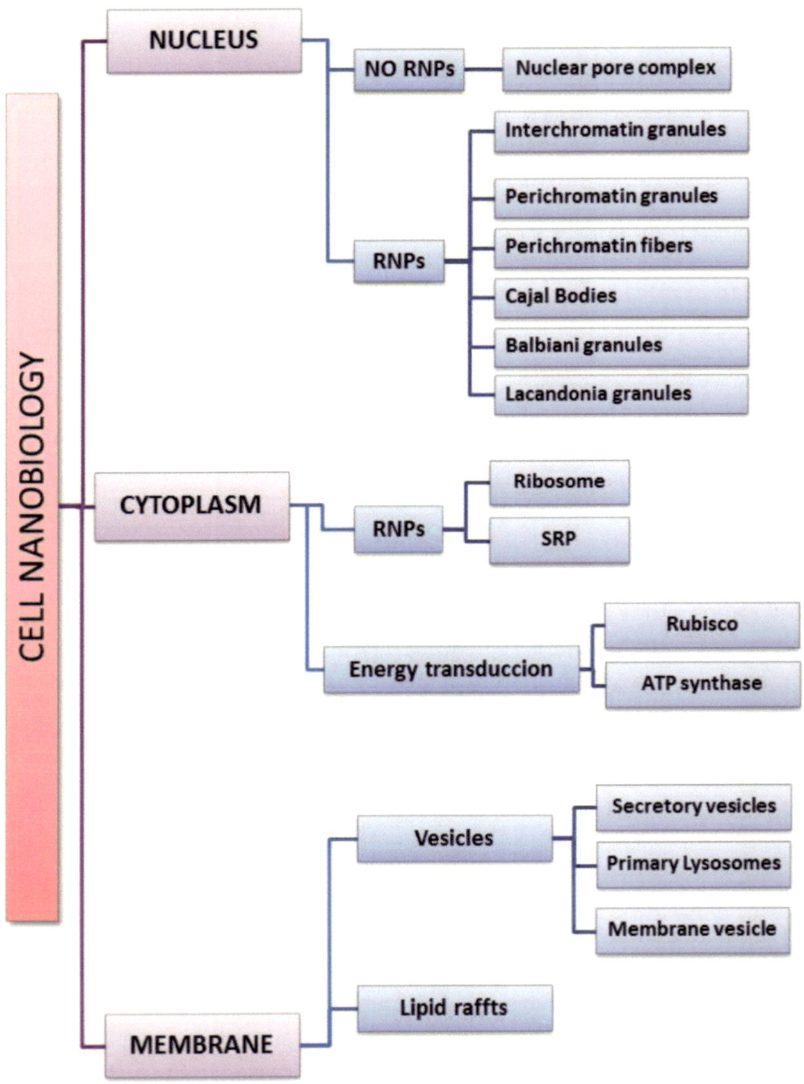

Figure 1. Cell nanobiology proposes to study cell structures using *in situ* high resolution microscopical approaches as electron and atomic force microscopy that could complement molecular and biochemical data to better understand a physiological role at the nanoscale.

http://dx.doi.org/10.5772/52003

a. Semenogelin

Semenogelin is the most abundant protein in the semen of mammals. It is a glycosylated protein that is responsible for properties such as density. As an example, the semenogelin of the tamarin *Saguinus oedipus* is used to show how the signals in the nucleic acids and proteins determine the intracellular pathways associated to that expression. Its expression includes intranuclear events as transcription by RNA polymerase II from a split gene consisting of 3 exones and 2 introns, processing of the transcript as 5′ end methylation, 3′ polyadenylation and splicing. All of them are associated to nuclear particles. Once in the cytoplasm, the mature transcript or mRNA associates to a ribosome that in turns translates the transcript. If the resulting protein contains a signal peptide, the signal recognition particle or SRP -a very well conserved RNA+protein complex- binds to it and associates to the rough endoplasmic reticulum, giving rise to the translocation process that introduces that protein to the lumen. Once there, N-glycosylation takes place at several asparagine residues following the basic rule of adjacent aminoacids showing a basic rule as Asn-X(except proline)-Ser or Asn. In *S. oedipus* semenogelin, there are 14 N-glycosylation sites. The protein then continues flowing through the Golgi apparatus or complex and at the TGN a secretory vesicle forms containing the protein that finally is secreted by the epithelial cell of seminal vesicle. The analysis of the gene sequences, as well as the transcription, processing, translation and post-translational products can predict the cell structures involved in the process [see 1].

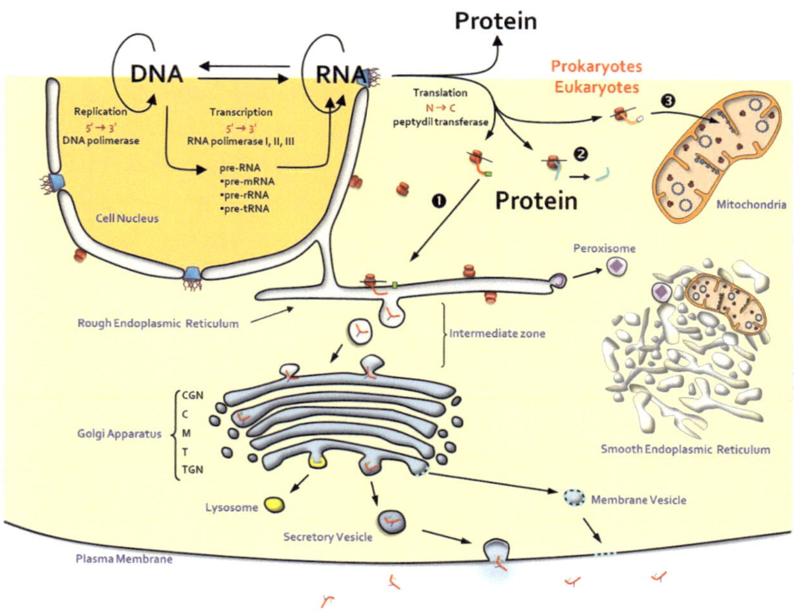

Figure 2. A general overview of the cell structure and function. The diagram illustrates the *in situ* flow of genetic information of a secretory protein encoded in the genome within the cell nucleus. A gene is copied into a pre-mRNA that is processed to mRNA within the nucleus. mRNA in the cytoplasm may contain a signal sequence that allows entrance to rough endoplasmic reticulum and further to Golgi complex. The protein inside a vesicle is secreted out of the cell.

1.1.2. Some nanoscale cell structures

There are many cell structures or products made by cells that could be analyzed under the present approach. Some of them are indicted in Figure 1, but there are others as extracellular matriz components, cytoskeleton elements, etc.; virus are also nanometric structures associated always to cell organelles. Here we will give an overview of some of the cell components, as examples.

a. Nuclear particles

In eukaryote cells, transcription and processing mainly takes place within the cell nucleus, associated to nuclear particles that are well known since a method for ribonucleoprotein (RNP) structures was described in 1969 [2]. These particles are few nanometers in diameter or lengh. To date, several nuclear RNPs have been described including involved in mRNA metabolism: perichromatin fibers, perichromatin granules, interchromatin granules in mammals. In insects, Balbiani ring granules are well known structures [3]. In 1992 Lacandonia granules were described for some plants [4]. In addition, other nuclear bodies around 300-400 nm in diameter have been described involved in gene expression. As for rRNA transcription and processing, the nucleolus is a nuclear organelle containing pre-ribosomes in the granular component that are about 10-20 nm in diameter.

b. Rough endoplasmic reticulum particles

i. The ribosome

Ribosomes are the universal ribonucleoprotein particles that translate the genetic code into proteins. The shape and dimensions of the ribosome were first visualized by electron microscopy [6-8]. Ribosomes have diameters of about 25 nanometers in size and are roughly two-thirds RNA and one-third protein. All ribosomes have two subunits, one about twice the mass of the other. The ribosome basic structure and functions are well-known. There are 70S ribosomes common to prokaryotes and 80S ribosomes common to eukaryotes. The bacterial ribosome is composed of 3 RNA molecules and more than 50 proteins. In humans, the small ribosome unit has 1 large RNA molecule and about 32 proteins; the large subunit has 3 RNA molecules, and about 46 proteins. Each subunit has thousands of nucleotides and amino acids, with hundreds of thousands of atoms. The small subunit (0.85 MDa) initiates mRNA engagement, decodes the message, governs mRNA and tRNA translocation, and controls fidelity of codon–anticodon interactions and the large subunit catalyzes peptide bond formation.

In 1980, the first three-dimensional crystals of the ribosomal 50S subunit from the thermophile bacterium *Geobacillus stearothermophilus* were reported [9]. Since then, ribosome crystallography advanced rapidly. To date crystal structures have been determined for the large ribosomal subunit from the archaeon *Haloarcula marismortui* at 2.4 Å [10] and the 30S ribosomal subunit from *Thermus thermophilus* [11]. The structure of the entire 70S ribosome in complex with tRNA ligands (at 5.5Å resolution) emerged shortly after the structures of the initial subunits [12].

http://dx.doi.org/10.5772/52003

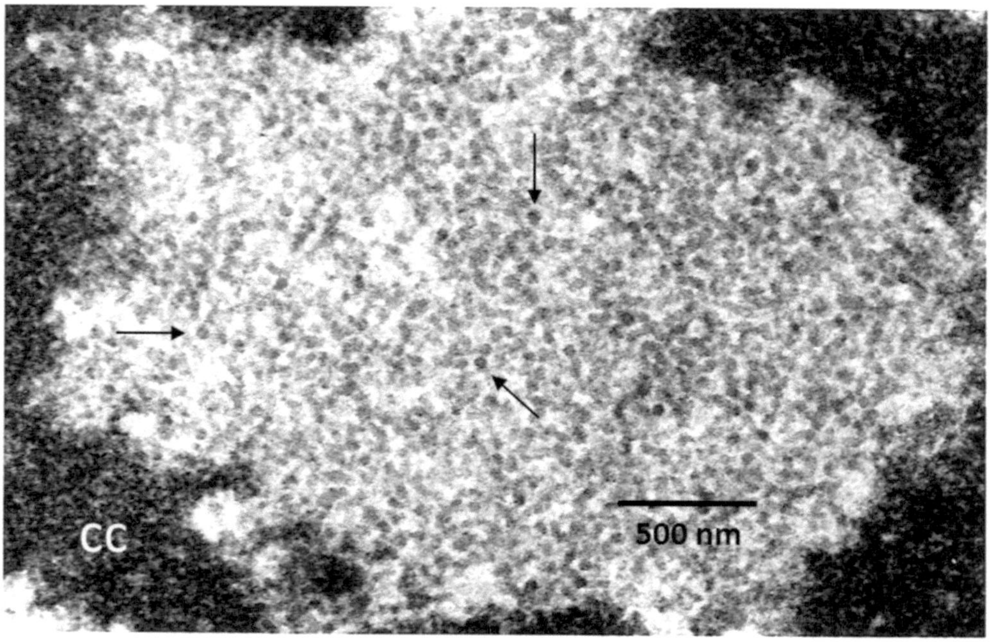

Figure 3. Lacandonia granules are nanoscale (32 nm in diameter) and abundant particles (arrows) in the nucleoplasm of the plant *Lacandonia schismatica*. Transmission electron microscopy image of a cell in the teguments of the flower. Cc, compact chromatin.

Snapshots of ribosome intermediates provided by cryo-EM and x-ray crystallography, associated translation factors, and transfer RNA (tRNA) have allowed dynamic aspects of protein translation to be reconstructed. For example, recent cryo-EM reconstructions of translating ribosomes allowed direct visualization of the nascent polypeptide chain inside the ribosomal tunnel at subnanometer resolution [13-15]. The dimension of the ribosomal tunnel in bacterial, archaeal, and eukaryotic cytoplasmic ribosomes is conserved in evolution [16 -18]. The ribosomal tunnel in the large ribosomal subunit is ~80 Å long, 10–20 Å wide, and predominantly composed of core rRNA [19]. The tunnel is clearly not just a passive conduit for the nascent chain, but rather a compartment in a dynamic molecular dialogue with the nascent chain. This interplay might not only affect the structure and function of the ribosome and associated factors, but also the conformation and folding of the nascent chain [20]. As the nascent polypeptide chain is being synthesized, it passes through a tunnel within the large subunit and emerges at the solvent side, where protein folding occurs.

Peptide bond formation on the bacterial ribosome and perhaps on the ribosomes from all organisms is catalyzed by ribosomal RNA as well as ribosomal protein and also by the 2′-OH group of the peptidyl-tRNA substrate in the P site. The high resolution crystal structures of two ribosomal complexes from T. thermophilus [21] revealed that ribosomal proteins L27 and L16 of the 50S subunit stabilize the CCA-ends of both tRNAs in the peptidyl-transfer reaction, suggesting that peptide chains from both these proteins take part in the catalytic mechanism of peptide bond formation.

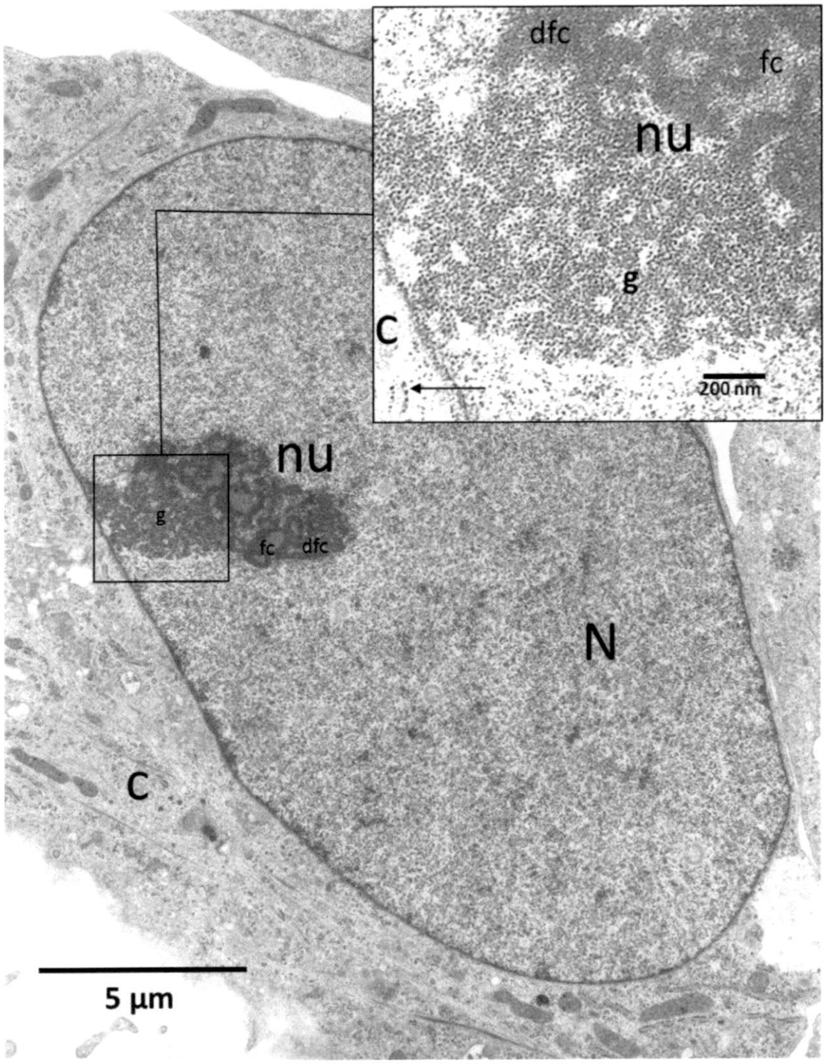

Figure 4. Nucleolus of a PtK2 cell. Within the cell nucleus (N), the nucleolus displays three different components named fibrillar center (fc), dense fibrillar component (dfc) and granular components (g). In the inset, a high magnification of the nucleolus shows granular particles or pre-ribosomes in the granular component (g). In the cytoplasm (c), ribosomes are also visible (arrow).

Ribosomes mediate protein synthesis by decoding the information carried by messenger RNAs (mRNAs) and catalyzing peptide bond formation between amino acids. When bacterial ribosomes stall on incomplete messages, the trans-translation quality control mechanism is activated by the transfer-messenger RNA bound to small protein B (tmRNA–SmpB ribo-

http://dx.doi.org/10.5772/52003

nucleoprotein complex). Trans-translation liberates the stalled ribosomes and triggers degradation of the incomplete proteins. The cryo-electron microscopy structures of tmRNA–SmpB accommodated or translocated into stalled ribosomes demonstrate how tmRNA–SmpB crosses the ribosome and how as the problematic mRNA is ejected, the tmRNA resume codon is placed onto the ribosomal decoding site by new contacts between SmpB and the nucleotides upstream of the tag-encoding sequence [22]. Recently, the crystal structure of a tmRNA fragment, SmpB and elongation factor Tu bound to the ribosome shows how SmpB plays the role of both the anticodon loop of tRNA and portions of mRNA to facilitate decoding in the absence of an mRNA codon in the A site of the ribosome [23].

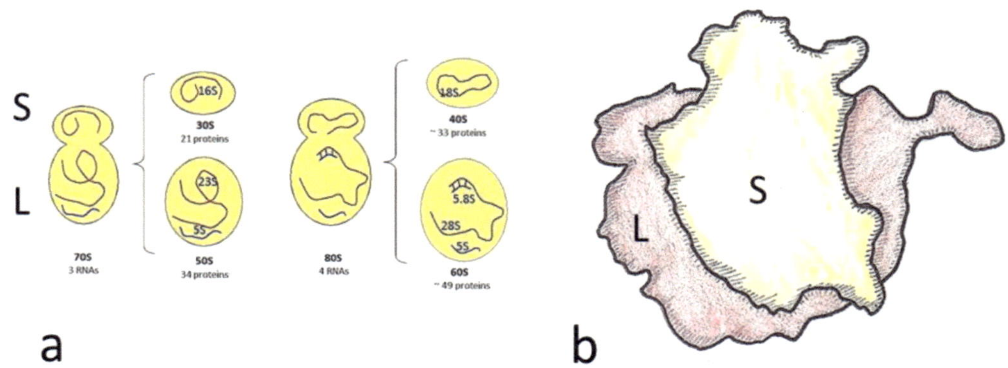

Figure 5. Representation of a prokaryotic (a) and a eukaryotic ribosome (a). Each one is an RNP is constituted by two subunits, each containing rRNA and proteins. b) a model to show the nanoscale morphology of a mammalian cytoplasmic ribosome (small [S] and large [L] subunits).

The structure of the ribosome at high resolution reveals the molecular details of the antibiotic-binding sites, explain how drugs exercise their inhibitory effects. Also, the crystal structures help us to speculate about how existing drugs might be improved, or novel drugs created, to circumvent resistance [24]. Recently, ribosome engineering has emerged as a new tool to promote new crystal forms and improve our knowledge of protein synthesis. To explore the crystallization of functional complexes of ribosomes with GTPase, a mutant 70S ribosomes were used to crystallize and solve the structure of the ribosome with EF-G, GDP and fusidic acid in a previously unobserved crystal form [25].

In contrast to their bacterial counterparts, eukaryotic ribosomes are much larger and more complex, containing additional rRNA in the form of so-called expansion segments (ES) as well as many additional r-proteins and r-protein extensions [26]. The first structural models for the eukaryotic (yeast) ribosome were built using 15-A° cryo–electron microscopy (cryo-EM) maps fitted with structures of the bacterial SSU [11] and archaeal LSU [10], thus identifying the location of a total of 46 eukaryotic r-proteins with bacterial and/or archaeal homologs as well as many ES [27].

Ribosome biogenesis is regulated by the conserved protein kinase TOR (target of rapamycin), a member of the ATM-family protein. TOR up-regulates transcription of rRNA and mRNA for ribosomal proteins in both yeast and mammals [28-30]. Recent results indicate that in yeast, conserved kinases of the LAMMER/Cdc-like and GSK-3 families function downstream of TOR complex 1 to repress ribosome and tRNA synthesis in response to nutrient limitation and other types of cellular stress [31].

ii. The signal recognition particle (SRP)

The Signal Recognition Particle (SRP) is an evolutionarily conserved rod-shaped 11S ribonucleoprotein particle, 5–6 nm wide and 23–24 nm long [32]. It comprises an essential component of the cellular machinery responsible for the co-translational targeting of proteins to their proper membrane destinations [33].

Although SRP is essential and present in all kingdoms of life maintaining its general function, structurally it shows high diversity. Vertebrates SRP consists of a single ~ 300-bp RNA (SRP RNA or 7S RNA) and six polypeptides designated SRP9, SRP14, SRP19, SRP54, SRP68 and SRP72. It can be divided into two major functional domains: the Alu domain (comprising the proteins SRP9 and -14) and the S domain (SRP19, -54, -68, and -72). The S domain functions in signal sequence recognition and SR interaction, whereas the Alu domain is required for translational arrest on signal sequence recognition [34]. In Archaea and Eucarya, the conserved ribonucleoproteic core is composed of two proteins, the accessory protein SRP19, the essential GTPase SRP54, and an evolutionarily conserved and essential SRP RNA [35]. SRP54, comprises an N-terminal domain (N, a four-helix bundle), a central GTPase domain [G, a ras-like GTPase fold, with an additional unique α-β- α insertion box domain (IBD)], and a methionine-rich C-terminal domain [36-37]. The N and G domains are structurally and functionally coupled; together, they build the NG domain that is connected to the M domain through a flexible linker [38]. The M domain anchors SRP54 to SRP RNA and carries out the principal function of signal sequence recognition [39-41]. The NG domain interacts with the SR in a GTP-dependent manner [43].

SRP is partially assembled in the nucleus and partially in the nucleolus. In agreement with that, nuclear localization for SRP proteins SRP9/14, SRP68, SRP72 and SRP19 has been determined [44]. After the transport into the nucleus the subunits bind SRP RNA and form a pre-SRP which is exported to the cytoplasm where the final protein, Srp54p, is incorporated [45-47]. Although this outline of the SRP assembly pathway has been determined, factors that facilitate this and/or function in quality control of the RNA are poorly understood [48]. SRP assembly starts during 7S RNA transcription by RNA polymerase III in the nucleolus, by binding of the SRP 9/14 heterodimer and formation of Alu-domain. Prior to transportation to the nucleus SRP9 and SRP14 form the heterodimer in the cytoplasm, a prerequisite for the binding to 7S RNA [49].

The signal recognition particle displays three main activities in the process of cotranslational targeting: (I) binding to signal sequences emerging from the translating ribosome, (II) pausing of peptide elongation, and (III) promotion of protein translocation through docking to the membrane-bound SRP receptor (FtsY in prokaryotes) and transfer of the ribosome nascent chain complex (RNC) to the protein-conducting channel [50]. Despite the diversity of signal sequences, SRP productively recognizes and selectively binds them, and this binding event serves as the critical sorting step in protein localization within the cell. The structural details that confer on SRP this distinctive ability are poorly understood. SRP signal sequences are characterized by a core of 8–12 hydrophobic amino acids that preferentially form an α-helix, but are otherwise highly divergent in length, shape, and amino acid composition [51-52]. This and the unusual abundance of methionine in the SRP54 M-domain led to the 'methionine bristle' hypothesis, in which the flexible side chains of methionine provide a hydrophobic environment with sufficient plasticity to accommodate diverse signal sequences [53].

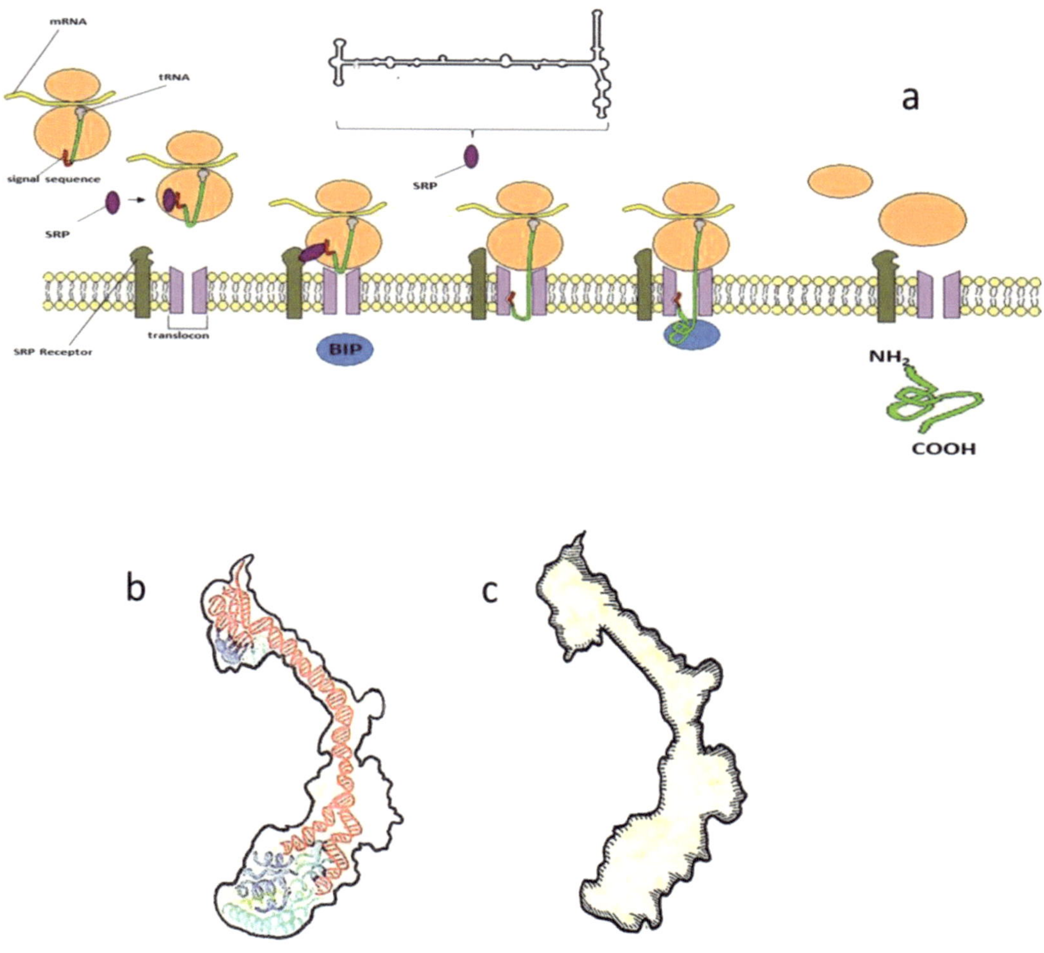

Figure 6. The signal recognition particle. During protein synthesis of the secretory pathway, the signal peptide binds to SRP, an RNP particle containing a small RNA and 6 different proteins (b, [modified from 60]). A model for SRP at nanoscale is shown in (c).

In the SRP pathway, SRP binds to the ribosome synthesizing the polypeptide, and subsequently also binds an SRP receptor, located next to the machinery that transfers proteins across the membrane and out of the cell. This process begins when a nascent polypeptide carrying a signal sequence emerges from the translating ribosome and is recognized by the SRP. The ribosome-nascent chain complex is delivered to the target membrane via the interaction of SRP with the SRP receptor. There, the cargo is transferred to the Sec61p (or secYEG in archaea and bacteria) translocon, which translocates the growing polypeptide across the membrane or integrates it into the membrane bilayer. SRP and SR then dissociate from one another to enter subsequent rounds of targeting.

During the last years, several structures have been solved by crystallography and cryo-electron microscopy that represent distinct functional states of the SRP cycle. On this basis, the first structure-based models can be suggested that explain important aspects of protein targeting, such as the SRP–ribosome [54], SRP-SRP receptor [55] and SRP–SR interactions. The snapshots obtained by single-particle EM reconstructions enable us to follow the path of a nascent protein from the peptidyl-transferase center, through the ribosomal tunnel, to and across the translocon in the membrane. With new developments in image processing techniques it is possible to sort a biological homogenous sample into different conformational states and to reach subnanometer resolution such that folding of the nascent chain into secondary structure elements can be directly visualized [56].

Molecular biology, biochemistry, and cryo-electron microscopy, have been combined to study the ribosome-protein complexes involved in protein assembly, folding and targeting. These approaches led to obtain structural snapshots of entire pathways by which proteins are synthesized and targeted to their final positions. The link between SRP and its receptor is usually transient and chemically unstable, for this reason, engineered SRP receptor bind more stably to SRP, then introduced to ribosomes and observed the resulting complexes using cryo-electron microscopy (cryo-EM). Cryo-EM can be performed in roughly physiological conditions, providing a picture that closely resembles what happens in living cells. This picture can then be combined with higher-resolution crystallography data and biochemical studies [57-58].

c. Peroxisomes

The oxidative stress (EO) is a disorder where reactive oxygen species (ROS) are produced. These compounds, that include free radicals and peroxides, play important roles in cell redox signaling. However, disturbances in the balance between the ROS production and the biological system can be particularly destructive. For example, the P450 oxide reductase activity produces H_2O_2 as a metabolite. This enormous family of enzymes is present in the mitochondrial and smooth endoplasmic reticulum (SER) membranes and catalyzes several reactions in the pathway of the biogenesis of steroid hormones [59] and in the detoxification process or in the first stage of drugs or xenobiotics hydrolysis, converting them in the SER, in water-soluble compounds for its excretion in the urine [60-62].

Peroxisomes are single membrane organelles present in practically every eukaryotic cell. Matrix proteins of peroxisomes synthesized in free polyribosomes in the cytoplasm and imported by a specific signal, are encoded in genes present in the cell nucleus genome.

http://dx.doi.org/10.5772/52003

Peroxisomal membrane-bound PEX proteins, also encoded in the nuclear genome, are synthetized by ribosomes associated to rough endoplasmic reticulum since they display signal peptide. Therefore the peroxisome as an organelle derives from the rough endoplasmic reticulum. These organelles participate in ROS generation, as H_2O_2, but also in cell rescue from oxidative stress by catalase activity. In several biological models for pathological processes involving oxygen metabolites, the role of peroxisomes in prevention of oxidative stress is strongly suggested by de co-localization of catalase and H_2O_2, and the induction of peroxisomes proliferation [63].

d. Mitochondrion and chloroplast particles

i. ATP synthase: A rotary molecular motor

To support life, cells must be continuously supplied with external energy in form of light or nutrients and must be equipped with chemical devices to convert these external energy sources into adenosine triphosphate (ATP). ATP is the universal energy currency of living cells and as such is used to drive numerous energy-consuming reactions, e.g., syntheses of biomolecules, muscle contraction, mechanical motility and transport through membranes, regulatory networks, and nerve conduction. When performing work, ATP is usually converted to ADP and phosphate. It must therefore continuously be regenerated from these compounds to continue the cell energy cycle. The importance of this cycle can be best illustrated by the demand of 50 Kg of ATP in a human body on average [64].

Prokaryotes use their plasma membrane to produce ATP. Eukaryotes use instead the specialized membrane inside energy-converting organelles, mitochondria and chloroplasts, to produce most of their ATP. The mitochondria are present in the cells of practically all eukaryotic organisms (including fungi, animals, plants, algae and protozoa), and chloroplasts occur only in plants and algae. The most striking morphological feature of both organelles, revealed by electron microscopy, is the large amount of internal membrane they contain. This internal membrane provides the framework for an elaborate set of electron-transport processes, mediated by the enzymes of the Respiratory Chain that are essential to the process of Oxidative Phosphorylation which generate most of the cell's ATP.

In eukaryotes, oxidative phosphorylation occurs in mitochondria and photophosphorylation in chloroplasts. In the mitochondria, the energy to drive the synthesis of ATP derive from the oxidative steps in the degradation of carbohydrates, fats and amino acids; whereas the chloroplasts capture the energy of sunlight and harness it to make ATP [60].

ii. The Chemiosmotic Model of Peter Mitchell

Our current understanding of ATP synthesis in mitochondria and chloroplasts is based on the chemiosmotic model proposed by Peter Mitchell in 1961 [60], which has been accepted as one of the great unifying principles of twentieth century. According with this model, the electrochemical energy inherent in the difference in proton concentration and the separation of charge across the inner mitochondrial membrane (the proton motive force) drives the synthesis of ATP as protons flow passively back into the matrix through a proton pore associated with the ATP synthase (Fig. 7).

Under aerobic conditions, the major ATP synthesis pathway is oxidative phosphorylation of which the terminal reaction is catalyzed by $F_O F_1$-ATP synthase. This enzyme is found widely in the biological world, including in thylakoid membranes, the mitochondrial inner membrane and the plasma membrane of bacteria, and is the central enzyme of energy metabolism in most organisms [65].

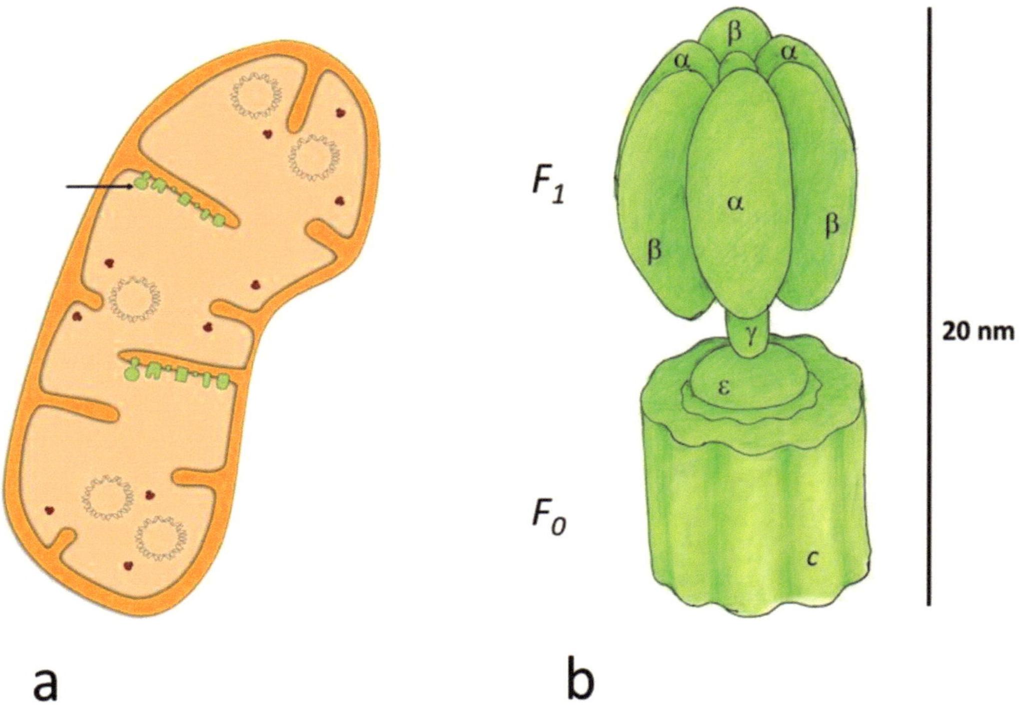

Figure 7. In the mitochondrion (a), ATP synthase (arrow in a; b) is part of the respiratory chain.

Like many transporters the $F_O F_1$-ATP synthase (or F-type ATPase) has been fascinating subject for study of a complex membrane-associated process. This enzyme catalyzes ATP synthesis from adenosine diphosphate (ADP) and inorganic phosphate (Pi), by using the electrochemical potential of protons (or sodium ions in some bacteria) across the membrane, i.e. it converts the electrochemical potential into its chemical form. ATP synthase also functions in the reverse direction (ATPase) when the electrochemical potential becomes insufficient: it catalyzes proton pumping to form an electrochemical potential to hydrolyze ATP into ADP and Pi. Proton translocation and ATP synthesis (or hydrolysis) are coupled by a unique mechanism, subunit rotation. Electrochemically energy contained in the proton gradient is converted into mechanical energy in form of subunit rotation, and back into chemical energy as ATP (Nakamoto RK, et al, 2008).

Mitochondrial ATP synthase is an F-type ATPase similar in structure and mechanism to the ATP synthases of chloroplasts and bacteria. This large complex of the inner mitochondrial membrane, also called Complex V, catalyzes the formation of ATP from ADP and Pi, accompanied by the flow of protons from P (positive) side to N (negative) side of the membrane [66].

iii. $F_O F_1$-ATP Synthase Structure and Function

ATP synthase is a supercomplex enzyme with a molecular weight of 500 kDal and consists of two rotary motors. One is F_1 subcomplex (~380 kDal), which is the water-soluble part of ATP synthase. F_1 was identified and purified by Efraim Racker and his colleagues in the early 1960s. When isolated from the membrane portion, it acts as an ATP-driven motor: it rotates its inner subunit to hydrolyze ATP and is therefore term F_1 ATPase. The other rotary motor of ATP synthase is the membrane-embedded Fo subcomplex (~120 kDal) through which the protons flow.

In the simplest form of the enzyme, in bacteria like *Escherichia coli*, F_1 is composed of five different subunits, in a stoichiometry of $\alpha_3\beta_3\gamma\delta\varepsilon$, and F_O consists of three distinct subunits in a stoichiometry of $ab_2c_{10\text{-}15}$. A newer more mechanically-based division differentiates between the "rotor" (in *E. coli*,$\gamma\varepsilon c_n$) and the "stator" ($\alpha_3\beta_3\delta ab_2$). The $\alpha_3\alpha_3$ ring of the stator contains the three catalytic nucleotide sites, on the β subunits at the interphase to the adjacent α subunit. The *a* subunit contains the static portion of the proton traslocator machinery. $\alpha_3\beta_3$ and *a* are held together by the "stator stalk" (or "peripheral stalk"), consisting of $b_2\gamma$ [65].

The crystallographic determination of the F_1 structure by John Walker and colleagues [67] revealed structural details very helpful in explaining the catalytic mechanism of the enzyme. The three α- and β- subunits that constitute the hexameric stator ring are alternately arranged like the sections of an orange. The rotor shaft is the γ-subunit, which is accommodated in the central cavity of the $\alpha_3\beta_3$-ring. The ε-subunit binds onto the protruding part of the γ-subunit and provides a connection between the rotor parts of F_1 and F_O. The δ-subunit acts as a connector between F_1 and F_O that connects the stator parts.

Catalytic reaction centers for ATP hydrolysis/synthesis reside at the three of the α-β interfaces, whereas the non-catalytic ATP-binding sites reside on the other α/β interfaces. While the catalytic site is formed mainly with amino acid residues from γ-subunit, the non- catalytic sites are primarily within the α-subunit. Upon ATP hydrolysis on the catalytic sites, F_1 rotates the γ-subunit in the anticlockwise direction viewed from the F_O side [68].

As mentioned before, F_O subcomplex (*o* denoting oligomycin sensitive) consists of $ab_2c_{10\text{-}15}$ subunits. The number of *c* subunits varies among the species and form a ring complex by aligning in a circle. It is widely thought that the *c*-ring and the *a* subunit form a proton pathway. With the downhill proton flow through the proton channel, the *c*-ring rotates against the ab_2 subunits in the opposite direction of the γ-subunit of the F_1 motor [69]. Thus, in the $F_O F_1$ complex, F_O and F_1 push each other in the opposite direction. Under physiological condition where the electrochemical potential of the protons is large enough to surpass the free energy of ATP hydrolysis, F_O forcibly rotates the γ-subunit in the clockwise direction and then F_1 catalyzes the reverse reaction, *i.e.* ATP synthesis which is the principle function of ATP synthase. In contrast, when the electrochemical potential is small or decreases, F_1 forces F_O to rotate the *c*-ring in the reverse direction to pump protons against the electrochemical potential.

The *c* subunit of the F_O complex is a small (Mr 8,000), very hydrophobic polypeptide, consisting almost entirely of two membrane-spanning α-helices, that are connected by a small loop extending from the matrix side of the membrane. The crystal structure of the yeast $F_O F_1$, solved in 1999, shows the arrangement of the subunits. The yeast complex has 10 *c* subunits, each with two transmembrane helices roughly perpendicular to the plane of the membrane and arranged in two concentric circles. The inner circle is made up of the amino-terminal helices of each *c* subunit; the outer circle, about 55 Å in diameter, is made up of the carboxyl-terminal helices. The ε andγ subunits of F_1 form a leg-and-foot that projects from the bottom (membrane) side of F_1 and stands firmly on the ring of *c* subunits. The *a* subunit is a very hydrophobic protein that in most models is composed of five transmembrane helices. Ion translocation takes place through subunit *a* and its interface with subunit *c*. The *b* subunits are anchored within the membrane by an N-terminal α-helix and extend as a peripheral stalk all the way to the head of the F_1 domain. According to cross-linking studies, the *b* subunits contact de C-terminal part of the c subunit and the loop between helices 4 and 5 of the *a* subunit at the periplasmic surface. The δ-subunit forms a strong complex with the α-subunit. In mitochondria, the peripheral stalk consists of more subunits named OSCP (Oligomycin Sensitive Conferring Protein), *b*, δ and F_6 [64].

iv. Structure of F_1 and binding-change mechanism for ATP Synthesis.

The classic working model for F1 is the "binding-change mechanism" proposed by Paul Boyer [70]. The early stage of this model postulated an alternating transition between two chemical states, assuming two catalytic sites residing on F_1. It was later revised to propose the cyclic transition of the catalytic sites based on the biochemical and electron microscopic experiments that revealed that F_1 has the three catalytic sites [71-73]. One important feature of this model is that the affinity for nucleotide in each catalytic site is different from each other at any given time, and the status of the three β-subunits cooperatively change in one direction accompanying γ rotation. This hypothesis is strongly supported by X-ray crystallographic studies performed by Walker's group [67] that first resolved crystal structure of F_1, which revealed many essential structural features of F_1 at atomic resolution. Importantly, the catalytic β-subunits differ from each other in conformation and catalytic state: one binds to an ATP analogue, adenosine 5′-(β,γ-imino)-triphosphate (AMP-PNP), the second binds to ADP and the third site is empty. Therefore, these sites are termed βTP, βDP and βEmpty, respectively. While βTP and βDP have a close conformation wrapping bound nucleotides on the catalytic sites, βEmpty has an open conformation swinging the C-terminal domain away from the binding site to open the cleft of the catalytic site. These features are consistent with the binding-change mechanism. Another important feature found in the crystal is that while the N-terminal domains of the α- and β-subunits form a symmetrical smooth cavity as the bearing for γ rotation at the bottom of the $\alpha_3\beta_3$-ring, the C-terminal domains of the β-subunit show distinct asymmetric interactions with the γ-subunit. Therefore, the most feasible inference is that the open-to-closed transition of the β-subunits upon ATP binding pushes γ, and the sequential conformational change among β- subunits leads the unidirectional γ rotation.

One strong prediction of the binding-change model of Boyer is that the γ subunit should rotate in one direction when $F_O F_1$ is synthesizing ATP and in the opposite direction when the enzyme is hydrolyzing ATP. This prediction was confirmed in elegant experiments in the laboratories of Masasuke Yoshida and Kazuhiko Kinosita Jr. [74]. The rotation of γ in a single F_1 molecule was observed microscopically by attaching a long, thin, fluorescent actin polymer to γ and watching it move relative to $\alpha_3\beta_3$ immobilized on a microscope slide, as ATP was hydrolyzed [see 75]. Lately the unidirectional γ rotation was visualized in simultaneous imaging of the conformational change of the β-subunit and the γ rotation.

v. New approaches for studying biological macromolecules.

The Atomic Force Microscope (AFM) is a powerful tool for imaging individual biological molecules attached to a substrate and place in aqueous solution. This technology allows visualization of biomolecules under physiological conditions. However, it is limited by the speed at which it can successively record highly resolved images. Recent advances have improved the time resolution of the technique from minutes to tens of milliseconds, allowing single biomolecules to be watch in action in real time. Toshio Ando and his coworkers at Kanazawa University have been leading innovators in this so-called High-Speed Atomic Force Microscope (HS-AFM) technology [76]. This technology allows direct visualization of dynamic structural changes and dynamic processes of functioning biological molecules in physiological solutions, at high spatial-temporal resolution. Dynamic molecular events appear in detail in an AFM movie, facilitating our understanding of how biological molecules operate to function.

In this regard, the Ando group showed a striking example of molecular motor action in their AFM movies of the isolated subcomplex of the rotary motor protein F_1-ATPase. Previous single-molecule experiments on parts of this enzyme had measured rotation, but they could only be done if at least one subunit of the rotor was attached. The AFM, however, could visualize the conformational change that the β subunits of the stator undergo when they bind ATP. By imaging at 12.5 frames/s, the authors followed the time dependence of these conformational changes, leading to the surprising conclusion that, contrary to what was widely assumed before, the catalysis on the enzyme maintains its sequential rotary order even in absence of the rotor subunits.

"To directly observe biological molecules at work was a holy grail in biology. Efforts over the last two decades at last materialized this long-quested dream. In high-resolution AFM movies, we can see how molecules are dynamically behaving, changing their structure and interacting with other molecules, and hence we can quickly understand in stunning detail how molecules operate to function. This new approach will spread over the world and widely applied to a vast array of biological issues, leading to a number of new discoveries. The extension of high-speed AFM to a tool for imaging live cells, which allows direct in situ observation of dynamic processes of molecules and organelles, remains an exciting challenge but will be made in the near future because it is a right and fruitful goal" [77].

e. Lipid rafts

Cell membranes are dynamic assemblies of a variety of lipids and proteins. They form a protective layer around the cell and mediate the communication with the outside world. The original fluid mosaic model [78] of membranes suggested a homogenous distribution of proteins and lipids across the two-dimensional surface, but more recent evidence suggests that membranes themselves are not uniform and that microdomains of lipids in a more ordered state exist within the generally disorder lipid milieu of the membrane. These clusters of ordered lipids are now referred to as lipid rafts [79] (Pike LJ 2009).

Lipid rafts (LRs), consist of cell membrane domains rich in cholesterol, sphingolipids and lipid-anchored proteins in the exoplasmic leaflet of the lipid bilayer. Because of their ability to sequester specific lipids and proteins and exclude others, rafts have been postulated to perform critical roles in a number of normal cellular processes, such as signal transduction [80], membrane fusion, organization of the cytoskeleton [81-83], lipid sorting, and protein trafficking/recycling, as well as pathological events [84].

LRs are too small to be resolved by standard light microscopy - they range from 10 to about 200 nm - with a variable life span in the order of milliseconds (msec). Detergent resistant membranes, containing clusters of many rafts, can be isolated by extraction with Triton X-100 or other detergents on ice. However, this method involves breaking up the membrane and has limitations in terms of defining the size, properties, and dynamics of intact microdomains [85-88].Thus, a variety of sophisticated techniques have recently been used to analyze in detail open questions concerning rafts in cell and model membranes including biochemical, biophysical, quantitative fluorescence microscopy, atomic force microscopy and computational methodologies [89-90].

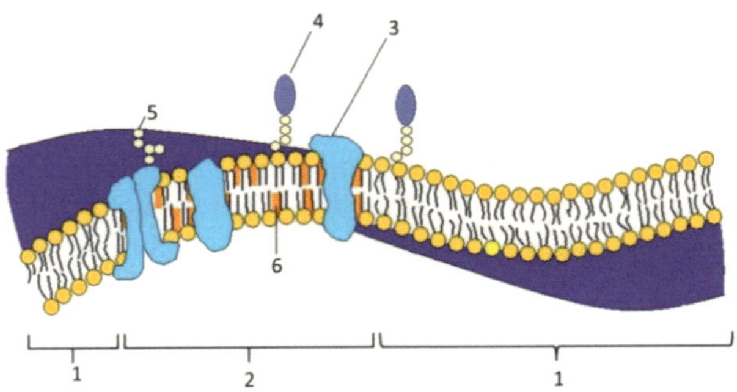

Figure 8. Components of a lipid raft. (1) non raft membrane, (2) lipid raft, (3) lipid raft associated transmembrane protein, (4) GPI-anchored protein, (5) glycosylation modifications (glycoproteins and glycolipids).

The raft affinity of a given protein can be modulated by intra- or extracellular stimuli. Saturated fatty acids are preferentially enriched in the side chains of the membrane phospholipids, which allows closer packing and thus increased rigidity, more order and less fluidity of the LRs compared to the surrounding membrane [91-92]. Proteins with raft affinity include glycosylphosphatidylinositol (GPI)-anchored proteins [93-94], doubly acylated proteins, such as Src-family kinases or the α-subunits of heterotrimeric G proteins8, cholesterol-linked and palmitoylated proteins such as Hedgehog9, and transmembrane proteins, particularly palmitoylated ones [92-95].

Different subtypes of lipid rafts can be distinguished according to their protein and lipid composition. Caveolae are types of rafts that are rich in proteins of the caveolin family (caveolin-1, -2 and -3) which present a distinct signaling platform [96]. The caveolae are enriched in cholesterol, glycosphingolipids, and sphingomyelin. They are the site of several important protein–protein interactions, for example, the neurotrophin receptors, TrkA and p75(NTR), whose respective interactions with caveolin regulates neurotrophin signaling in the brain. Caveolins also regulate G-proteins, MAPK, PI3K, and Src tyrosine kinases.

The most important role of rafts at the cell surface may be their function in signal transduction. Lipid rafts have been implicated as the sites for a great number of signaling pathways.They form concentrating platforms for individual receptors, activated by ligand binding [86]. If receptor activation takes place in a lipid raft, the signaling complex is protected from non-raft enzymes such as membrane phosphatases that otherwise could affect the signaling process. In general, raft binding recruits proteins to a new micro-environment, where the phosphorylation state can be modified by local kinases and phosphatases, resulting in downstream signalling. Individual signaling molecules within the raft are activated only for a short period of time.

Immobilization of signaling molecules by cytoskeletal actin filaments and scaffold proteins may facilitate more efficient signal transmission from rafts [97]. Current evidence supports a role for lipid rafts in the initiation and regulation of The B-cell receptor signaling and antigen trafficking [98-100]. The importance of lipid raft signalling in the pathogenesis of a variety of conditions, such as Alzheimer's, Parkinson's, cardiovascular and prion diseases, systemic lupus erythematosus and HIV, has been elucidated over recent years[101] and makes these specific membrane domains an interesting target for pharmacological approaches in the cure and prevention of these diseases [102]. Rafts serve as a portal of entry for various pathogens and toxins, such as human immunodeficiency virus 1 (HIV-1). In the case of HIV-1, raft microdomains mediate the lateral assemblies and the conformational changes required for fusion of HIV-1 with the host cell [103]. Lipid rafts are also preferential sites of formation for pathological forms of the prion protein (PrPSc) and of the β-amyloid peptide associated with Alzheimer's disease {104].

Plasma membranes typically contain higher concentrations of cholesterol and sphingomyelin than do internal membranous organelles [105-106]. Thus, along the secretion pathway, there are very low concentrations of cholesterol and sphingolipids in the endoplasmic reticulum, but the concentrations of these lipids increase from the cis-Golgi to the trans-Golgi and then to the plasma membrane [107-108]. On the contrary, recent evidence suggests that mitochondria do not contain lipid rafts, and lipid rafts do not contain mitochondrial proteins [109].

Lipid raft domains play a key role in the regulation of exocytosis [110]. The association of SNAREs protein complexes with lipid rafts acts to concentrate these proteins at defined sites of the plasma membrane that are of functional importance for exocytosis [111-114].

f. The nucleolus

The cell nucleus contains different compartments that are characterized by the absence of delineating membranes that isolate it from the rest of the nucleoplasm [5]. Due to the high concentration of RNA and proteins that form it, the nucleolus is the most conspicuous nuclear body in cycling cells observed by light and electron microscopy. Nucleoli are formed around nucleolar organizer regions (NORs), which are composed of cluster of ribosomal genes (rDNA) repeat units [115-121]. The number of NOR-bearing chromosomes varies depending on the species, can be found 1 in haploid yeast cells to 10 in human somatic cells (short arms of chromosomes 13, 14, 15, 21 and 22). Nucleolus is the organelle of rDNA transcription by RNA polymerase I, whose activity generates a long ribosomal precursor (pre-rRNA), this molecule is the target of an extensive process that includes removing or cutting the spacers and 2'-O-methylation of riboses and coversions of uridine residues into pseudouridines. The net result of these reactions is the release of mature species of ribosomal RNA (rRNA) 18S, 5.8S and 28S. These particles are assembled with approximately 82 ribosomal proteins and rRNA 5S (synthesized by RNA polymerase III) to form the 40S and 60S subunits; both of these subunits are then exported separately to the cytoplasm and are further modified to form mature ribosomal subunits. Currently, it is widely accepted that nucleolar transcription and early pre-rRNA processing take place in the fibrillar portion of nucleolus while the later steps of processing and ribosome subnits assembly occurs mainly in the granular zone. The architecture of the nucleolus reflects the vectorial maturation of the pre-ribosomes. The nucleolar structure is organized by three canonical subdomains that are morphologically and biochemically different. The fibrillar centers (FC), dense fibrillar component (DFC) and granular component (GC). The FCs are structures with a low electron density, often circular shape of ~0.1 to 1μm in diameter. The FCs are enriched with rDNA, RNA polymerase I, topoisomerase I and upstream binding factor (UBF). DFCs are a compact fibrillar region containing a high concentration of ribonucleoprotein molecules that confer a high electrodensity. This component entirely or partially surrounds the FCs. DFCs contains important proteins such as fibrillarin and nucleolin as well as small nucleolar RNAs, pre-rRNA and some transcription factors. FCs and DFCs are embedded in the GC, composed mainly of granules of 15 to 20nm in diameter with a loosely organized distribution. In the GC are located B23/nucleophosmin, Nop 52, r-proteins, auxiliary assembly factors, and the 40S and 60S subunits that the GC is itself composed of at least two distintc molecular domains. Considering the species, cell type and physiological state of the cell, there is considerable diversity in the prevalence and arrangement of the three nucleolar components.

On the other hand, the current eukaryotic nucleolus is involved in the ribosomal biogenesis but has been described as a multifunctional entity. Extra ribosomal functions include biogenesis and/or maturation of other ribonucleoprotein machines, including the signal recognition particle, the spliceosomal small nuclear RNPs and telomerase, processing or export of some mRNAs and tRNA, cell cycle and cell proliferation control, stress response and apoptosis [116]. The plurifunctional nucleolus hypothesis is reinforced by the description of nucleolar proteome of several eukaryotes. A proteomic analysis has identified more than 200 nucleolar proteins in Arabidopsis and almost 700 proteins in the nucleolus of HeLa cells. A comparison of nucleolar proteome from humans and budding yeast showed that ~90% of human nucleolar proteins have yeast homologues. Interestingly, only 30% of the human nucleolar proteome is intended for ribosomal biogenesis [120, 122].

1.1.3. *Microscopy*

Fundamental to approach the cell at the nanoscale in cell nanobiology are the classical and also remarkably new types of microscopy. Three different epochs characterize microscopy: 1) Light microscopy, developed since ca. 1500, where glass lenses and light as source of illumination are used to get resolution of up to 0.2 μm. Different types such as bright field, phase contrast, differential interference contrast (Nomarsky), dark field, polarization, fluorescence, confocal, and super-resolution, are variants of this type of microscopy. 2) Electron microscopy, developed since early 1930s, where electromagnetic lenses and electrons as source of illumination are used to get resolution of up to nm or A°. Transmission and scanning electron microscopy –including the environmental and high resolution modes- are the two forms of this microscopy. 3) Scanning probe microscopy, developed in the early 1980s, where no lenses or illuminations are used, but instead the microscope consists of a fine tip interacting with the samples to potentially obtaining atomic resolution. Scanning tunneling microscopy and atomic force microscopy are the major variants of this type of modern microscopy. Because atomic force microscopy may produce images at high resolution even under liquid, we have been using such microscopy for imaging the cell components. To test this approach, we used several cell types and generated images at low magnification (Figure 9a). Nuclear particles *i.e.* Lacandonia granules were already visualized using this approach (Figure 9b).

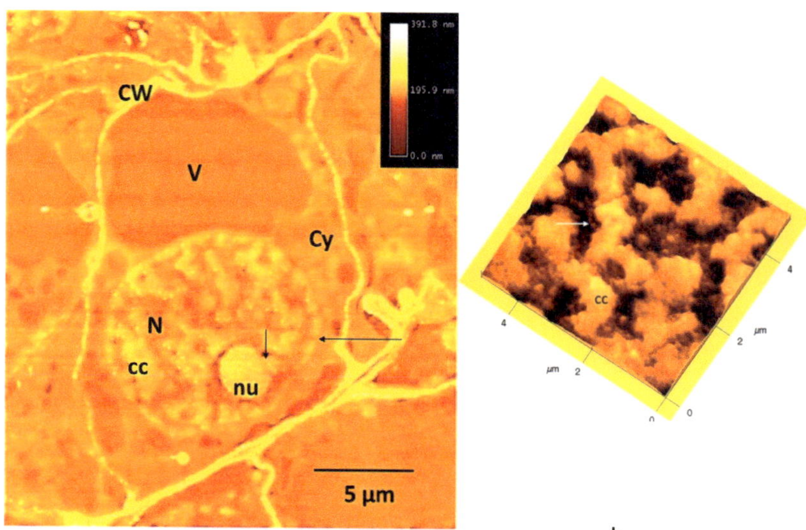

Figure 9. a) Atomic force microscopy image of a cell from the tegument of the plant *Lacandonia schismatica*. Cell wall (CW), vacuole (V), cytoplasm (Cy), Within the nucleus (N), compact chromatin (cc), nucleolus (nu), nucleolar organizer (small arrow) and nuclear pore (large arrow). b) Atomic force microscopy of Lacandonia granules within the nucleus of a tegument cell of the plant *Lacandonia schismatica*. Three dimensional displaying shows compact chromatin (cc) and associated particles (arrow).

1.1.4. Further research

Further research in our laboratory will focusing in visualizing the nanoscale cell structures involved in fundamental processes as ribosome biogenesis, at a high resolution *in situ* under liquid conditions to perform quantitative analysis.

2. Conclusion

A view of the cell emphasizing vertical resolution obtained by atomic force microscopy may represent a way to understand cell structure and function at the nanoscale, an interphase between molecular biology and cell biology.

Acknowledgements

DGAPA-UNAM PAPIIT IN-227810, PAPIME PE211412, CONACyT 180835.

Rogelio Fragoso-Soriano and Tomás Nepomuceno-Mejía are postdoctoral fellows from ICyTDF and DGAPA-UNAM at Faculty of Sciences-UNAM, respectively. Georgina Alvarez-Fernandez is on-a leave-of-absence from the Department of Biochemistry, Faculty of Medicine, UNAM. Luis and Teresa Jiménez Segura for SRP and ATP synthase figures.

Author details

María de Lourdes Segura-Valdez[1], Lourdes T. Agredano-Moreno[1],
Tomás Nepomuceno-Mejía[2], Rogelio Fragoso-Soriano[2], Georgina Álvarez-Fernández[2],
Alma Zamora-Cura[1], Reyna Lara-Martínez[1] and Luis F. Jiménez-García[1]

1 Laboratory of Cell Nanobiology and Electron Microscopy Laboratory (Tlahuizcalpan), Department of Cell Biology, Faculty of Sciences, National Autonomous University of Mexico (UNAM), Circuito Exterior, Ciudad Universitaria, Coyoacán 04510, México D.F, México

2 Visiting from the Department of Biochemistry, Faculty of Medicine, UNAM, México

References

[1] Jiménez García LFSegura Valdez ML. Biología celular del genoma. México: Las Prensas de Ciencias-UNAM: (2010).

[2] Monneron, A, & Bernhard, W. Fine structural organization in the interphase nucleus in some mammalian cells. Journal of Ultrastructural Research (1969). , 27-266.

[3] Vázquez-nin, G. H, & Echeverría, O. M. The polytene nucleus in morphological, cytochemical, and functional studies of Messenger RNA transcription, processing, and transportation. European Journal of Histochemistry (1996). , 40-7.

[4] Jiménez-garcía, J, Agredano-moreno, L. F, Segura-valdez, L. T, Echeverría, M. L, Martínez, O. M, Ramos, E, & Vázquez-nin, C. H. GH. The ultrastructural study of the interphase cell nucleus of Lacandonia schismatica (Lacandoniaceae:Triuridales) reveals a non typical extranucleolar particle. Biol. Cell (1992). , 75-101.

[5] Spector, D. L. Macromolecular domains within the cell nucleus. Annual Reviews in Cell Biology (1993). , 9-265.

[6] Tischendorf, G. W, Zeichhardt, H, & Stoffler, G. Architecture of the Escherichia coli ribosome as determined by immune electron microscopy. Proceedings of the National Academy of Sciences USA (1975). , 72-4820.

[7] Lake, J. A. (1976). Ribosome structure determined by electron microscopy of Escherichia coli small subunits, large subunits and monomeric ribosomes. Journal of Molecular Biology 1976; , 105-131.

[8] Boublik, M, Hellmann, W, & Kleinschmidt, A. K. Size and structure of Escherichia coli ribosomes by electron microscopy. Cytobiologie (1977). , 14-293.

[9] Yonath, A, Mussig, J, Tesche, B, Lorenz, S, Erdmann, V. A, & Wittmann, H. G. (1980). Crystallization of the large ribosomal subunits from Bacillus stearothermophilus. Biochemistry International 1980; , 1-428.

[10] Ban, P, Nissen, J, Hansen, P, Moore, B, & Steitz, T. A. The complete atomic structure of the large ribosomal subunit at 2.4 A resolution. Science (2000). , 289-905.

[11] Wimberly, B. T, Brodersen, D. E, Clemons, W. M, Morgan-warren, R. J, Carter, A. P, Vonrhein, C, Hartsch, T, & Ramakrishnan, V. Structure of the 30S ribosomal subunit. Nature (2000). , 407-327.

[12] Yusupov, M. M, Yusupova, G. Z, Baucom, A, Lieberman, K, Earnest, T. N, Cate, J. H, & Noller, H. F. Crystal structure of the ribosome at 5.5 A resolution. Nature (2001). , 292-883.

[13] Seidelt, B, Innis, C. A, Wilson, D. N, Gartmann, M, Armache, J. P, Villa, E, Trabuco, L. G, Becker, T, Mielke, T, Schulten, K, et al. Structural insight into nascent polypeptide chain-mediated translational stalling. Science (2009). , 326-1412.

[14] Bhushan, S, Gartmann, M, Halic, M, Armache, J. P, Jarasch, A, Mielke, T, Berninghausen, O, & Wilson, D. N. Beckmann R. α-helical nascent polypeptide chains visualized within distinct regions of the ribosomal exit tunnel. Nature Structural Molecular Biology (2010). , 17-313.

[15] Bhushan, S, Hoffmann, T, Seidelt, B, Frauenfeld, J, Mielke, T, Berninghausen, O, Wilson, D. N, & Beckmann, R. SecM-stalled ribosomes adopt an altered geometry at the peptidyl transferase center. PLoS Biol (2011). e1000581.

[16] Frank, J, Zhu, J, Penczek, P, Li, Y, Srivastava, S, Verschoor, A, Radermacher, M, Grassucci, R, Lata, R. K, & Agrawal, R. K. A model of protein synthesis based on cryo-electron microscopy of the E. coli ribosome. Nature (1995). , 376-441.

[17] Ben-shem, A. Garreau de Loubresse N, Melnikov S, Jenner L, Yusupova G, Yusupov M.The structure of the eukaryotic ribosome at 3.0A ° resolution. Science (2011). , 334-1524.

[18] Klinge, S, Voigts-hoffmann, F, Leibundgut, M, Arpagaus, S, & Ban, N. Crystal structure of the eukaryotic 60S ribosomal subunit in complex with initiation factor 6. Science (2011). , 334-941.

[19] Nissen, P, Hansen, J, Ban, N, Moore, P. B, & Steitz, T. A. The structural basis of ribosome activity in peptide bond synthesis. Science (2000). , 289-920.

[20] Kampmann, M. R, & Blobel, G. Biochemistry. Nascent proteins caught in the act. Perspectives in Biochemistry (2009). , 326, 1352-1353.

[21] Voorhees, R. M, Schmeing, T. M, Kelley, A. C, & Ramakrishnan, V. The mechanism for activation of GTP hydrolysis on the ribosome. Science (2010). , 330-835.

[22] Weis, F, Bron, P, Giudice, E, Rolland, J. P, Thomas, D, Felden, B, & Gillet, R. tmRNA-SmpB: a journey to the centre of the bacterial ribosome. The EMBO Journal (2010). , 29-3810.

[23] Neubauer, C, Gillet, R, Kelley, A. C, & Ramakrishnan, V. Decoding in the Absence of a Codon by tmRNA and SmpB in the Ribosome. Science (2012). , 335-1366.

[24] Poehlsgaard, J, & Douthwaite, S. The bacterial ribosome as target for antibiotics. Nature Reviews in Microbiology (2005). , 3-871.

[25] Selmer, M, Dunham, C. M, Murphy, F. V, Weixlbaumer, A, Petry, S, Kelley, A. C, Weir, J. R, & Ramakrishnan, V. Structure of the 70S ribosome complexed with mRNA and tRNA. Science (2006). , 313-1935.

[26] Wilson, D. N. Doudna Cate JH. The and function of the eukaryotic ribosome. Cold Spring Harbor Perspectives in Biology (2012).

[27] Spahn, C. M, Beckmann, R, Eswar, N, Penczek, P. A, Sali, A, Blobel, G, & Frank, J. Structure of the 80S ribosome from Saccharomyces cerevisiae-tRNA-ribosome and subunit- subunit interactions. Cell (2001). , 107-373.

[28] Powers, T, & Walter, P. Regulation of ribosome biogenesis by the rapamycin-sensitive TOR-signaling pathway in Saccharomyces cerevisiae. Mol Biol Cell (1999). , 10-987.

[29] Hannan, K. M, Brandenburger, Y, Jenkins, A, Sharkey, K, Cavanaugh, A, Rothblum, L, Moss, T, Poortinga, G, Mcarthur, G. A, Pearson, R. B, & Hannan, R. D. mTOR-dependent regulation of ribosomal gene transcription requires S6K1 and is mediated by phosphorylation of the carboxy-terminal activation domain of the nucleolar transcription factor UBF. Molecular Cellular Biology (2003). , 23-8862.

[30] Mayer, C, Zhao, J, Yuan, X, & Grummt, I. mTOR-dependent activation of the transcription factor TIF-IA links rRNA synthesis to nutrient availability. Genes & Development (2004). , 18-423.

[31] Lee, J, Moir, R. D, Mcintosh, K. B, & Willis, I. M. TOR signaling regulates ribosome and tRNA synthesis via LAMMER/Clk and GSK-3 family kinases. Molecular Cell (2012). , 45-836.

[32] Andrews, D. W, Walter, P, & Ottensmeyer, F. P. Structure of the signal recognition particle by electron microscopy. Proceedings of the National Academy of Sciences USA (1985). , 82-785.

[33] Walter, P, & Johnson, A. E. Signal sequence recognition and protein targeting to the endoplasmic reticulum membrane.Annual Reviews in Cell Biology (1994). , 10-87.

[34] Walter, P, & Blobel, G. Translocation of proteins across the endoplasmic reticulum III. Signal recognition protein (SRP) causes signal sequence-dependent and site-specific arrest of chain elongation that is released by microsomal membranes. Journal of Cell Biology (1981). , 91, 557-561.

[35] Andersen, E. S, Rosenblad, M. A, Larsen, N, Westergaard, J. C, Burks, J, Wower, I. K, Wewer, J, Gorodkin, J, Samuelsson, T, & Zwieb, C. The TmRDB and SRPDB resourdes. Nucleic Acid Research (1999). D , 163-188.

[36] Freymann, D. M, Keenan, R. J, Stroud, R. M, & Walter, P. Structure of the conserved GTPase domain of the signal recognition particle. Nature (1997). , 385-361.

[37] Montoya, G, Kaat, K, Moll, R, Schafer, G, & Sinning, I. The crystal structure of the conserved GTPase of SRP54 from the archaeon Acidianus ambivalens and its comparison with related structures suggests a model for the SRP-SRP receptor complex. Structure (2000). , 8-515.

[38] Keenan, R. J, Freymann, D. M, Walter, P, & Stroud, R. M. Crystal structure of the signal sequence binding subunit of the signal recognition particle. Cell (1998). , 94-181.

[39] Zopf, D, Bernstein, H. D, Johnson, A. E, & Walter, P. The methionine-rich domain of the 54 kd protein subunit of the signal recognition particle contains an RNA binding site and can be crosslinked to a signal sequence. The EMBO Journal (1990). , 9-4511.

[40] High, S, & Dobberstein, B. The signal sequence interacts with the methionine-rich domain of the kD protein of signal recognition particle. Journal of Cell Biology (1991). , 54.

[41] Tcke, L, High, H, Romisch, S, Ashford, K, & Dobberstein, A. J. B. The methionine-rich domain of the 54 kDa subunit of signal recognition particle is sufficient for the interaction with signal sequences. The EMBO Journal (1992). , 11-543.

[42] Egea, P. F, Shan, S. O, Napetschnig, J, Savage, D. F, Walter, P, & Stroud, R. M. Substrate twinning activates the signal recognition particle and its receptor. Nature (2004). , 427-215.

[43] Focia, P. J, Shepotinovskaya, I. V, Seidler, J. A, & Freymann, D. M. Heterodimeric GTPase core of the SRP targeting complex. Science (2004). , 303-373.

[44] Politz JC Yarovoi S Kilroy SM Gowda K Zwieb C and Pederson TSignal [44] recognition particle components in the nucleolus. Proceedings of the National Academy of Sciences USA (2000). , 97-55.

[45] Walter, P, & Blobel, G. Disassembly and reconstitution of signal recognition particle. Cell. (1983). , 34-525.

[46] Politz, J. C, Lewandowski, L. B, & Pederson, T. Signal recognition particle RNA localization in the nucleolus differs from the classical site of ribosome synthesis. Journal of Cell Biology (2002). , 159-411.

[47] Grosshans H Deinert K Hurt E and Simos GBiogenesis of the signal recognition particle (SRP) involves import of SRP proteins into the nucleolus, assembly with the SRP-RNA, and Xpo1p-mediated export. Journal of Cell Biology (2001). , 153-745.

[48] Leung, E, & Brown, J. D. Biogenesis of the signal recognition particle. Biochemistry. Society Transactions (2010). , 38-1093.

[49] Strub, K, & Walter, P. Assembly of the Alu domain of the signal recognition particle (SRP): dimerization of the two protein components is required for efficient binding to SRP RNA, Mol Cell Biol. (1990). , 10-777.

[50] Koch HG Moser M and Muller MSignal recognition particle-dependent protein targeting, universal to all kingdoms of life. Reviews of Physiology and Biochemical Pharmacology (2003). , 146, 55-94.

[51] Gierasch, L. M. Signal sequences. Biochemistry (1989). , 28-923.

[52] Zheng, N, & Gierasch, L. M. Signal sequences: the same yet different. Cell , 1996-86.

[53] Bernstein, H. D, Poritz, M. A, Strub, K, Hoben, P. J, Brenner, S, & Walter, P. Model for signal sequence recognition from amino-acid sequence of 54K subunit of signal recognition particle. Nature (1989). , 340-482.

[54] Halic, M, Becker, T, & Pool, M. R. Spahn CMT, Grassucci RA, Frank J and Beckmann R. Structure of the signal recognition particle ineracting with the elongation-arrested ribosome. Nature. (2004). , 427-808.

[55] Ataide, S. F, Schmitz, N, Shen, K, Ke, A, Shan, S, Doudna, J. A, & Ban, N. The crystal structure of the signal recognition particle in complex with its receptor. Science (2011). , 331-881.

[56] Knoops, K, Schoehn, G, & Schaffitzel, C. Cryo-electron microscopy of ribosomal complexes in cotranslational folding, targeting and translocation. Wiley Interdisciplinary Reviews in RNA. (2012). , 3-429.

[57] Schmeing, T. M, & Ramakrishnan, V. What recent ribosome structures have revealed about the mechanism of translation. Nature (2009). , 461-1234.

[58] Estrozi, L. F, & Boehringer, D. Shan Shu-ou,, Ban N, Schaffitzel C. Cryo-EM structure of the E. coli translating ribosome in complex with SRP and its receptor. Nat. Structural & Molecular Biology (2011). , 18-88.

[59] Repetto, M. Marcadores periféricos de estrés oxidativo en pacientes hipogonádicos. Revista Argentina de Endocrinología y Metabolismo (2005). , 42-26.

[60] Alberts, B, Johnson, A, Lewis, J, Raff, M, Roberts, K, & Walter, P. Molecular Biology of the Cell. New York: Garland; (2008).

[61] Chilo, N. H, & El Citocromo, p. y su rol en la hepatotoxicidad inducida por las drogas. Enfermedades del Aparato Digestivo (1999). , 2-34.

[62] Lodish, H, Berk, A, Kaiser, C. A, Kreiger, M, Scott, M, Bretscher, A, Ploegh, H, & Matsudaira, P. New York: Freeman: (2007).

[63] Schrader, M, & Fahimi, H. D. Peroxisomes and oxidative stress. Biochimica et Biophysica Acta (BBA)- Molecular Cell Research (2006). , 1763-1755.

[64] Von Ballmoos, C, Wiedenmann, A, & Dimroth, P. Essentials for ATP synthesis by F1F0 ATP synthases. Annual Review in Biochemistry (2009). , 78-649.

[65] Weber, J. ATP Synthase: Subunit-Subunit Interactions in the Stator Stalk. Biochimical Biophysical Acta (2006). , 1757-1162.

[66] Nelson, D. L, & Cox, M. Lehninger Principles of Biochemistry. New York: Freeman and Company; (2008).

[67] Abrahams, J. P, Leslie, A. G, Lutter, R, & Walker, J. E. Structure at 2.8 A resolution of F1ATPase from bovine heart mitochondria. Nature (1994). , 370-621.

[68] Okuno, D, Riota, I, & Noji, H. Rotation and structure of FOF1-ATP synthase. Journal of Biochemistry (2011). , 149-655.

[69] Diez, M, Zimmermann, B, Borsch, M, Konig, M, Schweinberger, E, Steigmiller, S, Reuter, R, Felekyan, S, Kudryavtsev, V, Seidel, , & Graber, P. . Proton-powered subunit rotation in single membrane-bound F0F1-ATP synthase. Nature Structural Molecular Biology 2004; 11-135.

[70] Boyer, P. D. The ATP synthase-a splendid molecular machine. Annual Reviews in Biochemistry (1997). , 66-717.

[71] Yoshida, M, Sone, N, Hirata, H, & Kagawa, Y. A highly stable adenosine triphosphatase from a thermophillie bacterium. Purification, properties, and reconstitution. Journal of Biological Chemistry (1975). , 250-7910.

[72] Kagawa, Y, Sone, N, Yoshida, M, Hirata, H, & Okamoto, H. Proton translocating ATPase of a thermophilic bacterium. Morphology, subunits, and chemical composition. Journal of. Biochemistry (1976). , 80-141.

[73] Wakabayashi, T, Kubota, M, Yoshida, M, & Kagawa, Y. Structure of ATPase (coupling factor TF1) from a thermophilic bacterium. Journal of Molecular Biology (1977). , 117-515.

[74] Noji, H, Yasuda, R, Yoshida, M, & Kinosita, K. Direct observation of the rotation of F1ATPase. Nature (1977). , 386-299.

[75] http://www.k2.phys.waseda.ac.jp/Movies.html for related experimental set up and video.

[76] Katan, A. J, & Dekker, C. High-Speed AFM Reveals the Dynamics of Single Biomolecules at the Nanometer Scale. Cell (2011). , 147-979.

[77] Ando, T. High-speed atomic force microscopy coming of age. Nanotechnology (2012)., 23-062001

[78] Singer, S. J, & Nicolson, G. L. The fluid mosaic model of the structure of cell membranes. Science. (1972). , 175-720.

[79] Pike, L. J. The challenge of lipid rafts. Journal of Lipid Research (2009). S , 323-328.

[80] Asano, S, Kitatani, K, Taniguchi, M, Hashimoto, M, Zama, K, Mitsutake, S, Igarashi, Y, Takeya, H, Kigawa, J, Hayashi, A, Umehara, H, & Okazaki, T. Regulation of cell migration by sphingomyelin synthases: sphingomyelin in lipid rafts decreases responsiveness to signaling by the CXCL12/CXCR4 pathway. Molecular Cell Biology (2012). , 27-35.

[81] Baumgart, T, Hammond, A. T, Sengupta, P, Hess, S. T, Holowka, D. A, Baird, B. A, & Webb, W. W. Large-scale fluid/fluid phase separation of proteins and lipids in giant plasma membrane vesicles. Proceedings of the National Academy of Sciences USA (2007). , 104-3165.

[82] Su, B, Gao, L, Meng, F, Guo, L. W, Rothschild, J, & Gelman, I. H. Adhesion-mediated cytoskeletal remodeling is controlled by the direct scaffolding of Src from FAK complexes to lipid rafts by SSeCKS/AKAP12. Oncogene (2012).

[83] Viola, A, & Gupta, N. Tether and trap: regulation of membrane-raft dynamics by actin-binding proteins. Nature Reviews in Immunology (2007). , 7-889.

[84] Mañes, S. del Real G, Martínez AC. Pathogens: raft hijackers. Nature Reviews in Immunology (2003). , 3-557.

[85] Douglass, A. D, & Vale, R. D. Single-molecule microscopy reveals plasma membrane microdomains created by protein-protein networks that exclude or trap signaling molecules in T cells. Cell (2005). , 121-937.

[86] Pike, L. J. Rafts defined: a report on the keystone symposium on lipid rafts and cell function. Journal of Lipid Research (2006). , 47-1597.

[87] Munro, S. Lipid rafts: elusive or illusive? Cell (2003). , 115-377.

[88] Lichtenberg, D, Goñi, F. M, & Heerklotz, H. Detergent-resistant membranes should not be identified with membrane rafts. Trends in Biochemical Sciences (2005). , 30-430.

[89] Owen, D. M (b, Williamson, D, Magenau, A, & Gaus, K. Optical techniques for imaging membrane domains in live cells (live-cell palm of protein clustering). Methods in Enzymology (2012). , 504-221.

[90] Anderton, C. R, Lou, K, Weber, P. K, Hutcheon, I. D, & Kraft, M. L. Correlated AFM and NanoSIMS imaging to probe cholesterol-induced changes in phase behavior and non-ideal mixing in ternary lipid membranes. Biochimical and Biophysical Acta (2011). , 1808-307.

[91] Simons, K, & Ikonen, E. Functional rafts in cell membranes. Nature (1997). , 387-569.

[92] Brown, D. A, & London, E. Structure and origin of ordered lipid domains in biological membranes. Journal of Membrane Biology (1998). , 164-103.

[93] Brown, D. A, & London, E. Structure and function of sphingolipid- and cholesterol-rich membrane rafts. Journal of Biological Chemistry (2000). , 275-17221.

[94] Hooper, N. M. Detergent-insoluble glycosphingolipid/cholesterol-rich membrane domains, lipid rafts and caveolae. Molecular Membrane Biology (1999). , 160-145.

[95] Simons, K, & Toomre, D. Lipid rafts and signal transduction. Nature Reviews in Molecular and Cell Biology (2000). , 1-31.

[96] Parton, R. G, & Simons, K. Digging into caveolae. Science (1995). , 269-1398.

[97] Kenichi, G, & Suzuki, N. Lipid rafts generate digital-like signal transduction in cell plasma membranes. Biotechnology Journal (2012). , 7-753.

[98] Cheng, P. C, Dykstra, M. L, Mitchell, R. N, & Pierce, S. K. A role for lipid rafts in BCR signaling and antigen targeting. Journal of Experimental Medicine (1999). , 190-1549.

[99] Chung, J. B, Baumeister, M A, & Monroe, J. G. Differential sequestration of plasma membrane-associated B cell antigen receptor in mature and immature B cells into glycosphingolipid-enriched domains. Journal of Immunology (2001). , 166-736.

[100] Pierce, S. K. Lipid rafts and B-cell activation.Nature Reviews in Immunology (2002). , 2-96.

[101] Hicks, D. A, Nalivaeva, N. N, & Turner, A. J. Lipids rafts and Alzheimer's disease: protein-lipid interactions and perturbation of signaling. Nature (2006). , 233-126.

[102] Michel, V, & Bakovic, M. Lipid rafts in health and disease. Biology of the Cell (2007). , 99-129.

[103] Brugger, B, Glass, B, Haberkant, P, Leibrecht, I, Wieland, F. T, & Krausslich, H. G. The HIV lipidome: A raft with an unusual composition. Proceedings of the National Academy of Sciences USA (2006). , 103-2641.

[104] Abad-rodriguez, J, Ledesma, M. D, Craessaerts, K, Perga, S, Medina, M, Delacourte, A, Dingwall, C, De Strooper, B, & Dotti, C. G. Neuronal membrane cholesterol loss enhances amyloid peptide generation. Journal of Cell Biology (2004). , 167-953.

[105] Keenan, T. W, & Morré, D. J. Phospholipid class and fatty acid composition of golgi apparatus isolated from rat liver and comparison with other cell fractions. Biochemistry (1997). , 9-19.

[106] Fridriksson, E. K, Shipkova, P. A, Sheets, E. D, Holowka, D, Baird, B, & Mclafferty, F. W. Quantitative analysis of phospholipids in functionally important membrane domains from RBL-2H3 mast cells using tandem high-resolution mass spectrometry. Biochemistry. (1999). , 38-8056.

[107] Van Helvoort, A, & Van Meer, G. Intracellular lipid heterogeneity caused by topology of synthesis and specificity in transport. Example: sphingolipids.FEBS Letters (1995). , 369-18.

[108] Gkantiragas, I, Brügger, B, Stüven, E, Kaloyanova, D, Li, X. Y, Löhr, K, Lottspeich, F, Wieland, F. T, & Helms, J. B. Sphingomyelin-enriched microdomains at the Golgi complex. Molecular Biology of the Cell (2001). , 12-1819.

[109] Zheng, Y. Z, Berg, K. B, & Foster, L. J. Mitochondria do not contain lipid rafts, and lipid rafts do not contain mitochondrial proteins Journal of Lipid Research (2009). , 50-988.

[110] Salaün, C, James, D. J, & Chamberlain, L. H. Lipid rafts and the regulation of exocytosis. Traffic (2004). , 5-255.

[111] Lang, T, Bruns, D, Wenzel, D, Riedel, D, & Holroyd, P. SNAREs are concentrated in cholesterol-dependent clusters that define docking and fusion sites for exocytosis. The EMBO Journal (2001). , 20-2202.

[112] Gil, C, Soler-jover, A, Blasi, J, & Aguilera, J. Synaptic proteins and SNARE complexes are localized in lipid rafts from rat brain synaptosomes. Biochemical and Biophysical Research Communications (2005). , 329-117.

[113] Predescu, S. A, Predescu, D. N, Shimizu, K, Klein, I. K, & Malik, A. B. Cholesterol dependent syntaxin-4 and SNAP-23 clustering regulates caveolar fusion with the endothelial plasma membrane. Journal of Biological Chemistry (2005). , 280-37130.

[114] Lang, T. SNARE proteins and ′membrane refts′. Journal of Physiology (2007). , 26-693.

[115] HernandezVerdun D. The nucleolus: a model for the organization of nuclear functions. Histochemistry and Cell Biology (2006). , 126-135.

[116] Pederson, T. The plurifunctional nucleolus. Nucleic Acid Research (1998). , 26-3871.

[117] Scheer, U, & Hock, R. Structure and function of the nucleolus. Curr Opin Cell Biol (1999). , 11-385.

[118] Thiry, M, & Lafontaine, D. Birth of a nucleolus: the evolution of nucleolar compartments. Trends Cell Biology (2005). , 15-194.

[119] Raska, I, Shaw, P. J, & Cmarko, D. Structure and function of the nucleolus in the spotlight. Current Opinion in Cell Biology (2006). , 18-325.

[120] Pederson, T. The nucleolus. Cold Spring Harb Perspect Biol (2010). pii: a000638. doi:cshperspect.a000638.

[121] Sirri, V, Urcuqui-inchima, S, Roussel, P, & Hernandez-verdun, D. Nucleolus: the fascinating nuclear body. Histochememistry and Cell Biology (2008). , 129-13.

[122] Andersen, J. S, Lam, Y. W, Leung, A. K, Ong, S. E, Lyon, C. E, Lamond, A. I, & Mann, M. Nucleolar proteome dynamics. Nature (2005). , 433-77.

Chapter 2

The Exogenous Antioxidants

Alejandro Chehue Romero,
Elena G. Olvera Hernández, Telma Flores Cerón and
Angelina Álvarez Chávez

Additional information is available at the end of the chapter

http://dx.doi.org/10.5772/52490

1. Introduction

One theory that was initially questioned was the proposal of Dr. Denhan Harman (1956) of the University of Nebraska. He was the first researcher to propose Free radicals (FR) as an important cause of cellular aging. Today this theory enjoys wide approval. FR are "disequilibrated" molecules that travel through our organism attempting to capture an electron of the stable molecules to obtain its electrochemical stability [1,2].

FR perform many useful functions in the organism (in fact, our own bodies manufacture these in moderate amounts to combat, for example, infections). When the increase of the intracellular contents of FR exceeds the cells' antioxidant defenses and are not efficient for inhibiting them, this causes organic damage known as Oxidative stress (OS), which leads to a variety of physiological and biochemical changes that induce damage to biological molecules such as nucleic acids, proteins, lipids, etc., which consequently cause deterioration and cell death. An FR comprises any atom or group that possesses one or more unpaired electrons; thus, FR are very reactive[3].

OS traditionally has been considered a static cell-damage process that derives from the aerobic metabolism, and its clinical importance has been recognized to the point of currently being considered a central component of any pathological process. OS in diverse pathological states affects a wide variety of physiological functions, contributing to or providing biofeedback on the development of a great number of human degenerative diseases, such as atherosclerosis, diabetes, cardiomyopathies, chronic inflammatory diseases (rheumatoid arthritis, intestinal inflammatory intestinal disease, and pancreatitis), neurological diseases, high blood pressure, ocular diseases, and pulmonary and hematological disease, cancer, and immunodepression, asthma, among others [4].

This implication does not mean that Reactive oxygen species (ROS) always play a direct role in the development of the disease. In fact, reactive species predispose the organism to diseases caused by other agents. In many cases, oxidative damage is to a greater degree the consequence of the tissue damage that the disease produces than a cause of the disease itself and therefore can contribute to worsening of the tissue damage generated [3].

While our own body produces FR in moderate amounts, amounts that decrease when we age, we must also bear in mind ROS-generated exogenous sources in organisms, such as antibiotics, drugs, alcohol, tobacco, stress, contaminants, chemotherapy, and exposure to Ultraviolet (UV) and ionizing radiation.

On the other hand, numerous epidemiological studies suggest that more persons could avoid the appearance of pathological processes if they consumed antioxidant-rich diets (fruits and vegetables). Thus, it would be possible to protect the organism more efficiently against OS, with the presentation of lesser risk of developing human degenerative diseases.

This has led to conducting experiments to identify the specific components responsible for the positive effects on health by the consumption of foods of plant origin. One explanation that has found great acceptance is that this is due to the presence of antioxidant nutrients such as vitamins C and E, carotenoids, flavonoids, selenium, etc., which would interfere with oxidative damage to the DNA, proteins, and lipids [3].

Antioxidants are synthetic or natural substances that present in low concentrations compared with the biomolecules that they should protect. Antioxidants protect by retarding or inhibiting the harmful effects of FR. They are classified as follows: endogens (glutathione, co-enzyme Q, etc.), which are manufactured by the cell itself; exogens, which enter the organism through the diet (existing in determined foods) or through supplements with antioxidant formulations, and co-factors (copper, zinc, manganese, iron, and selenium). The consumption of antioxidant exogens can increase protection of the body and aid antioxidant endogens in combating diseases [5].

Fortunately, numerous foods and supplements that we ingest are rich in the antioxidants that protect against damage to the cells. Vitamin C, which is found in abundance in citrics and vegetables, is perhaps the best known antioxidant. Vitamin E, which is liposoluble, can be found in nuts, unrefined vegetable oils such as corn, cotton seed, and wheat germ, and in whole grains. Beta carotene, which is converted into vitamin A in the organism, can be found in dark-leafed vegetables, carrots, and sweet potatoes.

In recent years, plant-derived natural antioxidants have been used frequently, given that they present activity that is comparable with the most frequently employed synthetic antioxidants. Antioxidants are also found in a variety of herbs and foods that are to a great extent unknown in and not easily available in our environment, such as green tea, cardo mariano, ginkgo biloba, pine bark, and red wine; however, we do have dulcamara, dragon's blood, cat's claw, anamu/guinea hen weed), garlic, onion, aloe vera, and others that are very rich in antioxidants.

Many benefits are conferred on antioxidants against diverse pathological states; in adition to this, an unequaled richness in natural foods is exhibited as well as our obligation to take advantage of and assess these.

In the present work, the description is performed of the characteristics of the exogenous antioxidants with regard to their employment in human health [6].

2. Vitamins

Vitamins are organic micronutrients that possess no energetic value, are biologically active, and with diverse molecular structure, which are necessary for humans in very small quantities (micronutrients) and which should be supplied by the diet because humans are unable to synthesize and which are essential for maintaining health [7].

The majority of vitamins are not synthesized by the organism, some can be formed in variable amounts in the organism (vitamin D and niacin are synthesized endogenously; the former forms in the skin by exposure to the sun, niacin can be obtained from tryptophan, and vitamins K2, B1, B2, and biotin are synthesized by bacteria). However, this synthesis is generally not sufficient to cover the organism's needs. [8,9].

The functions of the vitamins and the need of the organism for these are highly varied. Persons always need vitamins and at all life stages. However, during specific periods such as growth, pregnancy, lactancy, and disease, the needs are increased [8].

The majority of vitamins have a basic function in the maintenance of health (doing honor to their name: "vita" means life. The term vitamin, proposed for the first time (in 1912) by Polish Chemist Casimir Funk, is demonstrated by the appearance of deficiency or deficiency-related diseases that were caused by the lack of vitamins in the diet; for example, lack of vitamin A can produce blindness and the lack of vitamin D can retard bone growth; vitamins also facilitate the metabolic reactions necessary for utilization of proteins, fats, and carbohydrates.

In addition, today we know that their nutritional role extends beyond that of the prevention of deficiency or deficiency-associated diseases. They can also aid in preventing some of the most prevalent chronic diseases in developed societies. Vitamin C, for example, prevents scurvy and also appears to prevent certain types of cancer. Vitamin E, a potent antioxidant, is a protector factor in cardiovascular disease and folates help in preventing fetal neural tube defects [9].

Traditionally, vitamins have been classified into two large groups in terms of their solubility as follows:

Liposoluble vitamins: A (retinol); D (ergocalciferol); E (tocopherol), and K (filoquinone and menadione), which are soluble in lipids but not in water; thus, they are generally vehiculized in the fat found in foods. These vitamins can accumulate and cause toxicity when ingested in large amounts [9].

These are fat-soluble compounds and are found associated in foods with fats, mainly of animal origin, and are absorbed with them. Therefore, any problem with respect to the absorption of fats will be an obstacle to the absorption of liposoluble vitamins. The latter are stored in moderate amounts in the vital organs, especially in the liver [8].

Hydrosoluble vitamins: The following are vitamins of the B group: [B_1 (thiamin); B_2 (riboflavin; B3 (niacin); pantothenic acid; B_6 (pyridoxine); biotin; folic acid, and B_{12} (cyanocobalamin)], and vitamin C (ascorbic acid), contained in the aqueous compartments of foods. [9].

These are water-soluble compounds that are found in foods of animal and plant origin. Different from liposoluble vitamins, water-soluble vitamins are not stored in the body; thus, they should be ingested daily with food to avoid their supply becoming exhausted [8]. The hydrosoluble vitamins participate as co-enzymes in processes linked with the metabolism of organic foods: carbohydrates; lipids, and proteins.

One important difference between these two vitamin groups lies in their final destiny in the organism. An excess of water-soluble vitamins is rapidly excreted in the urine; on the other hand, liposoluble vitamins cannot be eliminated in this manner; they accumulate in tissues and organs. This characteristic is associated with a greater risk of toxicity, which means the ingestion of excessive amounts of liposoluble vitamins, especially vitamins A and E. Vitamin B12 constitutes an exception because it is stored in the liver in important quantities.

2.1. Vitamin C

Vitamin C, also known as ascorbic acid (enantiomer, L-ascorbic acid) is an antioxidant hydrosoluble vitamin, this due to that it is an electron donor, which explains its being a reducer that directly neutralizes or reduces the damage exercised by electronically disequilibrated and instable reactive species, denominated Free radicals (FR).

Action: The presence of this vitamin is required for a certain number of metabolic reactions in all animals and plants and is created internally by nearly all organisms, humans comprising a notable exception [10].Vitamin C is essential for the biosynthesis of collagen proteins, carnitine (which is a pro-catabolic transporter of fatty acids in the mitochondria), neurotransmitters (mediators of cell communications, primarily of nerve expression), neuroendocrine peptides, and in the control of angiogenesis; it aids in the development of teeth and gums, bone, cartilage, iron absorption, the growth and repair of normal connective tissue, the metabolism of fats, and the scarring of wounds; it promotes resistance to infections by means of the immunological activity of the leukocytes [11].

In addition to the biological functions mentioned, there are an infinite number of scientific and pseudoscientific reports that qualify this vitamin as an immunomodulator, an antiviral influenza protector, an antiatherogenic, an antiangiogenic, and as an anti-inflammatory, and debate continues on its activity in cancer and its antioxidant properties, given that there is information that lends support to its procancerigenenous and to its role as a pro-oxidant. Currently, this vitamin is the most widely employed vitamin in drugs, premedication, and nutritional supplements worldwide [11]. Various lines of experimental and epidemiological evidence suggest that vitamin C is a powerful antioxidant in biological systems, both *in vitro* as well as *in vivo*. Health benefits have been attributed to vitamin C, such as the anticancerigenous, immunoregulator, antiinflammatory, and neuroprotector effect. Vitamin C rapidly eliminates Reactive oxygen species (ROS), Reactive nitrogen species (ROS), or both, and reduces the transitional metallic ions of specific biosynthetic enzymes; thus, it can prevent biological oxidation (García G.A., et al., 2006). The damage exercised by electronically disequilibrated and instable Reactive oxygen-derived species (ROS) (Free radicals, FR), nitrogen-derived FR, NOS), and sulfa-derived or mixed FR harm through oxidation any of the cellular macromolecular components. If these are not neutralized, so-called "propagation" or "amplification" is produced and, in the case of oxidation, the peroxides are again oxidized into peroxyls [12].

Clinical Uses: Vascular diseases, cancer, cataracts, High blood pressure, acute pancreatitis, the common cold, iron fixation in blood hemoglobin, dermatological uses (photochronoaging, photoprotection, prevention of contact dermatitis, non-scarring of wounds, and hyperpigmentation) [9].

Foods are substances or products of any nature that due to their characteristics and components are utilized for human nutrition. Ascorbic acid, commonly known as vitamin C, promotes resistance to infections by means of the immunological activity of leukocytes; it is useful for preventing and curing the common cold, as well as improving iron absorption in the human body and diminishing the incidence of anemia caused by lack of this mineral, which presents a high incidence in Mexican population.

Chemical structure: Ascorbic acid is a 6-carbon ketolactone that has a structural relationship with glucose; it is a white substance, stable in its dry form, but in solution it oxidizes easily, even more so if exposed to heat. An alkaline pH (>7), copper, and iron also accelerate its oxidation. Its chemical structure is reminiscent of that of glucose (in many mammals and plants, this vitamin is synthesized by glucose and galactose).

Vitamin C is found mainly in foods of plant origin and can present in two chemically interchangeable forms: ascorbic acid (the reduced form), and dihydroascorbic acid (the oxidated form) (See Figure 1), with both forms biologically functional and maintaining themselves in physiological equilibrium. If dihydroascorbic acid is hydrated, it is transformed into diketogluconic acid, which is not biologically active, and with this an irreversible transformation. This hydration occurs spontaneously on neutral or alkaline dissolution.

Deficit: It is well known that a deficiency of vitamin C causes scurvy in humans, thus the origin of the name "ascorbic" given to the acid [10].Scurvy was recognized for the first time in the XV and XVI Centuries as a serious disease contracted by sailors on long sea journeys (it appeared in adults after a nutritional need had existed for >6 months, because sailors had no access to fresh foods, including fruits and vegetables). Prior to the era of research on vitamins, the British Navy established the practice of supplying lemons and other citric fruits to their sailors to avoid scurvy [13]. Scurvy is related with defective collagen synthesis, which manifests itself as the lack of scarring, progressive asthenia, gum inflammation, falling out of the teeth, joint inflammation and pain, capillary fragility, and esquimosis, thus the importance of the ingestion of vitamin C in the diet [11].

Obtaining Vitamin C: This is a nutrient that is localized, above all, in citric fruits and vegetables. All fruits and vegetables contain a certain amount of vitamin C. Foods that tend to be greater sources of vitamin C are, among others, the following: citrics (oranges, limes, lemons, grapefruit); guavas; pineapple; strawberries; kiwis; mangoes; melon; watermelon, and cantaloupe and, as examples of vegetables, green peppers, tomatoes, broccoli, cabbage, cauliflower, green peas, asparagus, parsley, turnips, green tea, and other green-leafed vegetables (spinach), potatoes or sweet potatoes, and yams.

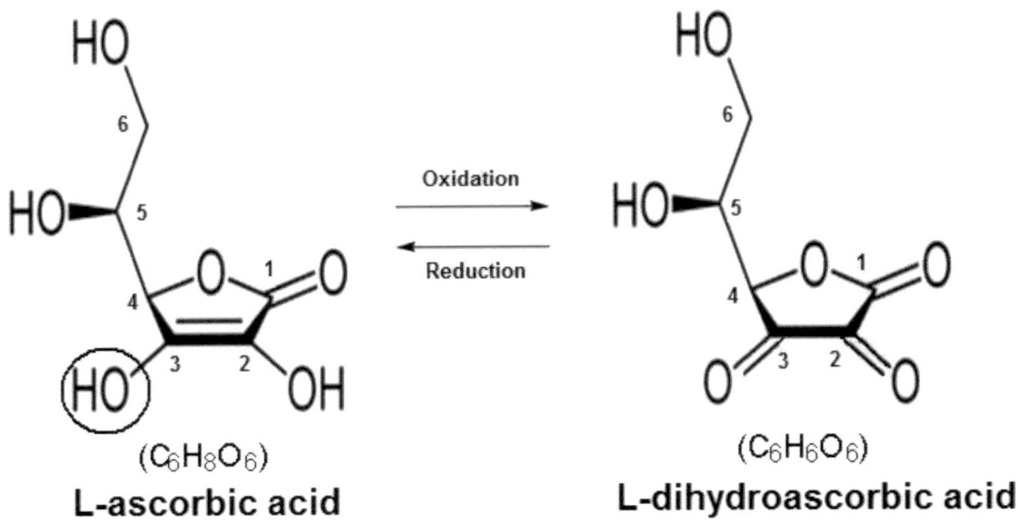

Figure 1. The oxidation-reduction (redox) reaction of vitamin C, molecular forms in equilibrium. L-dihydroascorbic acid also possesses biological activity, due to that in the body it is reduced to form ascorbic acid.

However, it is noteworthy that vitamin C diminishes on boiling, drying, or soaking foods; thus, it is convenient to consume these raw.

Daily recommended doses of ascorbic acid are 75 mg/day (for women) and 90 mg/day (for men). There are between 1.2 and 2 g (20 mg/kg body weight) of ascorbic acid available in the entire organism and its half-life ranges from 10–20 days [11–15].

Absorption: Vitamin C is easily absorbed in the small intestine, more precisely, in the duodenum. It enters the blood by active transport and perhaps also by diffusion. It would appear that the mechanism of absorption is saturable, due to that when large amounts of the vitamin are ingested, the percentage absorbed is much lower (Figure 2). In normal ingestions (30–180 mg), vitamin C is absorbed (bioavailability) at 70–90% vs. a 16% ingestion of 12 g. Its concentrations in plasma are 10–20 mcg/ml.

The vitamin C concentration in the leukocytes is in relation to the concentration of the vitamin in the tissues: therefore, by measuring the concentration of vitamin C in the leukocytes, we can know the real level of the vitamin in the tissues. The pool of vitamin C that humans possess under normal conditions is approximately 1,500 g. When this pool is full, vitamin C is eliminated at a high percentage by the urine in the form of oxalic acid (catabolite) or, if it is ingested in very high amounts, as ascorbic acid. If there are deficiencies, absorption is very high and there is no elimination by urine. Ascorbic acid is found at high concentrations in various tissues, for example, suprarenal, liver, spleen, and kidneys.Alcohol consumption diminishes absorption of the vitamin, and the smoking habit depletes the levels of the vitamin in the organism; thus, it is recommended that smokers and regular alcohol consumers supplement their diet with vitamin C.

The half-life of ascorbic acid in the organism is approximately16 days. Thus, the symptoms of scurvy do not appear for months in subjects with a diet deficient in vitamin C [7].

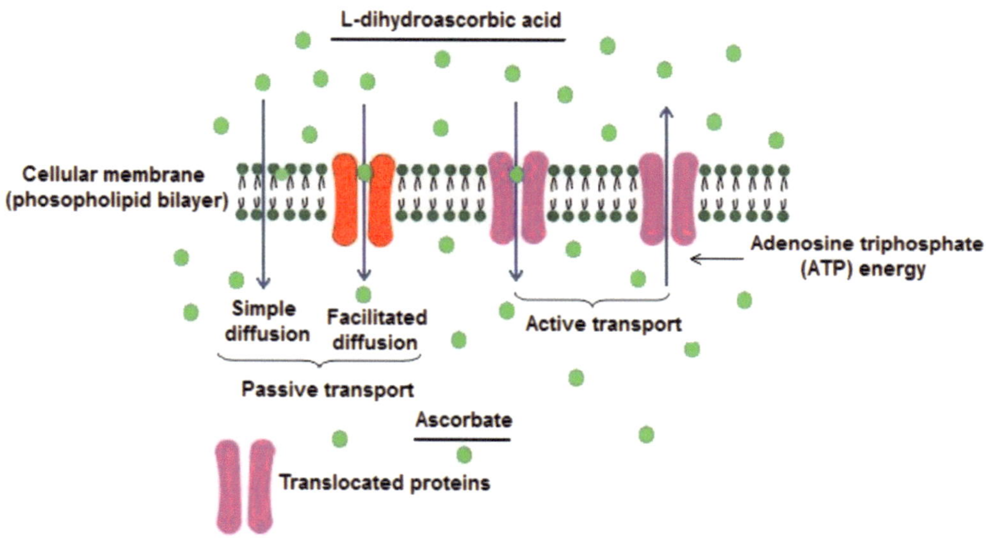

Figure 2. Mechanism of absorption of vitamin C. The L-dihydroascorbic acid molecule is better absorbed than that of L-ascorbic acid. Passive absorption is dependent on a glucose transporter and active absorption is dependent on Na^+.

Toxicity: It is scarcely probable for vitamin C intoxication (megadose) to occur because it is a hydrosoluble vitamin and excesses are eliminated through the urine. But if the daily dose of vitamin C exceeds 2,000 mg/day, the following can appear [16]:

- Diarrhea
- Smarting on urinating
- Prickling and irritation of the skin
- Important alterations of glucose in persons with diabetes
- Insomnia
- Excessive iron absorption
- Formation of oxalate and uric kidney stones.

Great controversy on the theme of Free radicals (FR) and antioxidants such as vitamin C continues, although there is conceptual dispute on whether these are the cause or consequence of the pathology. Biochemically, L-ascorbic acid donates two of its electrons from a double loop between the carbons in positions 2 and 3 (See Figure 1), and this donation is sequential, because the first molecular species generated after the loss of an electron is an FR denominated ascorbile acid. Similar to other FR with an unpaired electron, ascorbile is relatively stable and regularly non-reactive, with half-life of 10–15 seconds.

A great diversity of scientific works has allowed increasing the knowledge of the biological function of vitamin C, but this has also generated doubts, given that controversies have surfaced. One of these controversial points comprises the pro-oxidant activity of vitamin C [12,17]. In the meanwhile, there are starting points, such as that its pro-oxidant activity depends as much on the dose in the diet as on the presence of trace metals, such as iron and free copper, in order for these to produce Fenton-type reactions, and this is amplified by the additional presence of certain FR in the circulating medium [16,18]. This would also depend on the vitamin C-directed reaction.

2.2. Vitamin E

Discovered at the beginning of the 1920s in vegetable oils such as that of wheat germ by Herbert Evans and Katherine Bishop, vitamin E is also denominated tocopherol or the antisterile vitamin, due to its activity. Vitamin E is present in small amounts in all of the cells.

Vitamin E is a group of methylated phenolic compounds known as tocopherols and tocotrienols (a combination of the Greek words "τόκος" [birth] and "φέρειν" [possess or carry], which together mean "to carry a pregnancy"). Alpha-tocopherol is the most common of these and biologically that with the greatest vitaminic action. It is a lipophilic antioxidant that is localized in the cell membranes whose absorption and transport are found to be very highly linked with that of lipids. It is considered the most important lipid molecule protector because its action consists of protecting the polyunsaturated fatty acids of cell membrane phospholipids from cellular peroxidation, and also inhibiting the peroxidation of Low-density lipoproteins (LDL). It oxidizes the oxygen singlet, takes up hydroxyl FR, neutralizes peroxides, and captures the superoxide anion in order to convert it into less reactive forms [1].

Fortunately, the foods with the greatest amounts of Polyunsaturated fatty acids (PFA) also tend to have a high content of this vitamin. Sunflower seed oil, one of the foods richest in PFA, also has the highest content of vitamin E among all of the foods that we habitually consume. It is also found in other vegetable oils, in dry fruit, and in eggs. In the mean diet of Spaniards, vegetable oils furnish 79% of the vitamin E that they consume [9].

Ingestion that adequately covers the recommended allowance appears to behave as a factor of protection in cardiovascular disease, on protecting LDL from oxidation, one of the main risk factors of this pathology. Vitamin E acts jointly and synergically with the mineral selenium, another of the organism's antioxidants.

It can be easily destroyed by the action of heat and of oxygen in the air. Vitamin E is one of the least toxic liposoluble vitamins [9].

Action: It has been proposed that in addition to its antioxidant function, vitamin E can perform a specific physicochemical function in the ordering of the lipic membranes, especially of phospholipids rich in arachidonic acid (thus acting as a membrane stabilizer) [1].

In vivo, vitamin E acts when it breaks the chain of antioxidants, thus preventing the propagation of damage to the biological membranes that give rise to FR, something akin to a protective shield of the cells' membranes that allows them not to age or to deteriorate due to oxygen-containing FR, retarding cellular catabolism, impeding the chain reaction that can produce peroxides from ensuing. It participates in the hemo group and in vitamin E deficiency; hemolytic anemia appears as a result of damage by FR. It also exercises an antitoxic function, a protector in the face of various chemical agents, especially preventing the formation of peroxides from PFA, thus favoring the maintenance and stability of the biological membranes and of the lisosomes in erythrocytes, liver, and muscle [19].

Tocopherols act as intra- and extracellular liposoluble antioxidants within the body. In particular, the tocopherols protect the highly stored fatty acids (PFA) that are present in cellular and subcellular membranes, maintaining the integrity of the biological membranes, as well as other reactive compounds (e.g., vitamins A and C) from the oxidative damage that they could undergo on acting as FR traps. It has also been suggested that the tocopherols play an important role in cell respiration and in DNA and co-enzyme Q biosynthesis.

Tocopherols favor normal growth and development, act as an anticoagulant agent, stimulate the formation of red globules, stimulate the recycling of vitamin C, reduce the risk of the first mortal heart attack in males, protect against prostate cancer, improve immunity, and is a potent antioxidant against cancer in general, cardiac diseases, and FR, thus possessing a potent anti-aging function.

Vitamin E can reduce circulatory problems in the lower limbs, prevent coronary diseases, increase strength and muscular resistance (fostering achievement in sports), drive the sexual metabolism, and relieve menopausal symptoms. It can reduce the formation of scars (stimulating the curing of burns and wounds), could help in the treatment of acne, and is a potential treatment for diaper dermatitis and bee stings.

Chemical structure: The chemical formula for vitamin E ($C_{29}H_{50}O_2$) is utilized for designating a group of eight natural species (vitamers) of tocopherols and tocotrienols (α, β, γ, and δ). Together with vitamins A, D, and K, these constitute the group of liposoluble vitamins, characterized by deriving from the isoprenoid nucleus, soluble in lipids and organic solvents. They are essentials, given that the organism cannot synthesize them; therefore, their contribution is carried out through the diet in small amounts. For efficient absorption by the organism, these require the presence of fatty acids, bile, and lipolytic enzymes of the pancreas and intestinal mucosa [20].

Their structure comprises two primary parts: they contain a substitute aromatic ring denominated chromate and a long side chain (See Figure 3). These eight vitamers are divided into two basic groups: four tocopherols, and four tocotrienols, which are differentiated in the side-chain saturation; the tocopherols possess a saturated chain, and the tocotrienols, an unsaturated one with three double loops on carbons 3, 7, and 11 (Figure 4).

WIthin each group, the vitamers differ in the number and position of the methyl groups in the chromate ring, designating these as α, β, and δ (Figures 4 and 5) [19,20].

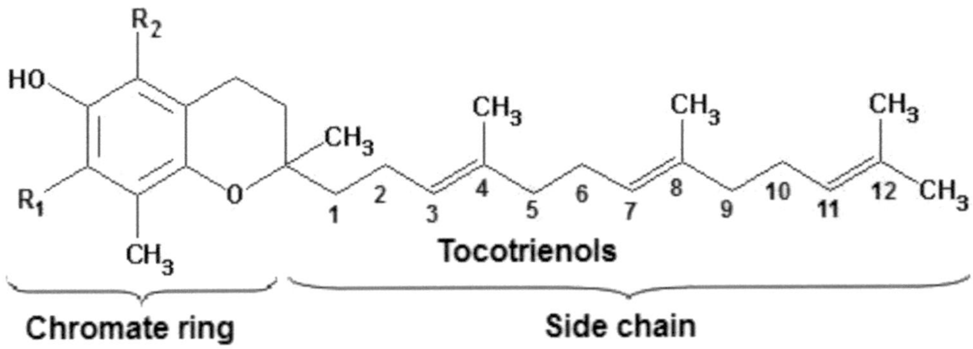

Figure 3. Components of the tocotrienol structure.

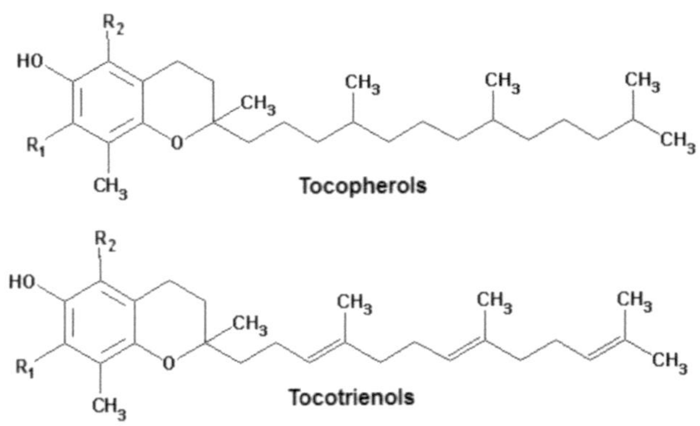

Tocopherol / Tocotrienol	R_1	R_2
α-	CH_3	CH_3
β-	H	CH_3
γ-	CH_3	H
δ-	H	H

Figure 4. Chemical structure of the possible stereoisomers of the tocopherols and tocotrienols that make up the natural vitamin E. The presence of the -CH_3 or -H groups in the chromate ring define that these substances as α, β, γ, and δ.

The presence of three chiral centers (position C2 of the chromate ring, positions C4 and C8 of the phytyl chain) allow there to be a total of eight configurations depending on the R or S orientation of the methyl group in each of the chiral centers (Figures 3 and 5) [19].

During vitamin E synthesis, equimolar amounts of these isomers (vitamers) are produced.

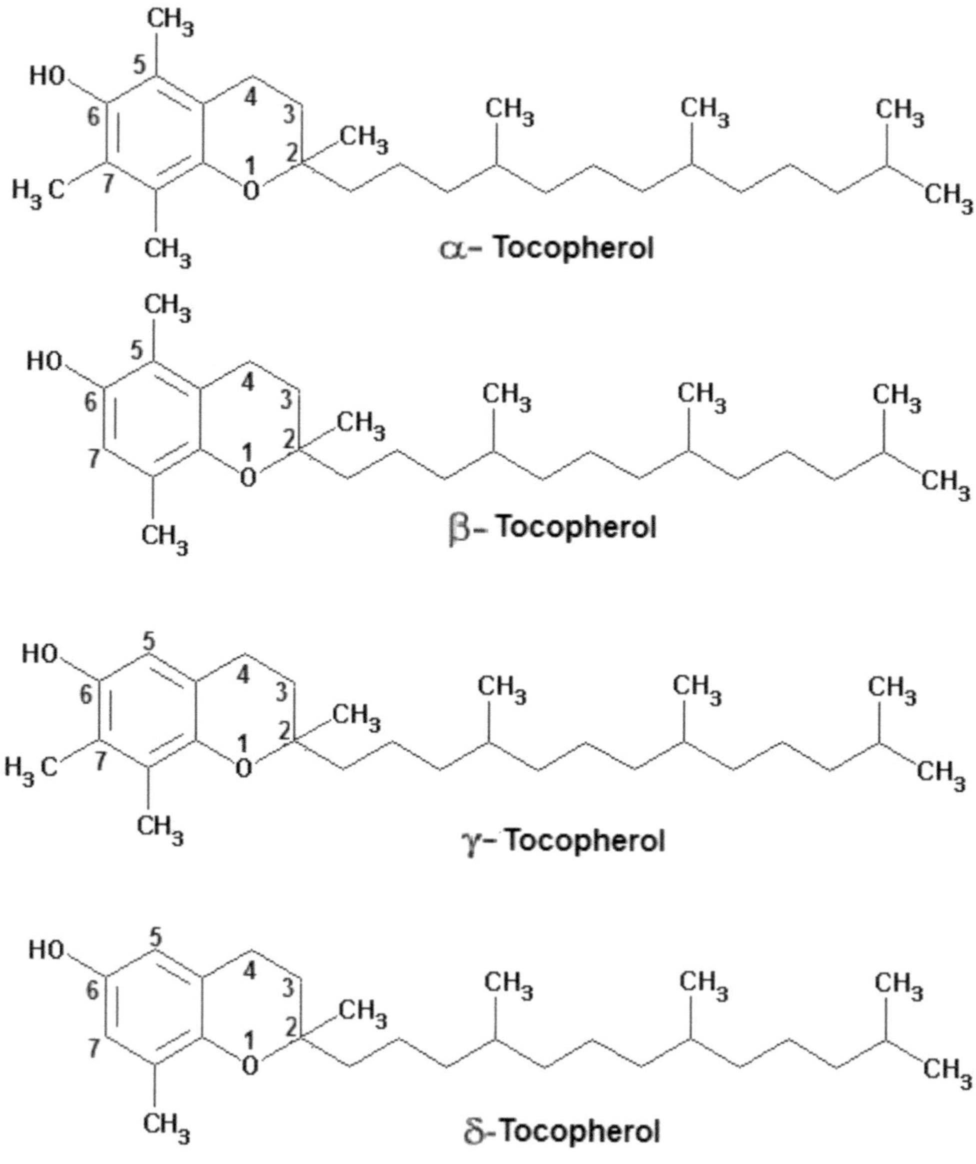

Figure 5. Chemical structure of the tocopherols.

Deficit: The deficiency of vitamin E can be due to two causes: not consuming a certain food that contains it, or poor fat absorption, due to that vitamin E is a liposoluble vitamin, that is, it is diluted in fats for its absorption in the intestine in the micelles.

Vitamin E is essential for humans. Its deficiency is not frequent even with persons who consume diets that are relatively poor in this vitamin, and could develop in cases of intense malabsorption of fats, cystic fibrosis, some forms of chronic liver disease, and congenital abetalipoproteinemia. The newborn, fundamentally the premature infant, is particularly vulnerable to vitamin E deficiency because of its deficient body reserves. The majority of vitamin E deficiency-associated sequelae are subclinical. Neuropathological alterations have been described in at-risk patients and the most frequent manifestations comprise diverse grades of areflexia, walk proprioception disorders, diminution of vibratory sensations, and ophthalmoplegia [1].

With regard to the relationship of vitamin E deficiency and the development of cardiovascular disease and cancer, there are no conclusive results to date [1,19].

The existence of a lack of vitamin E is rare. If this occurs, it is manifested in specific cases, that is, in the following three situations:

a. Persons with a difficulty of absorbing or secreting bile or who suffer from fat metabolism-related disease (celiac disease or cystic fibrosis)

b. Premature infants (with Very low birth weight, VLB) who weigh <1,500 grams at birth

c. Persons with genetic abnormalities in alpha-tocopherol transporter proteins.

Likewise, vitamin E levels can fall due to a zinc deficiency.

Lipid-absorption disorders can present in adults. From 3 years on, lack of absorption presents neurological conditions. The deficiency appears in less time due to the infants' not possessing so great a vitamin-E reserve.

2.2.1. Symptoms of vitamin E deficiency

Irritability, Fluid retention, Hemolytic anemia (destruction of red globules), Ocular alteration

Damage to the nervous system, Difficulty in maintaining equilibrium, Tiredness, apathy

Inability to concentrate, Alterations in the walk and Diminished immune response.

2.2.2. Vitamin E deficiency-related diseases

Encephalomalacia. This is due to the lack of vitamin E, which does not avoid PFA oxidation of the ration of the vitamin; consequently, hemorrhages and edema are produced in the cerebellum.

Exudative diathesis. This is due to deficient rations of vitamin E and selenium. The disease can be prevented with the administration of selenium, which acts on vitamin E as an agent that favors the storage of selenium in the organism.

Nutritional white muscle or muscular dystrophy. Rations with a scarcity of vitamin E, selenium, and azo-containing amino acids and a high content of polyunsaturated fats cause muscle degeneration in chest and thighs.

Ceroid pigmentation. This corresponds to the yellowish-brown coloration of adipose tissue in the liver due to the oxidation *in vivo* of lipids.

Erythrocytic hemolysis. The FR attack membrane and erythrocyte integrity; thus, these are also hemolysis-sensitive.

This produces sterility in some animals and certain disorders associated with reproduction, death, and fetal reabsorption in females and testicular degeneration in males.

The excess of vitamin E does not appear to produce noxious toxic effects.

Obtaining Vitamin E: Tocopherol-rich dietary sources include the following: alfalfa flour; wheat germ flour (125–100 mg/kg); hen's egg (egg yolk); polished rice (100–75 mg/kg); rice bran; mediator wheat (75–50 mg/kg); dry yeast; dry distillery solubles; barley grains; whole soy flour; corn grains; ground wheat residues (50–25 mg/kg); corn gluten flour; wheat bran; rye grains; sorghum; fish flour; oatmeal; sunflower seed flour; cotton seed flour (25–10 mg/kg); almonds; hazelnuts; sunflower seeds; nuts, and peanuts. Other sources include all vegetable oils and green vegetable harvests, above all those with green leaves, sweet chile peppers, avocado, fresh potatoes, celery, cabbage, fruits, chicken, fish, and butter [19, 20].

1 International unit (IU) of vitamin E = 1 mg αlpha-tocopherol, and 1 IU of vitamin E = 0.67 mg of vitamin E. In adults, the Minimum daily requirement (MDR) for vitamin E is 15 mg/day, and up to 200–600 mg/day would not cause any disorder.

The principal sources are vegetable oils and wheat germ. Hydrogenation of the oils does not produce a very important loss of tocopherols in terms of their content in the original oil; thus, margarine and mayonnaise contain this vitamin, in lesser amounts.

One hundred percent of the MDR of vitamin E can be covered with two tablespoons of sunflower seed or corn oil.

Absorption: The absorption of vitamin E in the intestinal lumen depends on the process necessary for the digestion of fats and uptake by the erythrocytes. In order to liberate the free fatty acids from the triglycerides the diet requires pancreatic esterases. Bile acids, monoglycerides, and free fatty acids are important components of mixed micelles. Esterases are required for the hydrolytic unfolding of tocopherol esters, a common form of vitamin E in dietary supplements. Bile acids, necessary for the formation of mixed micelles, are indispensable for the absorption of vitamin E, and its secretion in the lymphatic system is deficient. In patients with biliary obstruction, cholestasic disease of the liver, pancreatitis, or cystic fibrosis, a vitamin E deficiency presents as the result of malabsorption. Vitamin E is transported by means of plasma lipoproteins in an unspecific manner. The greater part of vitamin E present in the body is localized in adipose tissue [19, 20].

The four forms of tocopherol are similarly absorbed in the diet and are transported to the peripheral cells by the kilomicrons. After hydrolysis by the lipoprotein lipases, part of the tocopherols is liberated by the kilomicrons of the peripheral tissues [19].

Vitamin E accumulates in the liver as the other liposoluble vitamins (A and D) do, but different from these, it also accumulates in muscle and adipose tissue.

Toxicity: High doses of vitamin E can interfere with the action of vitamin K and also interfere with the effect of anticoagulants: hemorrhages.

Since 2001, it was calculated that 70% of the U.S. population occasionally consumes dietary supplements and that 40% do so on a regular basis. In 2002, Montuiler and collaborators informed, in a population of physicians, that 64% consumed doses of >400 IU/day of vitamin E and that the average obtained from food sources is 9.3 mg of αlpha-tocopherol per day (approximately 14 IU/day). In 2005, Ford and coworkers found that 11.3% of the U.S. population consumes at least 400 IU/day of vitamin E and that median daily ingestion is 8.8 IU/day.

Part of the potential danger of consuming high doses of vitamin E could be attributed to its effect on displacing other soluble antioxidants in fats and breaking up the natural balance of the antioxidant system. This can also inhibit the Glutathione S-transferase (GST) cytosolic enzymes, which contribute to the detoxification of drugs and endogenous toxins. In fact, one study on αlpha-tocopherol and β-carotene demonstrated a significant increase in the risk of hemorrhagic shock among study participants treated with vitamin E. Other data suggest that vitamin E could also affect the conversion of β-carotene into vitamin E and the distribution of the latter in animal tissues. Vitamin E possesses anticoagulant properties, possibly on interfering with the mechanisms mediated by vitamin K. In recent studies conducted *in vitro*, it was demonstrated that vitamin E potentiates the antiplatelet effects of acetylsalicylic acid; therefore, one should be alert to this effect when both substances are consumed [19].

2.3. Vitamin A

This is a term that is employed to describe a family of liposoluble compounds that are essential in the diet and that have a structural relationship and share their biological activity. It is an antioxidant vitamin that eliminates Free radicals (FR) and protects the DNA from their mutagenic action, thus continuing to halt cellular aging. Their oxygen sensitivity is due to the large amount of double loops present in their structure. Their biological activity is attributed to all-*trans* retinol, but from the nutritional viewpoint, they should be included in under the denomination of A provitamins, certain carotenoids, and similar compounds, the carotenals, which have the capacity to give rise to retinol from the organism.

Vitamin A is a hydrosoluble alcohol that is soluble in fats and organic solvents. It is stable when exposed to heat and light, but is destroyed by oxidation; thus, cooking in contact with the air can diminish the vitamin A content in foods. Its bioavailability increases with the presence of vitamin E and other antioxidants [21].

Function: In its different forms, vitamin A, also known as an antixerophthalmic, is necessary in vision, normal growth (its deficiency causes bone growth delay), reproduction, cellular proliferation, differentiation (which confers upon it a role in processes such as spermatogenesis, fetal development, immunological response, etc.), fetal development, and the integrity of the immune system. Others of these include its being an antioxidant, amino acid metabolism, the structure and function of other cells, reproduction, and epithelial tissues.

Vitamin A participates in the synthesis of glycoproteins, which contributes to maintaining the integrity of epithelial tissue in all of the body's cavities. Epithelial dissection especially affects the conjunctivae of the eye (xerophthalmia), which renders the cornea opaque and causes crevices, producing blindness and facilitating eye infections.

Sources: Retinol is only found in the lipidic part of foods of animal origin as follows; whole milk; lard; cream; cheese; egg yolk; eels, and fatty fish, due to their self-storage in the liver and in the oils extracted from the liver. The latter, as well as the oils extracted from the liver (veal and pork), comprise an important source of vitamin A. Cod liver oil constitutes source richest in vitamin A, although this cannot be considered a food in the strictest sense. In the case of skim/low-fat milk, this vitamin would be eliminated, but by law it is restituted to its original content; examples of these include manchego cheese, margarine, and butter.

Vegetables contain only provitamins or carotenes (all of these coloring pigments, such as alpha, beta, and gamma carotene). Garden vegetables (spinach and similar vegetables), carrots, sweet chile peppers, potatoes, tomatoes, and red and yellow fruits are the main suppliers. We must bear on mind that there are numerous carotenes that do not possess any provitamnic A activity, such as lycopene from the tomato, although it does act as a neutralizer of FR [21, 22].

Structure of Vitamin A: This vitamin is a diterpene ($C_{20}H_{32}$) that can present in the following various molecular forms:

Retinol (See Figure 6), when the side chain terminal is an alcohol group ($-CH_2OH$)

Retinal, the carbon terminal, is an aldehyde (–CHO)

Retinoic acid, when the terminal group is acidic (–COOH)

Retinyl-palmitate, in the case of lengthening of a side chain from esterification with palmitic acid ($-CH_2O\text{-}CO\text{-}(CH_2)_{14}\text{-}CH_3$).

Absorption: The metabolism of the vitamin responds to the same general mechanisms of digestion and absorption as those of other lipidic substances. Absorption is carried out in the form of carotenes or similar substances at the intestinal level within the interior of the micelles and quilomicrons, together with other fats.

Retinol esters are absorbed from 80–90%, while the beta-carotenes are absorbed at only 40–50%. Factors in the diet that affect carotene absorption include the origin and the concentration of the fat in the diet, the amount of carotenoids, and the digestibility of the foods. Vitamin A is first processed in the intestine, and afterward it arrives at the liver via portal, the liver being the main storage organ. In addition, the liver is responsible for regulating the secretion of the retinol bound to the retinoid-binding protein. Carotene absorption in particular is very inefficient in raw foods, and its content in lipids in the diet is low. The efficiency of conversion into retinol, which is quite variable and, in general, low, depends not only on the structure of the carotenoids, but also on their proteinic ingestion. Thus, when carotene ingestion is very high, those which have not been transformed into retinol in the retinal mucosa are absorbed unaltered, bind with the lipoproteins, and are deposited in the skin and the mucosa, on which they confer a typical yellowish color, constituting hypercarotenosis [21].

Toxicity: Both the deficiency as well as the excess of vitamin A causes fetal malformations. Ingestion of large amounts of this vitamin can give rise to skin alterations (scaling), hair fall, weakness, choking, vomiting, etc. In extreme cases, great amounts accumulate in the liver, producing hepatic disorders that end up as fatty liver.

It is noteworthy that the administration of vitamin A in chronic form and at doses higher than the recommended doses those can produce a clinical condition of toxicity characterized by fatigue, irritability, cephalea, febricula, hemorrhages in different tissues, and cutaneous alterations.

In children, this can trigger the early closing of the long bones, which causes the height to descend. Megadoses of vitamin A can produce acute intoxication that will be characterized by clinical features of sedation, dizziness, nausea, vomiting, erythema, pruritis, and generalized desquamation of the skin. We should also point out that in the elderly, the safety margin when we administer this vitamin is small; thus, we must be especially cautious and adjust the dose well [21].

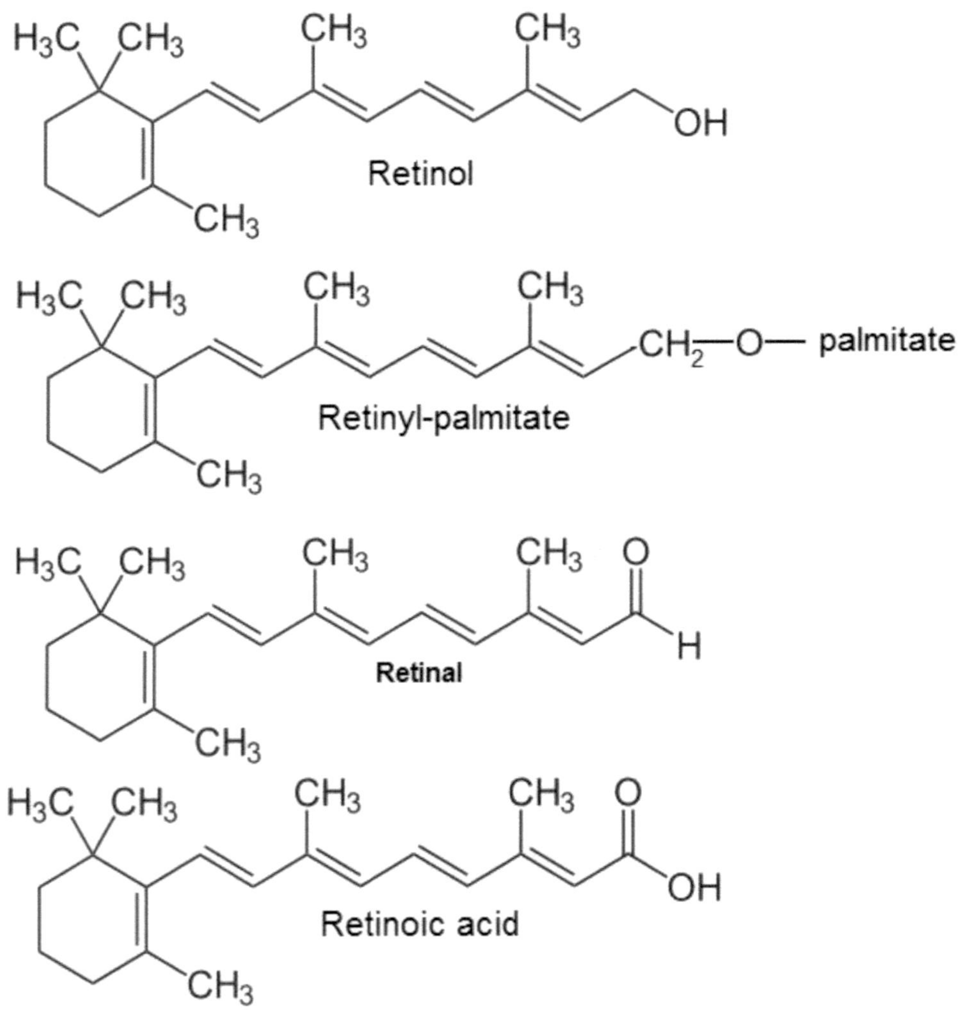

Figure 6. Molecular forms of vitamin A.

2.4. Flavonoids and their antioxidant actions

Flavo comes from the Latin *flavus* and means the color found between yellow and red, such as that of honey or of gold, and flavonoid refers to an aromatic group, with heterocyclic pigments that contain oxygen, which are widely distributed among plants, constituting the majority of yellow, red, and blue fruits. Consequently, flavonoids are found in abundance in grapes, apples, onions, cherries, and cabbage, in addition to forming part of the ginkgo biloba tree and *Camellia sinensis* (green tea). On consuming these, we obtain the anti-inflammatory, antimicrobial, antithrombotic, antiallergic, antitumor, anticancerigenous, and antioxidant properties. With regard to the latter properties, these lie within its function in the nervous system, because a protector relationship has been observed with regard to neurodegenerative diseases [22].

Flavonoids are Low-molecular-weight (LMW) compounds that share a common skeleton with diphenylpyrenes (C6-C3-C6); a flavonoid is a 2-phenyl-ring (A and B) compound linked through the pyrene C ring (heterocyclic). The carbon atoms in the C and A rings are numbered from 2–8, while those of the B ring are numbered from 2'–6'12 (Figure 7). The activity of flavonoids as antioxidants depends on the redox properties of their hydroxy phenolic groups and on the structural relationship among the different parts of their chemical structure[22].

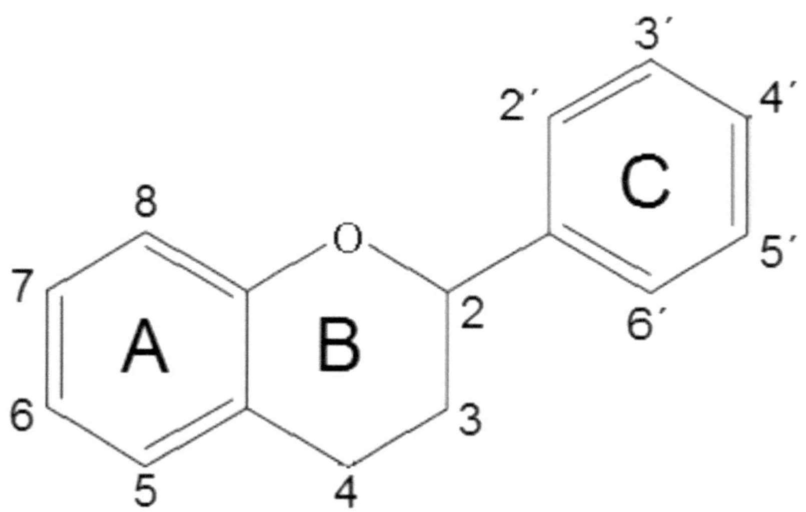

Figure 7. Base structure of the flavonoids

Thanks to the variations of pyrene, the flavonoids achieved classification, as shown in Table 1 (Antiatherogenic properties of flavonoids: Implications for cardiovascular health, 2010) [24].

Name	Structure	Description	Family members	Dietary sources
Flavanones	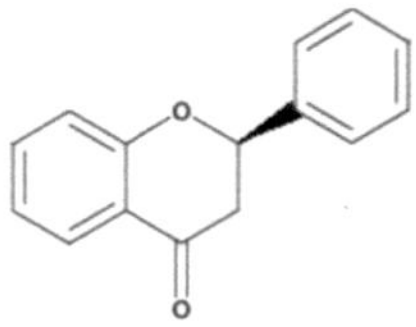	Carbonyl group at position 4 and an -OH group in position 3 of ring C	Quercetin, myricetin, isorhamnetin, kaempferol, pachypodol, rhamnazin	Onions, apples, broccoli, cranberries, berries, grapes, parsley, spinach
Flavan-3-ols	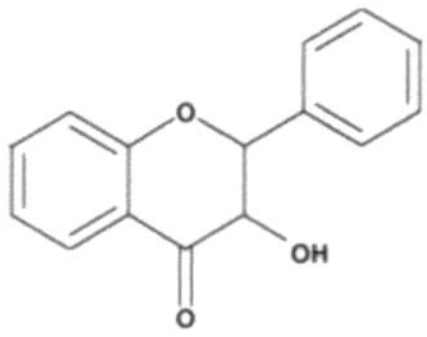	With an -OH group in position 3 of ring C	Catechins, epigallocatechin gallate, epicatechins, epicatechin gallate	Tea, red wine, cocoa, grapes, plums, fruits, legumes
Flavones	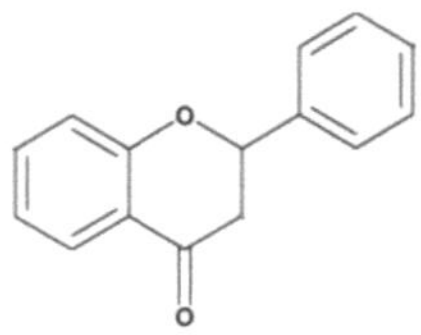	Have a carbonyl group in position 4 of ring C and lacking the hydroxyl group at position C3	Apigenin, nobiletin, tangeritin, luteolin	Celery, lettuce, parsley, citrus fruits, beets, bell peppers, spinach, Brussels sprouts, thyme
Anthocyanidins	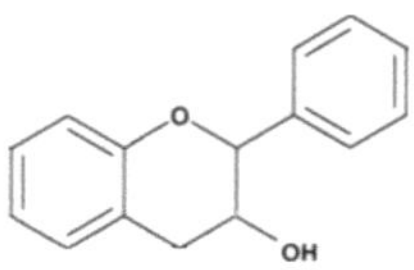	Carbonyl group at position 4 and an -OH group in position 3 of ring C	Cyanidin, delphinidin, peonidin, malvidin, pelargonidin	Red wine, blueberries cranberries, black currants, plums, red onions, red potatoes

Table 1. Classification of flavonoids

Distribution: The flavonoids are widely distributed among the higher plants, with the rutaceous, polygonaceous, compound, and umbelliferous plant families the principal ones containing flavonoids. Flavonoids abound, above all, in young, aerial plant parts and in those most exposed to the sun, such as the leaves, fruits, and flowers, because solar light favors their synthesis, controlling the levels of the auxins (vegetables hormones), which are growth regulators.

These compounds are important for the plant, similar to what occurs with the greater part of secondary metabolites, in addition to being responsible for the coloration of many flowers, fruits, leaves, and seeds, achieving >5,000 distinct flavonoids, because these can be found in the following groups:

a. Elegiac acid: found in fruits such as grapes and in vegetables

b. Anthocyanidines: the pigment responsible for the reddish-blue and red color of cherries

c. Catechins: found in black and green tea

d. Citroflavonoids: such as quercetin, lemonene, pyridine, rutin, and orangenine. The bitter flavor of the orange, lemon, and grapefruit confers orangenine on these fruits, while lemonene has been isolated from the lime and the lemon.

e. Isoflavonoids: such as genestein and daidzein, present in soy foods such as tofu, soy milk, soybeans, soy vegetable protein, tempeh/fermented soybeans, miso/soybean paste, and soy flour

f. Kaempherol: found in broccoli, leeks, endives, red beets, and radishes

g. Proanthocyanidines: these appear in grape seeds, sea pine bark extract, and in red wine.

These merit incorporation into the group of essential nutrients. The mean value of the ingestion of flavonoids is 23 mg/day. The main flavonoid consumed is quercetin, tea being its main source [22].

Properties: The flavonoids are white or yellowish solid, crystallized substances. Their heretosides are soluble in hot water, alcohol, and polar organic dissolvents, being insoluble in apolar organic dissolvents. However, when they are in their free state, they are scarcely hydrosoluble, but are soluble in more or less oxygenated organic substances, depending on their polarity.

On the other hand, these are easily oxidizable substances; thus, they exert an antioxidant effect because they are oxidized more rapidly than other types of substances [23].

Pharmacological activity: Pharmacologically, flavonoids are prominent due to their low toxicity, presenting in general activity on the vascular system with P vitaminic action (protector effect of the vascular wall due to the diminution of permeability and to the increase of capillary resistance). Likewise, they possess an antioxidant effect, can inhibit lipid peroxidation, have antimutagenic effects, and possess the capacity to inhibit diverse enzymes [23, 24].

Antioxidant functions: The flavonoids' antioxidant action depends mainly on their sequestering capacity of FR and on their chelant properties of metals such as iron, impeding the catalytic actions of FR, and they also act by inhibiting the enzyme systems related with vascular functionality, such as the following: Catechol-O-methyl transferase (COMT), with which it increases the duration of the action of the catecholamines, thus inciding in vascular resistance; histidine decarboxylase, thus affecting the histamine's action, and the phosphodiesterases, thus inhibiting platelet aggregation and adhesiveness, in addition to the following oxidases: lipo-oxygenase; cyclo-oxygenase; myeloperoxidase, and xanthinic oxide, therefore avoiding the formation of Reactive oxygen species (ROS) and organic hydroperoxides.

In addition to this, it has been observed that they also indirectly inhibit oxidative processes, such as phospholipase A_2, at the same time stimulating others with recognized antioxidant properties, such as catalase and SOD.

With respect to their structure, flavonoids are their hydroxylic constituents in positions 3′ and 4′; in the B ring, they demonstrate more action as antioxidants and this effect is potentiated by the presence of a double loop between carbons 2 and 3 and a free OH group in position 4. Additionally, the glycols show to be the most potent in their antilipoperoxidative actions than in their corresponding glycosidic actions.

As previously mentioned, quercetin is the flavonoid that unites the requisites for exercising an effective antioxidant function, because it is five times higher than vitamins A and C and additionally possesses a hydrosolubility similar to that of the latter. Therefore, rutin (quercetin-3-b-D-rutinoside) is, to date, the sole flavonoid with a pharmacological presence in Mexico.

There is a synergic effect with all of the vitamins to which we have alluded. This is due to that ascorbic acid reduces the oxidation of quercetin in such a way that combines with it and allows the flavonoids to maintain their functions for a longer time. For its part, quercetin protects vitamin E from oxidation.

The flavonoids remove reactive oxygen, especially in the form of SOD, hydroxyl radicals, hydroperoxides, and lipid peroxides, blocking the harmful effects of these substances on the cell, in which antioxidant protection of flavonoids has been corroborated in the following: queratinocytes; dermal fibroblasts; sensory lymph nodes; the endothelium; nervous tissue, and LDL.

On the other hand, the flavonoids exercise other actions as follows: diuretic; antispasmodic; anti-gastriculcerous, and anti-inflammatory.

In phytotherapy, the flavonoids are mainly employed in cases of capillary fragility as venotonics, although they are also utilized in proctology, metrorrhages, and retinopathies [22].

2.5. Pro-oxidant mechanisms

Due to the structural characteristics of some flavonoids, such as the anthocyanidines, these cause low oxidation potentials (EP/2), which permits them to reduce Fe^{3+} and Cu^{2+} in order for them to undergo auto-oxidation or even to become involved in the redox recycling process, acting in this manner as pro-oxidant agents, which explains the mutagenic and genotoxic effects of some flavonoids.

Some of these mechanisms include the temporary reduction of Cu (II) to CU (I), auto-oxidation of the aroxyl radical and generating the superoxide anion (O2–) that, on following its general sequence, becomes the harmful hydroxyl radical (HO.), as well as the affectation of the functions of the components of the nuclear antioxidant defense system: glutathione, and glutathione-S-transferase.

What determines the antioxidant or pro-oxidant character is the redox stability/lability of the radical compound forming part of the original flavonoid. The pro-oxidant actions only appear to be produced when the flavonoid doses are excessively high [25].

Under this heading, we will present a brief review of the remaining antioxidants present in our diet, their activity, and the foods that supply them.

2.6. Lycopene

Lycopene is the carotenoid that imparts the red color to the tomato and watermelon and that it not converted into vitamin A in the human organism, which does not impede it from possessing very high antioxidant properties.

The highest concentrations of lycopene are found in prostatic tissue. High consumption of lycopene has been related with the prevention of some cancer types, precisely that of the prostate.

Although the tomato is the greatest source of lycopene, there are also other vegetables and fruits that present intense colors, such as watermelons, papayas, apricots, and pink grapefruit. The tomato is the food that concentrates the greatest amount of lycopene, and it should be considered that there are factors that affect its assimilation into the organism, such as its maturity, the distinct varieties, or the manner of cooking, all of which exert an influence on the amount and degree of exploitation of lycopene.

Of all of these, the fried tomato is that which best assimilates this substance, frying being the best way of cooking because, in addition to the heat, there is a certain amount of fat involved, which renders better assimilation of lycopene (fat-soluble). In concrete fashion, its presence in the fried tomato is some 25 μg per 100 g, while in the fresh tomato, this is around 2 μg per 100 g [6, 26, 27].

3. Minerals

Other potent antioxidants include minerals such as copper, manganese, selenium, zinc, and iron. These minerals exercise their antioxidant function in diverse processes and metabolic steps in the organism [6, 26, 27].

3.1. Zinc

Zinc intervenes in >200 enzymatic reactions and its deficit increases the production of oxidant species and Oxidative stress (OS) [6, 26, 27].

3.2. Copper

Copper participates in functions with antioxidant features of the enzyme family denominated Superoxide dismutase (SOD), which is responsible for eliminating the superoxide anion.

It empowers the immune system, participates in the formation of enzymes, proteins, and brain neurotransmitters (cell renovation and stimulation of the nervous system) and is an anti-inflammatory and anti-infectious agent.

Similarly, it facilitates the synthesis of collagen and elastin (necessary constituents of the good state of the blood vessels, lungs, and the skin).

In addition, it acts as an antioxidant, protecting the cells from the toxic effects of FR, and it facilitates calcium and phosphorous fixation [6, 26, 27].

3.3. Manganese

Manganese also intervenes in this family of enzymes, concretely, in enzymes localized within the mitochondria [6, 26, 27].

3.4. Selenium

Selenium intervenes in the synthesis of enzymes related with the oxidative function, such as glutathione peroxidase, which, as its name indicates, eliminates peroxide groups, including oxygen peroxide.

This mineral is incorporated into proteins in the form of selenoproteins and, in this manner, aids in the prevention of cell damage. Epidemiological studies related the lack of selenium in the diet with the incidence of lung, colorectal, and prostate cancer.

The selenium content in the diet is directly related with the selenium content of the soil in which the food was grown. Thus, selenium-deficient soils give rise to a deficit of this element in the population, as in the case of China.

In this specific latter case, the method-of-choice comprises supplementing the diet with contributions of selenium, preferably in the form of selenomethionide, which is the analog, organic form of selenium and which easily increases selenium levels in the blood [6, 26, 27].

3.5. Iron

Iron forms part of the organism's antioxidant system because it contributes to eliminating the peroxide groups. However, its capacity to change valence with ease (2+/3+) renders that it can also intervene, depending on the environment, in the formation of Free radicals (FR) [6, 26, 27].

3.6. Co-enzyme Q

Co-enzyme Q10 or ubiquinone is a liposoluble compound that can be carried in many foods, although it can also be synthesized in the human organism. Co-enzyme Q10 diminishes with age; thus, the metabolic processes in which it has been found implicated are also co-enzyme Q10-sensitive.

Given its liposolubility, its absorption is very los, especially when the diet is poor in fats.

Its principal antioxidant activity resides in that, in its reduced form, it is a liposoluble antioxidant that inhibits lipid peroxidation in LDL. It is also found in the mitochondria, where it could protect protein membranes and the DNA from the oxidative damage that accompanies lipid peroxidation in these membranes.

It additionally acts as an immune system stimulant and through this stimulation also functions as an anticancerigen. In addition, it is capable of directly regenerating alpha-tocopherol [6, 26, 27].

4. Lipoic acid

Lipoic acid or thioctic acid is also a compound that forms part of the antioxidant capital of the organism.

Numerous studies have shown the protector effect of red globules and of the fatty acids of oxidative damage typical of intense exercise and excessive exposure to the sun's UV rays.

It is synthesized by plants and animals, as well as by the human organism, although in the latter case, in very small amounts. Lipoic acid is considered a very good regenerator of potent antioxidants such as vitamin C, vitamin E, glycation, and co-enzyme Q10. It is liposoluble and hydrosoluble, which means that it can act on any part of the organism.

It is found in spinach and similar green-leafed vegetables, broccoli, meat, yeast, and in certain organs (such as kidney and heart) [6, 26, 27].

4.1. Naringenine

The hypolipidemic and anti-inflammatory activities *in vivo* as well as *in vitro* of the flavonoids of citric fruits have been widely demonstrated. Among the flavonoids, naringenine, one of the compounds that causes the bitter taste of grapefruit, has been studied extensively in recent years. In a recently conducted clinical assay, it was found that naringenine reduced Low-density-lipoprotein (LDL) levels in the circulation of 17% of patients with hypercholesterolemia. Additionally, the reducer effects of cholesterol in rabbits and rats were demonstrated, in addition to the reducer effects of Very-low-density-lipoprotein (VLDL) levels through the inhibition of key proteins for their assembly. Other studies reported that naringenine activates enzymes that are important for the oxidation of fatty acids, such as CYP4A1 [28].

5. Conclusion

A good diet influences the development and treatment of diseases, it is increasingly evident. After that epidemiological studies have shown the association between moderate consumption of certain foods and reduced incidence of various diseases at the rate of these observations has attracted considerable interest in studying the properties of substances inherent in the chemical composition of food. Among the characteristics of these substances is the antioxidant activity, associated with the elimination of free radicals and therefore to the prevention of early stages which can trigger degenerative diseases. In this regard it is important to continue the study of dietary antioxidants on the activity may have on human diseases, paying attention to the substances primarily natural antioxidants of food and synthetic way to assess its protective effect on the body.

Author details

Alejandro Chehue Romero*, Elena G. Olvera Hernández, Telma Flores Cerón and Angelina Álvarez Chávez

*Address all correspondence to: chehue_alex@yahoo.com, chehuea@uaeh.edu.mx

Autonomous University of Hidalgo State, Mexico

References

[1] Criado D.C. y Moya M.M.S.(2009). Vitaminas y antioxidantes. Actualizaciones El Medico. Comisión Nacional de Formación Continuada del Sistema Nacional de Salud, Madrid.

[2] Harman D. Aging (1956): a theory based on free radical and radiation chemistry. J Gerontol 1956; 11(3):298-300.

[3] *Rosa Mayor-Oxilia R.* Estrés Oxidativo y Sistema de Defensa Antioxidante. *R. Rev. Inst. Med. Trop. 2010;5(2):23-29.*

[4] Olivares-Corichi IM, Guzmán-Grenfell AM, Sierra MP, Mendoza RS y Hicks Jj.Perspectivas del uso de antioxidantes como coadyuvantes en el tratamiento del asma. Rev Inst Nal Enf Resp Mex 2005; Volumen 18 - Número 2 , Páginas: 154-161.

[5] Carlsen MH. The total antioxidant content of more than 3100 foods, beverages, spices, herbs and supplements used worldwide. Nutrition Journal 2010; 9:3. http://www.nutritionj.com/content/9/1/3 (accessed 5 May 2012).

[6] Torrades S. Aportes extras de vitaminas. ¿Son realmente Necesarios? O F F A R M. Ámbito Farmacéutico Bioquimica 2005; Vol 24 Núm 6, Junio.

[7] Goodman, L.S; Gilman, A. Las Bases Farmacológicas da Terapéutica. 10ª ed., Editora McGraw Hill Interamericana, 1996. p. 1647.

[8] Gamboa C. Vitamina A, Guias alimentarias para la educacion nutricional en costa rica, http://www.ministeriodesalud.go.cr/index.php/inicio-menu-principal-comisiones-ms/140 (accessed 20 March 2012).

[9] Nutricion K, ABC de la Nutricion, Kellogg's, http://www.kelloggs.es/nutricion/index.php?donde=abc (accessed 5 April 2012).

[10] World Health Organization. Vitamin and mineral requirements in human nutrition. 2nd Edition.2004. http://www.who.int/nutrition/publications/micronutrients/9241546123/en/index.html (accessed 15 June 2012).

[11] Sandoval H, Sharon D. Cuantificación de Ácido Ascórbico (Vitamina C) en Néctares de Melocotón y Manzana Comercializados en Supermercados de la Ciudad Capital.Report of Thesis Presented, Universidad de San Carlos de Guatemala. Facultad de Ciencias Químicas y Farmacia; 2010.

[12] García GA, Cobos C, Rey CA, et al. Biología, patobiología y bioclínica de la actividad de oxidorreducción de la vitamina C en la especie humana. Universitas Médica 2006; VOL. 47 Nº 4.

[13] Latham M C. Carencia de vitamina C y escorbuto. In: Food and Agriculture Organization of the United Nations (ed.) Nutrición Humana en el Mundo en Desarrollo, *Colección FAO: Alimentación y nutrición N° 29,* Roma, 2002. Capitulo 19. http://www.fao.org/docrep/006/W0073S/w0073s0n.htm#TopOfPage (accessed 20 March 2012).

[14] Valdés F. Vitamina C. Actas Dermosifiliogr 2006 ; 97(9):557-68.

[15] *Taylor S.* Dietary Reference Intakes for vitamin C, vitamin E, Selenium and Carotenoids. Food and Nutrition Board, Institute of Medicine, National Academy Press 2000; Advance Copy, 3;6-7;97.

[16] World Health Organization and Food and Agriculture Organization of the United Nations. Vitamin and mineral requirements in human nutrition, Second edition. 2004.

[17] Anitra C. and Balz F. Does vitamin C act as A Pro-Oxidant?, The FASEB Journal. Vol. 13 June 1999.

[18] Proteggente vAR, Rehman A, Halliwell B, Rice-Evans CA. (2000) Potential problems of ascorbate and iron supplementation: pro-oxidant effect in vivo?. Biochem Biophys Res Commun 2000; 2;277(3):535-40.

[19] Blé-Castillo JL, Díaz-Zagoyab JC and Méndez JD. Suplementación con vitamina E, ¿benéfica o dañina?. Gac Méd Méx 2008; Vol. 144 No. 2.

[20] Sayago A, Marín MI, Aparicio R y Morales MT. Vitamina E y aceites vegetales, GRASAS Y ACEITES 2007; 58 (1), ENERO-MARZO, 74-86.

[21] Pérez RM y Ruano A. Vitaminas y salud. Aportación vitamínica. O F F A R M. Ámbito Farmacéutico Nutricion 2004; Vol 23 Núm 8, Septiembre.

[22] Christopher Isaac Escamilla Jiménez, E. Y. Flavonoides y sus acciones antioxidantes. *Rev. Fac. Med 2009; UNAM*, 73-75.

[23] Retta E D. . Marcela,A promising medicinal and aromatic plant from Latin America: A review. *Indusrial Crops and Products 2012; 38*, 27-38.

[24] Mulvihill, M. W. Antiatherogenic properties of flavonoids: Implications for cardiovascular health. *Can J Cardiol 2010; Vol 26 Suppl A March*, 17A-21A.

[25] Trueba GP. Los Flavonoides: Antioxidantes o Prooxidantes. *Rev. Cubana Biomed 2003; 22(1)*, 48-57.

[26] Vilaplana M. Antioxidantes presentes en los alimentos O F F A R M. Ámbito Farmacéutico Bioquimica 2007; Vol 26 Núm 10, Noviembre.

[27] García-Álvarez JL, Sánchez-Tovar T. y García-Vigil JL. Uso de antioxidantes para prevenir enfermedad cardiovascular. Metaanálisis de ensayos clínicos. Rev Med Inst Mex Seguro Soc 2009; 47 (1): 7-16.

[28] Goldwasser J, Cohen PY, Yang E, Balaguer P, Yarmush ML, et al. Transcriptional Regulation of Human and Rat Hepatic Lipid Metabolism by the Grapefruit Flavonoid Naringenin: Role of PPARα, PPARγ and LXRα. PLoS ONE 2010; 5(8). http://www.plosone.org/article/info:doi/10.1371/journal.pone.0012399 (accessed 15 January 2012)

Chapter 3

Chemistry of Natural Antioxidants and Studies Performed with Different Plants Collected in Mexico

Jorge Alberto Mendoza Pérez and
Tomás Alejandro Fregoso Aguilar

Additional information is available at the end of the chapter

http://dx.doi.org/10.5772/52247

1. Introduction

1.1. Antioxidants

Oxidation is the transfer of electrons from one atom to another and represents an essential part of both aerobic life and our metabolism, since oxygen is the ultimate electron acceptor in the electron flow system that produces energy in the form of ATP. However, problems may arise when the electron flow becomes uncoupled (transfer of unpaired single electrons), generating free radicals [1]. Antioxidants are important in living organisms as well as in food because they may delay or stop formation of free radical by giving hydrogen atoms or scavenging them. Oxidative stress is involved in the pathology of cancer, atherosclerosis, malaria and rheumatoid arthritis. An antioxidant can be defined in the broadest sense of the word, as any molecule capable of preventing or delaying oxidation (loss of one or more electrons) from other molecules, usually biological substrates such as lipids, proteins or nucleic acids. The oxidation of such substrates may be initiated by two types of reactive species: free radicals and those species without free radicals are reactive enough to induce the oxidation of substrates such as those mentioned. There are three main types of antioxidants:

1. Primary: Prevent the formation of new free radicals, converting them into less harmful molecules before they can react or preventing the formation of free radicals from other molecules. For example:

   - Enzyme superoxide dismutase (SOD) which converts $O_2 \bullet -$ to hydrogen peroxide (H_2O_2)

INTECH
open science | open minds

- Enzyme glutathione peroxidase (GPx), which converts H_2O_2 and lipid peroxides to harmless molecules before they form free radicals.
- Catalases
- Glutathione reductase.
- Glutathione S transferase.
- Proteins that bind to metals (ferritin, transferrin and ceruloplasmin) limit the availability of iron necessary to form the radical OH

2. Secondary: Capture free radicals, preventing the chain reaction (eg vitamin E or alpha-tocopherol, vitamin C or ascorbic acid, beta-carotene, uric acid, bilirubin, albumin, ubiquinol-10, methionine)
3. Tertiary: They repair damaged biomolecules by free radicals (eg DNA repair enzymes and methionine sulfoxide reductase) [2].

It also handles the classification based according to where they perform their activities, their background and their biochemical characteristics. So, antioxidants are also classified into two broad groups, depending on whether they are water soluble (hydrophilic) or lipid (hydrophobic). In general, water soluble antioxidants react with oxidants in the cell cytoplasm and blood plasma, whereas the liposoluble antioxidants protecting cell membranes against lipid peroxidation. In the metabolism it is a contradiction that while the vast majority of life requires oxygen for its existence, oxygen is a highly reactive molecule that damages living organisms by producing reactive oxygen species. Therefore, organisms possess a complex network of antioxidant metabolites and enzymes that work together to prevent oxidative damage to cellular components such as DNA, proteins and lipids. Usually antioxidant systems prevent these reactive species are formed or removed before they can damage vital components of the cell. Reactive oxygen species produced in cells include hydrogen peroxide (H_2O_2), hypochlorous acid (HClO), and free radicals such as hydroxyl radical (•OH) and superoxide radical (O_2•-). The hydroxyl radical is particularly unstable and reacts rapidly and non-specifically with most biological molecules. This species produces hydrogen peroxide redox reactions catalyzed by metals such as the Fenton reaction. These oxidants can damage cells starting chemical chain reactions such as lipid peroxidation or by oxidizing DNA or DNA damage proteins. These effects can cause mutations and possibly cancer if not reversed by DNA repair mechanisms, while damage proteins will cause enzyme inhibition, denaturation and degradation of proteins. The use of oxygen as part of the process for generating metabolic energy produces reactive oxygen species. In this process, the superoxide anion is produced as a byproduct of several steps in the electron transport chain. Particularly important is the reduction of coenzyme Q in the compound III as a highly reactive free radical is formed as intermediate (Q•-). This unstable intermediate can lead to loss of electrons when these jump directly to molecular oxygen to form superoxide anion instead of moving with well controlled series of reactions of electron transport chain. In a similar set of reactions in plants reactive oxygen species are also produced during photosynthesis under high light intensity. This effect is partly offset by the involvement of carotenoids in photoin-

http://dx.doi.org/10.5772/52247

hibition, which involves these antioxidants reacting with over-reduced forms of the photosynthetic reaction centers and thereby prevent the production of superoxide. Another process which produces reactive oxygen species is lipid oxidation that takes place following the production of eicosanoids. However, the cells are provided with mechanisms that prevent unnecessary oxidation. Oxidative enzymes of these biosynthetic pathways are coordinated and highly regulated [3].

1.2. Free radicals

A free radical from the chemical viewpoint, is any species (atom, molecule or ion) containing at least one unpaired electron and its outermost orbital, and which is in turn able to exist independently (Figure 1).

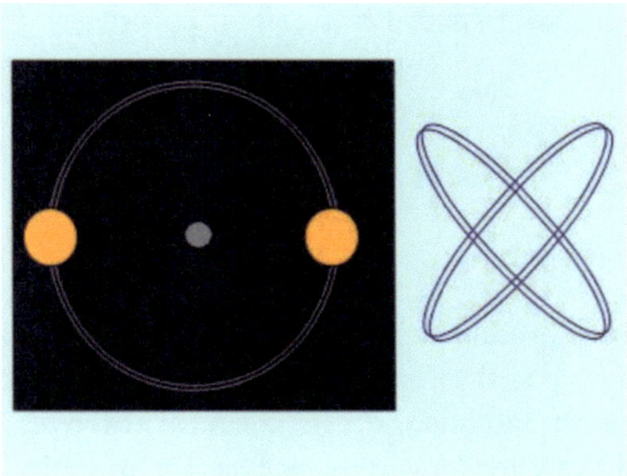

Figure 1. Atomic orbitals

The atoms arrange their electrons in regions called "atomic orbitals" in the form of pairs of electrons. The latter confers stability atom, or low chemical reactivity towards its environment. However, under certain circumstances, it may lose its parity orbital, either giving or capturing an electron. When this occurs, the resulting orbit exhibits an unpaired electron, making the atom in a free radical. The presence of an unpaired electron in an orbital outermost atom latter confers an increased ability to react with other atoms and / or molecules present in the environment, usually, lipids, proteins and nucleic acids (Figure 2). The interaction between free radicals and such substrates results in eventually structural and functional alterations [4].

Free radicals cause damage to different levels in the cell: Attack lipids and proteins in the cell membrane so the cell cannot perform its vital functions (transport of nutrients, waste disposal, cell division, etc.).

The superoxide radical, O_2, which is normally in the metabolism cause a chain reaction of lipid peroxidation of the fatty acids of phospholipids of the cell membrane. Free radicals attack DNA avoiding cell replication and contributing to cellular aging.

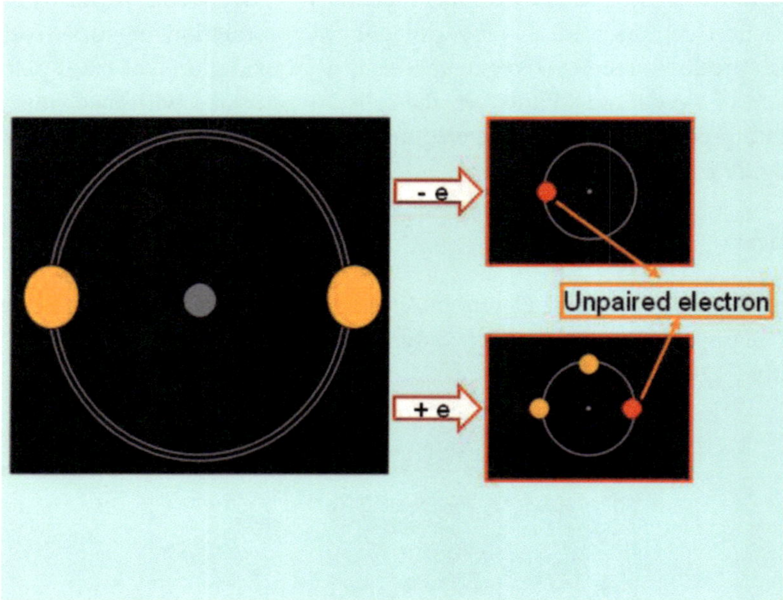

Figure 2. Unpaired electron in a free radical

The normal body processes produce free radicals that involve food metabolism, breathing and exercise. We are also exposed to environmental elements that create free radicals such as industrial pollution, snuff, radiation, drugs, chemical additives in processed foods and pesticides. Not all free radicals are dangerous because, for example, immune cells create free radicals to kill bacteria and viruses, but if there is sufficient control by antioxidants, healthy cell can be damaged.

The reactive oxygen species (ROS) is a collective term, widely used, comprising all the reactive species, whether or not free radicals, focus their reactivity in an oxygen atom. However, often under the designation ROS include other chemical species whose reactivity is focused on other than oxygen atoms [5].

1.3. Metabolites

In general, water soluble antioxidants react with oxidants in the cell cytoplasm and blood plasma, whereas the liposoluble antioxidants protecting cell membranes against lipid peroxidation. These compounds can be synthesized in the body or obtained from the diet (Table 1) [6].

1.4. Antioxidant metabolites in plants

Plants produce many different secondary metabolites some of them are potent antioxidants, some examples of these compounds is shown in Figure 3 [5,7].

http://dx.doi.org/10.5772/52247

Antioxidant metabolite	Solubility	Concentration in human serum (M)	Liver tissue concentration (mol / kg)
Ascorbic acid (vitamin C)	Water	50 to 60	260 (male)
Glutathione	Water	from 325 to 650	6,400 (male)
Lipoic acid	Water	from 0.1 to 0.7	4- 5 (rat)
Uric acid	Water	from 200 to 400	1,600 (male)
Carotene	Lipid	β-carotene: 0.5 to 1	retinol (vitamin A): 1 - 3 5 (male, total carotenoids)
α-tocopherol (vitamin E)	Lipid	10 to 40	50 (male)
Ubiquinol (coenzyme Q)	Lipid	5	200 (male)

Table 1. Biochemical properties of antioxidant metabolites

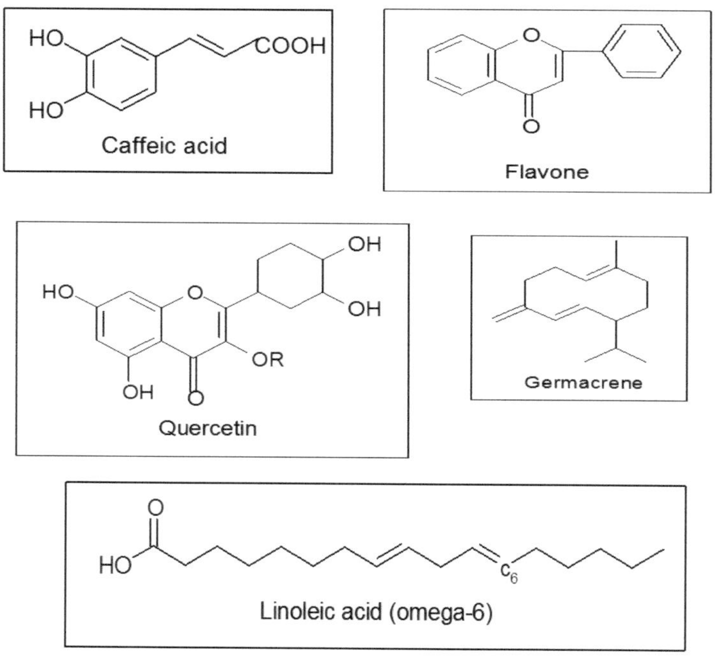

Figure 3. Different kinds of antioxidant metaboilites produced by plants

2. Enzyme systems

As with chemical antioxidants, cells are protected against oxidative stress by a network of antioxidant enzymes. Superoxide released by processes such as oxidative phosphorylation, is first converted into hydrogen peroxide and immediately reduced to give water. This route of detoxification is the result of multiple enzymes with superoxide dismutase catalyzing the first step and then catalases and peroxidases that eliminate several hydrogen peroxide [8].

2.1. Oxidative stress

Free radicals oxidize many biological structures, damaging them. This is known as oxidative damage, a major cause of aging, cancer, atherosclerosis, chronic inflammatory processes and cataracts, which are the most characteristic.

In certain circumstances, production of free radicals can increase uncontrollably, a situation known as oxidative stress. This means an imbalance between the speeds of production and destruction of toxic molecules, leading to an increase in cellular concentration of free radicals. Cells have mechanisms to protect against the harmful effects of free radicals based on a complex defense mechanism consisting of the antioxidants. Oxidative stress has been implicated in over one hundred human disease conditions, such as cancer, cardiovascular disease, aging and neurodegenerative diseases [9]. However, the innate defense in the human body may not be enough for severe oxidative stress. Hence, certain amounts of exogenous antioxidants are constantly required to maintain an adequate level of antioxidants in order to balance the ROS. As an example, epidemiological evidence indicates that the consumption of grapes reduces the incidence of coronary heart disease (CHD), atherosclerosis and platelet aggregation [10]. This greater protection may be due to the phenolic components of grapes, which are particularly abundant since they behave as reactive oxygen species-scavengers and metal-chelators. Polyphenolic substances in grapes and other red fruits are usually subdivided into two groups: flavonoids and nonflavonoids. The most common flavonoids are flavonols (quercetin, kaempferol, and myricetin), flavan-3-ols (catechin, epicatechin, and tannins), and anthocyanins (cyanin). Nonflavonoids comprise stilbenes, hydroxycinnamic acids and benzoic acids. Numerous papers have been published red fruits and their antioxidant properties have been correlated with their polyphenol contents [11,12,13,14,15].

2.2. Oxidative stress and disease

It is thought that oxidative stress contributes to the development of a wide range of diseases including Alzheimer's disease, Parkinson's disease, the pathologies caused by diabetes, rheumatoid arthritis, and neurodegeneration in motor neuron diseases. In many cases, it is unclear if oxidants trigger the disease, or occur as a result of this and cause the symptoms of the disease as a plausible alternative, a neurodegenerative disease may result from defective axonal transport of mitochondria that perform oxidation reactions. A case in which it fits is particularly well understood in the role of oxidative stress in cardiovascular disease. Here, the oxidation of low density lipoprotein (LDL) seems to trigger the process of atherogenesis, which leads to atherosclerosis, and ultimately to cardiovascular disease.

In diseases that have a high impact on the health sector Diabetes Mellitus is one of the most known. The World Health Organization (WHO) estimates that there are just over 180 million diabetics worldwide and likely to double this number for 2030 is quite high. Countries like China, India, United States of America and Mexico are at the top of this pathology [16]. In Mexico, this condition is a major cause of mortality and morbidity are estimated to be approximately 10 million individuals with diabetes, of whom 22.7% did not know they are sick, while 55% do not have good control netheir condition. This pathology is multifactorial, presenting various metabolic problems (polyuria, polyphagia, polydipsia, weight changes). The disorder is characterized by the inadequate use of glucose, due to insufficient production, insulin resistance and some without production of the hormone, resulting in unfavorable a high index of this monosaccharide in the blood. This causes abnormal function of some organs, tissues and systems that can cause kidney failure, vision loss, and amputation of a limb, diabetic coma and even death.

Different factors increase the likelihood of the individual to develop diabetes as are smoking, sedentary lifestyle, lack of exercise coupled with unbalanced diet causes both overweight and obesity. Naturally the body causes the formation of free radicals (highly unstable molecules), these chemical species are responsible for cellular aging, but when there is a greater concentration of these molecules may contribute to the development of various diseases and chronic degenerative neuro Parkinson's, Alzheimer's and diabetes. Obesity increases oxygen consumption and thus the production of free radicals, thus creating the phenomenon known as oxidative stress. Excess fat naturally stored in fat cells, causes the more than normal synthesis of substances called adiposines IL6 or leptines. These substances in higher concentrations also cause insulin resistance [17].

3. Alternative medicine

Due to the current problem in the health issue we propose the use of herbs as an option to improve the style of living of the people, not only for the adjuvant treatment, but because the use of plants offers great nutritional benefits somehow reducing the incidence of such chronic degenerative diseases. This is not intended to impair the option of preventive diagnosis by the health sector does not provide such benefits, but rather the use of plants known to have medicinal activity coupled with the clinical - pharmacology, could present better results, for the treatment of the various degenerative chronic diseases. Given the increasing scientific evidence that the etiology of several chronic degenerative diseases such as diabetes is influenced by factors such as metabolic redox imbalance. Is currently booming studying the formation of metabolites against free radicals that diverse plant species presents. An example of this has been widely documented, is the cranberry, a plant used for treating various diseases and, as has been discovered, is due to its potential antioxidant that has these properties beneficial to health [18, 19].

Similarly, Mexico has focused attention on other plants with potential antioxidant properties and for some years and was used in the treatment of diabetes. In this regard, since 2006, our research work focused on the task of describing the effects of plants such as Noni *(Morinda citrifolia)*, Moringa (*Moringa oleifera*), the Guarumbo (*Cecropia obtusifolia Bertolt*), the Musaro

(*Lophocereus sp.*) and Neem (*Azadirachta indica*) in murine models of chemically induced diabetes with streptozotocine. More recently, we began to evaluate the antioxidant properties of some of these plants through *in vitro* techniques [20].

3.1. Antioxidant effects in Mexican plants

The use of traditional medicine is widespread in Mexico and plants are indeed the first source for preparing remedies in this form of alternative medicine. Among the various compounds found in plants, antioxidants are of particular importance because they might serve as leads for the development of novel drugs. Several plants used as anti-inflammatory, digestive, antinecrotic, neuroprotective, and hepatoprotective properties have recently been shown to have and antioxidant and/or antiradical scavenging mechanism as part of their activity [21,22]. The search for natural sources of medicinal products that also have antioxidant and radical scavenging activity is on the rise [23,24]. Among the medicinal properties associated with them are the following: the fruitand bark of *Licania arborea* is used as a soap for hair infections, the latex from *Ficus obtusifolia* is employed as an anti parasitic and also for reducing fever, *Bunchosia cannesens* is prescribed as an antidiarrhoeic, *Sideroxylon capiri* is used for hiccups, as an antiseptic for cleaning wounds, and women use its leaves in a water bath after giving birth. The latex of *Sapium macrocarpum* is used against scorpion stings, fever and some skin problems such as warts; its use as an anti-coagulant is also widespread. The latex of *Ficus cotinifolia* is used in the treatments of urinary infections, vomiting, malaria and against inflammatory pathologies of the spleen. The leaves of *Annona squamosa* are used in cicatrisation of wounds, diarrhoea, ulcers, menstrual disorders, and also to help weight loss. The seeds of this plant are also employed as an insecticide. The leaves of *Vitex molli* are used to treat stomach ache, digestion disorders, nervous alterations, and also scorpion stings. *Piper leucophyllum* is employed for reducing fever and its dried leaves are used for cleaning eyes and as spice in cooking. The leaves and bark of *Gliricidia sepium* are used against high fever, skin infections, urine disorders, malaria, and headache. However, its seeds are reported to be toxic. *Hamelia paten is* used to accelerate wound cicatrisation. The Mexican and Central America native species of *Astianthus viminalis* is used for the curing of diabetes and malaria and to reduce hair loss. *Swietenia humilis* is used as anti parasitic, and it is also utilized for hair care as a shampoo. It is also used with other plants in mixed herbal teas, and used as home remedies. *Stemmandenia bella* is employed for curing wounds; *Rupechtia fusca* is used in some stomach disorders; *Bursera grandifolia* is used as a tooth paste and against digestive disorders; *Ziziphus amole* is prepared as infusion and it is applied for washing wounds and to treat gastric ulcers. The fruit and the latex of *Jacaratia mexicana* are used against ulcers in the mouth and digestive disorders. *Gyrocarpus jathrophifolius* leaves and bark are used as an analgesic. *Pseudobombax ellipticum* is used in respiratory disorders such as cough, and also against fever and as an anti microbial. The stems and flowers of *Comocladia engleriana* are toxic because they produce dermatitis. The flowers and the latex of *Plumeria rubra* can be used for stopping vaginal blood shed, and toothache, and the latex of the plant is used against earache. Infusions are used as an Eye-cleaning liquid [23, 24, 25 & 26].

Polyphenolic compounds are commonly found in both edible and inedible plants, and they have been reported to have multiple biological effects, including antioxidant activity [25]. Herbs are used in many domains, including medicine, nutrition, flavouring, beverages, dyeing, repellents, fragrances, cosmetics [26]. Many species have been recognized to have medicinal properties and beneficial impact on health, e.g. antioxidant activity, digestive stimulation action, antiinflammatory, antimicrobial, hypolipidemic, antimutagenic effects and anticarcinogenic potential [27,28]. Crude extracts of herbs and spices, and other plant materials rich in phenolics are of increasing interest in the food industry because they retard oxidative degradation of lipids and thereby improve the quality and nutritional value of food. The basic flavonoids structure is the flavan nucleus, which consists of 15 carbon atoms arranged in three rings (C6–C3–C6), labelled A, B, and C (Figure 3). Various clases of flavonoid differ in the level of oxidation and saturation of ring C, while individual compounds within a class differ in the substitution pattern of rings A and B. The differences in the structure and substitution will influence the phenoxyl radical stability and thereby the antioxidant properties of the flavonoids. Plant species belong to several botanical families, such as Labiatae, Compositae, Umbelliferae, Asteracae, Polygonacae and Myrtacae. Many spices have been investigated for their antioxidant properties for at least 50 years [29,30].

4. DPPH experiments with different plants collected in México

Herein it is presented a brief description of two experiments to evaluate the antioxidant properties of some plants collected in México, one conducted in the Moringa tree (*Moringa oleifera*) and the other in the Neem tree (*Azadirachta indica*).

1. Antioxidant properties of *Moringa oleifera*: In this experiment were collected fresh leaves of *M. oleifera* in the Municipality of Apatzingan in the state of Michoacan, Mexico, the leaves are brought to the facilities of the National School of Biological Sciences, IPN where allowed to air dry and then were macerated and placed in containers containing methanol. After one week the solvent was decanted and concentrated under reduced pressure using a rotary evaporator. The crude methanol extract was stable in distilled water obtained the following experimental concentrations: 50, 25, 12.5 and 6.25 mg / mL to which they evaluated the antioxidant capacity.

It was used as standard test the unstable radical 2,2-diphenyl-picrylhydrazyl (DPPH), which originally has a purple and when it is placed against a substance having antioxidant properties in a UV spectrophotometer at 517 nm, changes to yellow colour yellow (Figure 4). It was prepared a calibration curve of DPPH type in methanol at concentrations of 40, 120, 160 and 200 *ug*. Subsequently, aliquots of the above concentrations of methanol extract of *M. oleifera* and combined with DPPH in methanol for measuring its absorbance in the UV spectrophotometer at 0, 10, 30, 60 45 min., and then self-assess their antioxidant capacity [31].

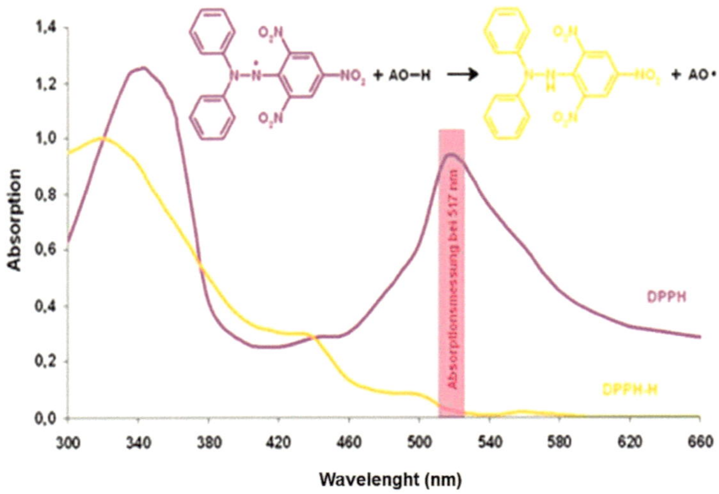

Figure 4. Graph showing the colour change of DPPH from purple to yellow when it is exposed to an antioxidant substance

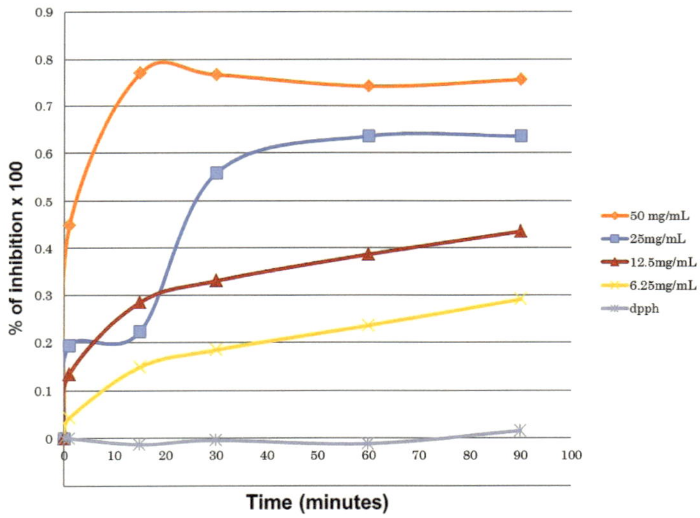

Figure 5. Antioxidant activity of different concentrations of methanol extract of *M. oleifera*

It was found that higher concentrations *of M. oleifera* antioxidant activity showed a concentration dependance, ie the higher the concentration of the extract metabolic higher antioxidant capacity (Figure 5). It was noted that the highest concentrations (50 mg / mL and 25 mg / mL) reached 50% of its antioxidant activity (assessed by the percentage of inhibition of purple fading to yellow against DPPH respective concentration of the methanol extract of *M. oleifera*) within 10 min. the reaction is initiated, which is indicative of the high antioxidant capacity of this plant and that could explain its therapeutic efficacy in the treatment of diabetes in mouse models that are currently underway [32].

2. Comparison of the antioxidant properties of *Azadirachta indica* with other species and a commercial product. This experiment followed a similar protocol to that described for *M. oleifera,* with the exception that the samples were evaluated for their antioxidant capacity at times 1, 15, 30, 60 and 90 minutes. The protocol is divided into two parts, one of which was evaluated for antioxidant activity from four different extracts from leaves of *Azadirachta indica*: a) methanol, b) infusion, c) ethyl acetate and d) ethanol. This was done by measuring the percent inhibition of loss of the color purple to yellow the respective front DPPH extract (Figure 6). It was found that infusion of Neem showed the highest antioxidant activity (80% inhibition) than the other extracts even from the first minute after initiating the reaction. This would correspond to the ethnomedical use that people from rural zones done with this tree, and then take it as a tea before the first food of the day. The methanol extract of Neem leaves also exhibited a high antioxidant capacity, as a percentage inhibition of the radical DPPH greater than 50% during the entire reaction time [33].

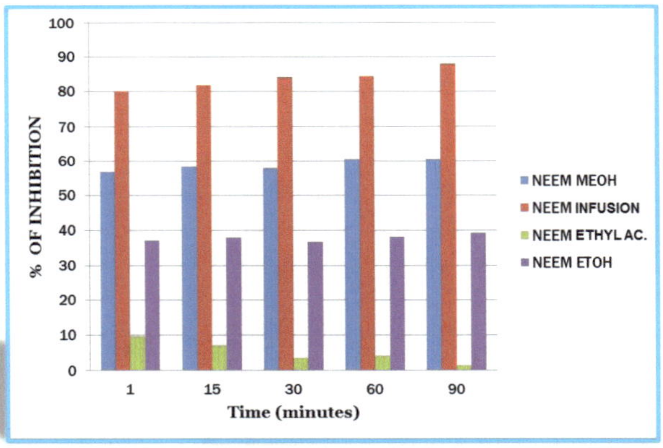

Sample	Time (min)	Inhibition (50%)	Concentrations (µg of DPPH)
Methanol	1	56	132,3478
Infusion	1	80	16,2609

Figure 6. Assessment of% inhibition of DPPH radical from 4 extracts of Neem (*Azadirachta indica*).

In the second part of this protocol, we chose a fraction of the methanol extract of Neem leaves and their antioxidant activity was compared against a commercial preparation, a kind of juice containing: Pomegranate (*Punica granatum*) green tea (*Camel sinensis*), cranberry (*Vaccinum myrtillus*), red grape (*Vitis vinifera*) and methanol extract of leaves, seed and fruit peel of passion fruit (*Passiflora edulis*). It was found that the commercial preparation showed the highest antioxidant activity throughout the reaction time (Figure 7), presented as a% inhibition greater than 50% from the first minute (68% inhibition), reaching end of the reaction with values higher than 85% inhibition of the presence of radical DPPH. Compared to the various organs of *Passiflora edulis*, the Neem leaf extract showed higher antioxidant capacity from the reaction started, it displayed a% inhibition of DPPH radical by 66% to reach 15 minutes of reaction, and reaching values greater than 85% inhibition at 60 minutes, falling a little activivty (63% approximately) at 90 min. the reaction is initiated. This would confirm previous data of different authors and from our laboratory (unpublished data), in the sense of the effectiveness of Neem tree leaves for the treatment of chronic degenerative diseases such as diabetes, it has proven effective in significantly reducing blood glucose levels in streptozotocin-treated mice, an effect that may be due to the presence of secondary metabolites of the steroid type saponins, flavonoids and phenols among others that seem to owe much its hypoglycemic action, thanks to its antioxidant properties. In this regard, our laboratory is conducting a more rigorous characterization of secondary metabolites found in these plants by using spectroscopic techniques (e.g, Infrared Spectroscopy with Fourier Transform, Proton Nuclear Magnetic Resonance, etc.) trying to isolate, purify these metabolites and test its therapeutic efficacy in animal models of diabetes. However, we should mention that, although the future is promising with regard to therapy options offered by these plants in Mexico (and other developing countries), yet to be made deeper studies and comparative scale phytochemical, toxicological and drug to confirm everything mentioned in this chapter [34].

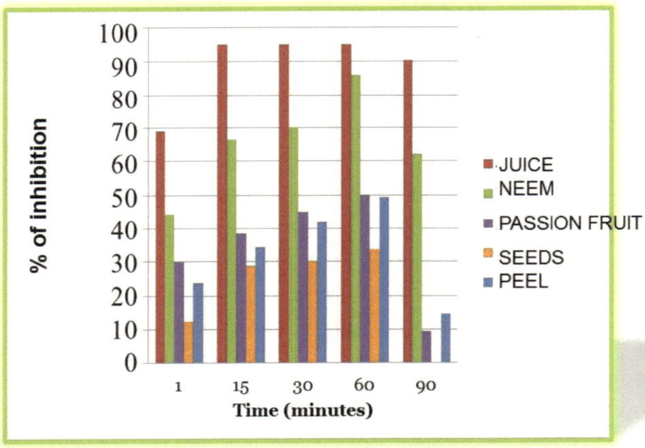

Sample	Time (min)	Inhibitión (50%)	Concentration (µg of DPPH)
Comercial juice	1	68%	56
Neem	15	66%	60

Figure 7. Assessment of% inhibition of DPPH radical of Neem (Azadirachta indica) with respect to a commercial preparation and different organs of Passion fruit (Passiflora edulis).

Another experiment with Noni (*Morinda citrifolia*) showed that the leaves of this plant also have a high antioxidant effect at low concentrations reaching a maximum effect at a dose of 5mg/mL (Figure 8). These results disagreed with the effect observed with the Noni fruit, where it was reported that the antioxidant effect of protection tends to increase with respect to higher concentrations in tests performed with lyophilized juice extracts [35].

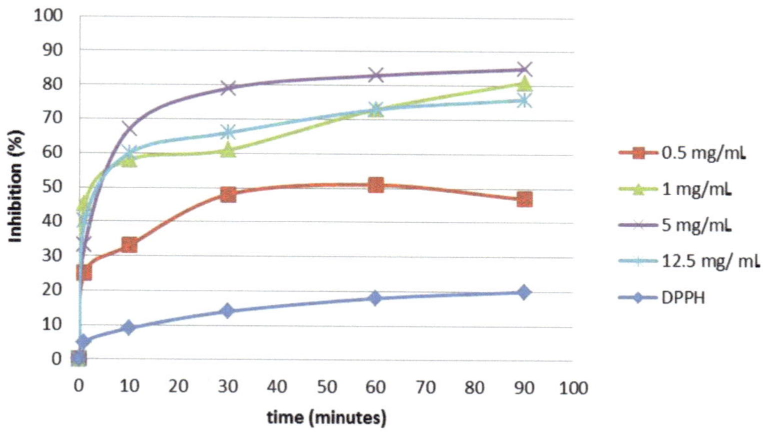

Figure 8. Results from DPPH method for testing the free-radical inhibition effect of different concentrations of methanolic extracts of Noni leaves.

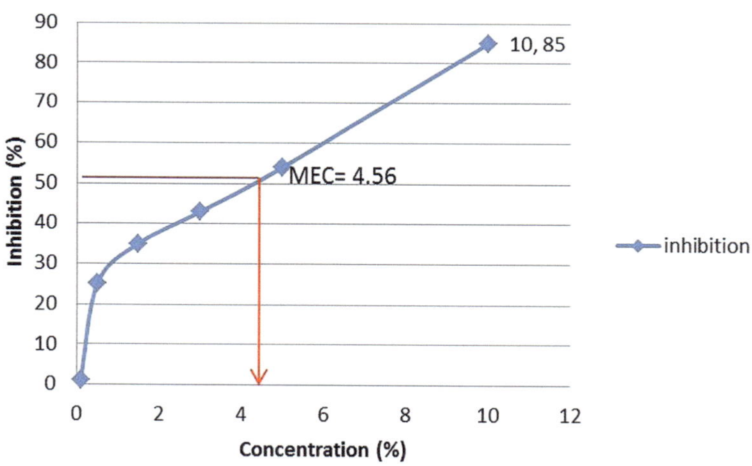

Figure 9. Shows the curve concentration vs. inhibition (%) of the radical 1,1-diphenyl-2-picrylhydrazyl (DPPH ●), in which it is observed that the mean effective concentration (MEC) of noni leaf extract was 4,56%, at this concentration is carried out the 50% inhibition of the unpaired radical.

Furthermore a methanolic solution was prepared with a concentration of 860 mmol / mL of DPPH. From this concentration the experiment was performed with a duplicated calibration curve at different concentrations, which was read at 517nm. An average concentration of those used for standard curve analysis was chosen for the initial concentration of the methanol extract from leaves of *Morinda citrifolia*. From this methanol extract different percentage dilutions were prepared. After obtaining the extract dilutions 1000μL of each one was mixed with 860μL methanol and 140μL of DDPH and after that it were allowed to stand for 10 minutes at 30°C and then absorbance was read (517nm) at time intervals of 10, 30, 45 and 60 minutes in a UV-Vis spectrophotometer HACH DR 5000. Through a mathematical analysis of linear regression curve was obtained concentration vs. inhibition (%), and finally determined the mean effective antioxidant concentration (Figure 9) [36].

5. Phytochemical results

5.1. Moringa oleifera

Extraction by maceration from *M. Oleifera* leaves was carried out in methyl alcohol for the phytochemical sieve, the results are shown in the following table (Table 2), and qualitative tests that were performed for each secondary metabolite are named in the same table.

	Reagent	Test
Alkaloids	Dragendorf	++
	Mayer	+
	Wagner	+
	Sonneshain	+
	Silicotungstine	+
Tanines	ferric chloride	-
phenols	Potassium ferricianide	+
reducing sugars	Fehling	-
	Benedic	-
Coumarins	ammonium hydroxide	-
	Erlich	+
Flavonoids	Sodium Hydroxide	+
Sesquiterpenlactones	hidroxylamine chlorehydrate	-
Saponines	Libermann Bouchard	+
cardiotonic Glycosides	Kedde, Baljet, legal	-
cyanogenic Glycosides	Guignar	-

Table 2. Phytochemical sieve of methanolic extract from *M. olerifera* leaves.

Metabolites	Reagent	Test			
		Azadirachta indica	***Cecropia obtusifolia* Bertolt**	***Lophocereus sp***	***Morinda citrifolia***
Alkaloids	Dragendorf	++	-	++	-
	Mayer	+	+	+	+
	Wagner	+	-	+	+
	Sonneshain	+	-	-	-
	Silicotungstine	+	-	-	-
Tanines	ferric chloride	-	+	-	-
phenols	Potassium ferricianide	+	+	+	+
reducing sugars	Fehling	-	-	+	+
	Benedic	-	-	+	+
Coumarins	ammonium hydroxide	-	-	+	+
	Erlich	+	+	+	+
Flavonoids	Sodium Hydroxide	+	+	-	+
Sesquiterpenlactones	hidroxylamine chlorehydrate	-	+	-	-
Saponines	Libermann Bouchard	+	+	+	+
cardiotonic Glycosides	Kedde, Baljet, legal	-	+	-	+
cyanogenic Glycosides	Guignar	-	+	+	-

Table 3. Phytochemical sieve of different plants collected in Mexico. All the tests were performed with methanolic extracts of the leaves.

The results of this phytochemical analysis presented in Table 2, shows qualitatively the secondary metabolites found in the methanol extract of leaves of *M. oleifera*. The more abundant compounds in the leaves of this tree and also the must reported in several articles are:

i. Alkaloids

ii. Coumarins

iii. Phenolics

iv. Flavonoids

v. Saponins

In the case of metabolites such as tannins, cardiac glycosides, cyanogenic glycosides, the tests results were negative for the methanol extract of leaves of *M. oleifera* [37, 38].

Phytochemical results for the extracts obtained from leaves of Guarumbo (Cecropia obtusifolia Bertolt), Musaro (Lophocereus sp.), Neem (Azadirachta indica) and Noni (Morinda citrifolia) are showed in Table 3. Several of those metabolites are highly active antioxidants.

6. Description of those plants collected in México and presenting high content of antioxidant compounds

6.1. *Azadirachta indica*

Active compounds of Neem have been identified while others have not, and analyzed the most common are: nimbin; nimbidin; ninbidol; gedunin; sodium nimbinate; queceretin; salannin and azadirachtin.

Parts used and their uses:

Neem bark is bitter, astringent, is used to treat diseases of the mouth, teeth, loss of appetite, fever, cough and intestinal parasites (Figure 10)

Figure 10. Stem of the Neem tree.

The leaves help to neuromuscular problems, eliminate toxins, purify the blood, are also used to treat snake bites and insect (Figure 11)

Figure 11. Branch and leaves of Neem or Azadirachta indica.

The Neem fruit is bitter, and it is used as a purgative and for hemorrhoids (Figure 12)

Figure 12. Fruits from Neem Tree

The flowers are astringent and expectorant and also Seed oil is extracted. (Figure 13).

Figure 13. Flowers from Neem Tree

Neem kills some infectious organisms, contributes to the immune response at various levels, this increases the possibility that the body fight bacterial infections alone, viral and fungal. Neem increases the production of antibodies, improves the response of immune cells that release mediators white blood cells. For Diabetes: Neem extract orally reduces insulin requirements by between 30% and 50% to people who are insulin dependent. With Cancer: polysaccharides and limonoids found in the bark, leaves and Neem oil reduces tumors and cancer [39].

6.2. *Passiflora edulis Sims*

Another plant which has been described with therapeutic applications is the genus Passiflora, whics comprises about 500 species and is the largest in the family Passifloraceae. *Passiflora edulis Sims* is native from the Brazilian Amazon, known by the common name for passion fruit [40, 41]. The word passion comes from the Portuguese- Brazilian passion fruit, which means food prepared in Totuma [41, 42].

Passiflora edulis (Figure 14) is a widely cultivated species in tropical and subtropical countries, there are two varieties: *Passiflora edulis Sims* var. *flavicarpa*, whose fruits are yellow and *Passiflora edulis Sims* var. purple, with purple fruits and adapts to higher ground [42].

Passion fruit (Figure 15) is a woody perennial, climbing habit and rapid development, which can reach up to 10 m long, the leaves are simple, alternate, and a tendril conestipules in the

http://dx.doi.org/10.5772/52247

armpit, with serrated margins, the flowers (Figure 16) are solitary and axillary, fragrant and showy, the fruit is a spherical berry, globose or ellipsoid, measuring 10 cm in diameter and weighs up to 190 g, yellow or purple, with a highly aromatic pulp [43].

Figure 14. Passion fruit

Figure 15. Leaves and flower of passion fruit

The ethnopharmacological information reveals that *Passiflora edulis Sims* has been used in traditional medicine around the world. In India, the fresh leaves of this plant are boiled in small amount of water and the extract is drunk to treat dysentery and hypertension, and the fruits are eaten to relieve constipation. In South America, native people drink the tea leaves and flowers as a sedative, infusion of the aerial parts is used in the treatment of tetanus, epilepsy, insomnia and hypertension is also indicated as a muscle relaxant, diuretic, to treat stomach aches, fever and intestinal tumors[44].

The phytochemical study of *Passiflora edulis Sims* (Passifloraceae) shows the presence of glycosides, including passiflorine, flavonoid glycosides: luteolin-6-Cchinovóside, cyanogenic glycosides, alkaloids harman, triterpenes and saponins, phenols, carotenoids, anthocyanins, L-ascorbic acid, γ-lactones, esters, volatile oils, eugenol, amino acids, carbohydrates and minerals [45].

Figure 16. Five ribs cladode from Musaro (*Lophocereus* sp)

6.3. *Lophocereus* sp

Gender *Lophocereus* sp. (Figure 17) develops as a succulent plant which presents a plus size and can reach 7m tall. These cacti have a columnar development right, has ribbed stems that branch with age. The flowers are nocturnal, appear only on copies of more than two meters high and are colored green, but in the four seasons can take a pink color, the stem proliferation and leads to masses of slender spines long they can get to cover it completely. Multiply by seeds in spring or summer. These plants are drought resistant so as to avoid daily watering and exposure to damp, besides being very sensitive to frost. But they need full sun and well drained soils. It is endemic to Baja California Sur, Baja California and Sonora in Mexico and Arizona in the United States It is a common species that has spread efficiently over the worldwide. The fruits are edible, but are hard to come by competition with birds and insects. This plant has curative properties. Southeastern Mexican indians prepare a tea from the pulp and skin of the cactus to relieve arthritis. Some members of this ethnic group say the plants with five ribs are very good for treating cancer. For diabetes: using the Musaro with a seven-pointed cladode, boiled in 1 liter of water, strain and drink three glasses of cooking, one before each meal. It is also widely used to treat ulcers, wounds, and stomach diseases.

Musaro or gooseberry is the tea made by slicing sections fifteen or twenty two inches long from the stems of cactus. These cuts are then placed in a container large enough to contain five gallons of water and boiled for eight to ten hours until the liquid is reduced to approximately one gallon. People take this treatment for serious stomach diseases but they should drink tea in large amount. From extracts of *Lophocereus sp* two compounds have been isolated: pylocereine and lophocine a dimeric alkaloid with cytotoxic activity [46, 47].

6.4. *Cecropia obtusifolia Bertolt*

Family: Cecropiaceae

Common name: The names that are known are "guarumbo", "chancarro", "hormiguillo", "chiflón" and "koochlle" among others. This plant is widely used by people suffering from type 2 diabetes. It is known mainly in rural areas.

Botanical description: A tree 20 m tall, commonly grows in tropical rain forests. Its trunk is straight and presents a cavity where you can find some ants inside it, with branches that grow along this horizontally. The leaves are placed in a spiral on the branches. They have a deep green on the upper surface of the leaf and gray on the back (Figure 18).

Distribution in Mexico: located along the coasts of Tamaulipas and San Luis Potosi to Tabasco on the side of the Gulf of Mexico and Sinaloa to Chiapas Pacific side. Traditionally the dried leaves (15 g) are heated in water (500 mL) and the result is an infusion which is then filtered and taken as "daily water", cold infusion is often consumed in hot weather. States where is often used as traditional medicine this plant are: Hidalgo, Guerrero, Veracruz, Yucatan, Campeche, Tabasco, Mexico State, Oaxaca and Chiapas. It is reported that this plant contains β-sitosterol, stigmasterol, 4-ethyl-5-(n-3 valeroil)-6-hexahydrocoumarins. From butanol extracts were isolated chlorogenic acid and isorientin [48, 49].

Figure 17. Guarumbo (*Cecropia obtusifolia Bertolt)* leaves.

7. Conclusion

The plant kingdom has been the best source of remedies for curing a variety of disease and pain. This is why medicinal plants have played a key role in the worldwide maintenance of health. Traditional herbal medicine is intimately related to the Mexican popular culture; its use has origins based on ancestral knowledge. Natural products of higher plants are an important source of therapeutic agents; therefore, many research groups are currently screening the different biological activities of plants. Mexico has an extensive variety of plants; it is the fourth richest country worldwide in this aspect. Some 25 000 species are registered, and it is thought that there are almost 30 000 not yet described. Natural antioxidants that are present in herbs and spices are responsible for inhibiting or preventing the deleterious consequences of oxidative stress. Spices and herbs contain free radical scavengers like polyphenols, flavonoids and phenolic compounds, having antioxidant activities, Indeed all these plant studied have several biological effects and they could also be used as a source of natural antioxidants. Further pharmacological studies are underway to identify the active constituents of the plant extracts responsible for the showed activities. As a final comment, compounds in plants are of great importance for the treatment of several chronic and degenerative diseases like diabetes and cancer, among others. For that reason the use in traditional medicine is of great interest in order to know the activity and the mechanism of action of these compounds which could be used for the treatment and prevention of that mentioned

http://dx.doi.org/10.5772/52247

diseases, which are associated nowadays with stress, fast food diets and lack of daily exercise, just to name a few factors.

Acknowledgments

The authors wants to thank the ENCB-IPN for the support received for this work through the SIP projects 20120789 and 20120899.

Author details

Jorge Alberto Mendoza Pérez[1] and Tomás Alejandro Fregoso Aguilar[2*]

1 Department of Environmental Systems Engineering at National School of Biological Sciences-National Polytechnic Institute. Mexico, D.F., Mexico

2 Department of Physiology at National School of Biological Sciences-National Polytechnic Institute. Mexico, D.F., Mexico

References

[1] PIETTA, P. G. (2000) Flavonoids as Antioxidants, *Journal Natural Product,* 63: 1035–1042.

[2] Rivera Arce E, Morales González J A, Fernández Sánchez A M, Bautista Ávila M, Vargas Mendoza N, Madrigal Santillán E O. (eds.) Chemistry of Natural Antioxidants and Studies Performed with Different Plants Collected in Mexico, (2009) 227-238

[3] Devasagaya T P, Tilak J C, Boloor K K. Free Radicals and Antioxidants in Human Health: Currens Status and Future Prospects. J. Assoc. Physicians India. 2004; 52:794-804.

[4] ACS. Chemistry. España:Reverte; 2010. p.460-65

[5] Camacho Luis A, Mendoza Pérez J A, The Ephemeral Nature of Free Radicals: Chemistry and Biochemistry of Free Radicals. In: Morales González J A, Fernández Sánchez A M, Bautista Ávila M, Vargas Mendoza N, Madrigal Santillán E O. (eds.) Antioxidants and Chronic Degenerative Diseases. México: Ciencia al Día; 2009. p. 27-76

[6] VanderJagt T J, Ghattas R, VanderJagt D J, Crossey M, Glew R H. Comparison of the Total Antioxidant Content of 30 Widely Used Medicinal Plants of New Mexico. Life Sci. 2002; 70: 1035-1040.

[7] Bors W, Heller W, Michael C, Saran, M. Radical Chemistry of Flavonoids Antioxidants. Advances in Experimental Medicine and Biology. 1990; 264: 165–170.

[8] Ballester M. Antioxidants, Free Radicals and Health. A Physic-Organic Chemistry View. *Med Clinc (Barc).*1996;107:509-515.

[9] Bagchi D, Bagchi M, Stohs S J, Das D K, Ray S D, Kuszynski C A. Free Radicals and Grape Seed Proanthocyanidin Extract: Importance in Human Health and Disease Prevention. Toxicology. 2000; 148:187–197.

[10] Tedesco I, Russo M, Russo P, Iacomino G, Russo G L, Carraturo A. Antioxidant Effect of Red Wine Polyphenols on Red Blood Cells. The Journal of Nutritional Biochemistry. 2000; 11(2):114–119.

[11] Fernandez-Pachon, M S, Villano D, Garcia-Parrilla M C, Troncoso A M. Antioxidant Activity of Wines and Relation with their Polyphenolic Composition. Analytica Chimica Acta.2004; 513(1):113–118.

[12] Fernandez-Pachon, M. S., Villano, D., Troncoso, A. M., Garcia-Parrilla, M. C. Determination of the Phenolic Composition of Sherry and Table White Wines by Liquid Chromatography and their Relation with Antioxidant Activity. Analytica Chimica Acta. 2006; 563(1–2):101–108.

[13] Cimino, F., Sulfaro, V., Trombetta, D., Saija, A., Tomaino, A. Radicalscavenging Capacity of Several Italian Red Wines. Food Chemistry. 2007;103(1):75–81.

[14] Arnous, A., Makris, D. P., Kefalas, P. Correlation of Pigment and Flavanol Content with Antioxidant Properties in Selected Aged Regional Wines from Greece. Journal Of Food Composition And Analysis. 2002; 15(6):655–665.

[15] Minussi, R. C., Rossi, M., Bologna, L., Cordi, L., Rotilio, D., Pastore, G. M. Phenolic Compounds and Total Antioxidant Potential of Commercial Wines. Food Chemistry. 2003; 82(3): 409–416.

[16] Nava Ch. G, Veras G. M. E. Epidemiology of diabetes In: Morales G.J.A., Madrigal S.E.O., Nava Ch.G., Durante M.I., Jonguitud F.A., Esquivel S. J. (eds.) Diabetes 2nd. ed. Universidad Autónoma del Estado de Hidalgo: México; 2010. p.57 – 63.

[17] Fernández S. A.M. (2009). Antioxidants and diseases: Obesity. In: Morales González J A, Fernández Sánchez A M, Bautista Ávila M, Vargas Mendoza N, Madrigal Santillán E O. (eds.) Antioxidants and Chronic Degenerative Diseases. México: Ciencia al Día; 2009. p.411-25

[18] Posmontier B. The Medicinal Qualities of Moringa Oleifera. USA:Lippincott Williams & Wilkin; 2011. p. 80-87

[19] Fragoso Antonio S. Antioxidant and Antigenotoxic Effects of Cranberry Juice. MsC Thesis. National School of Biological Sciences-National Polytechnic Institute; 2011.

[20] Castilla, D. Phytochemical and Biological Studies of Leaf Extract and Fruit of *Morinda citrifolia* (Noni) in a Mouse Model of Diabetes. Bachelor Thesis. National School of Biological Sciences-National Polytechnic Institute; 2012.

[21] Lin C C, Huang P C. Antioxidant and Hepatoprotective Effects of *Acanthopanax senticosus*. Phytother. Res. 2002; 14: 489-494.

[22] Perry E K, Pickering A T, Wang W W, Houghton P J, Perru N S Medicinal Plants and Alzheimer's Disease:from Ethnobotany to Phytotherapy . J. Pharm. Pharmacol. 1999; 51: 527-534.

[23] Schinella G R, Tournier H A, Prieto J M, Mordujovich de Buschiazzo P, Ríos J L. Antioxidant Activity of Antiinflammatory Plant Extracts. Life Sci. 2002; 70: 1023-1033.

[24] Velázquez E, Tournier HA, Mordujovich de Buschiazzo P, Saavedra G, Schinella GR. Antioxidant activity of Paraguayan plant extracts. Fitoterapia. 2003; 74: 91-97.

[25] Moure A, Franco D, Sineiro J, Domínguez H, Núñez JM, Lema MJ. Evaluation of Extracts from *Gevuina avellana* Hulls as Antioxidants. J. Agric. Food Chem. 2000; 48: 3890-3897.

[26] Singleton VL, Rossi JA Colorimetry of Total Phenolics with Phosphomolybdic-Phosphotungstic Acid Reagents. Am. J. Enol. Vitinicult. 1965; 16: 55-61.

[27] Takao T, Kitatani F, Watanabe N, Yagi A, Sakata K A Simple Screening Method for Antioxidants and Isolation of Several Antioxidants Produced by Marine Bacteria from Fish and Shellfish. Bioscience. Biotechnol. Biochem. 1994; 58: 1780-1783.

[28] Miller H E. A Simplified Method for the Evaluation of Antioxidants. J. Am. Oil Chem. Soc. 1971; 48: 92-97.

[29] Monroy-Ortíz C, Castillo E P. Medicinal Plants Used in Morelos State. UAEM: México, 2000. p. 20-50

[30] Morales-Cifuentes C, Gómez-Serranillos MP, Iglesias I, Villar del Fresno AM. Neuropharmacological profile of ethnomedicinal plants of Guatemala J. Ethnopharmacol. 2001; 76: 223-228.

[31] Robak J, Gryglewski R. J. Flavonoids are scavengers of superoxide anions.Biochemical Pharmacology.1988; 37(5):837–841.

[32] Brandwilliams W, Cuvelier M E, Berset C. Use of a Free-Radical Method to Evaluate Antioxidant Activity. Food Science and Technology-Lebensmittel- Wissenschaft & Technologie.1995; 28(1): 25–30.

[33] Cimino F, Sulfaro V, Trombetta D, Saija A, Tomaino A. Radicalscavenging Capacity of Several Italian Red Wines. Food Chemistry. 2007; 103(1): 75–81

[34] Sreelatha S, Padma P.R. Antioxidant Activity and Total Phenolic Content of Moringa Oleifera Leaves in Two Stages of Maturity. Plant Foods Hum Nutr. 2009; 64:303–311.

[35] De Las Heras B, Slowing K, Benedí J, Carreto E, Ortega T, Toledo C, Bermejo P, Iglesias I, Abad MJ, Gómez-Serranillos P, Liso PA, Villar A, Chiriboga X. Antiinflamatory and Antioxidant Activity of Plants Used in Traditional Medicine in Ecuador. J. Ethnopharmacol. 1998; 61: 161-166.

[36] Faustino R S, Clark T. A, Sobrattee S, Czubryl M. P, Pierce G. N. Differential Antioxidant Properties of Red Wine in Water Soluble and Lipid Soluble Peroxyl Radical Generating Systems. Molecular And Cellular Biochemistry. 2004; 263(1):211–215.

[37] Dinis T. C. P, Madeira V. M. C, Almeida L. M. Action of Phenolic Derivates (Acetoaminophen, Salicylate, And 5-Aminosalicylate) as Inhibitors of Membrane Lipid Peroxidation and as Peroxyl Radical Scavengers. Archive Of Biochemistry And Biophysics. 1994; 315:161–169.

[38] Huang D. J, Ou B. X, Hampsch-Woodill M, Flanagan J. A, Prior R. L. High-Throughput Assay of Oxygen Radical Absorbance Capacity (ORAC) Using a Multichannel Liquid Handling System Coupled with a Microplate Fluorescence Reader in 96-Well Format. Journal of Agricultural and Food Chemistry. 2002; 50(16):4437–4444.

[39] Abu Syed Md. Mosaddek and Md. Mamun Ur Rashid. A Comparative Study of the Anti-Inflammatory Effect of Aqueous Extract of Neem Leaf and Dexamethasone. Bangladesh J Pharmacol. 2008; 3: 44-47

[40] Arteaga S, Andrade-Cetto A, Cárdenas R. *Larrea tridentata* (Creosote bush), an Abundant Plant of Mexican and US-American Deserts and its Metabolite Nordihydroguaiaretic Acid. J. Ethnopharmacol. 2005; 98: 231 – 239.

[41] Mareck U, Herrmann K, Galensa R, Wray V. The 6-Cchinovoside and 6-C-Fucoside of Luteolin from *Passiflora edulis*. Phytochemistry. 1991; 30: 3486-7.

[42] Christensen J, Jaroszewski J. Natural Glycosides Containing Allopyranose from the Passion Fruit Plant Circular Dichroism of Benzaldehyde Cyanohydrin Glycosides. Org Lett. 2001; 3:2193-5.

[43] Seigler D, Pauli G, Nahrstedt A, Leen R. Cyanogenic Allosides and Glucosides from *Passiflora edulis* and *Carica papaya*. Phytochemistry. 2002; 60: 873-82.

[44] Slaytor M, McFarlane I. The Biosynthesis and Metabolism of Harman in *Passiflora edulis*. Phytochemistry. 1968; 7:605-11.

[45] Shawan K, Dhawan S, Sharma A. *Passiflora*: a Review Update. Journal of Ethnopharmacology. 2004; 94: 1-23.

[46] Andrade, C. A, Heinrich B M. Mexican Plants with Hypoglycemic Effect Used in the Treatment of Diabetes. Journal Of Ethnopharmacology. 2005; 99:325–348.

[47] Andrade, C. Hipoglicemic Effect of Cecropia Obtusifolia on Streptozotocin Diabetic Rats. Journal Of Ethnopharmacology. 2001; 78(2-3):145-149.

[48] Hicks S. Desert Plants and the People. The Naylor Company Book:San Antonio, Texas; 2009. p. 35-37

[49] Escalante J.S. Our Plants from the Sonora State http://www.apnsac.org. (Accesed 15 July 2012)

Chapter 4

Food Phenolic Compounds: Main Classes, Sources and Their Antioxidant Power

Maria de Lourdes Reis Giada

Additional information is available at the end of the chapter

http://dx.doi.org/10.5772/51687

1. Introduction

The natural phenolic compounds have received increasing interest in the last years, since a great amount of them can be found in plants and consumption of vegetables and beverages with a high level of such compounds may reduce the risk of development of several diseases due to their antioxidant power, among other factors.

It is known that the metabolism of plants is divided in primary and secondary. The substances that are common to living things and essential to cells maintenance (lipids, proteins, carbohydrates, and nucleic acids) are originated from the primary metabolism. On the other hand, substances originated from several biosynthetic pathways and that are restricted to determined groups of organisms are results of the secondary metabolism [1]. Phenolic compounds are constituted in one of the biggest and widely distributed groups of secondary metabolites in plants [2].

Figure 1 shows the inter-relationships between the primary and secondary metabolism in plants.

Biogenetically, phenolic compounds proceed of two metabolic pathways: the shikimic acid pathway where, mainly, phenylpropanoids are formed and the acetic acid pathway, in which the main products are the simple phenol [3]. Most plants phenolic compounds are synthesized through the phenylpropanoid pathway [4]. The combination of both pathways leads to the formation of flavonoids, the most plentiful group of phenolic compounds in nature [3].

Additionally, through the biosynthetic pathways to the flavonoids synthesis, among the not well elucidated condensation and polymerization phases, the condensed tannins or non-hydrolysable tannins are formed. Hydrolysable tannins are derivatives of gallic acid or hexahydroxydiphenic acid [5].

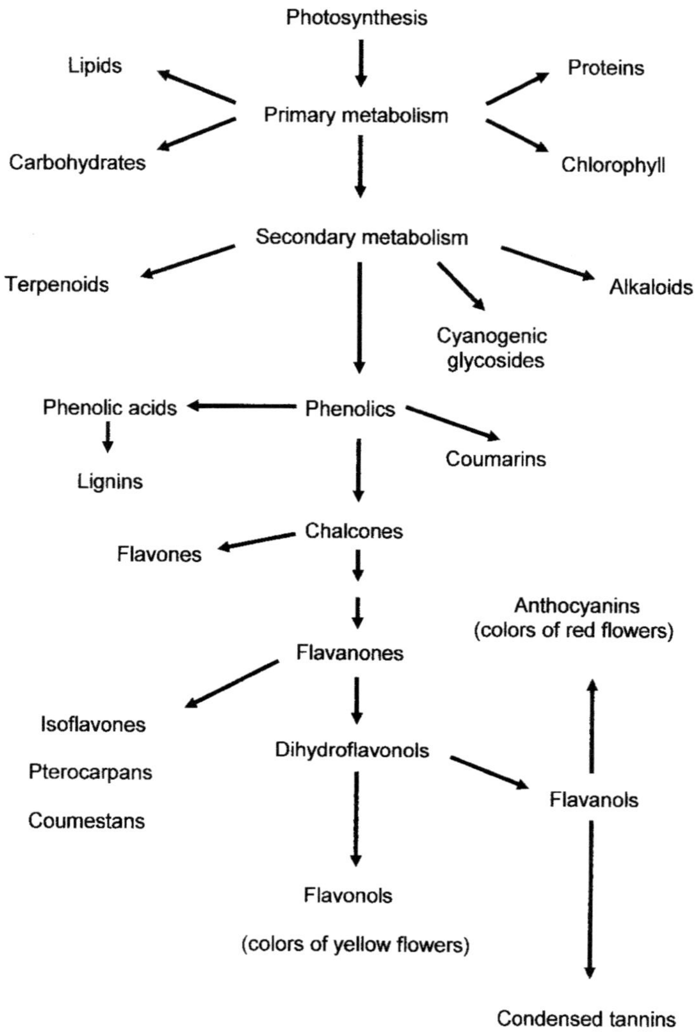

Figure 1. Inter-relationships between the primary and secondary metabolism in plants.

Therefore, phenolic compounds have, as a common characteristic, the presence of at least one aromatic ring hydroxyl-substituted [6]. Another characteristic of these substances is that they are presented commonly bound to other molecules, frequently to sugars (glycosyl residue) and proteins. The existence of phenolic compounds in free form also occurs in plant tissues. However, it is less common, possibly because they are toxic when present in the free state and detoxified, at least in part, when bound.

As a result, phenolic compounds play a role of protection against insects and other animals to the plants. The different types of bond between the glycosyl residue and the flavonoids,

http://dx.doi.org/10.5772/51687

such as anthocyanin, also lead to the different derivatives that add colors and color gradation to flowers [7].

This way, phenolic compounds are essential to the physiology and cellular metabolism. They are involved in many functions in plants, such as sensorial properties (color, aroma, taste and astringency), structure, pollination, resistance to pests and predators, germinative processes of seed after harvesting and growth as well as development and reproduction, among others [8].

Phenolic compounds can be classified in different ways because they are constituted in a large number of heterogeneous structures that range from simple molecules to highly polymerized compounds.

According to their carbon chain, phenolic compounds can be divided into 16 major classes [9].

The main classes of phenolic compounds regarding to their carbon chain can be seen in Figure 2.

On the other hand, as to their distribution in nature, phenolic compounds can be divided into three classes: *shortly distributed* (as simple phenols, pyrocatechol, hydroquinone, resorcinol, Aldehydes derived from benzoic acids that are components of essential oils, such as vanillin), *widely distributed* (divided in flavonoids and their derivatives, coumarins and phenolic acids, such as benzoic and cinnamic acid and their derivatives) and *polymers* (tannin and lignin) [10].

Finally, as to the location in the plant (free in the soluble fraction of cell or bound to compounds of cell wall), together with the chemical structure of these substances, phenolic compounds may also be classified as: *soluble* (such as simple phenol, flavonoids and tannins of low and medium molecular weight not bound to membranes compounds) and *insoluble* (essentially constituted by condensed tannins, phenolic acids and other phenolic compounds of low molecular weight bound to cell wall polysaccharides or proteins forming insoluble stable complexes). This classification is useful from the nutritional viewpoint, to the extent that the metabolic fate in the gastrointestinal tract and the physiological effects of each group will depend largely on their solubility characteristics. Insoluble phenolic compounds are not digested and may be partially or fully recovered quantitatively in the feces, while a part of the soluble can cross the intestinal barrier and be found in the blood, unchanged or as metabolites [3].

The antioxidant activity of food phenolic compounds is of nutritional interest, since it has been associated with the potentiation of the promoting effects of human health through the prevention of several diseases [11]. Additionally, in some cases, these compounds may also be used with therapeutic purposes due to their pharmacological properties [12]. Many phenolic compounds with low molecular weight, such as thymol, are used in medicine as antiseptic due to its toxicity [7].

However, the antioxidant activity of phenolic compounds depends largely on the chemical structure of these substances [2]. Among the phenolic compounds with known antioxidant activity, flavonoids, tannins chalcones and coumarins as well as phenolic acids are highlighted.

Class	Basic skeleton
Simple phenols	C_6
Benzoquinones	C_6
Phenolic acids	C_6-C_1
Acetophenones	C_6-C_2
Phenylacetic acids	C_6-C_2
Hydroxycinnamic acids	C_6-C_3
Phenylpropenes	C_6-C_3
Coumarins, isocoumarins	C_6-C_3
Chromones	C_6-C_3
Naphthoquinones	C_6-C_4
Xanthones	C_6-C_1-C_6
Stilbenes	C_6-C_2-C_6
Anthraquinones	C_6-C_2-C_6
Flavonoids	C_6-C_3-C_6
Lignans and neolignans	$(C_6$-$C_3)_2$
Lignins	$(C_6$-$C_3)_n$

Basic structure

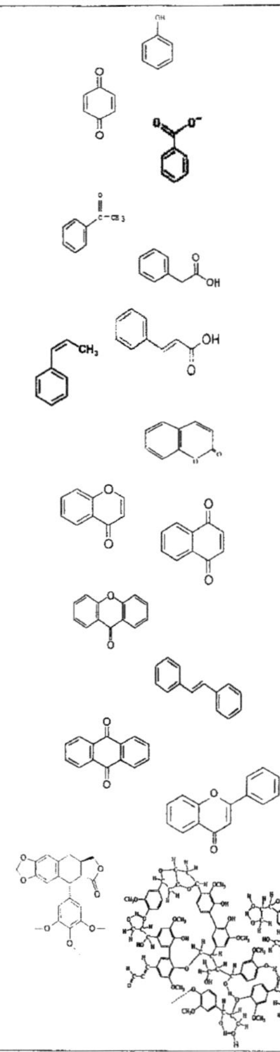

Figure 2. Main classes of phenolic compounds regarding to their carbon chain.

2. Main Classes

2.1. Flavonoids

According to the degree of hydroxylation and the presence of a C_2-C_3 double bond in the heterocycling pyrone ring, flavonoids can be divided into 13 classes [3], the most important being represented by the flavonols, flavanols, flavones, isoflavones, anthocyanidins or anthocyanins and flavanones [2]. Within these classes there are many structural varia-

http://dx.doi.org/10.5772/51687

tions according to the degree of hydrogenation and hydroxylation of the three ring systems of these compounds. Flavonoids also occur as sulfated and methylated derivatives, conjugated with monosaccharides and disaccharides and forming complexes with oligosaccharides, lipids, amines, carboxylic acids and organic acids, being known approximately 8000 compounds [13].

The basic chemical structures of the main classes of flavonoids are presented in Figure 3.

Flavonoid	Basic structure
Flavones	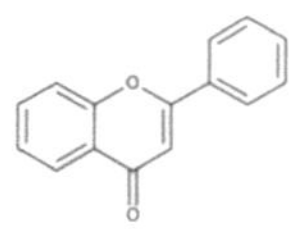
Flavonols	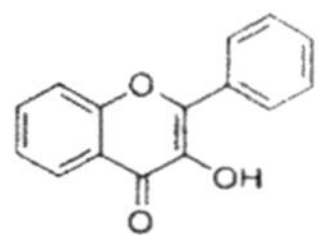
Flavanones	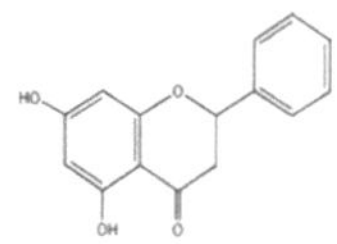
Flavanols	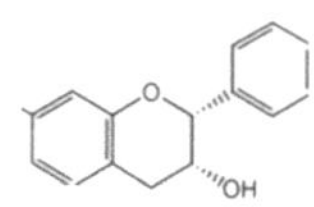
Anthocyanidins	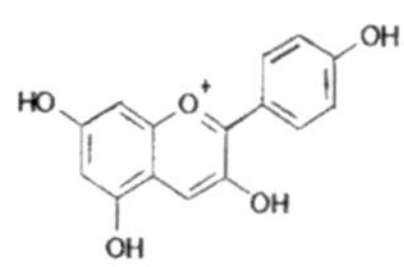
Isoflavones	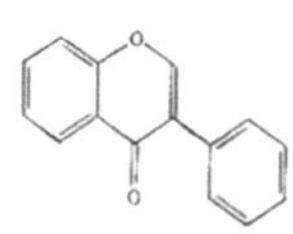

Figure 3. Chemical structures of the main classes of flavonoids.

While members of certain classes of flavonoids (eg., flavonones) are colorless, the other (eg, anthocyanins) are always colored, such as flowers pigments and other plant parts [7].

Flavonoids are important constituents of the human diet [14, 15], and are the most widely distributed phenolic compounds in plant foods and also the most studied ones [10].

It is known that flavonoids are among the most potent antioxidants from plants. The excellent antioxidant activity of these substances is related to the presence of hydroxyl groups in positions 3' and 4' of the B ring, which confer high stability to the formed radical by participating in the displacement of the electron, and a double bond between carbons C_2 and C_3 of the ring C together with the carbonyl group at the C_4 position, which makes the displacement of an electron possible from the ring B. Additionally, free hydroxyl groups in position 3 of ring C and in position 5 of ring A, together with the carbonyl group in position 4, are also important for the antioxidant activity of these compounds [16]. However, the effectiveness of the flavonoids decreases with the substitution of hydroxyl groups for sugars, being the glycosides less antioxidants than their corresponding aglycons [17].

2.2. Tannins

Tannins are phenolic compounds of molecular weight from intermediate to high (500-3000 D) [3] and can be classified into two major groups: hydrolysable tannins and non-hydrolysable or condensed tannins [18]. There is a third group of tannins, phlorotannins, which are only found in brown seaweeds and are not commonly consumed by humans [19].

The hydrolysable tannins have a center of glucose or a polyhydric alcohol partially or completely esterified with gallic acid or hexahydroxydiphenic acid, forming gallotannin and ellagitannins, respectively [20]. These metabolites are readily hydrolyzed with acids, bases or enzymes. However, they may also be oxidatively condensed to other galoil and hexahydroxydiphenic molecules and form polymers of high molecular weight. The best known hydrolysable tannin is the tannic acid, which is a gallotannin consisting of a pentagalloyl glucose molecule that can additionally be esterified with another five units of gallic acid [10].

The condensed tannins are polymers of catechin and/or leucoanthocyanidin, not readily hydrolyzed by acid treatment, and constitute the main phenolic fraction responsible for the characteristics of astringency of the vegetables. Although the term condensed tannins is still widely used, the chemically more descriptive term "proanthocyanidins" has gained more acceptance. These substances are polymeric flavonoids that form the anthocyanidins pigments. The proanthocyanidins most widely studied are based on flavan-3-ols (-)-epicatechin and (+)-catechin [5].

The chemical structures of casuarictin (hydrolysable tannin) and proanthocyanidins (non-hydrolysable or condensed tannins) are shown in Figure 4 A and 4B, respectively.

Although the antioxidant activity of tannins has been much less marked than the activity of flavonoids, recent researches have shown that the degree of polymerization of these substances is related to their antioxidant activity. In condensed tannins and hydrolysable (ellagitannins) of high molecular weight, this activity can be up to fifteen to thirty times superior to those attributed to simple phenols [16].

http://dx.doi.org/10.5772/51687

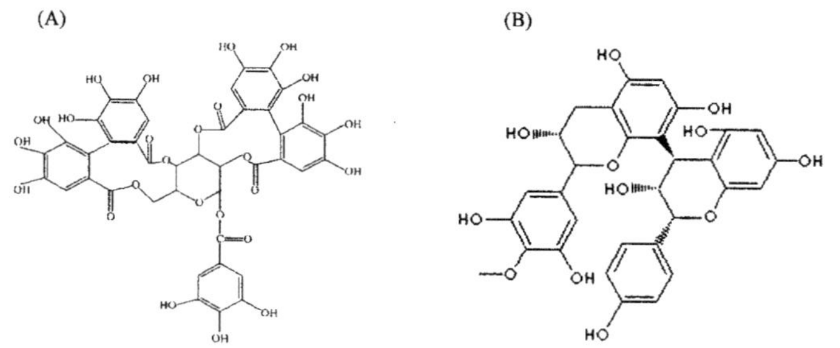

Casuarictin (ellagitannin)

Proanthocyanidins

Figure 4. Chemical structures of hydrolysable tannin (A) and non-hydrolysable or condensed tannins (B).

2.3. Chalcones and Coumarins

The chalcones are intermediate in the biosynthesis of flavonoids, being the phloretin and its glucoside phloridzin (phloretin 2'-*o*-glucose), as well as the chalconaringenin and the arbutin, the most frequently found in foods. The phloretin and phloridzin are characteristics of apples, as well as the chalconaringenin is characteristic of tomatoes and arbutin of pears. However, arbutin is also found in strawberries, wheat and its derivatives, as well as in trace amounts in tea, coffee, red wine and broccoli. In some species of plants, the main pigments of yellow flowers are chalcones [21].

Figure 5 shows the chemical estructures of the main chalcones.

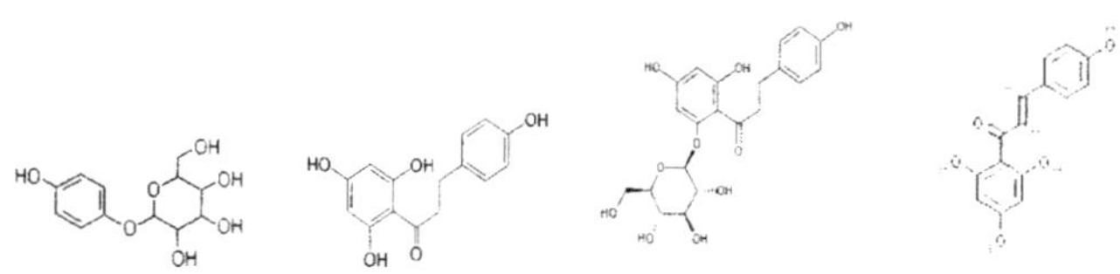

Arbutin | Phloretin | Phloridzin | Chalconaringenin

Figure 5. Chemical structures of the main chalcones.

Like the other phenylpropanoids, coumarins constitute a class of secondary metabolites of plants derivatives from cinnamic acid by cyclization of the side chain of the *o*-coumaric acid [22]. These substances are more common in nature in the form of glycosides, such as umbelliferone, esculetin and scopoletin, and are mainly found in olive oil, oats and spices [3].

The chemical structures of the main coumarins can be seen in Figure 6.

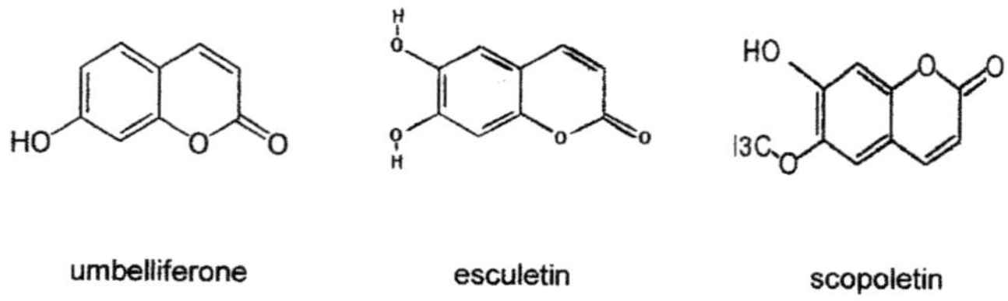

Figure 6. Chemical structures of the main coumarins.

Although the data are still limited, it is known that chalcones and coumarins have antioxidant activity [23].

2.4. Phenolic acids

Phenolic acids can be divided into two groups: benzoic acids and cinnamic acids and derivatives thereof. The benzoic acids have seven carbon atoms (C_6-C_1) and are the simplest phenolic acids found in nature. Cinnamic acids have nine carbon atoms (C_6-C_3), but the most commonly found in vegetables are with seven. These substances are characterized by having a benzenic ring, a carboxylic group and one or more hydroxyl and/or methoxyl groups in the molecule [24].

The general formulas and names of the main benzoic and cinnamic acids are found in Figures 7 and 8, respectively.

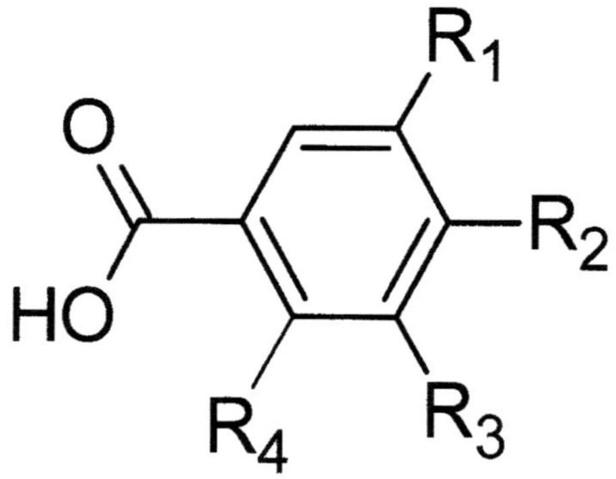

Salicylic acid (R_4 = OH, R_1, R_2, R_3 = H);
Gentisic acid (R_1, R_3 = OH; R_2, R_4 = H);
p-Hydroxybenzoic acid (R_2 = OH, R_1, R_3, R_4 = H);
Protocatechuic acid (R_1, R_2 = OH; R_3, R_4 = H);
Vanillic acid (R_1 = OCH_3, R_2 = OH; R_3, R_4 = H);
Gallic acid (R_1, R_2, R_3 = OH; R_4 = H);
Syringic acid (R_1, R_3 = OCH_3; R_2 = OH; R_4 = H)

Figure 7. The general formulas and names of the main benzoic acids.

In the group of benzoic acids the ones that stand out are protocatechuic acids, vanillic acids, syringic acid, gentisic acid, salicylic acid, p-hydroxybenzoic acid and gallic acid [3].

Among the cinnamic acids, *p*-coumaric, ferulic, caffeic and sinapic acids are the most common in nature [24].

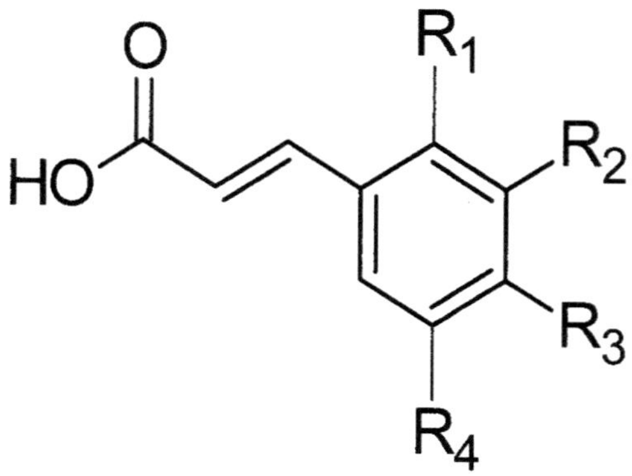

Ceramic acid ($R_1 = R_2 = R_3 = R_4 = H$)
o-Coumaric acid ($R_1 = OH$; $R_2, R_3, R_4 = H$)
m-Coumaric acid ($R_2 = OH$; $R_1, R_3, R_4 = H$)
p-Coumaric acid ($R_3 = OH$; $R_1, R_2, R_4 = H$)
Caffeic acid ($R_2 = R_3 = OH$; $R_1, R_4 = H$)
Ferulic acid ($R_2 = OCH_3$; $R_3 = OH$; $R_1, R_4 = H$)
Sinapic acid ($R_2 = R_4 = OCH_3$; $R_3 = OH$; $R_1 = H$)

Figure 8. The general formulas and names of the main cinnamic acids.

Cinnamic acids rarely found free in plants. They are generally in the form of esters, along with a cyclic alcohol-acid, such as quinic acid to form the isochlorogenic acid, neochlorogenic acid, cripto chlorogenic acid and chlorogenic acid, an caffeoyl ester, which is the most important combination [10].

Figure 9 shows the chemical structure of chlorogenic acid.

Phenolic acids may be about one-third of the phenolic compounds in the human's diet [24]. It is known that these substances and their esters have a high antioxidant activity, especially hydroxybenzoic acid, hydroxycinnamic acid, caffeic acid and chlorogenic acid, and although other characteristics also contribute to the antioxidant activity of phenolic acids and their esters, this activity is usually determined by the number of hydroxyl groups found in the molecule thereof. In general, the hydroxylated cinnamic acids are more effective than their benzoic acids counterparts [16].

Despite the antioxidant activity of phenolic compounds and their possible benefits to human health, until the beginning of the last decade, most studies on these substances occurred in relation to their deleterious effects. Tannins, one of the major components of this group, due to the large number of hydroxyl groups contained therein, among other functional groups (1 to 2 per 100 D), are capable of forming strong complexes with proteins, starch and other molecules, particularly digestive enzymes, reducing the digestibility of the feed. Likewise, by joining with their hydroxyl and carbonyl groups, tannins have the ability to chelate divalent cations, especially Fe and Zn, reducing the bioavailability of these minerals [10].

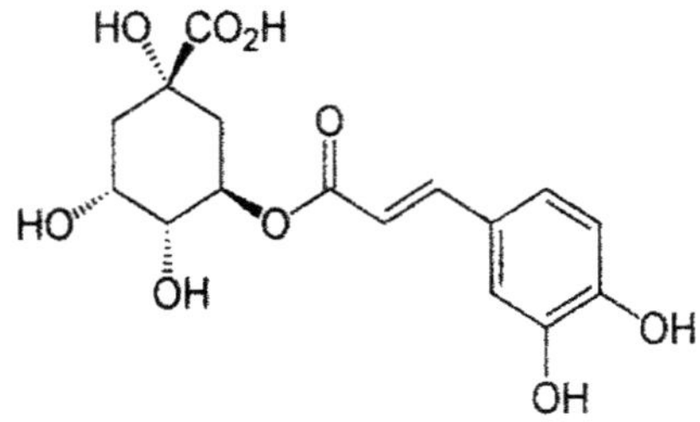

Figure 9. Chemical structure of chlorogenic acid.

Although phenolic compounds are traditionally considered antinutrients, and until the moment as non-nutrients because deficiency states are unknown for them, in recent years they have been seen as a group of micro-nutrients in the vegetable kingdom, which are important part of human and animal diet. The condensed and hydrolysable tannins (ellagitannins) of high molecular weight, since they are not absorbed by the mucosa, they have been regarded as insoluble antioxidants that may have high antioxidant activity in the gastrointestinal tract, protecting proteins, lipids and carbohydrates from oxidative damage during digestion [25].

Researches have also suggested that regular consumption of phenolic compounds directly from plant foods may be more effective in combating oxidative damage in our body than in the form of dietary supplement [26]. This can be explained by the possible synergistic interactions among food phenolic compounds, increasing the antioxidant capacity of these substances..

This way, the content of phenolic compounds and the antioxidant power of a wide variety of plant foods have been investigated.

3. Sources and their antioxidant power

Table 1 shows the mean content of total phenolic compounds (mg/ 100 g of sample) of some plant foods.

http://dx.doi.org/10.5772/51687

Source	Total phenolics (mg%)	Reference
Cereals and legumes		
Cowpea (*V. unguicuata*), brown	100	27
Soyabean	414	28
Oat	352	29
Wheat flour	184	30
Vegetables		
Black carrot	68	31
Broccoli	88	31
Brussels sprouts	69	31
Cabbage, white	76	32
Cabbage, red	186	32
Endive	92	32
Kale	136	33
Lettuce	107	32
Potato	150	31
Spinach	112	32
Tomato	68	32
Yam	92	31
Herbs and spices		
Basil	4425	34
Chilli, green	107	32
Chilli, red	277	32
Coriander	374	31
Garlic	145	31
Ginger	221	31
Leek	85	32
Mint	400	31
Onion, white	269	35
Onion, yellow	164	35
Onion, red	428	35
Pepper, black	1600	36
Pepper, white	800	36

Source	Total phenolics (mg%)	Reference
Shallot	1718	35
Sweet onion	142	35
Thyme	1646	37
Turmeric	176	31
Fruits		
Apple, green	118	38
Apple, red	125	38
Apple, yellow	100	38
Blueberry	362	39
Cherry, sour	156	40
Cherry,sweet	79	38
Grape, black	213	38
Grape, white	184	38
Grapefruit	893	41
Guava, pink flesh	247	42
Guava, white flesh	145	42
Kiwi	791	43
Lemon	843	41
Lime	751	41
Litchi	60	44
Nectarine, white flesh	38	45
Nectarine, yellow flesh	25	45
Orange, sweet	1343	41
Peach, white flesh	53	45
Peach, yellow flesh	35	45
Pear	125	38
Pineapple	94	44
Plum, black	88	44
Plum, red	73	44
Pomegranate	147	44
Pomelo	57	44
Raspberry, black	670	46

Source	Total phenolics (mg%)	Reference
Raspberry, red	342	46
Raspberry, yellow	426	46
Strawberry	199	47
Others		
Roasted cocoa bean	1305	48
Cocoa liquor	994	48
Alkalised cocoa powder	896	48
Baking chocolate	349	48
Red wine	242	49
Tea, black	62	50
Tea, green	83	50
Coffee	188	51

Table 1. Total phenolic compounds content of some plant foods.

As can be seen in Table 1, phenolic compounds are widely distributed in plant foods.

Cocoa, potato, yam, tomato, kale, Brussels sprouts, broccoli and others dark green leafy and brightly-colored vegetables as well as legumes and cereals, in addition to spices and fruits such as cherries and citrus, are particularly rich in phenolic compounds. Red wine also has a high concentration of phenolic compounds. It is known that the abundant phenolic compounds in red wine are anthocyanin [6, 52]. The green and black teas have been extensively studied, since they may contain up to 30% of their dry weight as phenolic compounds [53]. Coffee is also rich in phenolic compounds, especially chlorogenic acid. It has about 7% of the dry weight of the grains [24] and 15% of the dry instant coffee as phenolic compounds [54].

Although in some studies a few statistically significant correlations were found between the levels of total phenolic compounds and antioxidant power of foods, in others the total phenolics content of samples was highly correlated with the antioxidant capacity. On the other hand, there are still no standard methods and approved for determining the antioxidant power *in vitro*. The several available tests for this purpose involve different mechanisms of antioxidant defense system, from the chelation of metal ions to the measure of preventing oxidative damage to biomolecules, and offer distinct numerical results that are difficult to compare. Because of this, studies have used different methods to evaluate the antioxidant capacity of the studied sample, such as ABTS (2,2-azino-bis-3-ethylbenzothiazoline-6-sulfonic acid radical assay), DPPH (2,2-diphenyl-picrylhydrazyl radical assay), FRAP (Ferric Reducing/Antioxidant Power assay) and ORAC (Oxygen Radical Absorbance Capacity assay), among others tests.

By determining the content of total phenolic compounds and ability to reduce $FeCl_3$ as well as DPPH free radical of some commonly consumed and underutilized tropical legumes [27], it was concluded that one of the commonly consumed cowpea *Vigna unguiculata* (brown) as well as underutilized legumes C. cajan (brown) and *S. sternocarpa* could be considered as functional foods due to their relatively higher antioxidant power, which could be as a result of their relative higher total phenolics content. In a similar way, evaluating the antioxidant capacity of twenty soybean hybrids by DPPH assay and their total phenolics content [28], it was concluded that the two cultivars that showed the highest contents of total phenolics also showed the highest antioxidant powers.

Among cereals, milled oat groat pearlings, trichomes, flour, and bran were evaluated as to their antioxidant capacity against the oxidation of *R*-phycoerythrin protein in the ORAC assay, as well as against the oxidation of low density lipoproteins (LDL) [55]. In both the methods applied the antioxidant capacity of the fractions of oats was in the following order: pearlings > flour > trichome = bran. It was concluded through this study that a part of oat antioxidants, which is rich in phenolic compounds [29], is probably heat-labile because greater antioxidant power was found among the non-steam-treated pearlings. In another study, ten varieties of soft wheat were compared as to their content of total phenolic compounds and antioxidant capacity [30]. Important DPPH, oxygen, hydroxyl and ABTS radical removal capacity was found in all the studied varieties and the content of total phenolics of the samples showed correlation with their antioxidant power in DPPH, ORAC and ABTS assays.

On the other hand, searching the antioxidant capacity of vegetables in the genus Brassica and the best solvent (ethanol, acetone and methanol) for the extraction of their phenolic compounds [56], the results showed that the solvent used significantly affects the phenolics content and the properties of the studied extract. Methanolic extract showed the largest content of total phenolics of broccoli, Brussels sprouts, and white cabbage. In this study, the antioxidant power of the samples was confirmed by different reactive oxygen species and showed to be concentration-dependent. Kale extracts have also been evaluated as to their content of total phenolic compounds and antioxidant capacity [33]. It can be observed that all studied fractions (free and conjugated forms) were able to remove the DPPH radical and that the content of total phenolic compounds of fractions was highly correlated with their antioxidant power.

Herbs and spices are of particular interest, since they have been proved to have high content of phenolic compounds and high antioxidant capacity. The values of Trolox Equivalent Antioxidant Capacity (TEAC) and content of total phenolics were determined for 23 basil accessions [34]. A positive linear relationship was found between the content of total phenolic compounds and the antioxidant power of samples. This study concluded that basils have valuable antioxidant properties for culinary and possible medical application. The concentration of phenolic compounds in peppercorn (black and white), as well as the ability of hydrolyzed and nonhydrolyzed pepper extracts to remove DPPH, superoxide, and hydroxyl radicals [36] were also investigated. The results obtained showed that hydrolyzed and nonhydrolyzed extracts of black pepper contained significantly more phenolic compounds when compared with those of white pepper. For any of these peppers, the hydrolyzed ex-

http://dx.doi.org/10.5772/51687

tract contained significantly more phenolic compounds in comparison with the nonhydrolyzed extract. A dose-dependent effect was observed for all extracts concerning the power of removing free radical and reactive oxygen species, the black pepper extracts being the most effective. This study concluded that the pepper, especially black, which is an important component in the diet of many sub-Saharan and Eastern countries due to its nutritional importance, can be considered an antioxidant and radical scavenging. However, evaluating the content of phenolic compounds and antioxidant capacity of 14 herbs and spices [37], although a significant correlation has been obtained between the phenolics content and antioxidant capacity of samples, it was found that the trend of the antioxidant capacity was different according to the method applied. The leaves of the species Piper showed the highest antioxidant capacity in both methods studied (Folin-Ciocalteu reagent and FRAP method). Yet, the African mango showed the greatest content of free antioxidant by FRAP method, while by Folin method *Piper umbellatum* excelled followed by thyme. This study concluded that the antioxidant power of plant samples should be interpreted with caution when measured by different methods. In spite of that fact, regardless of the method used, the samples were rich in antioxidants.

In addition to the studies already mentioned, the antioxidant capacity of 36 plant extracts was evaluated by the β-carotene and linoleic acid model system [31] and the content of total phenolic compounds of the extracts was determined. Mint, black carrots, and ginger showed high content of total phenolics. The antioxidant capacity calculated as percentage of oxidation inhibition ranged from a maximum of 92% in turmeric extracts to a minimum of 12.8% in long melon. Other foods which have high antioxidant capacity (> 70%) were ginger, mint, black carrots, Brussels sprouts, broccoli, yam, coriander and tomato. The antioxidant power of the samples significantly and positively correlated with their content of total phenolic compounds, allowing the conclusion that the plant foods with high content of phenolic compounds can be sources of dietary antioxidants. In another study, 66 types of plant foods were analyzed as to their content of phenolic compounds and their antioxidant capacity in the ORAC assay [32]. The results showed that the antioxidants composition and concentration varied significantly among the different vegetables. The coriander, Chinese kale, water spinach and red chili showed high content of total phenolics and high antioxidant power.

Due to the growing recognition of their nutritional and therapeutic value, many fruits have also been investigated as to their content of phenolic compounds and antioxidant capacity. By evaluating the antioxidant capacity and total phenolics content, in addition to flavanol and monomeric anthocyanins, it was found from the flesh and peel of 11 apple cultivars [57] that the concentrations of the parameters investigated differed significantly among the cultivars and were higher in the peel in comparison to the flesh. The content of total phenolics and antioxidant capacity were significantly correlated in both flesh and peel. It was concluded that the contribution of phenolics to the antioxidant power in apple peel suggests that peel removal may induce a significant loss of antioxidants. It is also known that one of the most important sources of antioxidants among fruits is small red fruits. By determining the antioxidant capacity of four cultivars of blueberry through three different assays (DPPH, ABTS and FRAP), as well as the content of total phenolic compounds, in addition to flavonoids, anthocyanins and flavan-3-ols [39], it was found that all cultivars contained high content of total phenolics, flavonoids and anthocyanins and lower content of flavan-3-ols. However, significant differences were found in the total phenolics content among the different cultivars and growing seasons. Despite this, the studied cultivars showed high antioxidant power, which was highly correlated with the samples phenolic compounds. Similarly, by checking the content of total phenolics, in addition to flavonoids and anthocyanins, as well as the antioxidant capacity of three cultivars of sour cherries [58], a significant difference was observed in phenolics content among different cultivars and growing seasons. However, the cultivars analyzed showed high antioxidant capacity, which was correlated with the phenolic compounds found in them. In this study significant increases were also found in the content of total phenolic compounds and antioxidant power during the ripening of fruits. Additionally, different solvents were applied for comparing the antioxidant capacity and the yield of total phenolic compounds present in the extracts of sour and sweet cherries [40]. It was found that the solubility of phenolic compounds was more effective in extracts of sweet cherries with use of methanol at 50% and in extracts of sour cherries with the use of acetone at 50%. Extracts from lyophilized sour cherries (methanolic and acetone water-mixtures) presented in average twice as high phenolic compounds than ethanolic extracts. The DPPH antiradical efficiency values were higher in the extracts of sour cherries when compared with those of sweet cherries. It was concluded in this work that the strong antioxidant power of extracts of sour cherries is due to the substantial amount of total phenolic compounds present in them and that the fresh sour cherry can be considered as a good dietary source of phenolic compounds. The total phenolics content, total monomeric anthocyanins and antioxidant capacities of 14 wild red raspberry accessions were also examined [59]. In this study, more two cultivars were included in the investigation to determine the variation between wild and cultivated raspberries. Antioxidant capacity of fruits was evaluated by both FRAP and TEAC assays. Significant variability was found for total phenolics, total monomeric anthocyanins and antioxidant capacity of wild raspberries. Nevertheless, the results indicated that some of the wild accessions of red raspberries have higher antioxidant power and phytonutrients content than existing domesticated cultivars. Finally, two strawberry cultivars were studied as to their content of total phenolic compounds and antioxidant capacity in different ripeness stages [47]. It was concluded that despite the berries in general have better taste and be more appreciated at ripe stage, higher contents of total phenolic compounds and antioxidant power were observed at pink stage for both strawberry cultivars studied. Also with respect to the fruits, a less known snake fruit was compared with better known kiwi fruit regarding to their total phenolics content and four radical scavenging (FRAP, ABTS, DPPH and CUPRAC/Cupric Reducing Antioxidant Capacity) ability [43]. It was observed similarity between snake fruit and kiwi fruit in the contents of phenolic compounds as well as antioxidant power in DPPH assay. By this study, it was able to conclude that the two fruits can be applied as antioxidant supplements to the normal diet. Consumption of a combination of both fruits could be recommended in order to obtain the best results. In another study, 25 cultivars, 5 each of white-flesh nectarines, yellow-flesh nectarines, white-flesh peaches, yellow-flesh peaches, and plums at the ripe stage were studied for their total phenolics content and antioxidant capacity by the DPPH and FRAP assays [45]. In

http://dx.doi.org/10.5772/51687

descending order, the cultivars presenting higher contents of total phenolics were: white-flesh peaches, plums, yellow-flesh peaches and yellow-flesh nectarines. There was a strong correlation between total phenolics and antioxidant power of nectarines, peaches, and plums. By continuing to study the plum fruits, 20 genotypes of plums were investigated for their antioxidant capacity and total phenolics content [60]. Among the 20 genotypes, a strong correlation was observed between the total phenolics and antioxidant power of the samples, which was determined upon the FRAP assay. It was concluded that phenolic compounds seem to play a significant role in antioxidant value and health benefits of plums. Additionally, Mirabelle plums were examinated for their antioxidant capacity by different assays (DPPH, FRAP, ORAC) and total phenolics content [61]. The antioxidant power of the plum peels, flesh and pits reflected the total phenolics content of the samples with efficacy increasing of the order: peels < flesh < pits across the assays. Peel and flesh of six pear cultivars were also investigated for their antioxidant capacity by DPPH assay and total phenolics content [62]. The results obtained showed that the total phenolics content in the peel can be up to 25 times higher than in the flesh. The peel also showed higher antioxidant power. The pomegranate is another fruit that has been researched. Its peel, mesocarp and juice were evaluated for their antioxidant power by TEAC and FRAP assays as well as total phenolics content [63]. It was found not only high correlation between TEAC and FRAP values, but also with the total phenolics content, which was in the following order: mesocarp > peel > juice. This study demonstrated that selection of raw materials (co-extraction of arils and peel) and pressure, respectively, markedly affected the profile and content of phenolics in the pomegranate juices, underlining the necessity to optimise these parameters for obtaining products with well-defined functional qualities. Studies have also been carried out to quantify the total phenolics content and antioxidant capacity of citrus fruits. Comparing the antioxidant properties of peel (flavedo and albedo) and juice of grapefruit, lemon, lime and sweet orange, four different antioxidant assays (DPPH, Reducing Power, β-carotene–linoleate Model System and Thiobarbituric Acid Reactive Substances/TBARS) were applied to the volatile and polar fractions of peels and to crude and polar fraction of juices [41]. Phenolic compounds were among the two main antioxidant substances found in all extracts. Peels polar fractions showed the highest contents in phenolics, which probably contribute to the highest antioxidant power found in these fractions. However, peels volatile fractions showed the lowest antioxidant power. In another experiment, grapefruit and sour orange were extracted with five different polar solvents. The total phenolics content of the extracts was determined and the dried fractions were screened for their antioxidant capacity by four different assays (DPPH, Phosphomolybdenum method, Nitroblue tetrazolium/NBT Reduction and Reducing Power) [64]. All citrus extracts showed good antioxidant capacity. The best correlation between total phenolics and radical scavenging activity was observed by DPPH method. It was concluded that the data obtained clearly established the antioxidant power of the studied citrus fruit extracts. Studying the extraction efficiency of five different solvents on the total phenolics content and antioxidant capacities of pomelo and navel oranges by five antioxidant assays (DPPH, ORAC, ABTS, Phospomolybdenum method and Reducing Power) [65], it was found that the total phenolics content of extracts varied according to the solvent used. Significant differences were also found in antioxidant capacity values via the same method in different solvents, as well as on the antioxidant capacity of each extract via different methods. Nonetheless, the broad range of activity of the extracts led to the conclusion that multiple mechanisms are responsable for the antioxidant power of the samples and clearly indicated the potential application value of the citrus fruits studied. Finally, the study of the content of phenolic compounds and antioxidant power of tropical fruits such as guava has also been conducted. One white-fleshed and three pink-fleshed of guava were analyzed as to their content of total phenolics, in addition to ascorbic acid and total carotenoids, as well as to their antioxidant capacity [42]. The ABTS, DPPH and FRAP assays were used for determining the antioxidant capacity in methanol and dichloromethane extracts of the samples, while the ORAC assay was used only for determining it in methanol extracts. The results obtained showed that white pulp guava had more total phenolics and ascorbic acid than pink pulp guava. On the other hand, carotenoids were absent in the white pulp guava. In all antioxidant assays the methanol extracts showed good correlation with the content of total phenolics and ascorbic acid, as well as between them, but showed negative correlation with total carotenoids.

In addition to the aforementioned fruits, in the search for new foods rich in phenolic compounds and high antioxidant capacity, unconventional tropical fruits have been widely researched. Accordingly, the Antilles cherry, Barbados cherry or acerola (1063 mg/100 g), camu-camu (1176 mg/100 g), puçá-preto (868 mg/100 g), assai or açaí (454 mg/100 g) and jaboticaba (440 mg/100 g) showed to be rich in phenolic compounds. When testing the antioxidant capacity of these fruits fresh and dry matter by DPPH assay, it was found an association between their antioxidant power and total phenolics content [66]. Similarly, banana passion fruit (635-1018mg/100 g), cashew (445 mg/100 g) and guava apple (309 mg/100 g) also showed a high total phenolic content when evaluated by FRAP and ABTS assays. The antioxidant power of these fruits showed a strong correlation with their total phenolics content [67].

Other plant-originated foods studied for their content of phenolic compounds and antioxidant capacities are as follows. The cocoa and chocolate liquor antioxidant capacities as well as monomeric and oligomeric procyanidins were studied [68]. The results obtained showed that the procyanidins content was correlated with the antioxidant capacity, which was determined by the ORAC assay as an indicator for potential biological activity of the samples. However, following the changes in total and individual phenolics content as well as antioxidant capacity during the processing of cocoa beans [48], it can be noted that the loss of phenolic compounds and antioxidant capacity of cocoa vary according to the degree of technological processing. The roasting process and cocoa nib alkalization had the greatest influence on the content of phenolic compounds and antioxidant power. The antioxidant capacity of 107 different Spanish red wines, from different varieties of grapes, aging processes and vintages [69] was also investigated by different methods and the results showed that all samples had an important capacity of removing hydroxyl radical and were able to block the superoxide radical, but with 10 times lower intensity. The wines also showed important protective action on biomarkers of oxidative stress. However, few statistically significant correlations were found between the levels of total phenolics and antioxidant power of the wines and the values of these correlations were very low. In another investigation, the antioxidant capacities of three Argentine red wines were evaluated by TEAC and FRAP assays. The correlation between antioxidant capacity and content of phenolic compounds as well as between antioxidant capacity and phenolic profile of samples [49] was determined. It can be noted that the wines showed significant antioxidant capacity. However, no significant correlation was found between their antioxidant capacity and total phenolics content. Nevertheless, the canonical correlation and multiple regression analysis showed that the antioxidant capacity of the samples was highly correlated with their profile of phenolic compounds. The results obtained in this study showed the importance of analyzing the phenolic profile of the sample rather than total phenolics to help understand the differences in the antioxidant power of wines, which should be extended to other food products. Among the alcoholic beverages, antioxidant power has also been reported for whiskey, sake and sherries. [70]. In addition to alcoholic beverages, the free radical-scavenging activity and total phenolic content of commercial tea [50] were determined, finding that green tea contained higher content of phenolic compounds than black tea. The antioxidant capacity per serving of green tea was also much higher than that of black tea. However, comparing the content of total phenolics, flavonoids and antioxidant capacity of black tea, green tea, red wine and cocoa by ABTS and DPPH assays [71], it was found that cocoa contains much higher levels of total phenolics and flavonoids per serving than black tea, green tea and red wine. In the two methods applied, the antioxidant power of the samples per serving was found in the following descending order: cocoa, red wine, green tea and black tea. The content of total phenolic compounds and DPPH and ABTS radical removal capacity of coffee extracts obtained by continuous (Soxhlet 1 h and 3 h) and discontinuous (solid-liquid extraction and filter coffeemaker) methods, many solvents (water, methanol, ethanol and their mixtures), successive extractions and water with different pHs (4.5, 7.0 and 9.5) were also evaluated [72]. The coffee extracts with the highest antioxidant capacity were obtained after extraction with water neutral (pH 7.0) in the filter coffeemaker (24 g spent coffee per 400 mL water). In addition, the drink degreasing and lyophilization of the extract permitted to obtain coffee extract powder with high antioxidant power, which can be used as an ingredient or additive in the food industry with potential for preservation and functional properties.

It is also know that tamarind, canola, sesame, linseed and sunflower seeds are other possible sources of phenolic compounds [73] and have high antioxidant capacity. The antioxidant capacity of the striped sunflower seed cotyledon extracts, obtained by sequential extraction with different polarities of solvents, was determined by three in vitro methods: FRAP, DPPH and ORAC [74]. In the three methods applied, the aqueous extract showed higher antioxidant capacity than the ethanolic. When compared with the synthetic antioxidant Butylated Hydroxyl Toluene (BHT), the antioxidant power of the aqueous extract varied from 45% to 66%, according to the used method. It was concluded in this study that the high antioxidant power found for the aqueous extract of the studied sunflower seed suggests that the intake of this seed may prevent *in vivo* oxidative reactions responsible for the development of several diseases.

4. Conclusion

Phenolic compounds are widely distributed in plant foods (cereals, vegetables, fruits and others), stressing among them the flavonoids, tannins, chalcones, coumarins and phenolic acids. Although some studies have shown few statistically significant correlations between the levels of total phenolics and antioxidant capacity in foods, in others the content of total phenolic compounds was highly correlated with the antioxidant power of samples. Among the plant foods with a high content of phenolic compounds and antioxidant capacity, we can stand out the dark green leafy and brightly-colored vegetables, in addition to cocoa, soya-bean, spices and fruits such as cherries and citrus.

Author details

Maria de Lourdes Reis Giada*

Address all correspondence to: mlgiada@nutricao.ufrj.br

Department of Basic and Experimental Nutrition, Institute of Nutrition, Health Sciences Center, Federal University of Rio de Janeiro, Brazil

References

[1] Vickery, M. L., & Vickery, B. (1981). Secondary plant metabolism. *London: MacMillan.*

[2] Scalbert, A, & Williamson, G. (2000). Dietary intake and bioavailability of polyphenols. *Journal of nutrition*, 130, 2073S-2085S.

[3] Sánchez-Moreno, C. (2002). Compuestos polifenólicos: estructura y classificación: presencia en alimentos y consumo: biodisponibilidad y metabolismo. *Alimentaria*, 329, 19-28.

[4] Hollman, P. C. H. (2001). Evidence for health benefits of plant phenols: local or systemic effects? *Journal of the Science of Food and Agriculture*, 81(9), 842-852.

[5] Stafford, H. A. (1983). Enzymic regulation of procyanidin bisynthesis, lack of a flav-3-en-3-ol intermediate. *Phytochemistry*, 22, 2643-2646.

[6] Morton, L. W., Cacceta, R. A. A., Puddey, I. B., & Croft, K. D. (2000). Chemistry and biological effects of dietary phenolic compounds: relevance to cardiovascular disease. *Clinical and Experimental Pharmacology and Physiology*, 27(3), 152-159.

[7] Harborne, J. B. (1980). Plant phenolics. In: Bell EA, Charlwood BV, Archer B. (ed.) Secondary plant products. *Berlin: Springer-Verlag*, 330-402.

[8] Tomás-Barberán, F. A., & Espín, J. C. (2001). Phenolic compounds and related enzymes as determinants of quality in fruits and vegetables. *Journal of the Science of Food and Agriculture*, 81(9), 853-876.

[9] Harborne, J. B. (1989). Methods in plant biochemistry. In: Dey PM, Harborne JB. (ed.) Plant phenolics. *London: Academic Press*, 1.

[10] Bravo, L. (1998). Polyphenols: chemistry, dietary sources, metabolism and nutritional significance. *Nutrition Reviews*, 56(11), 317-333.

[11] Lampe, J. W. (1999). Health effects of vegetables and fruit: assessing mechanisms of action in human experimental studies. *The American Journal of Clinical Nutrition*, 70, 475S-490S.

[12] Percival, M. (1998). Antioxidants. *Clinical Nutrition Insights*, 10, 1-4.

[13] Duthie, G. G, Gardner, P. T, & Kyle, J. A. M. (2003). Plant polyphenols: are they the new magic bullet? *Proceedings of the Nutrition Society*, 62(3), 599-603.

[14] Hertog, M. G. L., Hollman, P. C. H., & Venema, D. P. (1992). Optimization of a quantitative HPLC determination of potentially anticarcinogenic flavonoids in vegetables and fruits. *Journal of Agricultural and Food Chemistry*, 40(9), 1591-1598.

[15] Jovanovic, S. V., Steenken, S., Tosic, M., Marjanovic, B., & Simic, M. G. (1994). Flavonoids as antioxidants. *Journal of the American Chemical Society*, 116(11), 4846-4851.

[16] Sánchez-Moreno, C. (2002). Compuestos polifenólicos: efectos fisiológicos: actividad antioxidante. *Alimentaria*, 329, 29-40.

[17] Rice-Evans, C., Miller, N. J., & Paganga, G. (1996). Structure-antioxidant activity relationships of flavonoids and phenolic acids. *Free Radical Biology and Medicine*, 20(7), 933-956.

[18] Chung, K. T., Wong, T. Y., Wei, C. I., Huang, Y. W., & Lin, Y. (1998). Tannins and human health: a review. *Critical Reviews in Food Science and Nutrition*, 38(6), 421-464.

[19] Ragan, M. A., & Glombitza, K. (1986). Phlorotannin: Brown algal polyphenols. *Progress in Physiological Research*, 4, 177-241.

[20] Okuda, T., Yoshida, T., & Hatano, T. (1995). Hidrolyzable tannins and related polyphenols. *Fortschritte der Chemie organischer Naturstoffe*, 66, 1-117.

[21] Karakaya, S. (2004). Bioavailability of phenolic compounds. *Critical Reviews in Food Science and Nutrition*, 44(6), 453-464.

[22] Matern, V., Lüer, P., & Kreusch, D. (1999). Biosynthesis of coumarins. In: Barton D, Nakanishi K, Meth-Cohn O, Sankawa V. (ed.) Comprehensive natural products chemistry: polyketides and other secondary metabolites including fatty acids and their derivatives. *Amsterdam: Elsevier Science*, 623-637.

[23] Pratt, D. E., & Hudson, B. J. F. (1990). Natural antioxidant no exploited commercially. In: Hudson BJF. (ed.) Food antioxidants. *London: Elsevier Applied sciences*, 171-180.

[24] Yang, C. S., Landau, J. M., Huang, M. T., & Newmark, H. L. (2001). Inhibition of carcinogenesis by dietary polyphenolic compounds. *Annual Review of Nutrition*, 21, 381-406.

[25] Martínez-Valverde, I., Periago, M. J., & Ros, G. (2000). Significado nutricional de los compuestos fenólicos de la dieta. *Archivos Latinoamericanos de Nutrición*, 50(1), 5-18.

[26] Martin, K. R., & Appel, C. L. (2010). Polyphenols as dietary supplements: A double-edged sword. *Nutritional and Dietary Supplements*, 2, 1-12.

[27] Oboh, G. (2006). Antioxidant properties of some commonly consumed and underutilized tropical legumes. *European Food Research and Technology*, 224(1), 61-65.

[28] Malencić, D., Popović, M., & Miladinović, J. (2007). Phenolic content and antioxidant properties of soybean (Glycine max (L.) Merr.) seeds. *Molecules*, 12(3), 576-581.

[29] Kovácová, M., & Malinová, E. (2007). Ferulic and coumaric acids, total phenolic compounds and their correlation in selected oat genotypes. *Czech Journal of Food Sciences*, 25(6), 325-332.

[30] Lv, J., Yu, L., Lu, Y., Nui, Y., Liu, L., Costa, J., & Yu, L. (2012). Phytochemical compositions, and antioxidant properties, and antiproliferative activities of wheat flour. *Food Chemistry*, doi: 10.1016/j.foodchem.2012.04.141.

[31] Kaur, C., & Kapoor, H. C. (2002). Anti-oxidant activity and total phenolic content of some Asian vegetables. *International Journal of Food Science and Technology*, 37(2), 153-161.

[32] Isabelle, M., Lee, B. L., Lim, M. T., Koh, W. P., Huang, D., & Ong, C. N. (2010). Antioxidant activity and profiles of common vegetables in Singapore. *Food Chemistry*, 120(4), 993-1003.

[33] Ayaz, F. A., Hayirhoglu-Ayaz, S., Alpay-Karaoglu, S., Grúz, J., Valentová, K., Ulrichová, J., & Strnad, M. (2008). Phenolic acid contents of kale (Brassica oleraceae L. var. acephala DC.) extracts and their antioxidant and antibacterial activities. *Food Chemistry*, 107(1), 19-25.

[34] Javanmardi, J., Stushnoff, C., Locke, E., & Vivanco, J. M. (2003). Antioxidant activity and total phenolic content of Iranian Ocimum accessions. *Food Chemistry*, 83(4), 547-550.

[35] Lu, X., Wang, J., Al-Qadiri, H. M., Ross, C. F., Powers, J. R., Tang, J., & Rasco, BA. (2011). Determination of total phenolic content and antioxidant capacity of onion (Allium cepa) and shallot (Allium oschaninii) using infrared spectroscopy. *Food Chemistry*, 129(2), 637-644.

[36] Agbor, G. A., Vinson, J. A., Oben, J. E., & Ngogang, J. Y. (2006). Comparative analysis of the in vitro antioxidant activity of white and black pepper. *Nutrition Research*, 26(12), 659-663.

[37] Agbor, G. A., Oben, J. E., Ngogang, J. Y., Xinxing, C., & Vinson, J. A. (2005). Antioxidant capacity of some herbs/spices from Cameroon: A comparative study of two methods. *Journal of Agricultural and Food Chemistry*, 53(17), 6819-6824.

[38] Marinova, D., Ribarova, F., & Atanassova, M. (2005). Total phenolics and total flavonoids in Bulgarian fruits and vegetables. *Journal of the University of Chemical Technology and Metallurgy*, 40(3), 255-260.

[39] Dragović-Uzelac, V., Savić, Z., Brala, A., Levaj, B., Kovacević, D. B., & Biško, A. (2010). Evaluation of phenolic content and antioxidant capacity of blueberry cultivars (Vaccinium corymbosum L.) grown in the Northwest Croatia. *Food Technology and Biotechnology*, 48(2), 214-221.

[40] Melichácová, S., Timoracká, M., Bystrická, J., Vollmannová, A., & Céry, J. (2010). Relation of total antiradical activity and total polyphenol content of sweet cherries (Prunus avium L.) and tart cherries (Prunus cerasus L.). *Acta Agriculturae Slovenica*, 95(1), 21-28.

[41] Guimarães, R., Barros, L., Barreira, J. C. M., Sousa, M. J., Carvalho, A. M., & Ferreira, I. C. F. R. (2010). Targeting excessive free radicals with peels and juices of citrus fruits: grapefruit, lemon, lime and Orange. *Food and Chemical Toxicology*, 48(1), 99-106.

[42] Thaipong, K., Boonprakob, U., Crosby, K., Zevallos, L. C., & Byrne, D. H. (2006). Comparison of ABTS, DPPH, FRAP, and ORAC assays for estimating antioxidant activity from guava fruit extracts. *Journal of Food Composition and Analysis*, 19(6), 669-675.

[43] Gorinstein, S., Haruenkit, R., Poovarodom, S., Park, Y. S., Vearasilp, S., Suhaj, M., Ham, K. S., Heo, B. G., Cho, J. Y., & Jang, H. G. (2009). The comparative characteristics of snake and kiwi fruits. *Food and Chemical Toxicology*, 47(8), 184-1891.

[44] Fu, L., Xu, B. T., Xu, X. R., Gan, R. Y., Zhang, Y., Xia, E. Q., & Li, H. B. (2011). Antioxidant capacities and total phenolic contents of 62 fruits. *Food Chemistry*, 129(2), 345-350.

[45] Gil, M. I., Tomás-Barberán, F. A., Hess-Pierce, B., & Kader, AA. (2002). Antioxidant capacities, phenolic compounds, carotenoids, and vitamin C contents of nectarine, peach, and plum cultivars from California. *Journal of Agricultural and Food Chemistry*, 50(17), 4976-4982.

[46] Gansch, H., Weber, C. A., & Lee, C. Y. (2009). Antioxidant capacity and phenolic phytochemiclas in black raspberries. *New York State Horticultural Society*, 17(1), 20-23.

[47] Pineli, L. L. O., Moretti, C. L., dos Santos, S. M., Campos, A. B., Brasileiro, A., Córdova, A. C., & Chiarello, M. D. (2011). Antioxidants and other chemical and physical characteristics of two strawberry cultivars at different ripeness stages. *Journal of Food Composition and Analysis*, 24(1), 11-16.

[48] Jolić, S. M., Redovniković, I. R., Marković, K., Šipušić, Đ. I., & Delonga, K. (2011). Changes of phenolic compounds and antioxidant capacity in cocoa beans processing. *International Journal of Food Science and Technology*, 46(9), 1793-1800.

[49] Baroni, M. V., Naranjo, R. D. D. P., García-Ferreyra, C., Otaiza, S., & Wunderlin, D. A. (2012). How good antioxidant is the red wine ? Comparison of some in vitro and in vivo methods to assess the antioxidant capacity of Argentinean red wines. *LWT-Food Science and Technology*, 47(1), 1-7.

[50] Lee, KW, Lee, HJ, & Lee, CY. (2002). Antioxidant activity of black te Lee a vs. green tea. *Journal of Nutrition*, 132(4), 785.

[51] Natella, F., Nardini, M., Belelli, F., & Scaccini, C. (2007). Coffee drinking induces incorporation of phenolic acids into LDL and increases the resistance of LDL to ex vivo oxidation in humans. *The American Journal of Clinical Nutrition*, 86(3), 604-609.

[52] Pellegrini, N., Serafini, M., Colombi, B., Del Rio, D., Salvatore, S., Bianchi, M., & Brighenti, F. (2003). Total antioxidant capacity of plant foods, beverages and oils consumed in Italy assessed by three different in vitro assays. *Journal of Nutrition*, 133(9), 2812-2819.

[53] Thiagarajan, G., Chandani, S., Sundari, C. S., Rao, S. H., Kulkarni, A. V., & Balasubramanian, P. (2001). Antioxidant properties of green and black tea, and their potential ability to retard the progression of eye lens cataract. *Experimental Eye Research*, 73(3), 393-401.

[54] King, A., & Young, G. (1999). Characteristics and occurrence of phenolic phytochemicals. *Journal of the American Dietetic Association*, 99(2), 213-218.

[55] Handelman, G. J., Cao, G., Walter, M. F., Nightingale, Z. D., Paul, G. L., Prior, R. L., & Blumberg, J. B. (1999). Antioxidant capacity of oat (Avena sativa L.) extracts. 1. Inhibition of low-density lipoprotein oxidation and oxygen radical absorbance capacity. *Journal of Agricultural and Food Chemistry*, 47(12), 4888-4893.

[56] Jaiswal, A. K., Abu-Ghannam, N., & Gupta, S. (2012). A comparative study on the polyphenolic content, antibacterial activity and antioxidant capacity of different solvent extracts of Brassica oleracea vegetables. *International Journal of Food Science and Technology*, 47(2), 223-231.

[57] Vieira, F. G. K., Borges, G. S. C., Copetti, C., Pietro, P. F., Nunes, E. C., & Fett, R. (2011). Phenolic compounds and antioxidant activity of the apple flesh and peel of eleven cultivars grown in Brazil. *Scientia Horticulturae*, 128(3), 261-266.

[58] Mitić, M. N., Obradović, M. V., Kostić, D. A., Micić, R. J., & Pecev, E. T. (2012). Polyphenol content and antioxidant activity of sour cherries from Serbia. *Chemical Industry and Chemical Engineering*, 18(1), 53-62.

[59] Çekiç, Ç., & Özgen, M. (2010). Comparison of antioxidant capacity and phytochemical properties of wild and cultivated red raspberries (RubusidaeusL.). *Journal of Food Composition and Analysis*, 23(6), 540-544.

[60] Rupasinghe, H. P. V., Jayasankar, S., & Lay, W. (2006). Variation in total phenolics and antioxidant capacity among European plum genotypes. *Scientia Horticulturae,* 108(3), 243-246.

[61] Khallouki, F., Haubner, R., Erben, G., Ulrich, C. M., & Owen, R. W. (2012). Phytochemical composition and antioxidant capacity of various botanical parts of the fruits of Prunus × domestica L. from the Lorraine region of Europe. *Food Chemistry,* 133(3), 697-706.

[62] Sánchez, A. C. G., Gil-Izquierdo, A., & Gil, M. I. (2003). Comparative study of six pear cultivars in terms of their phenolic and vitamin C contents and antioxidant capacity. *Journal of the Science of Food and Agriculture,* 83(10), 995-1003.

[63] Fischer, U. A., Carle, R., & Kammerer, D. R. (2011). Identification and quantification of phenolic compounds from pomegranate (Punica granatum L.) peel, mesocarp, aril and differently produced juices by HPLC-DAD-ESI/MS. *Food Chemistry,* 122(2), 807-821.

[64] Jayaprakasha, G. K., Girennavar, B., & Patil, B. S. (2008). Radical scavenging activities of Rio Red grapefruits and Sour orange fruit extracts in different in vitro model systems. *Bioresource Technology,* 99(10), 4484-4494.

[65] Jayaprakasha, G. K., Girennavar, B., & Patil, B. S. (2008). Antioxidant capacity of pummelo and navel oranges: Extraction efficiency of solvents in sequence. *Lebenson Wiss Technology,* 41(3), 376-384.

[66] Rufino, M. S. M., Alves, R. E., Brito, E. S., & Pérez-Jiménez, J. (2010). Bioactive compounds and antioxidant capacities of 18 non-traditional tropical fruits from Brazil. *Food Chemistry,* 121(4), 996-1002.

[67] Contreras-Calderón, J, Calderón-Jaimes, L, Guerra-Hernández, E, & García-Villanova, B. (2011). Antioxidant capacity, phenolic content and vitamin C in pulp, peel and seed from 24 exotic fruits from Colombia. *Food Research International,* 44(7), 2047-2053.

[68] Adamson, G. E., Lazarus, A. S., Mitchell, A. E., Prior, R. L., Cao, G., Jacobs, P. H., Kremers, B. G., Hammerstone, J. F., Rucker, R. B., Ritter, K. A., & Schmitz, H. H. (1999). HPLC method for the quantification of procyanidins in cocoa and chocolate samples and correlation to total antioxidant capacity. *Journal of Agricultural and Food Chemistry,* 47(10), 4184-4188.

[69] Rivero-Pérez, M. D., Muñiz, P., & González-Sanjosé, M. L. (2007). Antioxidant profile of red wines evaluated by total antioxidante capacity, scavenger activity, and biomarkers of oxidative stress methodologies. *Journal of Agricultural and Food Chemistry,* 55(14), 6476-5483.

[70] Moure, A., Cruz, J. M., Franco, D., Domínguez, J. M., Sineiro, J., Domínguez, H., Núñez, M. J., & Parajó, J. C. (2001). Natural antioxidants from residual sources. *Food Chemistry,* 72(2), 145-171.

[71] Lee, K. W., Kim, Y. J., Lee, H. J., & Lee, C. Y. (2003). Cocoa has more phenolic phytochemicals and a higher antioxidant capacity than teas and red wine. *Journal of Agricultural and Food Chemistry*, 51(25), 7292-7295.

[72] Bravo, J., Monente, C., Juániz, I., Peña, M. P., & Cid, C. (2011). Influence of extraction process on antioxidant capacity of spent coffee. *Food Research International*, doi: 10.1016/j.foodres.2011.04.026.

[73] Duthie, G. G., Duthie, S. J., & Kyle, J. A. M. (2000). Plant polyphenols in cancer an heart disease: implications as nutritional antioxidants. *Nutrition Research Reviews*, 13(1), 79-106.

[74] Giada, M. L. R., & Mancini-Filho, J. (2009). Antioxidant capacity of the striped sunflower (Helianthus annuus L.) seed extracts evaluated by three in vitro methods. *International Journal of Food Sciences and Nutrition*, 60(5), 395-401.

Chapter 5

Geranium Species as Antioxidants

Mirandeli Bautista Ávila,
Juan Antonio Gayosso de Lúcio,
Nancy Vargas Mendoza,
Claudia Velázquez González,
Minarda De la O Arciniega and
Georgina Almaguer Vargas

Additional information is available at the end of the chapter

http://dx.doi.org/10.5772/52002

1. Introduction

Complementary alternative medicine (CAM) has been widely used for a long time for the treatment of multiple diseases, despite the great advances in allopathic medicine. It is estimated that about 80% of the world population use some form of CAM.

CAM encompasses empirical knowledge and medical practice in which use is made of herbal medicinal plants, animals, minerals, manual therapy and exercise, alone or in conjunction for the treatment of diseases. In the early 1980's there emerged a strong interest in their study that has significantly influenced the pharmaceutical industry in developing technologies to identify new chemical entities and structures that are used for the synthesis of drugs. It has been shown that natural products play an important role in the discovery of compounds for drug development to treat multiple diseases.

Also, is important to recognize that use plants and their products have provided proven benefits to humanity, which falls into four areas: (i) food, (ii) essences and flavoring agents, (iii) perfumes and cosmetics, and (iv) biological and pharmaceutical agents [1]. Within the pharmaceutical area, the current outlook for natural products in drug discovery takes a central role, since at the beginning of this new millennium, only about 10% of 350,000 known species have been investigated from a phytochemical or pharmacology point of view [1].

A great examples of molecules that have hit the market as drugs by isolation from natural products metabolites are: taxol (1), an antitumor agent isolated from Taxus species [2] and camptothecin (2), isolated from the Chinese plant Camptotheca acuminate Decne (Nyssaceae), used to treat ovarian, breast and colorectal cancer, another example is ephedrine (3), which is isolated from the plant *Ephedra sinica*[3] and is used as a flu remedy. In drug discovery, researchers around the world use plants as an essential route in the search for new drugs leaders. One of the main objectives of the research laboratories is the preliminary meeting with isolated bioactive natural products, and its uses as anticancer, antiviral, antimalarial, antifungal and anti-inflammatory [3]. The search for active compounds in plants is an essential way for the development of new drugs, a process in which there is now more advanced and specific methodologies for the analysis of biological activities in particular.

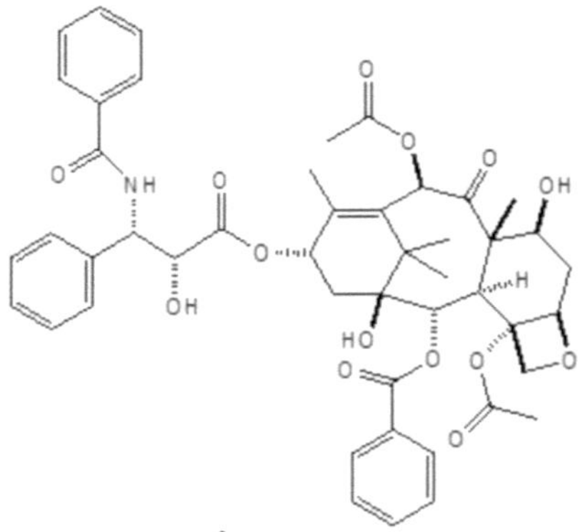

1

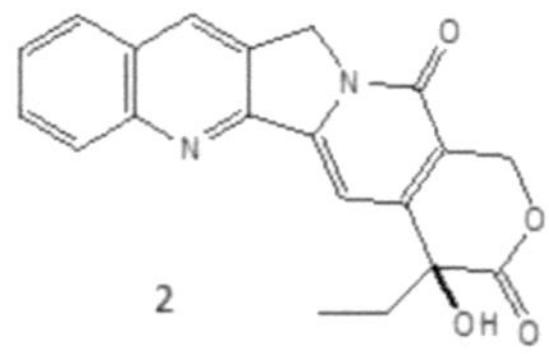

2

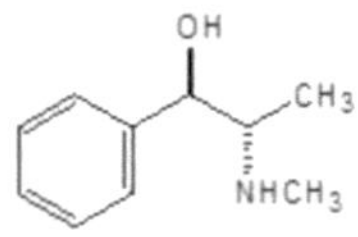

3

Documentary research from 1981 to 2006 showed that natural products have been a source of 5.7% of drugs produced in those years. The derivatives of natural products are most of the times, chemical molecules synthetized from natural products and contributed to the 27.6% of the total of the new molecule.

2. Characterization of *Geranium* genus

Geranium genus is taxonomically classified within the family *Geraniaceae Juss*, which includes five to eleven genuses, and in total near to 750 species. The genus best known are *Geranium* genus, as wild plants (Figure 1) and *Pelargonium* genus, as garden plants. The names of these genuses usually cause confusion because "geranium", is the common name for certain species of *Pelargonium*.

The names come from Greek and refer to the form that its fruits acquire, likes beaks. Thus, the word "Geranium" comes from "geranos" meaning crane, and "Pelargonium" derived from "Pelargos" meaning stork [4].

http://dx.doi.org/10.5772/52002

Figure 1. *Geranium* genus.

Subgenus	Section	Number of Species
Erodioidea	*Erodioidea*	3
	Aculeolata	1
	Subacaulia	15
	Brasiliensia	3
Geranium	*Geranium*	339
	Dissecta	4
	Tuberosa	19
	Neurophyllodes	6
	Paramensia	2
	Azorelloida	1
	Polyantha	7
Robertium	*Trilopha*	5
	Divaricata	2
	Batrachioidea	4
	Ungiculata	5
	Lucida	1
	Ruberta	4
	Anemonifolia	2

Table 1. *Geranium* genus clasification

Within the classification of *Geranium* genus are accepted 423 species, distributed in three subgenuses: Erodioidea, Geranium and Robertium. The following table (Table 1) shows the distribution of this classification.

Currently, in Hidalgo state, in Central Mexico, are classified 8 different species [5] and anyone has chemical or pharmacological studies.

2.1. Biological activities and compounds isolated from *Geranium* species

Some species of *Geranium* that have been studied has shown biological activity like: hypotensive, mild astringents, diuretics, hepatoprotection, antioxidants, anti-inflammatory and antiviral. *Geranium* species also are used as a remedy for tonsillitis, cough, whooping cough, urticarial, dysentery, kidney pain and gastrointestinal disorders [6-8]. It is probably that the species of this genus that growing in the State of Hidalgo possess a similar biological activities and metabolites. All phytochemicals studies described for these species, indicates the presence of polyphenolic compounds called tannins, which have been considered as water-soluble compounds of molecular weight between 500 and 30,000 g/mol with special properties such as the ability to precipitate alkaloids, gelatin and other proteins [9]. Nowadays tannins are well known because of its antioxidant properties. Tannin-protein complexes in the gastrointestinal tract provide persistent antioxidant activity.

One of the major components in *Geranium* species isgeraniin (4) [10] described by its discoverer as a crystallizable tannin. This substance first isolated from *Geranium thunbergi* Sieg. Et Zucc. by T. Okuda in 1976, has been evaluated showing an antihypertensive activity, geraniin inhibits the angiotensin converting enzyme [11,12] and reverse transcriptase of tumoral viruses RNA [13], inhibit HSV-1 and HSV-2 multiplication at different magnitudes of potency and also is an excellent antioxidant [14]. The corilagin (5) [15] is a derivative of geraniin, which has presented antimicrobial activity among other activities [16].

2.2. Different species of geraniums and its relevant compounds

The specie *Geranium* macrorizum presented a significant hypotensive activity in anesthetized cats [17], plus antioxidant activity. Of this specie germacrone (6) was isolated which is considered a precursor of pheromones.

Geranium robertianum L. well-known specie and one of the most variable in Britain has been used in conditions where increased diuresis is required, such as cystitis, urethritis, pyelonephritis, gout, hypertension and edema. Nowadays the phytochemistry of this geranium is relatively well known and its most studied active compounds are tannins, volatile oils, flavonoids and polyphenols (hyperoside, ellagic acid, isoquercitrin, quercitrine, kaempferols, caftaric acid, rutoside).Also infusions and decoctions prepared from leaves of this geranium: Robert herb or red Robin, are described as anti-hyperglycaemiant and commonly used in Portuguese herbal medicine [18]. In other hand G. robertianum extract treatment increased the efficiency of coupling between oxidative and phosphorylative systems, since RCR was considerably higher in GK rats consuming this plant extract [19].

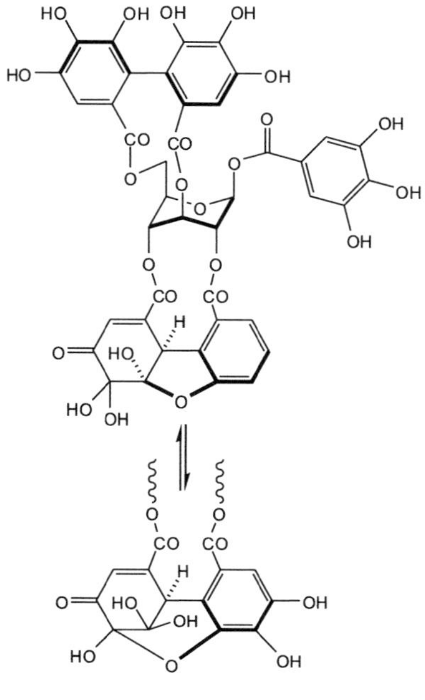

4

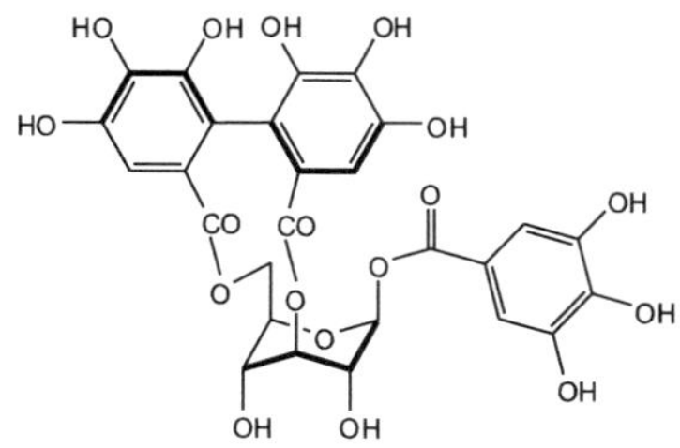

5

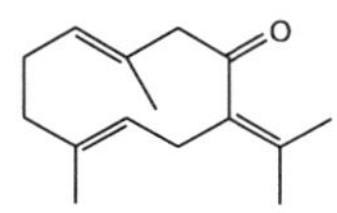

6

Recently the extracts of *G. sylvaticum* were studied [20] for antioxidant potential and all tested extracts had strong antioxidant activity and will be subject for further investigations. From flowers of *Geranium sylvaticum* was isolated 3-O-(6-O-acetyl-•-D-glucopyranoside)-5-O-•-D-glucopyranoside of malvidin (7) [21].*Geranium sanguineum L.* showed significant inhibitory activity of influenza virus and herpes simplex (8). The methanolic extract of *Geranium pratense* inhibited the action of the amylase enzyme in mouse plasma, isolated for first time the 3-O-(2-O-galloyl) -•-D-glucopyranoside myricetin(9) [22].

Geranium niveum, widely used by the Tarahumara Indians of Mexico. Is a specie rich in proanthocyanidins and other phenolics [23]. Previous in vitro assays have demonstrated that proanthocyanidins exhibited antiinflammatory, antiviral, antibacterial, enzyme-inhibiting, antioxidant, and radical-scavenging properties, the roots 25 of this species were isolated new proanthocyanidins named as geraniins A (9) and B (9a), latterlyin 2001 were found geraniins C (10) and D (10a) [24]. A recently study showed that geraniin A has antioxidant activity [25].

Geranium pusillum, commonly known as Small-flowered Cranesbill or (in North America) small Geranium, contains1-O-galloyl-3,6-hexahidroxibifenil-D-galactopyranoside (11) (pusilagin) a polyphenolic compound extracted from aerial parts [26]. The aqueous ethanolic extract of *Geranium wallichianum* showed antibacterial activity against *Staphylococcus aureus* [27] and the study of the chemical constituents of the whole plant has resulted in the isolation and characterization of six compounds. These six compounds were identified as ursolic acid, β-sitosterol, stigmasterol, β-sitosterol galactoside, herniarin, and 2,4,6-trihydroxyethyl-benzoate which were isolated for the first time from *Geranium wallichianum* [28].

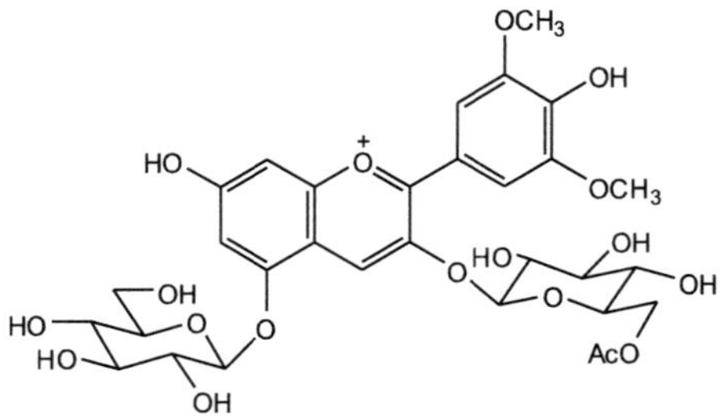

7

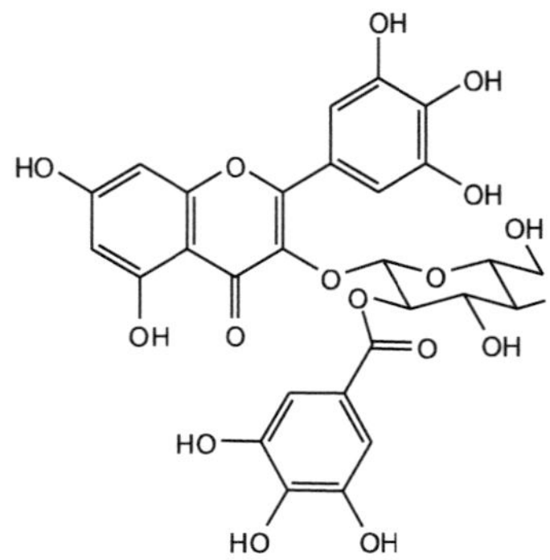

8

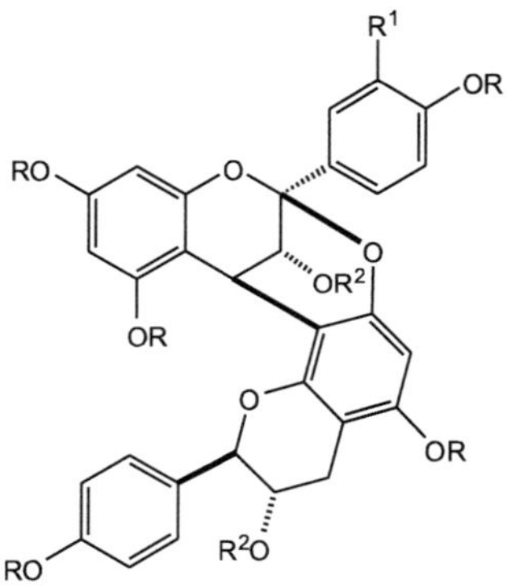

9: R, R1, R2 = H
9a: R, R2 = H; R1 = OH

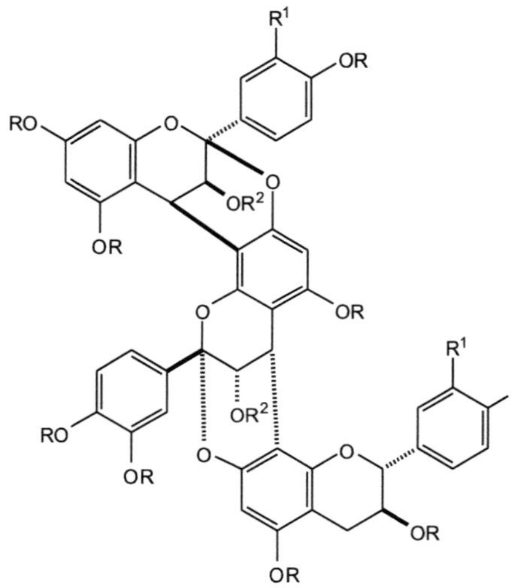

10: R, R1, R2 = H **10a**: R, R2 = H; R1=OH

Geranium caespitosum produces neohesperidoside (12) able to potentiate 10 to 100 times the action of drugs such as ciprofloxacin, norfloxacin, berberine and rhein, against bacterias such as *S. aureus, NorA S. aureus, B.* and *B. megaterium subtilis* [29]. Besides, *Geranium carolinianum* L., isa commonly used traditional Chinese medicine (TCM) with the efficacy of eliminating wind-damp and treating diarrhea. It is clinically used to treat the arthralgia due to wind-dampness, anaesthetization and muscular constriction. It has been reports that *Geranium carolinianum L.* as well of most of the congeneric plants contain significant amounts of tannins, flavonoids, organic acids, and volatile oils [30].Also, has shown that roots contain a substance that is extracted with water and can be a biological mechanism to control bacteria (*Ralstonia solanacearum*) which attacks potatoes [31].

From *Geranium pyrenaicum*, which showed antileishmanial activity [32],a new glycosylate flavonoid: 3-O-(2 ", 3"-di-O-galloyl)- •-D-glucopyranoside of kaempferol (13) was isolated, and anuncommon quercetin derivative: 3-O-(2 ", 3"-di-O-galloyl) - •-D-glucopyranoside of quercetin (13a) too. In *Geranium mexicanum* an antiprotozoal activity was assayed from its roots, where the most active compound founded was the flavan-3-ol-(-)-epicatechin (14), showing moderate activity (+)-catechin (14a), tyramine (15) and 3-O-β-D-glucopyranoside of β-sitosterol [33].

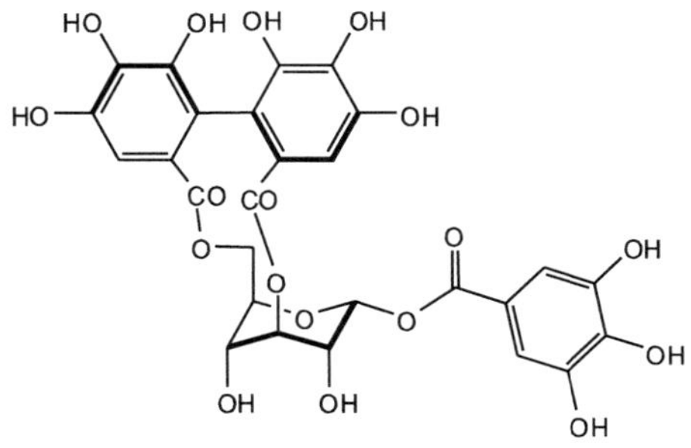

11

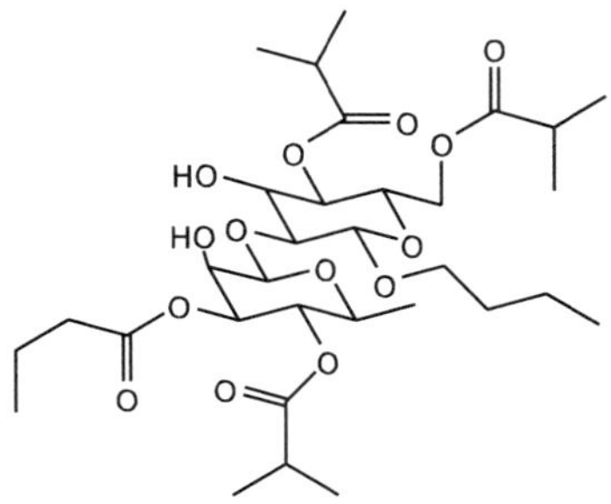

12

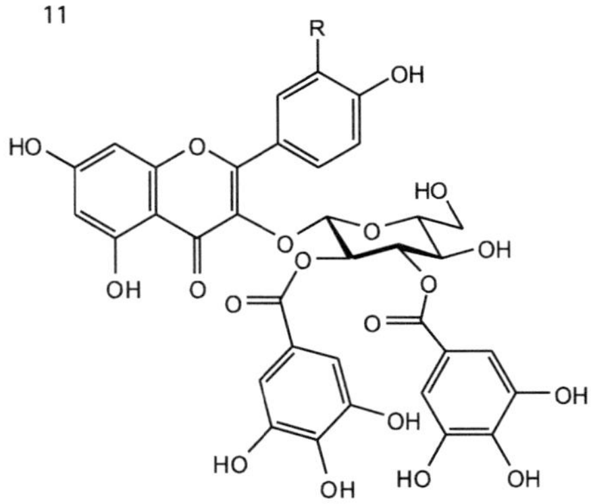

13: R=H
13a: R=OH

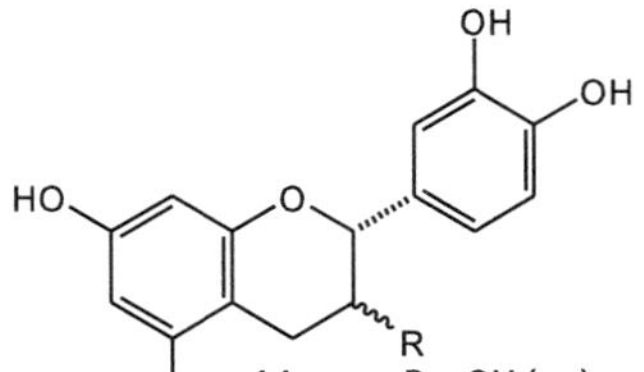

14: R = OH (α)
14a: R = OH (β)

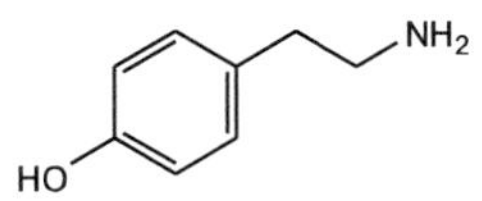

15

Geranium bellum Rose is a perennial plant with long roots, found in the grassy meadows bordering pine/oak forests in the mountains of Hidalgo State, Mexico, where it has the popular name "pata de león" and has been used as a traditional remedy for treatment of fevers, pain, and gastrointestinal disorders. Radical scavenging assay-guided fractionation of the antioxidant EtOAc and MeOH extracts from the aerial parts of *Geranium bellum* resulted in the isolation of bsitosterol 3-O-b-D-glucopiranoside, quercetin 3-O-a-L-(2″-O-acetyl)arabinofuranoside (16), quercetin 3-O-a-L-arabinofuranoside, quercetin, methyl gallate, gallic acid, methyl brevifolin carboxylate (17), and dehydrochebulic acid trimethyl ester (18). Compounds 2, 7 and 8 are isolated for the first time from *Geranium* genus [34]. Antioxidant activity of these extracts (both initial fractions and pure compounds), was tested by measuring their capacity to scavenge 2,2′-azino-bis(3-ethylbenzthiazoline-6-sulfonic acid) (ABTS) radicals, an assay widely used for screening of antioxidant activity of natural products [15].

Constituents from the aerial parts of *Geranium potentillaefoium* founded in certain studies were geraniin, corilagin, gallic acid, methyl gallate, methyl brevifolincarboxylate, quercetin, quercetin 3-O-β-Dglucopyranoside, quercetin 3-O-β-D- [6″-O-galloyl)glucopyranoside, kaempferol, β-sitosterol 3-O-β-D-glucopyranoside and β-sitosterol [35].

3. Study of *Geranium schiedeanum*

Geranium schiedeanum (Gs) (Figure 2), species that grows in Central Mexico, has been used as an antipyretic, anti-inflamatory and antiseptic. The use of other geranium species also has been reported a hypoglycemic, antihypertensive and cholesterol-lowering effect. However, scientific evidence does not exist in any literature to corroborate these targets or any other. In the present study the effect of Gs were studied in reference to postnecrotic liver damage induced by thioacetamide (TA).

Figure 2. *Geranium schiedeanum* [5]

3.1. Plant material

Specimen of *Geranium schiedeanum* was collected at Epazoyucan Municipality, in Hidalgo State, México, during June 2009. A voucher specimen (J. A. Gayosoo-de-Lucio) is preserved at the Herbarium of Biological Research Centre, Autonomous University of Hidalgo, Pachuca, Hidalgo, Mexico where Professor Manuel González Ledesma identified the plant material.

3.2. Extraction and purification

Air-dried aerial parts (1 kg) were extracted acetone-H_2O 7:3 (20 L) by maceration for 7 days. Vacuum evaporation of dissolvent give a 5 L residue Filtration give a fatty solid residue (12g) and complete evaporation of water give the acetone-water extract (115 g).

Lots of 3 g of acetone-water extract were purified on a Sephadex LH-20 (25 g) column using H_2O, H_2O-MeOH (9:1, 4:1, 7:3, 3:2, 1:1, 2:3, 3:7, 1:4, 1:9) and MeOH, as eluents. Fractions of 300 mL of each polarity were collected and marked "A–K". They were evaporated and analyzed by TLC and NMR. Fractions "B" gave 75 mg, and were purified over silica gel (10 g), using $CHCl_3$, $CHCl_3$-AcOEt(9:1, 4:1, 7:3, 3:2, 1:1, 2:3, 3:7, 1:4, 1:9) and AcOEt (10 mL of each polarity), as eluents and collecting fractions of 7 mL, fractions 13-16 give I 25 mg. Fractions "C" and "D" gave 56 mg, and were purified over silica gel (10 g), using $CHCl_3$-MeOH (50:7.0, 48:7, 45:7, 40:7, 35:7 and 30:7, 40 mL of each), as eluents and collecting fractions of 7 mL, fractions 33-66 give II 2 mg, (these procedure was repeated ten times to obtain 18 mg of compound), Fractions "F-I" gave 1.8 g, a portion of 500 mg were purified over silica gel C-18 (5 g) using H_2O, H_2O-MeOH (9:1, 4:1, 7:3, 3:2, 1:1, 2:3, 3:7, 1:4, 1:9) and MeOH (20 mL of each polarity), fractions of 10 mL were collected fractions 2-4 gave 325 mg of III, fraction "K" gave 90 mg were added 5 mL of (40°C) pyridine and were placed a room temperature for 72 h, filtrated of mixture give a yellow needles 60 mg of IV (Figure 3).

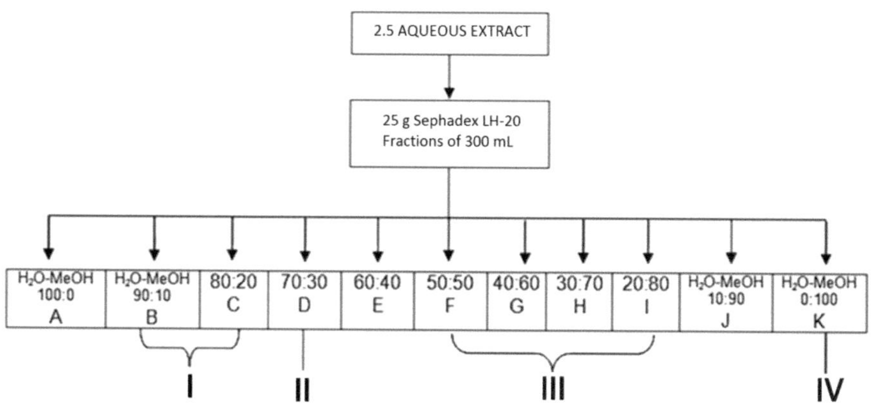

Figure 3. Extract fractions scheme.

3.3. Animals and treatment

Male adult Wistar rats 2 months old (200–220g) were obtained from UAEH Bioterio, and acclimated to our animal room for two weeks before use. Throughout these two weeks rats were supplied with food and water *ad libitum,* exposed to a 12 h light-dark cycle and given intraperitoneally a single necrogenic dose of thioacetamide (6.6 nmol/Kg body weights) freshly dissolved in 0.9% NaCl. The dose of thioacetamide was chosen as the highest dose with survival above 90% [36,37]. Wistar rats were intragastric pre-treated or not with a single dose of *Gs* extract (300 mg/kg) during 4 days, the fourth day of pretreatment were intraperitoneally injected with a single dose of TA (6.6 mmol/Kg). Samples of blood and liver were obtained from rats at 0, 24, 48,72 and 96 h following TA intoxication. Untreated animals received 0.5 ml of 0.9% NaCl. Experiments were performed on two different groups: rats treated with a single dose of thioacetamide (TA) and rats pre-treated with Gs and treated with a single dose of thioacetamide (Gs + TA). Each experiment was performed in duplicate from four different animals and followed the international criteria for the use and care of experimental animals outlined in *The Guiding Principles in the use of Animals in Toxicology* adopted by the Society of Toxicology in 1989.

3.4. Processing of samples

In order to clarify the sequential changes during the different stages of liver injury and the post-necrotic regenerative response, samples were obtained from control and at 24 and 48 h of TA intoxication in both Gs pre-treated or non pre-treated animals. Rats were sacrificed by cervical dislocation and samples of liver were obtained and processed as previously described. Blood was collected from hearts and kept at 4 °C for 24 h, centrifuged at 3000 rpm for 15 min, and serum was obtained as the supernatant.

3.5. Determination of AST

Enzymatic determination were carried out in serum in optimal conditions of temperature and substrate and cofactor concentrations. Aspartate aminotransferase (AST) activity were determined in serum. AST (EC 2.6.2.1) and was assayed following the method of Rej and Horder [38].

The activity of this enzyme was determined spectrophotometrically, by measuring the decrease in absorbance at 340 nm at 37 ° C, produced by the oxidation of NADH to NAD^+ in the coupled reaction of reduction of oxaloacetate to malate, catalyzed by malate dehydrogenase, according to the following process:

3.6. General

IR spectra measured in MeOH on a Perkin Elmer 2000 FT-IR spectrophotometer. Optical rotations were determined in MeOH on a Perkin Elmer 341 polarimeter. NMR measurements performed at 400 MHz for 1H and 100 MHz for 13C on a VARIAN 400 spectrometer from CDCl3, CD3OH, DMSO-d6 solutions. Column chromatography (CC) was carried out on Merck silica gel 60 (Aldrich, 230-400 mesh ASTM) and sephadex LH-20 Sigma Aldrich.

3.7. Statistical analysis

The results were calculated as the means ± SD of four experimental observations in duplicate (four animals). Differences between groups were analyzed by an ANOVA following Snedecor F ($\alpha = 0.05$). Students' test was performed for statistical evaluation as follows: (a) all values against their control; b) differences between two groups Gs + TA versus TA.

3.8. Results

3.8.1. Active compounds of Geranium schiedeanum

One kg of the aerial part of *G. schiedeanum* was extracted by maceration for 7 days with 20 L acetone-water (7:3), concentrated under reduced pressure to a volume of 3 L, which was extracted with $CHCl_3$ yielding 12.75 g of $CHCl_3$ phase and 125 g of aqueous phase. The phytochemical study of *Geranium schiedeanum* led to the isolation of hydrolysable tannins (I) gallic acid, (III) acetonylgeraniin and (IV) ellagic acid and to a lesser proportion of kaempferol glycosideflavonoid (II) (Figure 4 and 5). Is relevant to notice that is the first time discloses the compound II in the Geranium genus and further that the yield of compound III in the crude extract was 40%.

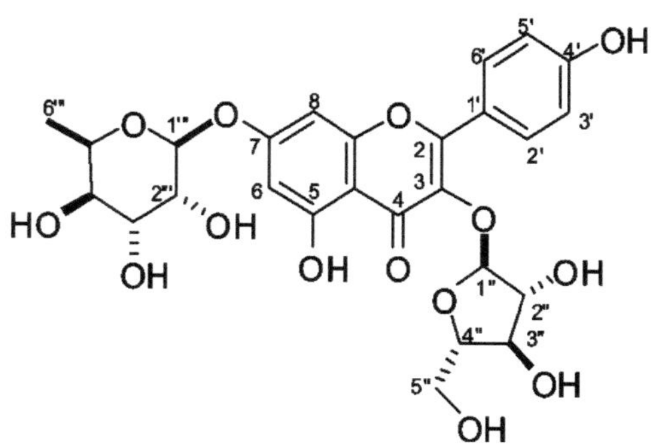

Figure 4. Compound (II) 3-O-α-L-arabinofuranoside-7-O-β-L-rhamnopyranoside de Kaempferol

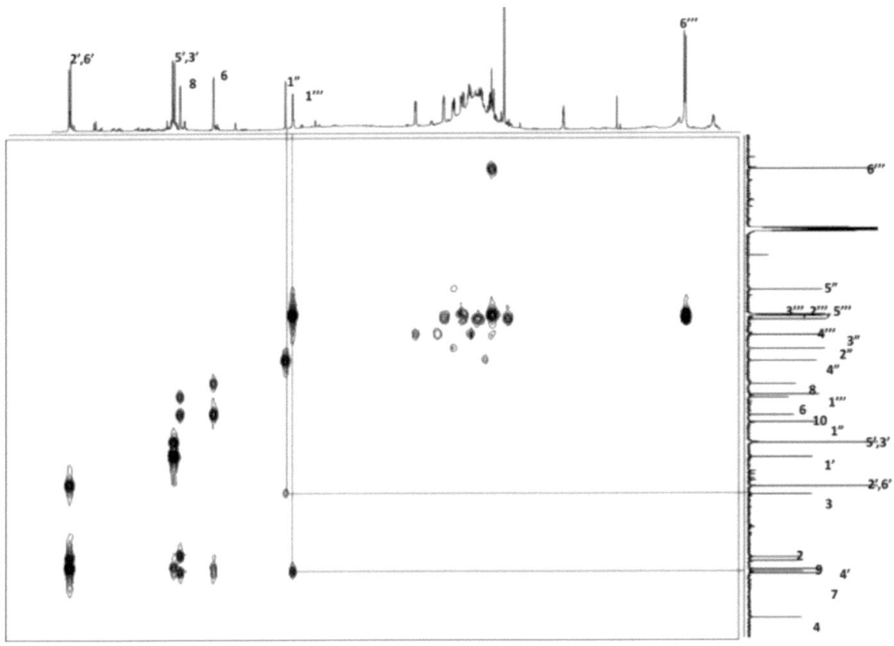

Figure 5. HMBC experiment of compound II

3.8.2. *Aspartate aminotransferase*

The acute liver injury induced by a necrogenic dose of thioacetamide (TA), a potent hepatotoxic agent, is characterized by a severe perivenous necrosis [39]. The necrosis develops as a consequence of the biotransformation of TA through the microsomal flavin-dependent monooxygenase [40]. The reactive metabolites responsible for TA hepatotoxicity are the radicals derived from thioacetamide-S-oxide and the reactive oxygen species derived as sub products in the process of microsomal TA oxidation, both of which can depleted reduced glutathione leading to oxidative stress [41, 42].

Liver damage induced by xenobiotics is characterized by the release in serum of hepatic enzymes due to necrosis of hepatocytes. AST is randomly distributed in the hepatic acinus, and is the enzyme activity used as marker of necrosis. Our results showed that *Gs* extract significantly reduced the level of liver injury. The levels of AST (Figure 6) were significantly lower in the rats pretreated with *Gs*.

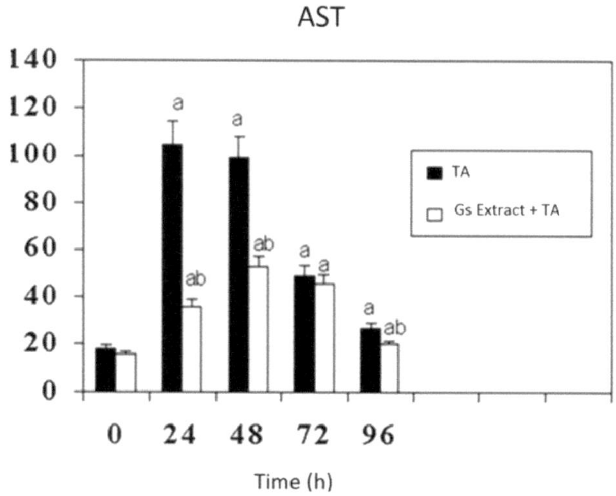

Figure 6. Effect of Gs pre-treatment on aspartate aminotransferase activity in serum of rats intoxicated with one sub-lethal dose of thioacetamide. Samples were obtained at 0, 24, 48, 72 & 96 h following thioacetamide (TA). The results, expressed as nmol per min per ml of serum, are the mean ± SD of four determinations in duplicate from four rats. Differences against the respective control are expressed as (a) and differences due to Gs extract are expressed as (b) $p<0.05$.

4. Conclusion

There is evidence that free radicals play a critical role in certain pathological conditions such as some cancers, multiple sclerosis, inflammation, arthritis and arterosclerosis [43]. For this reason, some research objectives directed toward the development or discovery of these compounds catchers of these radicals.

A large number of plant species, like *G. schiedeanum* contain chemical compounds that exhibit the ability to trap free radicals. The ability to trap free radicals has been called antioxidant activity. The phytochemical study of *Geranium shiedeanum* led to the isolation of hydrolysable tannins, well known as potent antioxidants: gallic acid, acetonylgeraniin and ellagic acid and a lesser proportion of kaempferol glycoside flavonoid (3-O-α-L-arabinofuranoside-7-O-β-D-rhamnoside de Kaempferol), notably is the first time discloses these compounds in the genus. Further the yield of acetonyl geraniin in the crude extract was 40%.

Also, in the present study TA-induced hepatotoxicity was used to investigate the effect of the pretreatment of *G. schiedeanum* on the events involved in liver regeneration. The results obtained in the present study provide evidence that *Gs*, when administered intravenously prior to TA, significantly reduce liver damage.

The pre-treatment with the crude extract in the model of thioacetamide-induced hepatotoxicity in rats, decreased and delayed liver injury by 66% at 24 h.

The data obtained indicate that the crude *Gs* extract pre-treatment has hepatoprotective and antioxidant effect in damage induced by TA. This result suggests that *Gs* extract may be used as an alternative for reduction of liver damage. However further investigation on the acute toxicity and on the mechanism of the hepatoprotective effect of the plant species needs to be carried out.

Acknowledgements

The authors would like to thank Teresa Vargas for her valuable technical Assistance. Supported by Grant PROMEP-MEXICO UAEHGO-PTC-454.

Author details

Mirandeli Bautista Ávila, Juan Antonio Gayosso de Lúcio, Nancy Vargas Mendoza,
Claudia Velázquez González, Minarda De la O Arciniega and Georgina Almaguer Vargas

Universidad Autónoma del Estado de Hidalgo, Mexico

References

[1] K-H Tan, Novel Compounds from Natural Products in the New Millennium: Potential and Challenges. 2004, World Scientific Publishing Company, Singapore.

[2] Dewick P. M., Medicinal Natural Products a Biosynthetic Aproach. 1998, John Wiley & Sons, New York, USA.

[3] Kuo-Hsiung L. *J. Nat. Prod.* 2004, 67(1), 273-283.

[4] Gómez, M. A., Borja y Tomé, López-Lomo V. M. A. M. Biotecnología aplicada a la mejora de "*Pelargonium*". 2005, Universidad Complutense de Madrid, España.

[5] Pérez Escandón B. E., Villavicencio M. A., Ramirez Aguirre A. Lista Floristica Del Estado De Hidalgo Recopilación Bibliografica, 1998, 1ª edición, Ed. UAEH. México.

[6] Calzada F, Cervantes-Martinez JA, Yepez-Mulia L. In vitro antiprotozoal activity from the roots of *Geranium mexicanum* and its constituents on *Entamoeba histolytica* and *Giardia lamblia*. *J. Ethnopharmacol.* 2005, 98: 191-193.

[7] Ercil D, Kaloga M, Redtke OA, Sakar MK, Kiderlen A, Kolodziej H. O-Galloyl flavonoids from *Geranium pyreniacum* and their *in vitro* antileishmanial activity. *Turk. Chem.* 2005, 29: 437-443.

[8] Küpeli E, Tatl I, Akdemir ZS, Yeflilada E. Estimation of antinociceptive and anti-inflammatory activity on *Geranium pratense* subsp. *finitimum*. *J. Ethnopharmacol.* 2007, 114: 234-240.

[9] Okuda T., Yoshida T. y Hatano T. *J. Nat. Prod.* 1989, 52(1), 1-31.

[10] Cheng J. T., Chang S. S., Hsu F. L. *J. of Pharm. and Pharmacol.* 1994, 46(1), 469

[11] Kameda K., Takaku T., Okuda H., Kimura Y., Okuda T., Hatano T., Agata I., Arichi S. *J. Nat. Prod.* 1987, 50(4), 680-683.

[12] Ueno H., Hoire S., Nishi Y., Shogawa H., Kawasaki M., Suzuki S., Hayashi, Shimizu A. M., Yoshizaki M., Morita N. *J. Nat. Prod.* 1988, 51(2), 357-359.

[13] Kakiuchi N., Hattori M., Namba T., Nisizahua M., Yamagishi T., Okuda T. *J. Nat. Prod.* 1985, 48(4), 614-621.

[14] Fujiki H., Sagunama M., Kurusu M., Okabe S., Imayoshi. Y., Tanigushi S., Yosida T. *Mutation Research.* 2003; 523-524, 119-25.

[15] Okuda T., Yoshida T., and Mori K. *Phytochemistry* 1975, 14, 1877–1878

[16] Shimizu M., Shiota S., Mizushima T., Ito H., Hatano T., Yoshida T., Tsuchiya T. *Antimicrobial Agents And Chemotherapy* 2001, 45, 3198–3201

[17] Chemical abstracts, vol. 95, 1981, 162140J.

[18] Cunha AP, Silva AP, Roque AR. Plantas e Produtos Vegetais em Fitoterapia. Fundação Calouste Gulbenkian. 2009. Lisboa, Portugal (in Portuguese).

[19] Ferreira FM, Peixoto FP, Nunes E, Sena C, Seiça R, Santos MS. *Vaccinium myrtillus* improves liver mitochondrial oxidative phosphorylation of diabetic Goto-Kakizaki rats. *J Med Plants.* 2010 Res 4: 692–696.

[20] Milena N, Reneta T, and Stephanka I. Evaluation of antioxidant activity in some Geraniacean species *Botanica Serbica.* 2010, 34 (2): 123-125

[21] Andersen M., Viksund R. I., Pedersen A.T. *Phytochemistry* 1995, 38(6), 1513-1517.

[22] Akdemir Z. S., Tatl J. J., Saracoglu J., Ismailoglu U. B., Sahin-ErdemLi I., Calis I., *Phytochemistry.* 2001, 56(2), 189-193.

[23] Maldonado PD, Rivero-Cruz I, Mata R, Pedraza-Chaverrí J. Antioxidant activity of A-type proanthocyanidins from *Geranium niveum* (Geraniaceae). *J Agric Food Chem.* 2005, 23;53(6):1996-200.

[24] Calzada F., García-Rojas C. M., Meches M., Rivera C. R., Bye R., Mata R. *J. Nat. Prod.* 1999, 62, 705-709.

[25] Maldonado P. D., Rivero-Cruz I., Mata R., Pedraza-Chaveri J. *J. Agric. Food Chem.* 2005, 53, 1996-2001.

[26] Kobakhidza K. B., Alaniya M. D. *Chem. of Nat. Comp.* 2004, 39(3), 262-264.

[27] Ahmad B., Ismail M., Iqbal Z., M. Iqbal Chaudhry. *Asian Journal of Plant Sciences* 2003, 2(13), 971-973.

[28] Mohammad I, Zafar Iqbaq, Javid H, Hidayat H, Manzoor Ahmed, Asma Ejaz, Muhammad I. C., Chemical Constituents and Antioxidant Activity of *Geranium wallichianum*. *Rec. Nat. Prod.* 2009, 3:4, 193-197

[29] Oshiro A., Takaesu K., Natsume M., Taba S., Nasu K., Uehara M., Muramoto Y. *Weed Biology and Management* 2004, 4, 187–194

[30] Pharmacopoeia of the People's Republic of China; Chemical Industry Press: Beijing, China, 2010; Vol 1, p.113.

[31] Oshiro A., Takaesu K., Natsume M., Taba S., Nasu K., Uehara M., Muramoto Y. *Weed Biology and Management* 2004, 4, 187–194.

[32] Ercil D., Kaloga M., Ratke O. A., Sakar M. K., Kiderlen F.A., Kolodziej H. Turk *J. Chem.* 2005, 29, 437-443.

[33] Calzada F., Cervantes-Martíneza J. A., Yépez-Muliab L. *Journal of Ethnopharmacology* 2005, 98(1-2), 191-193

[34] Camacho-L A, J Gayosso-De-Lucio, J. Torres-Valencia, J Muñoz-Sánchez, E Alarcón-Hernández, Rogelio L, Blanca L. Barrón. Antioxidant Constituents of *Geranium bellum* Rose. *J. Mex. Chem.* Soc. 2008, 52(2), 103-107

[35] J.A. Gayosso-De-Lucio, J.M. Torres-Valencia, C.M. Cerda-García-Rojas and P. Joseph-Nathan. Ellagitannins from *Geranium potentillaefolium* and *G. bellum*. *Nat. Prod. Comm.*, 2010; 5, 531-534

[36] Sanz N, Diez-Fernández C, Andrés D, Cascales M. Hepatotoxicity and aging: endogenous antioxidant systems in hepatocytes from 2-, 6-, 12-, 18- and 30-month-old rats following a necrogenic dose of thioacetamide. *Biochim Biophys Acta*. 2002; 1587: 12-20.

[37] Zaragoza A, Andrés D, Sarrión D y Cascales M. Potentiation of thioacetamide hepatotoxicity by phenobarbital pretreatment in rats. Inducibility of FAD monooxygenase system and age effect. *Chem Biol Interact*. 2000, 124: 87-101.

[38] Rej R y Horder M. Aspartate aminotransferase. L-aspartate: 2-oxoglutarate aminotranferase, EC 2.6.2.1. Routine U.V. method. En: Bergmeyer HU Editor. Methods of Enzymatic Analysis. 3rd ed., vol III. Weinheim. *Verlag Chemie*, pp. 416-24, (1984).

[39] Cascales M., Martin-Sanz P, Craciunescu DC, Mayo I, Aguilar A, Robles-Chillida EM, Cascales C. Alterations in hepatic peroxidation mechanisms in thioacetamide-induced tumors in rats. Effect of a rhodium complex. *Carcinogenesis* 1991; 12: 233-240.

[40] Dyroff MC y Neal RA. Studies of the mechanism of metabolism of thioacetamide-S-oxide by rat liver microsomes. *Cancer Res.* 1981; 41: 3430-3445.

[41] Sanz N, Diez-Fernandez C, Andres D, Cascales M. Hepatotoxicity and aging: endogenous antioxidant systems in hepatocytes from 2-, 6-, 12-, 18- and 30-month-old rats following a necrogenic dose of thioacetamide. *Biochim Biophys Acta* 2002; 1587: 12-20.

[42] Sanz N, Díez-Fernández C, Alvarez AM, Cascales M. Age-dependent modifications in rat hepatocyte antioxidant defense systems. *J Hepatol* 1997; 27: 525-534.

[43] Latté P. K., Kolodziej H. *J. Agric. Food Chem*. 2004, 52, 4899-4902.

Chapter 6

Foods or Bioactive Constituents of Foods as Chemopreventives in Cell Lines After Simulated Gastrointestinal Digestion: A Review

Antonio Cilla, Amparo Alegría, Reyes Barberá and
María Jesús Lagarda

Additional information is available at the end of the chapter

http://dx.doi.org/10.5772/51504

1. Introduction

Epidemiological studies on the relationship between dietary habits and disease risk have shown that food has a direct impact on health. Indeed, our diet plays a significant role in health and well-being, since unbalanced nutrition or an inadequate diet is known to be a key risk factor for chronic age-related diseases [1]. An example that illustrates this fact is the protective effect of the so-called Mediterranean diet. The lower occurrence of cancer and cardiovascular disease in the population located around the Mediterranean sea has been linked to the dietary habits of the region, in which the components of the diet contain a wide array of molecules with antioxidant and antiinflammatory actions [2].

Many diseases with a strong dietary influence include oxidative damage as an initial event or in an early stage of disease progression [3]. In fact, Western diets (typically dense in fat and energy and low in fiber) are associated with disease risk [4]. Therefore, dietary modification, with a major focus on chronic age-related disease prevention through antioxidant intervention, could be a good and cost-effective strategy [5]. The intake of whole foods and/or new brand developed functional foods rich in antioxidants would be suitable for this purpose. In this sense, dietary antioxidants such as polyphenols, carotenoids and peptides, as well as other bioactive chemopreventive components such as fiber and phytosterols have been regarded to have low potency as bioactive compounds when compared to pharmaceutical drugs, but since they are ingested regularly and in significant amounts as part of the diet, they may have noticeable long-term physiological effects [6].

For decades, the beneficial role of antioxidants was related to the reduction of unwanted and uncontrolled production of reactive oxygen species (ROS), leading to a situation referred to as oxidative stress [7]. Nowadays, the term "antioxidant" has become ambiguous, since it has different connotations for distinct audiences. For instance, for biochemists and nutritionists, the term is related to the scavenging of metabolically generated ROS, while for food scientists the term implies use in retarding food oxidation or for the categorization of foods or substances according to *in vitro* assays of antioxidant capacity, such as the ORAC and TEAC tests [8]. The antioxidant values provided by these assays sometimes have been misinterpreted by both food producers and consumers due to the fact that health claims advertised on the package labeling are directly associated with benefits that include slowing of the aging process and decreasing the risk of chronic disease. Nevertheless, contemporary scientific evidence indicates that total antioxidant capacity measured by currently popular chemical assays may not reflect the actual activity *in vivo,* since none of them take biological processes such as bioavailability, uptake and metabolism into account [9]. Therefore, no *in vitro* assay that determines the antioxidant capacity of a nutritional product describes *in vivo* outcomes, and such testing should not be used to suggest such a connection. In this sense, it is currently recognized that the mechanisms of action of antioxidants *in vivo* might be far more complex than mere radical scavenging - involving interactions with specific proteins central to intracellular signaling cascades [10], and in the specific case of cancer cells there might be a direct antioxidant effect, antiproliferation and anti-survival action, the induction of cell cycle arrest, the induction of apoptosis, antiinflammatory effects and the inhibition of angiogenesis and metastasis [11].

In order to determine and verify the action of these bioactive compounds, it is clear that data from human intervention studies offer the reference standard and the highest scientific evidence considering the bioavailability and bioactivity of a food component, while *in vitro* methods are used as surrogates for prediction [12]. From a physiological perspective, food after consumption undergoes a gastrointestinal digestion process that may affect the native antioxidant potential of the complex mixture of bioactive compounds present in the food matrix before reaching the proximal intestine. *In vitro* methods which apply human simulated digestion models (including or not including colonic fermentation) are considered valuable and useful tools for the estimation of pre-absorptive events (i.e., stability, bioaccessibility) of different food components from distinct food sources, and also for determining the effect which processing may have upon food components bioavailability [13]. In addition, *in vitro* assays combining a simulated gastrointestinal digestion process and cell cultures as pre-clinical models can be useful for unraveling mechanisms of action and for projecting further *in vivo* assays [9]. Nevertheless, in most cases these *in vitro* studies are unrealistic, because they involve single compounds used at high concentrations (pharmacological and not dietary concentrations) far from the low micromolar or nanomolar concentrations detected *in vivo*, or use the bioactive compounds "as they are in food" versus the metabolites or derivatives considered to be the true bioactive compounds, over an extended period of time (up to 120 h). As a result, biological activity may be overestimated, since no account is taken of the possible transformation of these compounds during gastrointestinal digestion with or without colonic fermentation [6]. Likewise, the use of single or crude compounds instead of whole foods impedes the detection of synergistic and/or antagonistic actions among bioactive chemopreventive compounds [14, 15].

Taking this background together, and in order to obtain a more precise view of the *in vivo* situation, we propose the use of whole foods or related target bioactive constituents subjected to a human simulated gastrointestinal digestion including or not including colonic fermentation, depending on the nature of the studied compounds, in order to gain better insight from a nutritional/functional point of view of the chemopreventive action derived from foods and bioactive compounds in cell models of disease.

This review introduces the main features of the different *in vitro* gastrointestinal digestion (solubility and dialysis) and colonic fermentation procedures (batch, continuous and continuous with immobilized feces) for studying the bioaccessibility and further bioavailability and bioactivity of nutrients and bioactive compounds. It also includes a definition of the terms: bioavailability including bioaccessibility and bioactivity. Likewise, the main advantages and disadvantages of these *in vitro* methods versus *in vivo* approaches, the improvement of these models with the inclusion of cell lines, and a short comment on the main effects that digestion and/or fermentation have on bioactive compounds are included. On the other hand, a short description is provided of the studies involving the use of human simulated gastrointestinal digestion and/or colonic fermentation procedures, and of the subsequent bioactivity-guided assays with cell line models.

2. Simulated gastrointestinal digestion assays

Bioavailability is a key concept for nutritional effectiveness, irrespective of the type of food considered (functional or otherwise). Only certain amounts of all nutrients or bioactive compounds are available for use in physiological functions or for storage.

The term bioavailability has several working conditions. From the nutritional point of view, bioavailability is defined as the proportion of a nutrient or bioactive compound can be used for normal physiological functions [16]. This term in turn includes two additional terms: bioaccessibility and bioactivity. Bioaccessibility has been defined as the fraction of a compound that is released from its food matrix in the gastrointestinal tract and thus becomes available for intestinal absorption. Bioaccessibility includes the sequence of events that take place during food digestion for transformation into potentially bioaccessible material, absorption/assimilation through epithelial tissue and pre-systemic metabolism. Bioactivity in turn includes events linked to how the bioactive compound is transported and reaches the target tissue, how it interacts with biomolecules, the metabolism or biotransformation it may undergo, and the generation of biomarkers and the physiologic responses it causes [12]. Depending on the *in vitro* method used, evaluation is made of bioaccessibility and/or bioactivity.

In vitro methods have been developed to simulate the physiological conditions and the sequence of events that occur during digestion in the human gastrointestinal tract. In a first step, simulated gastrointestinal digestion (gastric and intestinal stages, and in some cases a salivary stage) is applied to homogenized foods or isolated bioactive compounds in a closed system, with determination of the soluble component fraction obtained by centrifugation or dialysis of soluble components across a semipermeable membrane (bioaccessible fraction). Simulated gastrointestinal digestion can be performed with static models where the products of digestion remain largely immobile and do not mimic physical processes such as shear, mixing, hydration. Dynamic models can also be used, with gradual modifications in pH and enzymes, and removal of the dialyzed components – thereby better simulating the actual *in vivo* situation. All these systems evaluate the aforementioned term "bioaccessibility", and can be used to establish trends in relative bioaccessibility.

The principal requirement for successfully conducting experimental studies of this kind is to achieve conditions which are similar to the *in vivo* conditions. Temperature, shaking or agitation, and the chemical and enzymatic composition of saliva, gastric juice, duodenal and bile juice are all relevant aspects in these studies. Interactions with other food components must also be taken into account, since they can influence the efficiency of digestion [12, 17]. A recent overview of the different *in vitro* digestion models, sample conditions and enzymes used has been published by Hur et al. [13]. En lipophilic compounds such as carotenoids and phytosterols, it is necessary to form mixed micelles in the duodenal stage through the action of bile salts, phospholipases and colipase. This allows the compounds to form part of the micelles, where they remain until uptake by the enterocytes [18]. In the case of lycopene, during digestion isomerization of trans-lycopene may occur with the disadvantage that trans-isomers are less soluble in bile acid micelles [19]. Salivary and gastric digestion exert no substantial effect on major phenolic compounds. However, polyphenols are highly sensitivity to the mild alkaline conditions in pancreatic digestion, and a good proportion of these compounds can be transformed into other unknown and/or undetected forms [20].

Bioactive compounds such as dietary fiber, carotenoids, polyphenols and phytosterols undergo very limited absorption, and may experience important modifications as a result of actions on the part of the intestinal microbiota. Small intestine *in vitro* models are devoid of intestinal microbes, and are designed to only replicate digestion and absorption processes; as a result, they are unable to provide information on intestinal fermentation processes. The incorporation of colonic/large intestine fermentation offers a better approximation to the *in vivo* situation, and allows us to study the effect/interaction between these compounds and the intestinal microbiota.

In vitro colonic fermentation models are characterized by the inoculation of single or multiple chemostats with fecal microbiota (of rat or human origin) and operated under physiological temperature, pH and anaerobic conditions. There are two types of colonic fermentation models: batch culture and continuous cultures. Batch culture describes the growth of pure or mixed bacterial suspensions in a carefully selected medium without the further addition of nutrients in closed systems using sealed bottles or reactors containing suspensions of fecal material under anaerobic conditions. The advantages of batch fermentation are that the technique is inexpensive, easy to set up, and allows large number of substrates of fecal samples to be tested. However, these models have their weakness in microbiological control and the need to be of short duration in order to avoid the selection of non-representative microbial populations. The technique is useful for fermentation studies, for the investigation of metabolic profiles of short chain fatty acids arising from the active metabolism of dietary compounds by the gut microbiota, and especially for substrate digestion evaluation studies [21, 22]. Several of the publications in this field are based on a European interlaboratory study for estimation of the fermentability of dietary fiber *in vitro* [23].

Continuous cultures allow us to control the rate and composition of nutrient feed, bacterial metabolism and the environmental conditions. These models simulate proximal (single-state models) or proximal, transverse and distal colonic regions (multistage models). Continuous cultures are used for performing long-term studies, and substrate replenishment and toxic product removal are facilitated - thereby mimicking the conditions found *in vivo*. The most variable factor in these models is the technique used for fecal inoculation. The use of liquid fecal suspension as inoculum, where the bacterial populations are in the free-cell state, produces rapid washout of less competitive bacteria; as a result, the operation time is less than four weeks. The formation of fecal beads from the immobilization of fecal microbiota in a porous polysaccharide matrix allows release of the microbiota into the culture medium, with better reproduction of the *in vivo* flora and longer fermentation times [21, 22].

Artificial continuous models including host functions/human digestive functions have been developed. Models of this kind control peristaltic movement, pH and gastrointestinal secretions. The SHIME model (Simulated Human Intestinal Microbial Ecosystem) comprises a 5-step multi-chamber reactor simulating the duodenum and jejunum, ileum, cecum and the ascending colon, transverse colon and descending colon [24]. In turn, TIM-1 is an intestinal model of the stomach and small intestine, while TIM-2 is a proximal colon simulator model developed by TNO (*Netherlands Organization for Applied Scientific Research*). These models have been validated based on human and animal data [25]. They incorporate some host functions; however, they do not reproduce immune modulating and neuroendocrine responses. A remaining challenge is the difficulty of establishing a representative human gut microbiota *in vitro*. Other difficulties are the availability of the system, its cost, the prolonged time involved, its laboriousness, the use of large working volumes, and long residence times.

Combined systems that include the fractions obtained from simulated human digestion (gastrointestinal and/or colonic fermentation) and the incorporation of cell culture-based models allow us to evaluate bioaccessibility (estimate the amount of bioactive compounds assimilated from the bioaccessible fraction by cell culture) and to conduct bioactivity studies. The Caco-2 cell model is the most widely used and validated intestinal epithelium or human colon carcinoma cell model. Although colonic in origin, Caco-2 cells undergo spontaneous differentiation in cell culture to form a monolayer of well-polarized cells at confluence, showing many of the functional and morphological properties of mature human enterocytes (with the formation of microvilli on the brush border membrane, tight intercellular junctions and the excretion of brush border-associated enzymes) [26]. However it must be mentioned that this cell line differs in some aspects from *in vivo* conditions. For example, it does not reproduce the different populations of cells in the gut, such as goblet, Paneth and crypt cells, which are less organized and therefore leakier. Likewise, the model lacks regulatory control by neuroendocrine cells and through the blood [27].

The advantage of these systems versus those which only evaluate the influence of digestion is their greater similarity to the *in vivo* conditions. The combination of *in vitro* human intestinal cell models with *in vitro* digestion models in turn creates an advanced *in vitro* model system where samples obtained from host responses lacking in *in vitro* digestion models can be directly applied to monolayer cell models for host function studies [21].

3. Bioactivity of digested/fermented foods or related target bioactive compounds in cell lines

The chemopreventive properties of bioactive compounds have been investigated in cultured cells exposed to individual compounds. However, gut epithelial cells are more likely to be exposed to complex food matrixes containing mixtures of bioactive and antioxidant *in vivo* compounds [6]. In addition, food matrixes undergo a digestion process that may affect the structure and properties of the bioactive compounds. Therefore, the *in vitro* protective effects of antioxidant bioactive compounds do not necessarily reflect *in vivo* chemoprotection, which is more likely due to the combined effects of all the bioactive components present in the food [28].

A potential cell culture model for cancer or cardiovascular chemoprevention research involving dietary antioxidants (polyphenols, carotenoids and peptides) and other bioactive chemopreventive components such as phytosterols, should include some of the proposed mechanisms of action: inhibition of cell proliferation, induction of tumor suppressor gene expression, induction of cell cycle arrest, induction of apoptosis, antioxidant enzyme induction, and enhanced detoxification, antiinflammatory activities and the inhibition of cholesterol absorption [9, 15, 29, 30]. In addition, other mechanisms of chemoprevention could involve protection against genotoxic compounds or reactive oxygen species [31].

It recently has been stated that the measurement of cellular bioactivity of food samples coupled to *in vitro* digestion can provide information close to the real-life physiological situation [32]. In this sense, we surveyed more than 30 studies conducted in the past 10 years, involving human simulated gastrointestinal digestion and/or colonic fermentation procedures and subsequent bioactivity-guided assays with cell line models. These studies are presented in Tables 1, 2 and 3, which correspond to the mechanism of action related to chemoprevention of digested, fermented or digested plus fermented foods or bioactive constituents in cell lines, respectively.

The chemopreventive effect of digested foods or bioactive constituents in cell lines is summarized in Table 1. From the 22 studies surveyed, and according to the digestion method used, it can be seen that most of them involve solubility (n = 17) versus dialysis (n = 5). Samples used are preferably of vegetal origin (n = 15), the target compounds responsible for the chemopreventive action being polyphenols, antioxidants (in general), antioxidant peptides, lycopene and phytosterols. Furthermore, these compounds are mainly studied in colon-derived cells (as a cancer model when not differentiated, or as an intestinal epithelial model when differentiated). Concentrations tested are physiologically achievable in colon cells, since the bioaccessible fractions obtained after digestion are considered to be fractions that can pass through the stomach and small intestine reaching the colon, where they can exert antioxidant activity *in situ* [33]. In addition, polyphenols are studied in neuronal cells, liver-derived cells and lymphocytes. In the case of neuronal cells, the concentrations used (0-6 μM polyphenols) are similar to those reported for dietary polyphenolic-derived metabolites found in plasma (0-4 μM) [34], but for lymphocytes and liver, the concentrations are unknown or higher than expected *in vivo*, respectively. Another aspect to bear in mind is the time of cell exposure to the digested food or bioactive constituents. The range found in these studies is from 30 min to 120 h (this latter time-point not being expectable from a physiological standpoint).

Bioactive compounds of digested foods present four different but in some cases complementary modes of action: (1) inhibition of cholesterol absorption (phytosterols), and (2) antiproliferative, (3) cytoprotective and (4) antiinflammatory activities (polyphenols and general antioxidants).

1. The inhibition of cholesterol absorption has been reported to be mainly due to competition between phytosterols and cholesterol for incorporation to the micelles as a previous step before absorption by the intestinal epithelial cells [35].
2. Antiproliferative activity has been linked to cell growth inhibition associated to polyphenols [28, 32, 36-38] and lycopene [39], which is mainly regulated by two mechanisms: cell-cycle arrest and apoptosis induction. The cell cycle can be halted at different phases: G_0/G_1 with down-regulation of cyclin D_1 [39], S with down-regulation of cyclins D_1 and B_1 [28, 37] and G_2/M [36]. Apoptosis induction in turn occurs as a result of caspase-3 induction and down-regulation of the anti-apoptotic proteins Bcl-2 and Bcl-xL [39].
3. The cytoprotective effect of polyphenols, peptides and antioxidants against induced oxidative stress is related to the preservation of cell viability [40-47], an increase in the activity of antioxidant enzymes (such as catalase, glutathione reductase or glutathione peroxidase) [41, 43, 47, 48], the prevention of reduced glutathione (GSH) depletion [46, 47, 49], a decrease in intracellular ROS content [46, 50, 51], the maintenance of correct cell cycle progression [41, 43, 47, 52], the prevention of apoptosis [43], and the prevention of DNA damage [42, 51, 52].
4. The antiinflammatory action of peptides and polyphenols is derived from the decrease in the release of proinflammatory cytokines such as IL-8 when cells are stimulated with stressors such as H_20_2 or TNFα [53, 54].

Studies on the chemopreventive effect of foods or isolated bioactive constituents following colonic fermentation or gastrointestinal digestion plus colonic fermentation in cell lines are shown in Tables 2 and 3, respectively. The colonic fermentation procedure used in these assays has always been a batch model, except for one study combining batch and dynamic fermentation. In turn, when gastrointestinal digestion is involved, dialysis has been the method used. Foods of plant origin rich in fiber, and short chain fatty acids (mainly butyrate) and polyphenols as the target compounds have been used in such studies. The use of colon-derived cell lines is common in these assays, which have been performed using physiologically relevant concentrations and time periods of exposure of samples to cells ranging between 24 h and 72 h.

The mechanism of action underlying the treatment of cells with colonic fermented foods or isolated bioactive constituents (see Table 2) mainly comprises antiproliferative activity (i) and/or cytoprotective action (ii). In the first case, antiproliferative activity (i) has been attributed to cell growth inhibition [55-59], mainly due to apoptosis induction [58-59] and/or the up-regulation of genes involved in cell cycle arrest (*p21*) and apoptosis (*WNT2B*) [59]. Studies referred to a cytoprotective effect against oxidative damage (ii) in turn have been linked to the prevention of DNA damage [55, 56] and to the induction of antioxidant enzymes such as glutathione-S-transferase (GST) [56].

The bioactivity observed with the incubation of cells lines with foods or isolated bioactive constituents following gastrointestinal digestion plus colonic fermentation (see Table 3) is derived from antiproliferative activity (i) regulated by cell growth inhibition [60-62], cell cycle arrest [60] and/or apoptosis induction [60, 62], or by a cytoprotective effect against induced oxidative stress (ii) as a result of preservation of cell viability [63], protection against DNA damage [31, 61, 63] and/or induction of antioxidant enzymes such as CAT, GST and sulfotransferase (SULT2B1) [31].

4. Conclusions and future perspectives

From the data here reviewed in disease cell models, it can be concluded that gastrointestinal digestion/colonic fermentation applied to whole foods or isolated bioactive constituents may have potential health benefits derived from cell growth inhibition through the induction of cell-cycle arrest and/or apoptosis, cytoprotection against induced oxidative stress, antiinflammatory activity and the reduction of cholesterol absorption.

Studies conducted with single bioactive compounds are unrealistic from a nutritional and physiological point of view, since they do not take into account physicochemical changes during digestion and possible synergistic activities. Thus, a combined model of human simulated digestion including or not including colonic fermentation (depending on the nature of the studied compounds) with cell lines should be carried out if *in vitro* bioactivity assays with whole foods or bioactive chemopreventive compounds for the prevention of oxidative stress-related diseases are planned.

Although digested/fermented bioactive compounds appear as promising chemopreventive agents, our understanding of the molecular and biochemical pathways behind their mechanism of action is still limited, and further studies are warranted. In addition, the need for harmonization of the *in vitro* methods: (i) conditions of the gastrointestinal procedure, (ii) cell line used, (iii) concentrations of bioactive compounds used (usually much higher than those achievable in the human body when the digestion process is not considered), and (iv) time of cell exposure to the bioactive compounds (more than 24 h is unlikely to occur *in vivo*), should be considered for improved study designs more similar to the *in vivo* situation and for allowing comparisons of results among laboratories. This task is currently being carried out at European level within the project "Improving health properties of food by sharing our knowledge on the digestive process (INFOGEST) (2011-2015) (FAO COST Action FA 1005) (http://www.cost-infogest.eu/ABOUT-Infogest)".

Sample (Target compound/s)	Cell type	Cell treatment (Concentrations and time)	Cellular mechanism	References
		Gastrointestinal digestion (dialysis)		
(Polyphenols)				
Chokeberry juice	Caco-2 (human colon carcinoma)	85 to 220 (μM total polyphenols) 2 h a day for a 4-day period	Cell growth inhibition Viability decrease Cell cycle arrest at G_2/M phase Up-regulation of tumor suppression gene *CEACAM1*	Bermúdez-Soto et al. (2007) [36]
Raspberries	HT29, Caco-2 and HT115 (human colon carcinoma)	3.125 to 50 (μg/mL) 24 h	Prevention of H_2O_2 (75μM/5min)-induced DNA damage and decrease in G_1 phase of cell cycle (HT29 cells) No effect on epithelial integrity (Caco-2 cells) Inhibition of colon cancer cell invasion (HT115 cells)	Coates et al. (2007) [52]
Green tea	Differentiated PC12 (model of neuronal cells)	0.3-10 μg/mL (for H_2O_2) and 0.03-0.125 μg/mL (for $A\beta_{(1-42)}$) Pretreatment 24 h and stressed 24 h	Protection against H_2O_2 and $A\beta_{(1-42)}$ induced cytotoxicity (only at low concentrations)	Okello et al. (2011) [44]
Blackberry (*Rubus sp.*)	SK-N-MC (neuroblastoma cells)	1.5-6 μM total polyphenols 24 h	Preservation of cell viability against H_2O_2 (300 μM- 24 h) –induced oxidative stress (not related to modulation of ROS nor GSH levels)	Tavares et al. (2012a) [45]

CECAM1: Carcinoembryonic antigen-related cell adhesion molecule 1. Aβ(1-42): β-amyloid peptide 1-42. ROS: reactive oxygen species. GSH: reduced glutathione.

Table 1. Mechanisms involved in the chemopreventive effect of *in vitro* digested foods or bioactive constituents in cell lines.

The *in vitro* simulation of the conditions of gastrointestinal digestion represents an alternative to *in vivo* studies for evaluating the bioavailability and/or functionality of bioactive components of foods. *In vitro* studies do not replace *in vivo* studies; rather, both complement each other. *In vitro* methods need to be improved and validated with more *in vivo* studies. Thus, caution is mandatory when attempting to extrapolate observations obtained *in vitro* in cell line studies to humans.

Sample (Target compound/s)	Cell type	Cell treatment (Concentrations and time)	Cellular mechanism	References
Wild blackberry species	SK-N-MC (neuroblastoma cells)	0-6 μM total polyphenols 24 h	Preservation of cell viability and mitochondrial membrane potential against H_2O_2 (300 μM -24 h)-induced oxidative stress Decrease of intracellular ROS against H_2O_2 (200 μM -1 h)-induced oxidative stress (only *R. brigantines*) Prevention of GSH depletion against H_2O_2 (300 μM -24 h)-induced oxidative stress Induction of caspase 3/7 activity against H_2O_2 (300 μM -24 h)-induced oxidative stress (preconditioning effect)	Tavares et al. (2012b) [46]
			Gastrointestinal digestion (solubility)	
(Polyphenols)				
Fruit beverages with/without milk and/or iron	Caco-2 (human colon carcinoma)	2%, 5% and 7.5% (v/v) in culture medium (3.4-22.7 mg/mL total polyphenols) 4 hours-4 days or 24 h	Cell growth inhibition (no clear dose-response) Cell cycle arrest at S phase (7.5%) Down-regulation of cyclins D_1 and B1 No apoptosis (cytostatic effect)	Cilla et al. (2009) [28]
Zinc-fortified fruit beverages with/without iron and/or milk	Caco-2 and HT-29 (human colon carcinoma)	7.5% (v/v) in culture medium (~50 μM total polyphenols) 24 h	Cell growth inhibition (without citotoxicity) Cell cycle arrest at S phase No apoptosis and resumption of cell cycle after digest removal (cytostatic effect)	Cilla et al. (2010) [37]
Fruit juices enriched with pine bark extract	Caco-2 (human colon carcinoma)	4% (v/v) in culture medium 24-120 h	Cell growth inhibition	Frontela-Saseta et al. (2011) [38]

ROS: reactive oxygen species. GSH: Reduced glutathione.

Table 1. (continued-I).

Sample (Target compound/s)	Cell type	Cell treatment (Concentrations and time)	Cellular mechanism	References
Feijoada-traditional Brazilian meal	HepG2 (human liver cancer cells)	10-100 mg/mL 72 h (antiproliferation) and 1 h (antioxidant)	Antiproliferative activity ("/ 80 mg/mL) Increase in cellular antioxidant activity (0.6 μM quercetin equivalents)	Kremer-Faller et al. (2012) [32]
Culinary herbs: rosemary, sage and thyme	PBL (peripheral blood lymphocytes) and Differentitated Caco-2 (model of intestinal epithelium)	1:10 (v/v) in culture medium. Stressors (H_2O_2 2 mM and TNFα 100 μg/mL) Co-incubation 24 h or pre-incubation 3h then stress 24 h	PBL: significant decrease in IL-8 release when co-incubation with H_2O_2 and pre-incubation prior H_20_2 and TNFα Caco-2: significant decrease in IL-8 release only when co-incubation with TNFα	Chohan et al. (2012) [54]
(Antioxidants)				
Fruit beverages with/without milk and/or iron/zinc	Differentiated Caco-2 (model of intestinal epithelia)	1:1 (v/v) in culture medium	Preservation of cell viability No alteration of SOD	Cilla et al. (2008) [40]
Fruit beverages with/without milk or CPPs	Differentiated Caco-2 (model of intestinal epithelia)	1:1 (v/v) in culture medium or CPPs (1.4 mg/mL)	Preservation of cell viability (only fruit beverages)	Laparra et al. (2008) [41]
Beef patties enriched with sage and oregano	Caco-2 (human colon carcinoma)	10-100% (v/v) 24 h	Increase in cell viability at low concentrations (20-40%) but slight decrease at high concentrations (80-100%) Increase in GSH (only sage-enriched samples at 10%) Protection against H_20_2 (200 μM/1h)-induced GSH depletion (at 10%)	Ryan et al. (2009) [49]

IL-8: Proinflammatory interleukin-8. TNFα: tumor necrosis factor α. SOD: Superoxide dismutase. CPPs: caseinophosphopeptides. GSH-Rd. glutathione reductase. GSH: reduced glutathione.

Table 1. (continued-II).

Sample (Target compound/s)	Cell type	Cell treatment (Concentrations and time)	Cellular mechanism	References
Ellagic acid-, lutein- or sesamol-enriched meat patties	Caco-2 (human colon carcinoma)	0-20% (v/v) in culture medium 24 h	Viability maintenance against H_2O_2 (500 µM/ 1h)-induced stress Prevention of H_2O_2 (50 µM/30 min)-induced DNA damage	Daly et al. (2010) [42]
Pacific hake fish protein hydrolysates	Caco-2 (human colon carcinoma)	0.625-5 mg/mL 2 h	Inhibition (at non cytotoxic doses) of intracellular oxidation induced by AAPH (50 µM/1-2 h)	Samaranayaka et al. (2010) [50]
Human breast milk	Co-culture of Caco-2 BBE and HT29-MTX (model of human intestinal mucosa)	1:3 (v/v) in culture medium 30 min	Decrease of H_2O_2 (1 mM/30 min)-induced ROS Prevention of H_2O_2 (500 µM/30 min)-induced DNA damage	Yao et al. (2010) [51]
Fruit beverages with/without milk and/or iron/zinc	Differentiated Caco-2 (model of intestinal epithelia)	1:1 (v/v) in culture medium Pre-incubation 24 h then stressed 2h with H_2O_2 5 mM	Preservation of cell viability Increase in GSH-Rd activity (only Fe or Zn with/without milk samples) Prevention of G_1 cell cycle phase decrease induced by H_2O2 Prevention of apoptosis (caspase-3) induced by H202	Cilla et al. (2011) [43]
Purified milk hydrolysate peptide fraction from digested human milk	Caco-2 and FHs 74 int (human colon carcinoma and primary fetal enterocytes)	0.31-1.25 g/L (peptide) and 150 µM (tryptophan) 2 h (peptide) and 1-12 h (tryptophan)	Exacerbation of AAPH (50 µM/1-2 h)-induced oxidative stress (peptide) Up-regulation of Nrf-2 and subsequent up-regulation of GSH-Px2 gene as adaptive response to stress (tryptophan)	Elisia et al. (2011) [48]

AAPH: 2,2′-azobis (2-amidinopropane) dihydrochloride. ROS: reactive oxygen species. GSH-Rd: glutathione reductase. Nrf-2: nuclear response factor 2. GSH-Px2: glutathione peroxidase.

Table 1. (continued-III).

Sample (Target compound/s)	Cell type	Cell treatment (Concentrations and time)	Cellular mechanism	References
CPPs from digested cow's skimmed milk	Differentiated Caco-2 (model of intestinal epithelia)	1, 2 and 3 mg/mL Pre-incubation 24 h then stressed 2h with H_2O_2 5 mM	Preservation of cell viability Increase in GSH content and induction of CAT activity Decrease in lipid peroxidation Maintenance of correct cell cycle progression	García-Nebot et al. (2011) [47]
Purified hen egg yolk-derived phosvitin phosphopeptides	Differentiated Caco-2 (model of intestinal epithelia)	0.05-0.5 mg/mL 2 h	Reduced IL-8 secretion in H_2O_2 (1 mM/6 h)-induced oxidative stress	Young et al. (2011) [53]
(Lycopene)				
Tomatoes	HT29 and HCT-116 (human colon carcinoma)	20-100 mL/L 24 h	Cell growth inhibition Cell cycle arrest at G_0-G_1 phase and apoptosis induction (caspase-3) Down-regulation of cyclin D_1 and anti-apoptotic proteins Bcl-2 and Bcl-xL	Palozza et al. (2011) [39]
(Phytosterols)				
Orange juice enriched with fat-free phytosterols	Differentiated Caco-2 (model of intestinal epithelia)	2 mL test medium/well 4 h	Reduced micellarization of cholesterol Decrease in cholesterol accumulation by Caco-2 cells	Bohn et al. (2007) [35]

GSH: reduced glutathione. CAT: catalase. IL-8: proinflammatory interleukin-8.

Table 1. (continued-IV).

Sample (Target compound/s)	Cell type	Cell treatment (Concentrations and time)	Cellular mechanism	References
(SCFA)				
Fibre sources: linseed, watercress, kale, tomato.soya	HT29 (human colon carcinoma)	2.5-25% (v/v) in culture medium 72 h	Cell growth inhibition (all samples except watercress) Prevention of HNE (150μM/30 min)-induced DNA damage (only soya flour)	Beyer-Sehlmeyer et al. (2003) [55]

Sample (Target compound/s)	Cell type	Cell treatment (Concentration s and time)	Cellular mechanism	References
flour, chicory inulin and wheat				
Wheat bran-derived arabinoxylans	HT29 (human colon carcinoma)	0.01-50% (v/v) in culture medium 24-72 h	Cell growth inhibition Prevention of HNE (200μM/30 min)-induced DNA damage (at 25-50%) Induction of GST activity (at 10%)	Glei et al. (2006) [56]
Inulin-type fructans	LT97 and HT29 (human colon adenoma and carcinoma)	1.25-20% (v/v) in culture medium 24-72 h	Cell growth inhibition (at 5-10%) Apoptosis induction (cleavage of PARP) only in LT97 cells (at 5-10%)	Munjal et al. (2009) [58]
Wheat aleurone	LT97 and HT29 (human colon adenoma and carcinoma)	5-10% (v/v) in culture medium 24-72 h	Cell growth inhibition Apoptosis induction (caspase-3) Up-regulation of genes *p21* (cell cycle arrest) and *WNT2B* (apoptosis)	Borowicki et al. (2010a) [59]
(polyphenols)				
Apples	LT97 and HT29 (human colon adenoma and carcinoma)	100-900 μg/mL 24-48 h	Cell growth inhibition (LT97 more sensitive than HT29 cells)	Veeriah et al. (2007) [57]

SCFA: short chain fatty acids. GST: Glutathione-S-Transferase. HNE: 4-Hydroxynonenal. PARP: Poly (ADP-ribose) polymerase. WNT2B: Wingless-type MMTV integration site family member 2.

Table 2. Mechanisms involved in the chemopreventive effect of *in vitro* colonic fermented (in batch) of foods or bioactive constituents in cell lines.

Sample (Target compound/s)	Cell type	Cell treatment (Concentrations and time)	Cellular mechanism	References
(SCFA)				
Resistant starches	Differentiated Caco-2 (model of intestinal epithelia)	10% (v/v) in culture medium 24 h	Preservation of cell viability Prevention of H_2O_2 (75 μM/5 min)-induced DNA damage Maintenance of barrier function integrity (TEER)	Fässler et al. (2007) [63]
Wheat aleurone	HT29 (human colon carcinoma)	10% (v/v) in culture medium 24-72 h	Cell growth inhibition Cell cycle arrest in G_0-G_1 phase Apoptosis induction (caspase-3)	Borowicki et al. (2010b) [60]
Wheat aleurone	HT29 (human colon carcinoma)	5-10% (v/v) in culture medium 24-72 h	Induction of antioxidant enzymes (CAT and GST) Up-regulation of genes *CAT*, *GSTP1* and *SULT2B1* Prevention of H_2O_2 (75 μM/5 min)-induced DNA damage	Stein et al. (2010) [31]
(SCFA and polyphenols)				
Bread	HT29 (human colon carcinoma)	2.5-5% (v/v) in culture medium 24-72 h	Cell growth inhibition Prevention of H_2O_2 (75 μM/5 min)-induced DNA damage	Lux et al. (2011) [61]
(butyrate)				
Bread	LT97 (human colon adenoma)	5-20% (v/v) in culture medium 24-72 h	Up-regulation of genes from DNA repair, biotransformation, differentiation and apoptosis Increase in GST activity, GSH content and AP activity (differentiation) Cell growth inhibition Apoptosis induction (caspase-3)	Schölrmann et al. (2011) [62]

SCFA: Shot chain fatty acids. TEER: Trans Epithelial Electrical Resistance. GST: Glutathione-S-Transferase. GSH: Glutathione. CAT: catalase. SULT: Sulfotransferase. AP: Alkaline phosphatase.

Table 3. Mechanisms involved in the chemopreventive effect of *in vitro* digested (dialysis) plus colonic fermented (batch) foods or bioactive constituents in cell lines.

Acknowledgements

This work was partially supported by Consolider Fun-C-Food CSD2007-00063 and the Generalitat Valenciana (ACOMP 2011/195).

Author details

Antonio Cilla*, Amparo Alegría, Reyes Barberá and María Jesús Lagarda

*Address all correspondence to: antonio.cilla@uv.es

Nutrition and Food Science Area. Faculty of Pharmacy, University of Valencia, Avda. Vicente Andrés Estellés s/n, 46100 - Burjassot, Valencia, Spain

References

[1] Millen, B. E., Quatromoni, P. A., Pencina, M., Kimokoti, R., Nam-H, B., Cobain, S., Kozak, W., Appagliese, D. P., Ordovas, J., & D'Agostino, R. B. (2005). Unique dietary patterns and chronic disease risk profiles of adult men: The Framinghan nutrition studies. *J. Am. Diet. Assoc.*, 105, 1723-1734.

[2] Puawels, E. K. J. (2011). The protective effect of the Mediterranean diet: focus on cancer and cardiovascular risk. *Med. Princ. Pract.*, 20, 103-111.

[3] Valko, M., Leibfritz, D., Moncol, J., Cronin, M. T., Mazur, M., & Telser, J. (2007). Free radicals and antioxidants in normal physiological functions and human disease. . Int. J. Biochem. Cell Biol. , 39, 44-84.

[4] Johansson, I., Nilsson, L., Stegmayr, B., Boman, K., Hallmans, G., & Winkvist, A. (2012). Associations among 25-year trends in diet, cholesterol and BMI from 140,000 observations in men and women in Northern Sweden. *Nutr. J.*, 11, 1-40.

[5] Bruce, W. R., Giacca, A., & Medline, A. (2000). Possible mechanisms relating diet and risk of colon cancer. *Cancer Epidemiol. Biomarkers Prev.*, 9, 1271-1279.

[6] Espín, J. C., García-Conesa, M. T., & Tomás-Barberán, F. A. (2007). Nutraceuticals: facts and fiction. *Phytochemistry*, 68, 2986-3008.

[7] Holst, B., & Williamson, G. (2008). Nutrients and phytochemicals: from bioavailability to bioefficacy beyond antioxidants. *Curr. Opin. Biotechnol.*, 19, 73-82.

[8] Finley, J. W., Kong, N.-A., Hintze, K. J., Jeffery, E. H., Ji, L. L., & Lei, X. G. (2011). Antioxidants in foods: state of the science important to the food industry. *J. Agric. Food Chem*, 59, 6837-6846.

[9] Liu, R. H., & Finley, J. (2005). Potential cell culture models for antioxidant research. *J. Agric. Food Chem.*, 53, 4311-4314.

[10] Stevenson, D. E., & Hurst, R. D. (2007). Polyphenolic phytochemicals- just antioxidants or much more? *Cell. Mol. Life Sci.*, 64, 2900-2916.

[11] Ramos, S. (2008). Cancer chemoprevention and chemotherapy: dietary polyphenols and signaling pathways. *Mol. Nutr. Food Res.*, 52, 507-526.

[12] Fernández-García, E., Carvajal-Lérida, I., & Pérez-Gálvez, A. (2009). In vitro bioaccessibility assessment as a prediction tool of nutritional efficiency. *Nutr. Res.*, 29, 751-760.

[13] Hur, S. J., Lim, B. O., Decker, E. A., & Mc Clements, D. J. (2011). In vitro human digestion models for food applications. *Food Chem.*, 125, 1-12.

[14] Liu, R. H. (2003). Health benefits of fruit and vegetables are from additive and synergistic combinations of phytochemicals. *Am. J. Clin. Nutr.*, 78, 517S-520S.

[15] de Kok, T. M., van Breda, S. G., & Manson, M. M. (2008). Mechanisms of combined action of different chemopreventive dietary compounds: a review. *Eur. J. Nutr.*, 47, 51-59.

[16] Fairweather-Tait, S. J. (1993). Bioavailability of nutrients. . Macrae R, Robinson RK, Sadler MJ, editors. Encyclopaedia of food science, food technology and nutrition. London: Academic Press. , 384-388.

[17] Ekmekcioglu, C. (2002). A physiological approach for preparing and conducting intestinal bioavailbility studies using experimental systems. *Food Chem.*, 76, 225-230.

[18] Yonekura, L., & Nagao, A. (2007). Intestinal absorption of dietary carotenoids. . Mol. Nutr. Food Res. , 51, 107-115.

[19] Parada, J., & Aguilera, J. M. (2007). Food microstructure affects the bioavailability of several nutrients. *J. Food Sci.*, 72, R 21-R32.

[20] Bermúdez-Soto, M. J., Tomás-Barberán, F. A., & García-Conesa, M. T. (2007). Stability of polyphenols in chokeberry (Aronia melanocarpa) subjected to in vitro gastric and pancreatic digestion. *Food Chem.*, 102, 865-874.

[21] Payne, A. N., Zihler, A., Chassard, C., & Lacroix, C. (2012). Advances and perspectives in in vitro human gut fermentation modeling. *Trends Biotech.*, 30, 17-25.

[22] Macfarlane, G. T., & Macfarlane, S. (2007). Models for intestinal fermentation: association between food components, delivery sustems, bioavailability and functional interactions in the gut. *Curr. Opin. Biotechnol.*, 18, 156-162.

[23] Barry, B. J. L., Hoebler, C., Macfarlane, G. T., Macfarlane, S., Mathers, J. C., Reed, K. A., Mortensen, P. B., Norgaard, I., Rowland, I. R., & Rumney, C. J. (1995). Estimation of the fermentability of dietary fibre in vitro: a European interlaboratory study. *Br. J. Nutr.*, 74, 303-322.

[24] Molly, K., Woestyne, M. V., & Verstraete, W. (1993). Development of a 5-step multi-chamber reactor as a simulation of the human intestinal microbial ecosystem. *Appl. Microbiol. Biotechnol.*, 39, 254-258.

[25] Minekus, M., Marteau, P., Havenaar, R., & Huis in't Veld, J. H. J. (1995). A multi compartmental dynamic computer-controlled model simulating the stomach and small intestine. *ATLA.*, 23, 197-209.

[26] Pinto, M., Robine-Leon, S., Appay, M. D., Kedinger, M., Triadou, N., Dussaulx, E., Lacroix, B., Simon-Assmann, P., Haffen, K., Fogh, J., & Zweibaum, A. (1983). Enterocyte-like differentiation and polarization of the human colon carcinoma cell line Caco-2 in culture. *Biol. Cell*, 47, 323-330.

[27] Ekmekcioglu, C., Pomazal, K., Steffan, I., Schweiger, B., & Marktl, W. (1999). Calcium transport from mineral waters across Caco-2 cells. *J. Agric. Food Chem.*, 47, 2594-2599.

[28] Cilla, A., González-Sarrías, A., Tomás-Barberán, F. A., Espín, J. C., & Barberá, R. (2009). Availability of polyphenols in fruit beverages subjected to in vitro gastrointestinal digestión and their effects on proliferation, cell-cycle and apoptosis in human colon cancer Caco-2 cells. *Food Chem.*, 114, 813-820.

[29] Bradford, P. G., & Awad, A. B. (2010). Modulation of signal transduction in cancer cells by phytosterols. *Biofactors*, 36, 241-247.

[30] Brüll, F., Mensik, R. P., & Plat, J. (2009). Plant sterols: functional lipids in immune function and inflammation? *Clin Lipidol.*, 4, 355-365.

[31] Stein, K., Borowicki, A., Scharlau, D., & Glei, M. (2010). Fermented wheat aleurone induces enzymes involved in detoxification of carcionogens and in antioxidative defence in human colon cells. *Br. J. Nutr.*, 104, 1101-1111.

[32] Kremer-Faller, A. N., Fialho, E., & Liu, R. H. (2012). Cellular antioxidant activity of Feijoada whole meal coupled with an in vitro digestion. *J. Agric. Food Chem.*, 60, 4826-4832.

[33] Halliwell, B., Rafter, J., & Jenner, A. (2005). Health promotion by flavonoids, tocopherols, tocotrienols, and other phenols: direct or indirect effects? Antioxidant or not? *Am. J. Clin. Nutr.*, 81, 268S-276S.

[34] Manach, C., Williamson, G., Morand, C., Scalbert, A., & Remesy, C. (2005). Bioavailability and bioefficacy of polyphenols in humans. I. Review of 97 bioavailability studies. *Am. J. Clin. Nutr.*, 81, 230S-242S.

[35] Bohn, T., Tian, Q., Chitchumroonchokchai, C., Failla, M. L., Schwartz, S. J., Cotter, R., & Waksman, J. A. (2007). Supplementation of test meals with fat-free phytosterol products can reduce cholesterol micellarization during simulated digestion and cholesterol accumulation by Caco-2 cells. . J. Agric. Food Chem. , 55, 267-272.

[36] Bermúdez-Soto, Larrosa. M., García-Cantalejo, J. M., Espín, J. C., Tomás-Barberán, F. A., & García-Conesa, M. T. (2007). Up-regulation of tumor supresor carcinoembryon-

ic antigen-related cell adhesión molecule 1 in human colon cancer Caco-2 cells following repetitive exposure to dietery levels of a polyphenol-rich chokeberry juice. *J. Nutr. Biochem.*, 18, 259-271.

[37] Cilla, A., Lagarda, Barberá. R., & Romero, F. (2010). Polyphenolic profile and antiproliferative activity of bioaccessible fractions of zinc-fortified fruit beverages in human colon cancer cell lines. *Nutr. Hosp.*, 25, 561-571.

[38] Frontela-Saseta, C., López-Nicolás, R., González-Bermúdez, C. A., Peso-Echarri, P., Ros-Berruezo, G., Martínez-Graciá, C., Canalli, R., & Virgili, F. (2011). Evaluation of antioxidant activity and antiproliferative effect of fruit juices enriched with Pycnogenol® in colon carcinoma cells. The effect of in vitro gastrointestinal digestion. *Phytoter. Res.*, 25, 1870-1875.

[39] Palozza, P., Serini, S., Bonisegna, A., Bellovino, D., Lucarini, M., Monastra, G., & Gaetani, S. (2007). The growth-inhibitory effects of tomatoes digested in vitro in colon adenocarcinoma cells occur through down regulation of cyclin D1, Bcl-2 and Bcl-xL. *Br. J. Nutr.*, 98, 789-795.

[40] Cilla, A., Laparra, J. M., Alegría, A., Barberá, R., & Farré, R. (2008). Antioxidant effect derived from bioaccessible fractions of fruit beverages against H2O2-induced oxidative stress in Caco-2 cells. *Food Chem.*, 106, 1180-1187.

[41] Laparra, J. M., Alegría, A., Barberá, R., & Farré, R. (2008). Antioxidant effect of casein phosphopeptides compared with fruit beverages supplemented with skimmed milk against H202-induced oxidative stress in Caco-2 cells. *Food Res. Int.*, 41, 773-779.

[42] Daly, T., Ryan, E., Aherne, S. A., O'Grady, M. N., Hayes, J., Allen, P., Kerry, J. P., & O'Brien, N. M. (2010). Bioactivity of ellagic acis-, lutein- or sesamol-enriched meat patties assessed using an in vitro digestion in Caco-2 cell model system. *Food Res. Int.*, 43, 753-760.

[43] Cilla, A., Laparra, J. M., Alegría, A., & Barberá, R. (2011). Mineral and/or milk supplementation of fruit beverages helps in the prevention of H202-induced oxidative stress in Caco-2 cells. *Nutr. Hosp.*, 26, 614-621.

[44] Okello, E. J., Mc Dougall, G. J., Kumar, S., & Seal, C. J. (2011). In vitro protective effects of colon-available extract of Camellia sinensis (tea) against hydrogen peroxide and beta-amyloid (Aβ(1-42)) induced cytotoxicity in differentiated PC12 cells. *Phytomedicine*, 18, 691-696.

[45] Tavares, L., Figueira, I., Macedo, D., Mc Dougall, G. J., Leitao, M. C., Vieira, H. L. A., Stewart, D., Alves, P. M., Ferreira, R. B., & Santos, C. N. (2012). Neuroprotective effect of blackberry (Rubus sp.) polyphenols is potentiated after simulated gastrointestinal digestion. *Food Chem.*, 131, 1443-1452.

[46] Tavares, L., Figueira, I., Mc Dougall, G. J., Vieira, H. L., Stewart, D., Alves, P. M., Ferrerira, R. B., & Santos, C. N. (2012). Neuroprotective effects of digested polyphenols from wild blackberry species. *Eur. J. Nutr.*, Doi: 10.1007/s00394-012-0307-7.

[47] García-Nebot, M. J., Cilla, A., Alegría, A., & Barberá, R. (2011). Caseinophosphopeptides exert partial and site-specific cytoprotection against H202-induced oxidative stress in Caco-2 cells. *Food Chem.*, 129, 1495-1503.

[48] Elisia, I., Tsopmo, A., Friel, J. K., Diehl-Jones, W., & Kitts, D. D. (2011). Tryptophan from human milk induces oxidative stress and upregulates Nrf-2-mediated stress response in human intestinal cell lines. *J. Nutr.*, 141, 1417-1423.

[49] Ryan, E., Aherne-Bruce, S. A., O'Grady, M. N., Mc Govern, L., Kerry, J. P., & O'Brien, N. M. (2009). Bioactivity of herb-enriched beef patties. *J. Med. Food*, 12, 893-901.

[50] Samaranayaka, A. G. P., Kitts, D. D., & Li-Chan, E. C. Y. (2010). Antioxidative and angiotensin-I-converting enzyme inhibitory potential of Pacific hake (Merluccius productus) fish protein hydrolysate subjected to simulated gastrointestinal digestion and Caco-2 cell permeation. *J. Agric. Food Chem.*, 58, 1535-1542.

[51] Yao, L., Friel, J. K., Suh, M., & Diehl-Jones, W. L. (2010). Antioxidant properties of breast milk in a novel in vitro digestion/enterocyte model. *J. Pediatr. Gastroenterol. Nutr.*, 50, 670-676.

[52] Coates, E. M., Popa, G., Gill, C. I. R., Mc Cann, M. J., Mc Dougall, G. J., Stewart, D., & Rowland, I. (2007). Colon-available raspberry polyphenols exhibit anti-cancer effects on in vitro models of colon cancer. *J. Carcinog.*, 6(4).

[53] Young, D., Nau, F., Pasco, M., & Mine, Y. (2011). Identification of hen egg yolk-derived phosvition phosphopeptides and their effects on gene expression profiling against oxidative-stress induced Caco-2 cells. *J. Agric. Food Chem.*, 59, 9207-9218.

[54] Chohan, M., Naughton, D. P., Jones, L., & Opara, E. I. (2012). An investigation of the relationship between the anti-inflammatory activity, polyphenolic content, and antioxidant activities of cooked and in vitro digested culinary herbs. Oxid. Med. Cell Longev. 627843.

[55] Beyer-Sehlmeyer, G., Glei, M., Hartmann, E., Hughes, R., Persin, C., Böhm, V., Rowland, I., Schubert, R., Jahreis, G., & Pool-Zabel, B. L. (2003). Butyrate is only one of several growth inhibitors produced during gut flora-mediated fermentation of dietary fibre sources. *Br. J. Nutr.*, 90, 1057-1070.

[56] Glei, M., Hofmann, T., Küster, K., Hollmann, J., Lindhauer, M., & Pool-Zabel, B. L. (2006). Both wheat (Triticum aestivum) bran arabinoxylans and gut flora-mediated fermentation products protect human colon cells from genotoxic activities of 4-hydroxynonenal and hydrogen peroxide. *J. Agric. Food Chem.*, 54, 2088-2095.

[57] Veeriah, S., Hofmann, T., Glei, M., Dietrich, H., Will, F., Schreier, P., Knaup, B., & Pool-Zabel, B. L. (2007). Apple polyphenols and products formed in the gut differently inhibit survival of human cell lines derived from colon adenoma (LT97) and carcinoma (HT29). *J. Agric. Food Chem.*, 55, 2892-2900.

[58] Munjal, U., Glei, M., Pool-Zabel, B. L., & Scharlau, D. (2009). Fermentation products of inulin-type fructans reduce proliferation and induce apoptosis in human colon tumour cells of different stages of carcinogenesis. *Br. J. Nutr.*, 102, 663-671.

[59] Borowicki, A., Stein, K., Scharlau, D., & Glei, M. (2010). Fermentation supernatants of wheat (Triticum aestivum L.) aleurone beneficially modulate cancer progression in human colon cells. *J. Agric. Food Chem.*, 58, 2001-2008.

[60] Borowicki, A., Stein, K., Scharlau, D., Scheu, K., Brenner-Weiss, G., Obst, U., Hollmann, J., Lindhauer, M., Wachter, N., & Glei, M. (2010). Fermented wheat aleurone inhibits growth and induces apoptosis in human HT29 colon adenocarcinoma cells. *Br. J. Nutr.*, 103, 360-369.

[61] Lux, S., Scharlau, D., Schlörmann, W., Birringer, M., & Glei, M. (2011). In vitro fermented nuts exhibit chemopreventive effects in HT29 colon cancer cells. *Br J. Nutr.*, 15, 1-10.

[62] Schlörmann, W., Hiller, B., Janhs, F., Zöger, R., Hennemeier, I., Wilheim, A., Lindhauer, M. G., & Glei, M. (2011). Chemopreventive effects of in vitro digested and fermented bread in human colon cells. *Eur. J. Nutr*, 10.1007/s00394-011-0262-8.

[63] Fässler, C., Gill, C. I. R., Arrigoni, E., Rowland, I., & Amadò, R. (2007). Fermentation of resistant starches: influence of in vitro models on colon carcinogenesis. *Nutr. Cancer*, 58, 85-92.

Section 2

Cell Damage by Free Radicals - Oxidative Stress in Disease

Chapter 7

The Chemoprevention of Chronic Degenerative Disease Through Dietary Antioxidants: Progress, Promise and Evidences

Eduardo Madrigal-Santillán,
Eduardo Madrigal-Bujaidar, Sandra Cruz-Jaime,
María del Carmen Valadez-Vega,
María Teresa Sumaya-Martínez,
Karla Guadalupe Pérez-Ávila and
José Antonio Morales-González

Additional information is available at the end of the chapter

http://dx.doi.org/10.5772/52162

1. Introduction

1.1. Epidemiology of chronic degenerative diseases in Mexico and the world

During the last 30 years relevant changes in the public health field have arisen worldwide, among which the most representative are observed in developed countries where a big deal of infectious diseases have been reduced and controlled as a result of the creation and introduction of powerful antibiotics [1].

In countries such as Australia, Austria, Belgium, Canada, Denmark, Finland, France, Germany, Greece, Ireland, Italy, Japan, Luxembourg, Netherlands, New Zealand, Norway, Portugal, Spain, Sweden, Switzerland, the United Kingdom, and the United States, incorporated to the OECD (Organization for Economic Cooperation and Development), mortality due to those diseases has diminished up to 38% in people between 35 and 69 years old. Likewise, the risk of mortality before age 70 has diminished up to the 23%. Those reductions have been the result of social changes and of the improvement of preventive methods of infectious diseases. However, in recent years the prevalence of chronic degenerative diseases has increased [1].

Chronic degenerative diseases (CDDs) represent a problem of public health for they have become the cause of death worldwide both in adolescents and adults. Among the most prevalent CDDs worldwide is obesity, the cardiovascular diseases (such as hypertension, atherosclerosis), heart diseases, diabetes, chronic respiratory diseases, and cancer; which have caused the 60% of the 58 million yearly deaths, which are approximately 35 million people death for these diseases between 2005 and 2007 [2].

Prevalence in chronic degenerative diseases results from different factors, among which the technologic advance and modernization affect life styles where an increase in processed foods consumption with a high level of fat content, a sedentary lifestyle since childhood, alcohol and tobacco, stress and a lack of culture in terms of damage prevention and health risks [1,2].

Mexico does not escape this situation as a result of specific factors to our country such as economic development, concentration of population in urban areas, lack of support to improve the health services and the limitations in preventive programs, particularly in the population under 10 years. Besides, there is a transformation of the population pyramid due to a reduction in mortality and a decrease in birth rate; both phenomena are identified as epidemiologic and demographic transitions [2]. In México, the morbidity data produced by the CDDs are taken from the statistics of the healthcare sector and published by the healthcare ministry. Although in those reports not all the existing cases are included (not all patients request healthcare services), they are a good help to understand the damage behavior along with other indicators of prevalence that estimate the number of cases in the population within a specific period of time. Such indicators are obtained from the national healthcare survey and from the national healthcare and nutrition survey 2006 [2]. On the other hand, the mortality statistics are considered as more reliable due to the permanent job in updating the database. The information is obtained from the records of the national institute of statistics, geography and informatics (INEGI) and the general bureau of health information, in conjunction with the epidemiological AVAD index, which is a measure that combines years of healthy life lost due to premature mortality and years of life lost due to disability [3].

As mentioned above, the epidemiological and demographic transitions are important factors for the prevalence of chronic degenerative diseases and indicate changes in the behavior of population dynamics, as well as damage to health which are the result of the low socioeconomic development and the impact of government policies on public health. The demographic transition shows the change in a steady state population with high fertility and mortality associated with the low socioeconomic development process and/or modernization. This process is irreversible and was constructed from the first countries reaching socioeconomic development in Europe such as France and England. In recent years it has made rapid changes affecting the world population [2].

According to data from INEGI and the national population council (CONAPO), Mexico has experienced an accelerated process of demographic transition, which has influenced the economic development and migration, leading to a reduction in mortality and a parallel high birth rates, as well as the consequent population growth, so it is estimated that between 2010 and 2050 the proportion of elderly people in Mexico will grow from 7% to 28% and with it the possibility of an chronic degenerative disease is greater [2,3].

In the case of the epidemiological transition, this is characterized by a reduction of morbidity and mortality from transmissible diseases and an increase in chronic degenerative diseases. In recent years, this parameter has shown that in both developed and developing countries, the proportion of infectious diseases in individuals over age 15 is stable, but unfortunately the CDDs are increasing, showing that they occupy almost half of value of morbility globally. A relevant fact is observed in developed countries (like France, Germany, Japan, United Kingdom, and United States) where the greatest impact of transmissible diseases remains the HIV/AIDS; but the cerebrovascular diseases and the ischemic heart disease are among the main causes of morbidity and mortality (Table 1) in individuals over age 15, both diseases represent more than 36% of deaths worldwide [2].

In the specific case of Mexico, it is well-known that infectious diseases made up the profile of mortality in the fifties, since half of the deaths were caused by diarrhea and respiratory infections, for reproductive problems and associated malnutrition conditions. Nowadays, these diseases (classified as lag diseases) are concentrated in less than 15% of deaths [2].

In the last 10 years, there has been an overlap between lag diseases and the so-called emerging diseases. Thus, the epidemiological transition has ranked the chronic degenerative diseases among the 10 leading causes of death, highlighting the type 2 diabetes, obesity, cardiovascular diseases, malignant neoplasms and cerebrovascular diseases [4].

Mortality (individuals between 15 and 50 years)			Mortality (over 60 years)		
	Cause	Deaths (thousands)		Cause	Deaths (thousands)
1	HIV / AIDS	2279	1	Ischemic heart disease	5825
2	Ischemic heart disease	1332	2	Cerebrovascular diseases	4689
3	Tuberculosis	1036	3	Chronic obstructive pulmonary disease	2399
4	Injuries from traffic accidents	814	4	Infections of lower respiratory	1396
5	Cerebrovascular diseases	783	5	Trachea and lung cancer	928
6	Self-harm	672	6	Diabetes	754
7	Violence	473	7	Hypertensive heart disease	735
8	Liver cirrhosis	382	8	Stomach Cancer	605
9	Infections of lower respiratory	352	9	Tuberculosis	495
10	Chronic obstructive pulmonary disease	343	10	Colorectal cancer	477

Table 1. Leading causes of death in people over 15 years in the world (as a function of AVAD index)

2. Definition, importance and control of oxidative stress

The term "oxidative stress" was first introduced in the eighties by Helmut Sies (1985), defining it as a disturbance in the prooxidant-oxidant balance in favor of the first. From that time, a great number of researchers have studied this phenomenon; so, the concept has evolved and now, has been defined as "A situation when steady-state ROS concentration is transiently or chronically enhanced, disturbing cellular metabolism and its regulation and damaging cellular constituents" [5,6].

However, oxidative stress is a phenomenon not entirely detrimental for the organism; also, free radicals (FR) have an important function in several homeostatic processes. They act as intermediate agents in essential oxidation-reduction (redox) reactions. Some examples are the destruction of microorganisms through phagocytosis, synthesis of inflammatory mediators and detoxification. Therefore, FR in low concentrations are useful and even essential [7].

FR represents any chemical species that exists independently and has one or more unmatched (odd) electrons rotating in its external atomic orbits. This highly unstable configuration causes this chemical species to be very aggressive and to have a short life span. Once generated, FR interact with other molecules through redox reactions to obtain a stable electronic configuration [8-10].

Several authors have classified FR according to the functional group in their molecule, being the most frequent the reactive oxygen species (ROS) and reactive nitrogen species (RNS). Thiol radicals are less important, their reactive group contains sulfur; well as those containing carbon or phosphorus in their reactive center. ROS are constituted by superoxide anion (O_2•−), hydroxyl radical (•OH), hydrogen peroxide (H_2O_2), and singlet oxygen. While, that the RNS are nitric oxide (NO), nitrogen dioxide (NO_2•−) and peroxynitrite (OONO−) [11,12].

Due to the constant production of ROS and RNS during metabolic processes, the organism has developed a powerful, complex defense system that limits its exposure to these agents these are the so-called antioxidants (AO). Several antioxidants are enzymes or essential nutrients, or include these in their molecular structure. An essential nutrient is a compound that must be eaten because the organism is unable to synthesize it. Based on this characteristic, some authors classify AO as non-enzymatic and enzymatic [12-14].

2.1. Enzimatic antioxidants

Some researchers state that the AO function performed by enzymes has advantages compared to AO compounds, for this activity is regulated according to cellular requirements: they can be induced, inhibited or activated by endogenous effectors [15]. Ho and colleagues (1998) showed evidence of the importance of AO enzymes in protection against oxidant agents. When using transgenic mice designed to overexpress the activity of some AO enzymes, it was noticed that there is a notorious tolerance of certain tissues when they are exposed to toxics and pathologic conditions that would promote ROS action [9,16].

Enzymatic AO catalyze electron transference from a substrate towards FR. Later, the substrates or reducing agents used in these reactions are regenerated to be used again, they achieve this by using the NADPH produced in different metabolic pathways [14]. A prolonged exposure to ROS can result in diminished NADPH concentration, which is needed in other important physiologic processes, even though some enzymatic AO do not consume cofactors. Superoxide dismutase (SOD), catalase (CAT) and glutathione peroxidase (GSH-Px) belong to this group [12,14,17].

2.2. Non-enzymatic antioxidants

Non-enzymatic antioxidants constitute a heterogeneous group of hydrophobic and hydrophilic molecules that trap FR and create chemical species that are less noxious to cell integrity [18]. Essentially, they give an electron to a FR to stabilize it. Hydrophilic non-enzymatic antioxidants are located mainly in the cytosol, mitochondrial and nuclear matrixes and in extracellular fluids. They are vitamin C, glutathione, uric acid, ergothioneine and polyphenolic flavonoids [9,18].

3. Role of oxidative stress in the development and pathogenesis of the chronic degenerative diseases

Currently, studies related to reactive oxygen species (ROS) and reactive nitrogen species (RNS) have become a relevant issue in research with the main purpose of understanding their functions and effects in the organism. Studies developed throughout the 20th century have explained the action mechanisms of ROS and the operation of the systems responsible for their elimination. These evidences have shown the existence of enzyme systems that produce ROS (cytochrome P450, xanthine oxidase, respiratory chain) of the Fenton reaction, catalase, peroxidase, and superoxide dismutase [19].

All researches lead to the same conclusions so far: a) the evidence that the cells have specialized systems to convert ROS into less reactive compounds, and b) if those systems would fail, the ROS could be preexisting compounds for the development of diseases. Thus, several researchers agree in the relevance of "oxidative stress" in medical problems, specifically in pathogenesis and/or complications of chronic degenerative diseases [9,12,20,21].

Different observations suggest that these pathologies could be originated when reactive species are formed and suffer alterations, or when they are eliminated, or both. However, the situation is real and much more complicated, for it is difficult to determine the crucial event that originates this disease due to the diversity of forms of oxidative stress (Figure 1). Different researches indicate that mutations produced in genes are responsible of the metabolic unbalance of ROS, while others suggest that environmental changes and common habits weigh on human metabolic processes. However, doubt remains, if oxidative stress is the primary event that leads to the disease or the oxidative phenomenon is developed throughout the disease [22].

Whatever the means by which oxidative stress is induced and pathology is developed, the majority of evidences coincide in the relevance of alterations or enzyme deficiencies. These deficiencies are often caused by mutations in genes coding antioxidant or related enzymes, for example, by genetic polymorphism.

This concept is frequently related to large number of pathologies. Enzymes involved in defence against ROS are not an exception. All enzymes contributing to antioxidant defence can be classified to really antioxidant ones, dealing directly with ROS as substrates, and auxiliary ones. The latter enzymes respond for reparation or degradation of oxidatively modified molecules, maturation and posttranslational modification of antioxidant enzymes and metabolism of low molecular mass antioxidants. As a rule, genetic polymorphisms of enzymes may lead to oxidative stress and consequent diseases, among which cancer, neurodegeneration, cardiovascular disorders, and diabetes are most frequently mentioned. Among the most studied enzymes with genetic polymorphism is the glucose-6-phosphate dehydrogenase, catalase, superoxide dismutase, glutathione peroxidase and those involved in reparation of oxidized molecules and the disease progression [22].

3.1. Glucose-6-phosphate dehydrogenase deficiency

The most striking example among polymorphisms of genes coding enzymes related to antioxidant defence is well-known deficiency in glucose-6-phosphate dehydrogenase (G6PDH) which leads to favism; genetic disease characterized by the lysis of erythrocytes when consumed broad beans and other substances which are harmless to the general population [23].

Other pathologies which are related to the same deficiency are diabetes [24,25], vascular diseases [24], and cancer [26]. In these cases, oxidative stress is induced in specific cells; it was shown that GSH may react with superoxide anion radical providing partial defence against this ROS [27]. When decreasing GSH concentration in G6PDH-deficient individuals enhances their sensitivity to redox-active compounds, producing superoxide. Superoxide is able to react also with nitric oxide, leading to the formation of rather harmful oxidant peroxynitrite. However, relation of this reaction to diabetes and vascular diseases is not because of peroxynitrite production and subsequent oxidative damage, but rather because of decrease in nitric oxide level [28].

The latter is an important second messenger in certain signalling pathways particularly related to vasodilation [29]. There is some probability also that individuals with G6PDH-deficiency may fail to regulate properly blood pressure [30]. Despite possible impairment in nitric oxide production, there is also other way to connect G6PDH deficiency with vascular diseases. It is known, that development of vascular diseases depends on the levels of homocysteine and folate, intermediates in metabolism of sulfur-containing amino acids [31]. Production of two these metabolites depends on GSH and NADPH levels in cells [32].

Data regarding association of G6PDH deficiency with cancer are controversial, because some studies demonstrated that G6PDH-deficient patients may additionally suffer from cancer [33], while others state opposite [34]. Nevertheless, both situations are possible. In particular, there is a large data body indicating that different cancer types are developed at increased DNA damage. It often happens under polymorphism in enzymes contributing to DNA repair, what will be discussed below.

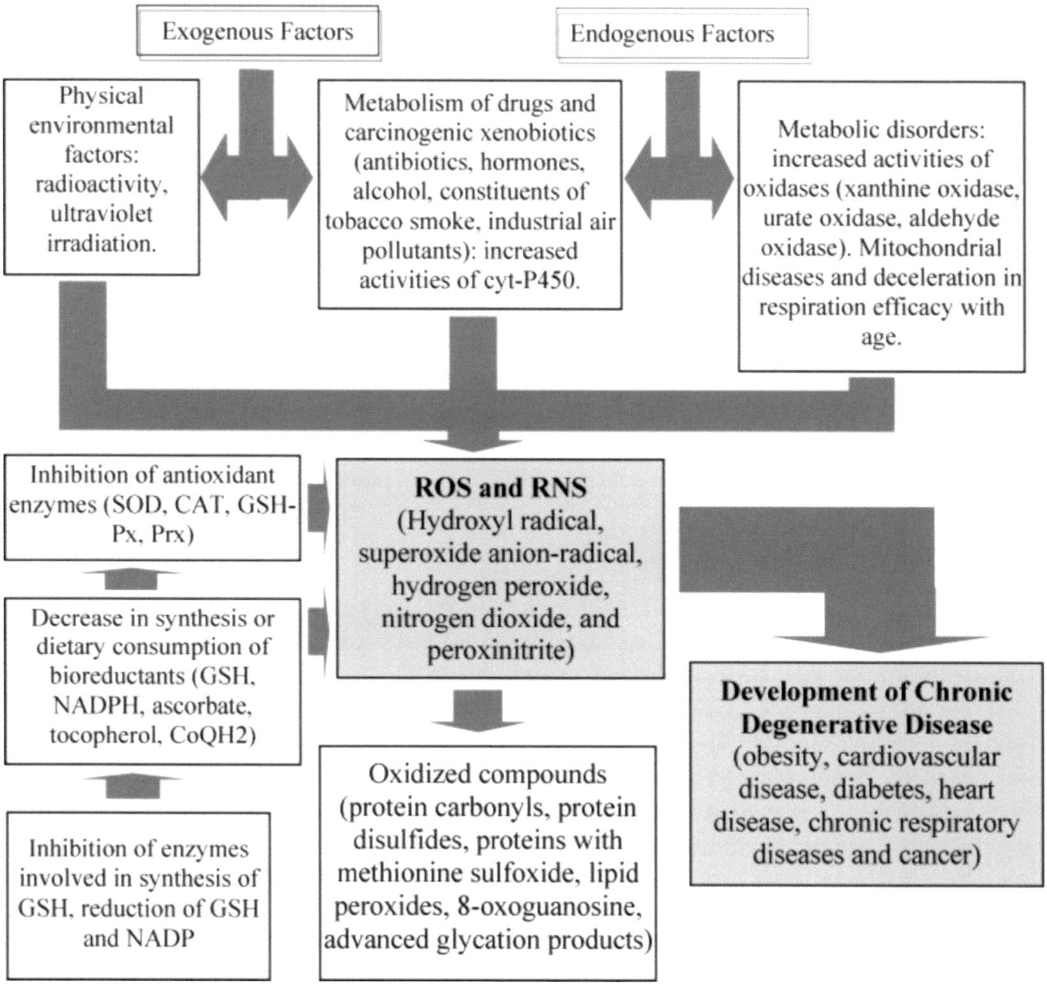

Figure 1. Scheme of the different ways that produce oxidative stress and stimulate the development of chronic degenerative diseases. *Abbreviations*: ROS – reactive oxygen species, RNS – reactive nitrogen species, SOD – superoxide dismutase, CAT - catalase, GSH-Px – glutathione-peroxidase, Prx - peroxyredoxin, GSH – reduced glutathione, CoQH2 – ubiquinol.

On the other hand, NADPH supply at certain conditions may be even harmful leading to enhanced oxidative damage and cancer development. Indeed, it was shown that G6PDH was particularly responsible for cell growth and frequently correlated with cell growth [26]. Tian and colleagues (1998) found that cancer cells possessed several times higher G6PDH activity. The positive correlation between tumour progression and G6PDH activity was found also for humans [35,36].

Increased NADPH supply resulting from G6PDH overexpression can lead to so-called "reductive stress" [37]. Enhanced activity of G6PDH, a lipogenic enzyme, was found at diabetes and obesity [38]. In humans, G6PDH is regulated by many transcription factors, in particular, SREBP-1a (sterol regulatory element binding protein) [39], AP-1 [40] and Sp1 [41]. It was shown that elevation of G6PDH activity might lead to enhanced lipid synthesis [42] and to possible reductive stress [43].

3.2. Catalase deficiency

The first case of catalase deficiency was described by Shigeo Takahara (1947) in a child with cold sores and called acatalasemia to the patology [44]. The cause of this patology is related to ability of oral Streptococci to produce hydrogen peroxide which may promote death of mouth mucosa cells in acatalasemic patients [45]. Catalase deficiency is also associated with diabetes mellitus [46]. This association is attributed for Hungarian hypocatalasemic patients. They were shown to possess higher levels of homocysteine and lower levels of folate [32]. It hints, on one hand, to abnormalities of sulfur metabolism, but on the other hand, it is commonly known that higher homocysteine levels are related to cardiovascular diseases [47], the fact we mentioned above in the context of G6PDH deficiency.

3.3. Polymorphism of Cu,Zn-SOD and protein aggregation

In recent years, the main attention has focused on the polymorphism of genes coding the enzyme superoxide dismutase. More than 100 nucleotide substitutions for the gene SOD1 coding human cytosolic copper- and zinc containing SOD (Cu,Zn-SOD) were described [48]. It is known that several mutations in SOD1 gene are associated with cases of familial amyotrophic lateral sclerosis (ALS), a neurodegenerative disease which is characterized by paralysis and subsequent death [49]. Mechanisms of the disease development are still unknown, but there are many evidences that oxidative stress, developed in neurons, is rather caused by unexpected pro-oxidative activity of SOD than by the loss of the activity at all [50]. It was found that the aggregates cause harm to the cells not only via oxidative stress, but also via inhibition of glutamate receptors [51] and induction of apoptosis [52].

Irwin Fridovich presented some examples of unusual activities of SOD, such as oxidase-like or reductase-like ones [53]. His works and data of other authors suggest that SOD, being mutated or placed in specific conditions, may produce more harmful ROS tan hydrogen peroxide, i.e. hydroxyl radical [54, 55]. Some studies suggested that SOD aggregation can be triggered by higher susceptibility to oxidation of mutated protein [56,57]. Indeed, Cu,Zn-SOD is considered to be rather stable, resistant to many, deleterious to other proteins, compounds [48]. These evidences suggest that Alzheimer, Huntington, and Parkinson diseases are other pathologies related to this enzymatic alteration [22].

3.4. Polymorphism of Mn-SOD, extracellular SOD and glutathione peroxidase

Unlike Cu,Zn-SOD, less mutations were found in the gene coding human manganese containing superoxide dismutase (SOD2). Substitution of alanine-16 to valine (so called "Ala variant") is the most known mutation [58]. This mutation has recently been associated with cancers of breast, prostate, ovaries and bladder, as well as non-Hodgkin lymphoma, mesothelioma and hepatic carcinoma [58]. Mammals possess also extracellular Cu,Zn-SOD (EC-

SOD) encoded in humans by gene SOD3. The enzyme is a homotetramer presenting in plasma, lymph, and synovial fluid [59]. Extracellular SOD is abundant particularly in the lung, blood vessels, and the heart. Consequently, polymorphism of SOD3 gene is associated with pulmonary and cardiovascular diseases [60].

Polymorphism of glutathione peroxidase (GSH-Px) was found to be associated with some cancers. Four GSH-Px isoforms have been described in humans. It was found that mutations in exon 1 of human GSH-Px-1 gene lead to appearance of polyalanine tract at N-terminus of the protein [59]. These tracts themselves are not connected with diminished enzyme activity. Another polymorphism, substitution of proline-198 to leucine, was found in Japanese diabetic patients and associated with intima-media thickness of carotid arteries [61]. The same substitution for adjacent proline-197 was associated with lung and breast cancers, as well as with cardiovascular diseases [59].

3.5. Polymorphism of enzymes involved in reparation of oxidized molecules

Mutations may also affect enzymes involved in DNA reparation. The enzyme 8-hydroxy-2'-deoxyguanosine glycosylase (hOGG) encoded in human genome by the gene hOGG1 is probably the most known example. Recent studies associate mutations in hOGG1 with different cancer types, such as lung, stomach and bladder cancers [62]. Most of the mutations in this gene affect exon 7 and cause serine-to-cysteine substitution. It was demonstrated that substitution S326C in hOGG1 protein confers susceptibility to oxidation and makes the enzyme prone to form disulfide bond between different polypeptide chains [63].

Hydrolase MTH1 is other important enzyme preventing incorporation of oxidized purine nucleotide triphosphates in DNA [64]. Knockout of this enzyme in mice resulted in increased frequency of lung, stomach and liver tumours with age [65].

Other important antioxidant enzymes are glutathione S-transferases (GSTs). Its main function is to conjugation of different electrophilic compounds with glutathione [66]. Oxidatively modified compounds as well as lipid oxidation products, like 4-hydroxy-2-nonenal, are subjected to conjugation with glutathione. In general, GSTs are belong to xenobiotic-elimitating system. Some of them, namely GSTs of μ class, are known well by their ability to eliminate polycyclic aromatic hydrocarbons, oxidized previously by cytochrome P450 monooxygenases. To date, eight classes of GSTs have been described: α, κ, μ,σ, ξ, π, θ, and ω. Cytosolic enzymes belong to classes α, μ, π and θ [67]. The gene coding GSTM1 (GST of μ class, isoform 1) is appeared to be highly polymorphic and found inactivated in half of human population.

Some studies associate polymorphism of GSTM1 with lung cancer [59,68], although reports are controversial. For example, meta-analysis conducted by [69] found no association of GSTM1 null genotype with lung cancers as well as with smoking. Other authors found such association and reported increased susceptibility to cancerogens among Caucasian and African-American populations [70]. Polymorphism of GSTM1 was also found to be associated with head and neck carcinomas [67]. The need in GSTM1 and its role in prevention of lung cancer are explained by the ability of the enzyme to detoxify constituents of cigarette smoke, such as mentioned above polycyclic aromatic hydrocarbons. Some studies also associate lung cancer with polymorphism of GSTT1 (GST of θ class) which participates in catabolism of tobacco smoke constituents, such as halomethanes and butadione [70].

3.6. Role of oxidative modifications of antioxidant and related enzymes in disease progression

Many disorders related to the metabolism of transition metals, amino acids or low molecular mass reductants are known to be connected with activities of antioxidant enzymes. Particularly, impairement in selenium uptake or synthesis of selenocysteine needed for glutathione peroxidase may lead to GSH-Px deficiency and subsequent disorders such as cardiovascular ones [47]. Disruption of iron-sulfur clusters by superoxide anion radicals or peroxynitrite leads frequently to impairment of many metabolic pathways. Indeed, aconitase, NADH-ubiquinone-oxidoreductase (complex I of mitochondrial electron transport chain), ubiquinol-cytochrome c oxidoreductase (complex III), ribonucleotide reductase, ferredoxins possess iron-sulfur clusters, susceptible to oxidation. Owing to this, aconitase is used as one of oxidative stress markers [71]. On the other hand, iron is a component of haem, a prosthetic group in catalase holoenzyme. Susceptibility to oxidative modification is described for catalase, glutathione peroxidase, Cu,Zn-SOD, and G6PDH. The latter is believed to be one of the most susceptible to oxidation enzymes [22]. Thus, oxidative stress induced by exogenous factors, like carcinogens, certain drugs, ions of transition metals, etc., or by metabolic disorders, like diabetes, can be exacerbated by oxidative modification of antioxidant enzymes. These assumptions demonstrate the potential of antioxidant therapy in particular cases. At some pathological states, whatever the cause of the disease, oxidative stress is seen to be a powerful exacerbating factor. Type II diabetes, cardiovascular diseases and neurodegenerative diseases, associated with protein aggregation are among such pathologies. Indeed, enhanced level of glucose results in higher probability of protein glycation [72].

4. Impact of chemopreventive agents in the chronic degenerative diseases

The available evidences indicates that individuals with chronic degenerative diseases are more susceptible to oxidative stress and damage because they have elevated levels of oxidants and/or reduced antioxidants. Therefore, it has been posited that antioxidant supplementation in such individuals may be beneficial. Different research has confirmed that many common foods contain nonnutritive components that may provide protection against chronic degenerative diseases, however, the most studies have had impact on the cancer [20,21].

The "chemoprevention" seeks to eliminate precancerous cells in order to avoid the necessity of chemotherapy. It can be further classified as primary, secondary, or tertiary prevention. Primary chemoprevention focuses on preventing the development of precancerous lesions, secondary chemoprevention focuses on preventing the progression of these lesions to cancer, and tertiary chemoprevention aims to prevent the recurrence or spread of a primary cancer [73].

It has been known for some time that dietary factors play a role in the development of some human cancers [73,74] and that some foods contain mutagens and carcinogens [74,75]. Investigations of last decades, has focused on the existence of a number of non nutritional components in our regular diet that possess antimutagenic and anticarcinogenic properties, these compounds have been called as chemopreventers [76,77].

The chemopreventers are classified as food entities that can prevent the appearance of some long-term diseases like cancer or cardiovascular disorders. It has been suggested that chemoprevention should be considered as an inexpensive, easily applicable approach to cancer control and "may become a major weapon in the anticancer arsenal" [76,78,79]. These compounds can be found in all food categories, but mainly in fruits, vegetables, grains and tea [78,79]. Chemopreventers belong to different classes of chemicals but the most recognized are some vitamins, food polyphenols, flavonoids, catechins, and some components in spices [78, 79].

The mechanisms of action of the chemopreventers are complex and can be categorized according to the site of action or by the specific type of action. It appears that most chemopreventers act primarily as antioxidants. As such, they may scavenge free radicals formed during the preparation of food or as a normal biological process in the body. Recall, that the free radicals can react with DNA, lipids, or cell membranes, leading to aging, injuries of the organ, and greater susceptibility to develop the chronic degenerative disease. Therefore, any event that removes free radicals in the human body is considered beneficial for human health. In addition to their antioxidative activities, there are other mechanisms that show in the Table 2 [80-82].

5. Chemopreventive evidence of some fruits and food supplements evaluated by our research group

5.1. Cactus pears

Plants from the genus *Opuntia* are the most abundant of the Cactaceae family, grown throughout the Americas as well as the central area of the Mediterranean, Europe, Asia, Africa, and Australia. *Opuntia* species display flattened stems called "pencas" or cladodes. The cactus pear (also called prickly pears) is the fruit of this plant (*Opuntia* spp.). The fruit is an oval berry with a large number of seeds and a semi-hard rind with thorns, which may be grouped by fruit colors: red, purple, orange-yellow and white. The fruits with white pulp and green rind are preferred for consumption as food, and their domestic production corresponds to almost 95% of the total production. Mexico is the main producer of cactus pears and accounts for more than 45% of the worldwide production; however, only 1.5% of this production is exported [83,84].

Mechanism	Action	Examples
Inhibition of carcinogen formation	Agents that block or inhibit to the enzymes responsible for the biotransformation of procarcinogens to carcinogen form	Dithiocarbamates, isothiocyanates, diallyl sulfide, and ellagic acid
Inducing agents	Agents that induce or enhance enzyme activity (e.g., glutathione S-transferase, GST) for detoxify and reduce the level of mutagenic/carcinogenic species of the body	Isothiocyanates, sulfaraphane, d-limonene, terpinoids, turmeric, and curcurains
Suppressing agents	Agents that may react on different processes (e.g., inhibition of arachidonic acid metabolism, activity of protease or protein kinase C) involved in tumor promotion/progression	isoflavones, phytoestrogens, selenium
Immune activity and modulation	Since the immune system can influence on growth either via effects on the inflammation status or by causing apoptosis. Some chemopreventers can act on the early stages in neoplasia or have effects on frank malignancies	Carotenoid, flavonoids, lactoferrin
Signal transduction pathways and their regulation	Some chemopreventers may alter signaling pathways of receptors for hormones and others factors responsible for cell regulation and can be modified the potential for growth, either by acting to increase mitosis or alter the level of apoptosis.	d-limonene, sulfur compounds, lactoferrin, retinoids

Table 2. Other mechanisms of chemoprevention

A viable strategy to increase the competitiveness of the Mexican cactus pear in national and international markets is the innovation and creation of new high value-added products. This could be achieved by determining the nutritional and functional properties that differentiate the Mexican cactus pear from analogous products. In addition, providing functional products for a market in constant growth would offer a key competitive advantage and would allow the producers to diversify its commercialization, not as fresh fruit only, but also as an ingredient or high-value additive for the food industry. A commercialization of the cactus-pear based on its antioxidant properties could generate competitive advantages that may turn into business opportunities and the development of new products [85].

Different studies with the varieties of European and Asian cactus pears have shown notable antioxidant activities that significantly reduce oxidative stress in patients and may help in preventing chronic pathologies (as diabetes and cancer) [85-87]. For this reason, the cactus pear is considered a functional food; this feature is attributed to its bioactive compounds such as vitamin C and vitamin E, polyphenols, carotenoids, flavonoid compounds (e.g., kaempferol, quercetin, and isorhamnetin), taurine and pigments [88,89].

Betalains are water-soluble pigments. Two betalain derivatives are present in cactus-pears: betacyanin, which gives the red-purple color, and betaxanthin, which gives a yellow-orange color. These pigments have shown beneficial effects on the redox-regulated pathways involved in cell growth and inflammation, and have not shown toxic effects in humans [90,91].

In addition, a neuroprotector activity against oxidative damage induced in cultures of rat cortical cells has been attributed to the cactus pear flavonoids [92]. Another beneficial effect of the fruit was observed in the prevention of stomach ulcers through the stimulation of prostaglandin production: cactus pear promoted mucous secretion of bicarbonate, involved in the protection of gastric mucosa [93]. On the other hand, their contents of natural antioxidants has raised interest in the use of cactus pears as substitute for synthetic antioxidants, such as butylhydroxytoluene (BHT), butylhydroxyanisole (BHA) [88].

In the Institute of Health Sciences (Autonomous University of Hidalgo State) have been performed studies to demostrate the chemopreventive capacity of the cactus pear. The first studies were developed by Hernández-Ceruelos et al. (2009) with the main objective to evaluate the antioxidant effect of three varieties of prickly pear juice (red-purple, white-green and yellow-orange) in four different concentrations (25, 50, 75 and 100%) by the technique of DPPH (1,1-diphenyl-2-picrylhydrazyl). Their results indicated that the juice of princkly pear variety red-purple (PPRP) had the highest antioxidant capacity in all concentrations in comparison with the positive control (vitamin E). Subsequently, researchers evaluated the anticlastogenic potential of PPRP by micronucleus assay against of methyl methane sulfonate (MMS) in mice. This experiment had a duration of 2 weeks, was included a negative control (animals treated with water), a positive control of MMS (40 mg/kg), a group of mice treated with princkly pear variety red-purple (25mL/Kg), and three groups with PPRP (in doses of 25, 16.5 and 8.3 mL/Kg) plus the mutagen. The PPRP was administered by oral gavage and the mutagen was injected intraperitoneally 5 days before the end of the experiment (single dose). Finally, blood samples were obtained in four times (0, 24, 48 and 72 hours) to determine the frequency of micronucleated polychromatic erythrocytes (MNPE). The result indicated that PPRP is not a genotoxic agent, on the contrary, may reduce the number of micronucleated polychromatic erythrocytes. In this regard, the princkly pear variety red-purple showed an anticlastogenic effect directly proportional to the concentrations. The highest protection was obtained with the concentration of 25 mL/Kg (approximately, 80%) after 48 hours of treatment [94].

In the second study was evaluated the antioxidant activities [with three assays: a)1,1-diphenyl-2-picrylhydrazyl radical-scavenging, b) protection against oxidation of a β-carotene-linoleic acid emulsion, and c) iron(II) chelation], the content of total phenolic compounds, ascorbic acid, betacyanin, betaxanthins and the stability of betacyanin pigments in presence of Cu(II)-dependent hydroxyl radicals ($OH^{\bullet}$), in 18 cultivars of purple, red, yellow and white cactus pear from six Mexican states (Hidalgo, Puebla, Guanajuato, Jalisco, Zacatecas and the State of Mexico). The results indicated that the antiradical activities from yellow and white cactus pear cultivars were not significantly different and were lower than the average antiradical activities in red and purple cultivars. The red cactus pear from the state of Zacatecas showed the highest antioxidant activity. The free radical scavenging activity for red cactus pears was significantly correlated to the concentration of total phenolic compounds ($R2$ = 0.90) and ascorbic acid ($R2$ = 0.86). All 18 cultivars of cactus pears studied showed significant chelating activity of ferrous ions. The red and purple cactus pears showed a great stability when exposed to $OH^{\bullet}$ [88].

5.2. Cranberries

Among small soft-fleshed colorful fruits, berries make up the largest proportion consumed in our diet. Berry fruits are popularly consumed not only in fresh and frozen forms, but also as processed and derived products including canned fruits, yogurts, beverages, jams, and jellies. In addition, there has been a growing trend in the intake of berry extracts as ingredients in functional foods and dietary supplements, which may or may not be combined with other colorful fruits, vegetables, and herbal extracts [95]. Berry fruits commonly consumed in America include blackberries, black raspberries, red raspberries and strawberries, blueberries, and cranberries.

Other "niche-cultivated" berries and forest/wild berries, for example, bilberries, black currant, lingonberry, and cloudberry, are also popularly consumed in other regions of the World [95]. The North American cranberry (*Vaccinium macrocarpon*) is of a growing public interest as a functional food because of potential health benefits linked to phytochemicals of the fruit. Cranberry juice has long been consumed for the prevention of urinary tract infections, and research linked this property to the ability of cranberry proanthocyanidins to inhibit the adhesion of *Escherichia coli* bacteria responsible for these infections [96]. These studies, which brought to light the unique structural features of cranberry proanthocyanidins [97], have sparked numerous clinical studies probing a cranberry´s role in the prevention of urinary tract infections and targeted the nature of the active metabolites. Further antibacterial adhesion studies demonstrated that cranberry constituents also inhibit the adhesion of *Helicobacter pylori*, a major cause of gastric cancer, to human gastric mucus [98]. The earliest report of potential anti-carcinogenic activity appeared in 1996 in the University of Illinois [99].

Extracts of cranberry and bilberry were observed to inhibit ornithine decarboxylase (ODC) expression and induce the xenobiotic detoxification enzyme quinonereductase *in vitro* [99]. Subsequent studies with cranberry and other berries in cellular models have focused on some cancers such as breast, colon, liver, prostate and lung [100-102]. This biological activity of berries are partially attributed to their high content of a diverse range of phytochemicals such as flavonoids (anthocyanins, flavonols, and flavanols), tannins (proanthocyanidins, ellagitannins, and gallotannins), quercetin, phenolic acids, lignans, and stilbenoids (e.g., resveratrol) [100]. With respect to his genotoxic and/or antigenotoxic potential, there are few reports in the literature that demonstrate this effect and the majority of studies were performed *in vitro* cell culture models [101,103,104]. Boateng *et al.* demonstrated that consumption of some juices of berries (as blueberries, blackberries, and cranberry) can reduce the formation of aberrant crypt foci (ACF) induced by azoxymethane in Fisher male rats [105]. Another study, in which it was administrated a lyophilized extract of *Vaccinium ashei* berries in male Swiss mice during 30 days, showed to have improved the performance on memory tasks and has a protective effect on the DNA damage in brain tissue evaluated with the comet assay [106].

Although the types of berry fruits consumed worldwide are many, the experiment executed in our laboratory is focuses on cranberries that are commonly consumed in Mexico, especially in the states of Tlaxcala, Hidalgo, and Puebla. The purpose of our study was to determine whether cranberry ethanolic extract (CEE) can prevent the DNA damage produced by benzo[a]pyrene (B[a]P) using an in vivo mouse peripheral blood micronucleus assay. The experimental groups were organized as follows: a negative control group (without treatment), a positive group treated with B[a]P (200 mg/kg), a group administered with 800 mg/kg of cranberry ethanolic extract, and three groups treated with B[a]P and cranberry ethanolic extract (200, 400, and 800 mg/kg) respectively. The CEE and benzo[a]pyrene were administered orally for a week, on a daily basis. During this period the body weight, the feed intake, and the determination of antigenotoxic potential were quantified. At the end of this period, we continued with the same determinations for one week more (recovery period) but anymore administration of the substances. The animals treated with B[a]P showed a weight increase after the first week of administration. The same phenomenon was observed in the lots combined with B[a]P and CEE (low and medium doses). The dose of 800 mg/kg of CEE showed similar values to the control group at the end of the treatment period. In the second part of the assay, when the substances were not administered, these experimental groups regained their normal weight. The dose of CEE (800 mg/kg) was not genotoxic nor cytotoxic. On the contrary, the B[a]P increases the frequency of micronucleated normochromatic erythrocytes (MNNE) and reduces the rate of polychromatic erythrocytes (PE) at the end of the treatment period. With respect to the combined lots, a significant decrease in the MN rate was observed from the sixth to the eighth day of treatment with the two high doses applied; the highest protection (60%) was obtained with 800 mg/kg of CEE. The same dose showed an anticytotoxic effect which corresponded to an improvement of 62.5% in relation to the animals administered with the B[a]P. In the second period, all groups reached values that have been seen in the control group animals. Our results suggest that the inhibition of clastogenicity of the cranberry ethanolic extract against B[a]P is related to the antioxidant capacity of the combination of phytochemicals present in its chemical composition [107].

5.3. Grapefruit juice and naringin

The grapefruit is a subtropical citrus tree known for its bitter fruit. These evergreen trees usually grow around 6 meters tall. The leaves are dark green, long and thin. His fruit (called toronja in Spanish) has become popular since the late 19th century, is yellow-orange skinned and largely an oblate spheroid and generally, is consumed in form of juice [108].

The grapefruit juice is an excellent source of many nutrients and phytochemicals that contribute to a healthy diet. Is a good source of vitamin C, contains the fiber pectin, and the varieties pink and red contain the beneficial antioxidant lycopene [108]. But, the main flavonoid, existing in highest concentration in grapefruit juice is naringin, which in humans is metabolized to naringenin [109].

Since grapefruit juice is known to inhibit enzymes necessary for the clearance of some drugs and hormones, some researchers have hypothesized that grapefruit juice and the naringin may play an indirect role in the development of hormone-dependent cancers. A study found a correlation between eating a quarter of grapefruit daily and a 30% increase in risk for breast cancer in post-menopausal women. The study points to the inhibition of CYP3A4 enzyme by grapefruit, which metabolizes estrogen [110]. However, an investigation conducted in 2008 has shown that grapefruit consumption does not increase breast cancer risk and found a significant decrease in breast cancer risk with greater intake of grapefruit in women who never used hormone therapy [111].

In the case of naringin, this compound exerts a variety of pharmacological effects such as antioxidant activity, blood lipid lowering, anticancer activity, and inhibition of selected drug-metabolizing cytochrome P450 enzymes, including CYP3A4 and CYP1A2, which may result in drug-drug interactions in vivo. Ingestion of naringin and related flavonoids can also affect the intestinal absorption of certain drugs, leading to either an increase or decrease in circulating drug levels [112].

This evidence has motivated to our research group to develop various studies with grapefruit juice (GJ) and the naringin (Nar) to assess his chemoprotective ability.

Our first experience was with naringin in 2001. On that occasion, the study was designed for three main purposes: (1) to determine whether Nar has a genotoxic effect in mouse in vivo. This was evaluated by measuring the rate of micronucleated polychromatic erythrocytes (MNPE); (2) to determine its antigenotoxic and its anticytotoxic potential on the damage produced by ifosfamide. The first study was done by scoring the rate of MNPE, and the second one by establishing the index polychromatic erythrocytes/normochromatic erythrocytes (PE/NE); and (3) to explore whether its antigenotoxic mechanism of action is related to an inhibitory effect of Nar on the expression of the CYP3A enzyme, an effect which could avoid the biotransformation of ifosfamide.

A single oral administration was used for all groups in the experiment: three groups were given different doses of Nar (50, 250, and 500 mg/kg), other groups received the same doses of Nar plus an administration of ifosfamide (60 mg/kg), another group treated with distilled water and another with ifosfamide (60 mg/kg) were used as negative and positive controls, respectively. The micronuclei and the cell scoring were made in blood samples taken from the tail of the animals at 0, 24, 48, 72, and 96 h. The results showed that Nar was neither genotoxic nor cytotoxic with the doses tested, but ifosfamide produced an increase in the rate of MNPE at 24 and 48 h. The highest value was 24+/-1.57 MNPE per thousand cells at 48 h. The index PE/NE was significantly reduced by ifosfamide at 24 and 48 h. Concerning the antigenotoxic capacity of Nar, a significant decrease was observed in the MNPE produced by ifosfamide at the three tested doses. This effect was dose-dependent, showing the highest reduction in MNPE frequency (54.2%) at 48 h with 500 mg/kg of Nar. However, no protection on the cytotoxicity produced by ifosfamide was observed. Immunoblot analysis was used to assess the CYP3A expression in liver and intestinal microsomes from mouse exposed orally to Nar. An induction in the CYP3A protein was observed in both intestinal and hepatic microsomes from treated mice. This induction correlated with an increase in erythromycin N-demethylase activity. These data suggest that other mechanism(s) are involved in the antigenotoxic action of naringin [113].

With regard to grapefruit juice (GJ), we performed two experiments which are summarized below. The first evaluated the capacity of GJ to inhibit the micronucleated polychromatic erythrocytes (MNPE) produced by daunorubicin in an acute assay in mice, as well as to determine its antioxidant potential in mouse hepatic microsomes, and its capacity to trap free radicals in vitro.

The results showed that GJ is not toxic or genotoxic damage; on the contrary, it generated a significant reduction of the MNPE formed by daunorubicin. The effect was found throughout the examined schedule (from 24 to 96 h). The two high doses produced inhibition of about 60% at 48 h, 86% at 72 h and 100% at 96 h after the treatment. With respect to the grapefruit juice antioxidant potential, a 50% decrease in liver microsomal lipid peroxidation produced by daunorubicin was found by quantifying malondialdehyde formation. Finally, a strong GJ scavenging activity evaluated with the 1,1-diphenyl-2-picryl-hydrazyl (DPPH) was observed, giving rise to a concentration-dependent curve with a correlation coefficient of 0.98. Overall, our results established an efficient anticlastogenic potential of grapefruit juice, probably related to its antioxidant capacity, or to alterations of daunorubicin metabolism [114].

Based on this background; recently, we finished another study in which using the comet assay was demonstrated a strong effect by hydrogen peroxide (HP) and no damage by grapefruit juice (GJ) in human lymphocytes. Cells exposed to HP and treated with GJ was shown an increase of DNA damage by HP over the control level, and a decrease of such damage by GJ. With the comet assay plus formamidopyrimidine-DNA-glycosylase we found the strongest increase of DNA damage by HP over the control level, and the strongest reduction of such damage by GJ. By applying the comet/FISH method we determined 98% of the p53 gene signals in the comet head of control cells along the experiment, in contrast with about 90% signals in the comet tail of cells exposed to HP. Cells treated with both agents showed a significant, concentration/time dependent return of p53 signals to the head, suggesting enhancement of the gene repair. Finally, with the annexin V assay we found an increase in apoptosis and necrosis by HP, and no effect by GJ; when GJ was added to HP treated cells no modification was observed in regard to apoptosis, although a decrease of necrosis was observed [115].

5.4. Chamomile

Chamomile (*Matricaria chamomilla* or *Chamomilla recutita*) is an asteraceae plant native to Europe and distributed around the world, except in tropical and polar regions. This plant has been used for its curative properties since ancient Egyptian and Greek times, and at present is frequently used as an antiseptic, antiflogistic, diuretic, expectorant, febrifuge, sedative, anti-inflammatory and anticarcinogen [116]. Pharmacological activities of various components of the plant have been reported, for example, the anti-inflammatory capacity and the modulating effects of the heat shock protein on apigenin and quercetin flavonoids, as well as the anti-inflammatory, antioxidant, and antiseptic activities detected on α-bisabolol, guargazulene, and chamazulene [117, 118]. The essential oil extracted from the chamomile flower varies from 0.42 to 2%, and consists of compounds such as bisabolol, chamazulene, cyclic sesquiterpenes, bisabolol oxides, and other azulenes and terpenes [119].

With respect to his genotoxic and/or antigenotoxic potential, there are few reports in the literature that demonstrate this effect. Therefore, our laboratory performed two investigations with the main purpose to evaluate the chemoprotection capacity of chamomile. Initially, we obtained the chamomile essential oil (CEO) from flowers of *Chamomilla recutita* by steam distillation, and then it was analyzed by gas chromatography to identify the chemical species. Thirteen compounds were determined with this assay, including bisabolol and its oxides, β-farnecene, chamazulene, germacrene, and sesquiterpenes (Table 3).

Compound	**RT**[a]	**Area (%)**
(*E*)-β-Farnecene	38.46	28.17
Germacrene-D	39.23	2.19
Unidentified sesquiterpene	40.07	1.40
Unidentified sesquiterpene	41.17	0.78
(Z,E)–α–Farnecene	41.35	1.59
Unidentifiedsesquiterpene	48.52	0.71
α–Bisabolol oxide A	54.46	41.77
α–Bisabolol oxide B	49.28	4.31
α-Bisabolol oxide	50.65	5.30
α–Bisabolol	51.18	2.31
Chamazulene	52.80	2.39
1,6-Dioxaspiro[4,4]non-3-ene,2-(2,4hexadyn-1-ylidene)	60.73	2.19
Hexatriacontane	67.49	0.50

RT[a], Retention time obtained with gas chromatography.

Table 3. Components of the tested chamomile essential oil

The first work was to determine the inhibitory effect of the CEO, on the sister chromatid exchanges (SCEs) produced by daunorubicin and methyl methanesulfonate (MMS) in mouse bone marrow cells.

The authors performed a toxic and genotoxic assay of chamomile essential oil; both showed negative results. To determine whether CEO can inhibit the mutagenic effects induced by daunorubicin, one group of mice was administered corn oil, another group was treated with the mutagen (10 mg/kg), a third group was treated with 500 mg/kg of CEO; three other groups were treated first with CEO (5, 50 and 500 mg/kg) and then with 10 mg/kg of daunorubicin. In the case of MMS, the experimental groups consisted of the following: the negative control group which was administered corn oil, a group treated with 25 mg/kg of MMS, a group treated with 1000 mg/kg of CEO, and three groups treated first with CEO (250, 500 and 1000 mg/kg) and then with MMS (25 mg/kg). The results indicated a dose-dependent inhibitory effect on the SCEs formed by both mutagens. In the case of daunorubicin, a statistically significant result was observed in the three tested doses: from the lowest to the highest dose, the inhibitory values corresponded to 25.7, 63.1 and 75.5%. No alterations were found with respect to the cellular proliferation kinetics, but a reduction in the mitotic index was detected. As regards MMS, the inhibitory values were 24.8, 45.8 and 60.6%; no alterations were found in either the cellular proliferation kinetics or in the mitotic indices [120]. This results suggested that CEO may be an effective antimutagen and was the reason for develop the second study.

The aim of the second investigation was to determine the inhibitory potential of CEO on the genotoxic damage produced by daunorubicin (DAU) in mice germ cells. We evaluated the effect of 5, 50, and 500 mg/kg of essential oil on the rate of sister chromatid exchange (SCE) induced in spermatogonia by 10 mg/kg of the mutagen. We found no genotoxicity of CEO, but detected an inhibition of SCE after the damage induced by DAU; from the lowest to the highest dose of CEO we found an inhibition of 47.5%, 61.9%, and 93.5%, respectively. As a possible mechanism of action, the antioxidant capacity of CEO was determined using the 1,1-diphenyl-2-picrylhydrazyl (DPPH) free radical scavenging method and ferric thiocyanate assays. In the first test we observed a moderate scavenging potential of the oil; nevertheless, the second assay showed an antioxidant capacity similar to that observed with vitamin E. In conclusion, we found that CEO is an efficient chemoprotective agent against the damage induced by DAU in the precursor cells of the germinal line of mice, and that its antioxidant capacity may induce this effect [116].

5.5. Silymarin

Silybum marianum is the scientific name of milk thistle or St. Mary's thistle. It is a Mediterranean native plant belonging to the Asteraceae family. It is characterized by thorny branches, a milky sap, with oval leaves that reach up to 30 centimeters, its flowers are bright pink and can measure up to 8 cm to diameter [121].

Milk thistle (Mt) grows of wild form in the southern Europe, the northern Africa and the Middle East but is cultivated in Hungary, China and South American countries as Argentina, Venezuela and Ecuador. In México, is consumed as supplement food for many years ago [122].

In the sixties years, German scientists performed a chemical investigation of his fruits, isolating a crude extract formed by active compounds with hepatoprotective capacity; this group of compounds was called silymarin. In 1975, it was found that the principal components of silymarin are silybin A, silybin B, isosilybin A, isosilybin B, silychristin A, silychristin B and silydianin [123]. Currently it is known that the chemical constituents of silymarin are flavonolignans, ie, a combination conformed by flavonoids and lignins structures [124].

Mt is one of the most investigated plant extracts with known mechanisms of action for oral treatment of toxic liver damage. Silymarin is used as a protective treatment in acute and chronic liver diseases [125]. His protective capacity is related with different mechanisms as suppress toxin penetration into the hepatic cells, increasing superoxide dismutase activity, increasing glutathione tissue level, inhibition of lipid peroxidation and enhancing hepatocyte protein synthesis. The hepatoprotective activity of silymarin can be explained based on antioxidant properties due to the phenolic nature of flavonolignans. It also acts through stimulating liver cells regeneration and cell membrane stabilization to prevent hepatotoxic agents from entering hepatocytes [126].

Silymarin is also beneficial for reducing the chances for developing certain cancers [127]. The molecular targets of silymarin for cancer prevention have been studied. Milk thistle interfere with the expressions of cell cycle regulators and proteins involved in apoptosis to modulate the imbalance between cell survival and apoptosis. Sy-Cordero et al. (2010) isolated four key flavonolign and diastereoisomers (silybin A, silybin B, isosilybin A and isosilybin B) from *S. marianum* in gram scale. These compounds and other two related analogues, present in extremely minute quantities, were evaluated for antiproliferative/cytotoxic activity against human prostate cancer cell lines. Isosilybin B showed the most potent activity [126]. The isolation of six isomers afforded a preliminary analysis of structure-activity relationship toward prostate cancer prevention. The results suggested that an *ortho* relationship for the hydroxyl and methoxy substituents in silybin A, silybin B, isosilybin A and isosilybin B was more favorable than the *meta* relationship for the same substituents in the minor flavonolignans. Silymarin suppressed UVA-induced oxidative stress that can induce skin damage. Therefore, topical application of silymarin can be a useful strategy for protecting against skin cancer [128].

In our laboratory, we evaluated the antigenotoxic effect of two doses of silymarin (200 and 400 mg/Kg) administered by oral gavage against the chronic consumption of ethanol (solution: 92 mL of water/8 mL of ethanol) during a week with alkaline single cell electrophoresis (comet) assay.

Figure 2 shows the comet measurements obtained in our assay. To summarize, at the 24 hours of the schedule we found no significant DNA damage induced in the control group (only water) and the silymarin group (400 mg/kg), both groups had a mean T/N index of 1.1. On the contrary, the mice (strain CD-1) that consumed the solution of ethanol showed a slight comet increase during this same time. But at 48, 72 and 96 hours, this group showed a T/N index increase of about four times as much. During the last times (120, 144, 168 and 192) there is a decrease of DNA damage, suggesting that hepatocytes are in the process of cell regeneration. With respect to the groups treated with the combination of chemicals, a clear antigenotoxic effect was found with the two doses of silymarin; particularly with 400 mg/kg, the prevention of DNA damage was about 70% during the 48, 72, 96 and 120 hours of treatment. At the end of the experiment, these groups reached similar values to the negative control [129].

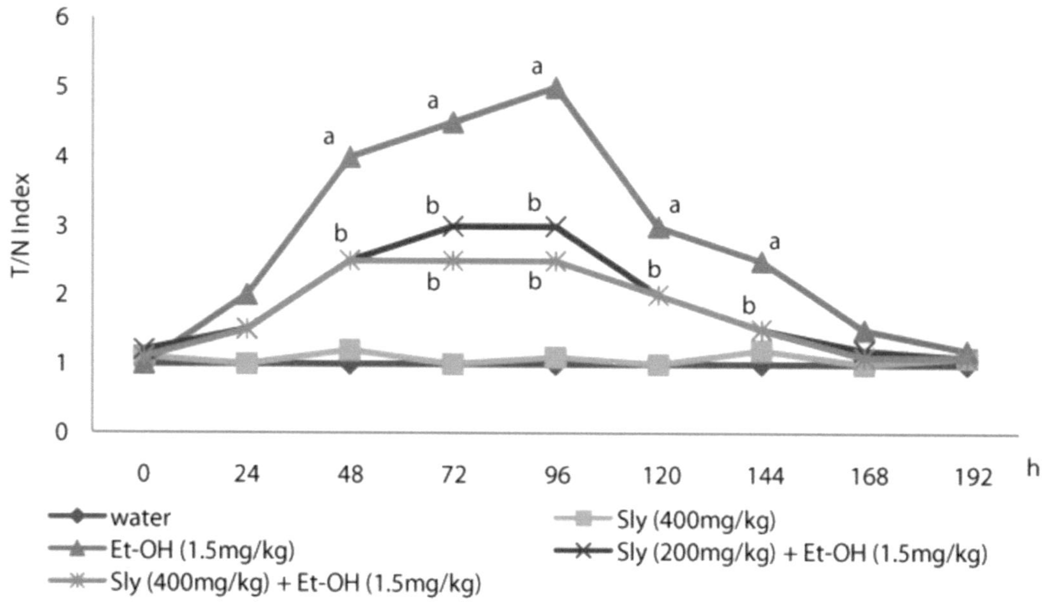

Figure 2. Antigenotoxic effect of silymarin (Sly) against the DNA damage induced by the chronic consumption of etanol (Et-OH). Results are the mean ± SD of 5 mice per group (100 nuclei per doses) [a] statistically significant difference with respect to the value of the control group and, [b] with respect to the value obtained in mice treated with Et-OH only. ANOVA and Student-Newman Keuls tests, $p \leq 0.05$.

Author details

Eduardo Madrigal-Santillán[1], Eduardo Madrigal-Bujaidar[2], Sandra Cruz-Jaime[1], María del Carmen Valadez-Vega[1], María Teresa Sumaya-Martínez[3], Karla Guadalupe Pérez-Ávila[1] and José Antonio Morales-González[1]

1 Instituto de Ciencias de la Salud, UAEH, México

2 Escuela Nacional de Ciencias Biológicas, IPN, México

3 Universidad Autónoma de Nayarit, Tepic, México

References

[1] Doll R. Chronic and degenerative disease: Major causes of morbidity and death. TheAmerican Journal of ClinicalNutrition.1995; 62(6) 1301S-1305S.

[2] Nava-Chapa G., Ortiz-Espinosa RM., Reyes-Gómez D. Epidemiologia de las enfermedades crónico degenerativas. In: Morales-González JA., Fernández-Sánchez AM.,

Bautista-Ávila M., Madrigal-Santillán E. (ed.) Los antioxidantes y las enfermedades crónico degenerativas. México: UAEH; 2009. p269-310.

[3] Ugalde A., Jackson, JT. The World Bank and international health policy: a critical review. Journal of International Development 1995; 7(3) 525-41.

[4] Mejía-Median JI., Hernández-Torres I., Moreno-Aguilera F., Bazan-Castro M. Asociación de factores de riesgo con el descontrol metabólico de diabetes mellitus en pacientes de la clínica oriente del ISSSTE. Revista de Especialidades Médico Quirúrgicas 2007; 12(2) 25-30.

[5] Sies, H. Oxidative stress: Introductory remarks, In: Sies H.(ed.) Oxidative stress. London: Academic Press; 1985. p1-8.

[6] Lushchak VI. Environmentally induced oxidative stress in aquatic animals. Aquatic Toxicology 2011; 101(1) 13-30.

[7] Halliwell B. Antioxidants and human disease: a general introduction. Nutrition Reviews. 1997; 55(1Pt2) S44-S52.

[8] Yu BP. Cellular defenses against damage from reactive oxygen species. Physiological Reviews 1994; 74(1) 139-162.

[9] ChihuailafRH., Contreras PA., Wittwer FG. Pathogenesis of oxidative stress: Consequences and evaluation in animal health. Veterinaria México 2002; 33(3) 265-283.

[10] Halliwell, B. Free radicals and antioxidants-quo vadis? Trends in Pharmacological Sciences 2011; 32(3) 125–130.

[11] Halliwell B, Gutteridge JMC. Oxygen toxicity, oxygen radical, transition metals and disease. The Biochemical Journal 1984; 219(1) 1-14.

[12] Erejuwa OO., Sulaiman SA., AbWahab MS. Honey: a novel antioxidant. Molecules 2012; 17(4) 4400-4423.

[13] Halliwell B., GutteridgeJMC., Cross CE. Free radicals, antioxidants, and human disease: Where are we now?. The Journal of Laboratory and Clinical Medicine 1992; 119(6) 598-620.

[14] Chaudière J., Ferrari-Iliou R. Intracellular antioxidants: from chemical to biochemical mechanisms. Food and Chemical and Toxicology 1999; 37(9-10) 949-962.

[15] Harris ED. Regulation of antioxidant enzymes. FASEB Journal 1992; 6(9) 2675-2683.

[16] Ho YS., Magnenat JL., Gargano M., Cao J. The nature of antioxidant defense mechanism: a lesson from transgenic studies. Environmental Health Perspective 1998; 106(5) 1219-1228.

[17] Maxwell SRJ. Prospects for the use of antioxidants therapies. Drugs 1995; 49(3) 345-361.

[18] Bandyopadhyay U, Das D, Banerjee RK. Reactive oxygen species: oxidative damage and pathogenesis (review). Current Science 1999; 77(5) 658-666.

[19] Lushchak VI., Gospodaryov DV. Introductory Chapter. In: Lushchak V &Gospodaryov D (ed.) Oxidative Stress and Diseases.Rijeka: InTech; 2012. p3-10.

[20] Shibata N., Kobayashi M. The role for oxidative stress in neurodegenerative diseases. Brain and Nerve 2008; 60(2) 157-170.

[21] Kadenbach B., Ramzan R., Vogt S. Degenerative diseases, oxidative stress and cytochrome c oxidase function. Trends in Molecular Medicine 2009; 15(4) 139-147.

[22] GospodaryovDV.,Volodymyr IL. Oxidative Stress: Cause and Consequence of Diseases. In: Lushchak V. &Gospodaryov D (ed.) Oxidative Stress and Diseases.Rijeka: InTech; 2012. p13-38.

[23] Beutler E. (). Glucose-6-phosphate dehydrogenase deficiency: a historical perspective. Blood. 2008; 111(1) 16-24.

[24] Gaskin RS.,Estwick D. Peddi R. G6PDH deficiency: its role in the high prevalence of hypertension and diabetes mellitus. Ethnicity & Disease. 2001; 11(4) 749-754.

[25] Carette C., Dubois-Laforgue D., Gautier JF. Timsit J. Diabetes mellitus and glucose-6-phosphate dehydrogenase deficiency: from one crisis to another. Diabetes & Metabolism. 2011; 37(1) 79-82.

[26] Ho HY., Cheng ML., Chiu D TY. G6PDH-an old bottle with new wine. Chang Gung Medical Journal. 2005; 28(9) 606-612.

[27] Winterbourn CC., Metodiewa D. The reaction of superoxide with reduced glutathione. Archives of Biochemistry and Biophysics. 1994; 314(2) 284-290.

[28] Lubos E., Handy DE., Loscalzo J. Role of oxidative stress and nitric oxide in atherothrombosis. Frontiers in Bioscience. 2008; 13 5323-5344.

[29] Förstermann U. Nitric oxide and oxidative stress in vascular disease, PflügersArchiv: European Journal of Physiology 2010; 459(6) 923-939.

[30] Matsui R., Xu S., Maitland KA., Hayes A., Leopold J.A., Handy DE., Loscalzo J., Cohen RA. Glucose-6 phosphate dehydrogenase deficiency decreases the vascular response to angiotensin II. Circulation. 2005; 112(2) 257-263.

[31] Rimm EB., Stampfer MJ. Folate and cardiovascular disease: one size does not fitall. Lancet. 2011; 378(9791) 544-546.

[32] Leopold JA. Loscalzo J. Oxidative enzymopathies and vascular disease. Arteriosclerosis, Thrombosis, and Vascular Biology. 2005; 25(7) 1332-1340.

[33] Pavel S., Smit NP., van der Meulen H., Kolb RM., de Groot AJ., van der Velden PA., Gruis NA., Bergman W. Homozygous germline mutation of CDKN2A/p16 and glucose-6-phosphate dehydrogenase deficiency in a multiple melanoma case. Melanoma Research. 2003; 13(2) 171-178.

[34] Cocco P., Ennas M.G., Melis MA., Sollaino C., Collu S., Fadda D., Gabbas A., Massarelli G., Rais M., Todde P., Angelucci E. Glucose-6-phosphate dehydrogenase polymorphism and lymphoma risk. Tumori 2007; 93(2) 121-123.

[35] TianWN.,Braunstein LD., Pang J., Stuhlmeier KM., Xi QC., Tian X., Stanton RC. Importance of glucose-6-phosphate dehydrogenase activity for cell growth. The Journal of Biological Chemistry. 1998; 273(17) 10609-10617.

[36] Batetta B., Pulisci D., Bonatesta RR., Sanna F., Piras S., Mulas MF., Spano O., Putzolu M., Broccia G., Dessì S. G6PDH activity and gene expression in leukemic cells from G6PDH-deficient subjects. Cancer Letters. 1999; 140(1-2) 53-58.

[37] Lushchak VI. Adaptive response to oxidative stress: Bacteria, fungi, plants and animals. Comparative Biochemistry and Physiology. Toxicology & Pharmacology CBP. 2011; 153(2) 175-190.

[38] Gupte SA. Targeting the Pentose Phosphate Pathway in Syndrome X-related Cardiovascular Complications. Drug Development Research. 2010; 71(3) 161-167.

[39] Amemiya-Kudo M., Shimano H., Hasty AH., Yahagi N., Yoshikawa T., Matsuzaka T., Okazaki H., Tamura Y., Iizuka Y., Ohashi K., Osuga J.-ichi., Harada K., Gotoda T., Sato R., Kimura S., Ishibashi S., Yamada N. Transcriptional activities of nuclear SREBP-1a, -1c, and -2 to different target promoters of lipogenic and cholesterogenic genes. Journal of Lipid Research. 2002; 43(8) 1220-1235.

[40] Kletzien RF., Harris PK., Foellmi LA. Glucose-6-phosphate dehydrogenase: a "housekeeping" enzyme subject to tissue-specific regulation by hormones, nutrients, and oxidant stress. The FASEB Journal: Official Publication of the Federation of American Societies for Experimental Biology. 1994; 8(2) 174-181.

[41] Franzè A., Ferrante MI., Fusco F., Santoro A., Sanzari E., Martini G., Ursini MV. Molecular anatomy of the human glucose 6-phosphate dehydrogenase core promoter. FEBS Letters. 1998; 437(3) 313-318.

[42] Lee JW., Choi AH., Ham M., Kim JW., Choe SS., Park J., Lee GY., Yoon KH., Kim JB. G6PDH up-regulation promotes pancreatic beta-cell dysfunction. Endocrinology. 2011; 152(3) 793-803.

[43] Ralser M., Benjamin IJ. Reductive stress on life span extension in C. elegans, BMC Research Notes. 2008; 1, 19.

[44] KirkmanHN.,Gaetani GF. Mammalian catalase: a venerable enzyme with new mysteries, Trends in Biochemical Sciences. 2007; 32(1) 44-50.

[45] Ogata M., Wang DH., Ogino K. Mammalian acatalasemia: the perspectives of bioinformatics and genetic toxicology. Acta Medica Okayama. 2008; 62(6) 345-361.

[46] Góth L. Catalasedeficiency and type 2 diabetes. Diabetes Care. 2008; 31(12) e93.

[47] Lubos E., Loscalzo J., Handy DE. Homocysteine and glutathione peroxidase-1, Antioxidants & Redox Signaling. 2007; 9(11) 1923-1940.

[48] Valentine JS., Doucette PA., Zittin Potter S. Copper-zinc superoxide dismutase and amyotrophic lateral sclerosis, Annual Review of Biochemistry. 2005; 74 563-593.

[49] Vucic S., Kiernan MC. Pathophysiology of neurodegeneration in familial amyotrophic lateral sclerosis. Current Molecular Medicine. 2009; 9(3) 255-272.

[50] Liochev SI., Fridovich I. Mutant Cu, Zn superoxide dismutases and familial amyotrophic lateral sclerosis: evaluation of oxidative hypotheses. Free Radical Biology & Medicine, 2003; 34(11) 1383-1389.

[51] Tortarolo M., Grignaschi G., Calvaresi N., Zennaro E., Spaltro G., Colovic M., Fracasso C., Guiso G., Elger B., Schneider H., Seilheimer B., Caccia S., Bendotti C. Glutamate AMPA receptors change in motor neurons of SOD1 G93A transgenic mice and their inhibition by a noncompetitive antagonist ameliorates the progression of amytrophic lateral sclerosis-like disease. Journal of Neuroscience Research 2006; 83(1) 134-146.

[52] Beckman JS., Estévez AG., Crow JP., Barbeito L. Superoxide dismutase and the death of motoneurons in ALS, Trends in Neurosciences. 2001; 24(11) S15-20.

[53] Liochev SI., Fridovich I. Copper- and zinc-containing superoxide dismutase can act as a superoxide reductase and a superoxide oxidase. The Journal of Biological Chemistry. 2000; 275(49) 38482-38485.

[54] YimMB., Chock PB, Stadtman ER. Copper, zinc superoxide dismutase catalyzes hydroxyl radical production from hydrogen peroxide. Proceedings of the National Academy of Sciences of the United States of America. 1990; 87(13) 5006-5010.

[55] Kim KS., Choi SY., Kwon HY., Won MH., Kang TC., Kang JH. Aggregation of alpha-synuclein induced by the Cu,Zn-superoxide dismutase and hydrogen peroxide system. Free Radical Biology & Medicine. 2002; 32(6) 544-550.

[56] Rakhit R., Cunningham P., Furtos-Matei A., Dahan S., Qi XF., Crow JP., Cashman NR., Kondejewski LH. Chakrabartty A. Oxidation-induced misfolding and aggregation of superoxide dismutase and its implications for amyotrophic lateral sclerosis. The Journal of Biological Chemistry. 2002; 277(49) 47551-47556.

[57] Poon H.F., Hensley K., Thongboonkerd V., Merchant ML., Lynn BC., Pierce WM., Klein J. B., Calabrese V., Butterfield DA. Redox proteomics analysis of oxidatively modified proteins in G93A-SOD1 transgenic mice-a model of familial amyotrophic lateral sclerosis. Free Radical Biology & Medicine. 2005; 39(4) 453-462.

[58] Lightfoot TJ., Skibola CF., Smith AG., Forrest MS., Adamson PJ., Morgan GJ., Bracci PM., Roman E., Smith MT., Holly EA. Polymorphisms in the oxidative stress genes superoxide dismutase, glutathione peroxidase and catalase and risk of non-Hodgkin's lymphoma. Haematologica. 2006; 91(9) 1222-1227.

[59] Forsberg L., de Faire U., Morgenstern R. Oxidative stress, human genetic variation, disease. Archives of Biochemistry and Biophysics. 2001; 389(1) 84-93.

[60] GongoraMC., Harrison DG. Sad heart from no SOD. Hypertension. 2008; 51, 28-30.

[61] Hamanishi T., Furuta H., Kato H., Doi A., Tamai M., Shimomura H., Sakagashira S., Nishi M., Sasaki H., Sanke T., Nanjo K. Functional variants in the glutathione peroxidase-1 (GPx-1) gene are associated with increased intima-media thickness of carotid arteries and risk of macrovascular diseases in japanese type 2 diabetic patients. Diabetes. 2004; 53(9) 2455-2460.

[62] Sun LM., Shang Y., Zeng, YM., Deng YY., Cheng JF. hOGG1 polymorphism in atrophic gastritis and gastric cancer after Helicobacter pylori eradication. World Journal of Gastroenterology. 2010; 16(35) 4476-4482.

[63] Bravard A., Vacher M., Moritz E., Vaslin L., Hall J., Epe B., Radicella JP. Oxidative status of human OGG1-S326C polymorphic variant determines cellular DNA repair capacity. Cancer Research. 2009; 69, 3642-3649.

[64] Nakabeppu Y., Kajitani K., Sakamoto K., Yamaguchi H., Tsuchimoto D. MTH1, an oxidized purine nucleoside triphosphatase, prevents the cytotoxicity and neurotoxicity of oxidized purine nucleotides. DNA Repair. 2006; 5(7) 761-772.

[65] Halliwell B. Oxidative stress and cancer: have we moved forward?.Biochemical Journal. 2007; 401, 1-11.

[66] Hayes JD., Flanagan JU., Jowsey IR. Glutathione transferases. Annual Review of Pharmacology and Toxicology. 2005; 45, 51-88.

[67] Konig-Greger D., Riechelmann H., Wittich U., Gronau S. Genotype and phenotype of glutathione-S-transferase in patients with head and neck carcinoma, Otolaryngology-Head and Neck Surgery: Official Journal of American Academy of Otolaryngology-Head and Neck Surgery. 2004; 130(6) 718-725.

[68] Mohr LC., Rodgers JK., Silvestri GA. Glutathione S-transferase M1 polymorphism and the risk of lung cancer. Anticancer Research. 2003; 23(3A) 2111-2124.

[69] Benhamou S., Lee WJ., Alexandrie AK., Boffetta P., Bouchardy C., Butkiewicz D., Brockmöller J., Clapper ML., Daly A., Dolzan V., Ford J., Gaspari L., Haugen A., Hirvonen A., Husgafvel-Pursiainen K., Ingelman-Sundberg M., Kalina I., Kihara M., Kremers P., Le Marchand L., London SJ., Nazar-Stewart V., Onon-Kihara M., Rannug A., Romkes M., Ryberg D., Seidegard J., Shields P., Strange RC., Stücker I., To-Figueras J., Brennan P., Taioli, E. Meta- and pooled analyses of the effects of glutathione S-transferase M1 polymorphisms and smoking on lung cancer risk, Carcinogenesis. 2002; 23(8) 1343-1350.

[70] Cote M.L., KardiaSLR.,Wenzlaff AS., Land SJ., Schwartz AG. Combinations of glutathione S-transferase genotypes and risk of early-onset lung cancer in Caucasians and African Americans: a population-based study. Carcinogenesis. 2005; 26(4) 811-819.

[71] Lushchak VI. Oxidative stress and mechanisms of protection against it in bacteria. Biochemistry. 2001; 66(5) 592-609.

[72] Wautier JL. Schmidt Ann Marie. Protein glycation: a firm link to endothelial cell dysfunction. Circulation Research. 2004; 95(3) 233-238.

[73] Davis JS., Wu X. Current state and future challenges of chemoprevention. Discovery Medicine. 2012; 13(72), 385-90.

[74] Stavric B. Role of chemopreventers in human diet. Clinical Biochemistry. 1994; 27(5) 319-32.

[75] Ferguson L. Dietary influences on mutagenesis--where is this field going?.Enviromental and Molecular Mutagenesis. 2010; 51(8-9) 909-918.

[76] Stavric B. Antimutagens and anticarcinogens in foods. Food and Chemical Toxicology. 1994; 32(1) 79-90.

[77] Tanaka T., Shnimizu M., Moriwaki H. Cancer chemoprevention by carotenoids. Molecules. 2012; 17(3) 3202-3242.

[78] Tsuda H., Ohshima Y., Nomoto H., Fujita K., Matsuda E., Iigo M., Takasuka N., Moore MA. Cancer prevention by natural compounds. Drug Metabolism and Pharmacokinetics. 2004; 19(4) 245-263.

[79] Madrigal-BujaidarE., Viveros Martha E. La prevención química del cáncer. Revista del Instituto Nacional de Cancerología México. 1996; 42(1) 37-41.

[80] Gullett NP., RuhulAmin AR, Bayraktar S, Pezzuto JM, Shin DM, Khuri FR, Aggarwal BB, Surh YJ, Kucuk O. Cancer prevention with natural compounds. Seminars in Oncology. 2010; 37(3) 258-281.

[81] Steele VE. Current mechanistic approaches to the chemoprevention of cancer. Journal of Biochemistry and Molecular Biology. 2003; 36(1) 78-81.

[82] Ferguson LR, Bronzetti G, De Flora S. Mechanistic approaches to chemoprevention of mutation and cancer. Mutation Research. 2005; 591(1-2) 3-7.

[83] de Wit M., Nel P., Osthoff G., Labuschagne, MT. The effect of variety and location on cactus pear (Opuntiaficus-indica) fruit quality. PlantFoodsfor Human Nutrition.2010; 65(2) 136-145.

[84] Jolalpa-Barrera JL., Aguilar-Zamora A., Ortiz-Barreto O., García-López L. Producción y comercialización de tuna en fresco bajo diferentes modalidades en Hidalgo. Revista Mexicana de Agronegocios. 2011; 28, 605-614.

[85] Sumaya-Martínez MT., Suárez-Diéguez T., Cruz-Cansino NS., Alanís-García E., Sampedro JG. Innovación de productos de alto valor agregado a partir de la tuna mexicana. Revista Mexicana de Agronegocios.2010; 27, 435-441.

[86] Becerra-Jiménez J., Andrade-Cetto A. Effect of OpuntiastreptacanthaLem. on alpha-glucosidase activity. Journal of Ethnopharmacology. 2012; 139(2) 493-496.

[87] Hahm SW., Park J., Son YS. Opuntiahumifusa stems lower blood glucose and cholesterol levels in streptozotocin-induced diabetic rats. Nutrition Research. 2011; 31(6) 479-87.

[88] Sumaya-Martínez MT., Cruz-Jaime S., Madrigal-Santillán EO., García-Paredes JD.,Cariño-Cortés R., Cruz-Cansino N., Valadez-Vega C., Martínez-Cardenas L., Alanís-García E. Betalain, Acid Ascorbic, Phenolic Contents and Antioxidant Propierties of Purple, Red, Yellow and White Cactus Pears. International Journal of Molecular Science. 2011; 12(10) 6452-6468.

[89] Fernández-López J., Almela L., Obón J., Castellar R. Determination of Antioxidant Constituents in Cactus Pear Fruits. Plant Foods for Human Nutrition. 2010; 65(3) 253-259

[90] Castellar R., Obón JM., Alacid M., Fernández-López JA. Color properties and stability of betacyanins from Opuntiafruits. Journal of Agricultural and Food Chemistry. 2003, 51(9) 2772-2776.

[91] Livrea MA., Tesoriere L. Antioxidative effects of cactus pear [Opuntiaficus-indica(L) Mill] fruits from Sicily and bioavailability of betalain components in healthy humans. ActaHorticulturae.2009; 811, 197-204.

[92] Dok-Go H., Lee KH., Kim HJ., Lee EH., Lee J., Song YS., Lee YH., Jin Ch., Lee YS., Cho J. Neuroprotective effects of antioxidative flavonoids, quercetin, (+)-dihydroquercetin and quercetin 3-methyl ether, isolated from Opuntiaficus-indicavar. Saboten. Brain Research. 2003; 965(1-2) 130-136.

[93] Galati EM., Mondello MR., Giuffrida D., Dugo G., Miceli N., Pergolizzi S., Taviano MF. Chemical characterization and biological effects of Sicilian Opuntiaficus-indica(L.) Mill. Fruit juice: Antioxidant and antiulcerogenic activity. Journal of Agricultural and Food Chemistry.2003; 51(17) 4903-4908.

[94] García-Melo LF. Degree Thesis. Evaluación de la capacidad quimioprotectora del jugo de tuna mediante la técnica de micronúcleos. Institute of Health Sciences, Autonomous University of Hidalgo State, México. 2009.

[95] Seeram NP. Berry fruits for cancer prevention: Current status and future prospects. Journal of Agricultural and Food Chemistry. 2008; 56(3) 630-635.

[96] Howell AB., Vorsa N., Der Marderosian A., Foo L. Inhibition of adherence of P-fimbricatedEscherischia coli to uroepithelial-cell surfaces by proanthocyanidin extracts from cranberries. The New England Journal of Medicine. 1998; 339(15) 1085-1086.

[97] Foo LY., Lu Y., Howell AB., Vorsa N. A-type proanthocyanidintrimers from cranberry that inhibit adherence of uropathogenic P-fimbriatedEscherichia coli. Journal of Natural Products. 2000; 63(9) 1225-1228.

[98] Burger O., Ofek I., Tabak M., Weiss EI., Sharon N., Neeman I. A high molecular mass constituent of cranberry juice inhibits Helicobacter pylori adhesion to human gastric mucus. FEMS Immunology and Medical Microbiology. 2000; 29(4) 295-301.

[99] Bomser J., MadhaviDL., Singletary K., Smith MA. In vitro anticancer activity of fruit extracts from Vaccinium species. PlantaMedica. 1996; 62(3) 212-216.

[100] Seeram NP., Adams LS., Zhang Y., Sand D., Heber D. Blackberry, black raspberry, blueberry, cranberry, red raspberry and strawberry extracts inhibit growth and stimulate apoptosis of human cancer cells in vitro. Journal of Agricultural and Food Chemistry. 2006; 54(25) 9329-9339.

[101] Sun J., Liu RH. Cranberry phytochemical extracts induce cell cycle arrest and apoptosis in human MCF-7 breast cancer cells. Cancer Letters. 2006; 241(1) 124-134.

[102] Neto CC., Amoroso JW., Liberty AM. Anticancer activities of cranberry phytochemicals: An update. Molecular Nutrition and Food Research. 2008; 52(1) S18-S27.

[103] Coates EM., Popa G., Gill CI., McCann MJ., McDougall GJ., Stewart D., Rowland I. Colon-available raspberry polyphenols exhibit anti-cancer effects on in vitro models of colon cancer. Journal of Carcinogenesis. 2007; 6, 4.

[104] Schmidt BM., Erdman JW., Lila MA. Differential effects of blueberry proanthocyanidins on androgen sensitive and insensitive human prostate cancer cell lines. Cancer Letters. 2006; 231(2) 240-246.

[105] Boateng J., Verghese M., Shackelford L., Walker LT., Khatiwada J., Ogutu S., Williams DS., Jones J., Guyton M., Asiamah D., Henderson F., Grant L., DeBruce M., Johnson A., Washington S., Chawan CB. Selected fruits reduce azoxymethane (AOM)-induced aberrant crypt foci (ACF) in Fisher 344 male rats. Food and Chemical Toxicology. 2007; 45(5) 725-732.

[106] Barros D., Amaral OB., Izquierdo I., Geracitano L., do CarmoBassolsRaseira M., Henriques AT., Ramírez MR. Behavioral and genoprotective effects of Vaccinium berries intake in mice. Pharmacology, Biochemistry and Behavior. 2006; 84(2) 229-234.

[107] Madrigal-Santillán E., Fragoso-Antonio S., Valadez-Vega C., Solano-Solano G., Pérez CZ., Sánchez-Gutiérrez M., Izquierdo-Vega JA., Gutiérrez-Salinas J., Esquivel-Soto J., Esquivel-Chirino C., Sumaya-Martínez T., Fregoso-Aguilar T., Mendoza-Pérez J., Morales-González JA. Investigation on the protective effects of cranberry against the DNA damage induced by benzo[a]pyrene. Molecules. 2012; 17(4) 4435-4451.

[108] Fellers PJ, Nikdel S, Lee HS. Nutrient content and nutrition labeling of several processed Florida citrus juice products. Journal of the American Dietetic Association. 1990; 90(8) 1079-84.

[109] Kumar A. Dogra S. Prakash A. Protective effect of naringin, a citrus flavonoid, against colchicine-induced cognitive dysfunction and oxidative damage in rats. Journal of Medicinal Food. 2010; 13(4) 976-984.

[110] Monroe KR., Murphy SP., Kolonel LN., Pike MC. Prospective study of grapefruit intake and risk of breast cancer in postmenopausal women: the Multiethnic Cohort Study. British Journal of Cancer. 2007; 97(3) 440-5.

[111] Kim EH., Hankinson SE., Eliassen AH., Willett WC. A prospective study of grapefruit and grapefruit juice intake and breast cancer risk.British Journal of Cancer. 2008; 98(1) 240-241.

[112] Bressler R. Grapefruit juice and drug interactions. Exploring mechanisms of this interaction and potential toxicity for certain drugs. Geriatrics.2006; 61(11) 12-18.

[113] Alvarez-González I., Madrigal-Bujaidar E., Dorado V., Espinosa-Aguirre JJ. Inhibitory effect of naringin on the micronuclei induced by ifosfamide in mouse, and evaluation of its modulatory effect on the Cyp3a subfamily. Mutation Research. 2001; 1(480-481) 171-178.

[114] Alvarez-González I., Madrigal-Bujaidar E., Martino-Roaro L., Espinosa-Aguirre JJ. Antigenotoxic and antioxidant effect of grapefruit juice in mice treated with daunorubicin. Toxicology Letters. 2004; 152(3) 203-211.

[115] Razo-Aguilera G., Baez-Reyes R., Alvarez-González I., Paniagua-Pérez R., Madrigal-Bujaidar E. Inhibitory effect of grapefruit juice on the genotoxicity induced by hydrogen peroxide in human lymphocytes. Food and Chemical Toxicology. 2011; 49(11) 2947-2953.

[116] Hernández-Ceruelos A., Madrigal-Santillán E., Morales-González JA., Chamorro-Cevallos G., Cassani-Galindo M., Madrigal-Bujaidar E. Antigenotoxic Effect of Chamomillarecutita (L.) Rauschertssential Oil in Mouse Spermatogonial Cells, and Determination of Its Antioxidant Capacity in Vitro. International Journal of Molecular Sciences. 2010; 11(10) 3793-3802.

[117] Jakolev V., Issac O., Flaskamp E. Pharmacological investigation with compounds of chamazulene and matricine. PlantaMedica. 1983; 49(10) 67-73.

[118] Viola H., Wasowski C., Levi de Stein M., Wolfman C., Silveira R., Dajas F., Medina JH., Paladini AC. Apigenin, a component of Matricariarecutitaflowers is a central benzodiazepine receptor-ligand with anxiolytic effects. PlantaMedica. 1995; 61(3) 213-216.

[119] McKay DL., Blumberg JB. A review of the bioactivity and potential health benefits of chamomile tea (MatricariarecutitaL.). Phytotherary Research. 2006; 20(7) 519-530.

[120] Hernández-Ceruelos A., Madrigal-Bujaidar E., de la Cruz C. Inhibitory effect of chamomile essential oil on the sister chromatid exchanges induced by daunorubicin and methyl methanesulfonate in mouse bone marrow. Toxicology Letters. 2002; 135(1-2) 103-110.

[121] Hamid S., Sabir A., Khan S., Aziz P. Experimental cultivation of Silybummarianumand chemical composition of its oil. Pakistan Journal of Scientific and Industrial Research. 1983; 26, 244-246

[122] Morazzoni P., Bombardelli E. Silybummarianum(Cardusarianum). Fitoterapia. 1995; 66(1) 3-42.

[123] Lee DY-W., Liu Y. Molecular structure and stereochemistry of silybinA, silybin B, isosilybin A, and isosilybin B, isolated from Silybummarianum(Milk thistle). Journal of Natural Products. 2003; 66(9) 1171-1174.

[124] Ligeret H., Brault A., Vallerand D., Haddad Y., Haddad PS. Antioxidant and mitochondrial protective effects of silibinin in cold preservation-warm reperfusion liver injury. Journal of Ethnopharmacology. 2008; 115(3) 507-514.

[125] Shaker E., Mahmoud H., Mnaa S. Silymarin, the antioxidant component and Silybummarianumextracts prevents liver damage. Food and Chemical Toxicology 2010; 48(3) 803-806.

[126] AbouZid S. Silymarin, Natural Flavonolignans from Milk Thistle. In: RaoVenketeshwer (ed.) Phytochemicals-A Global Perspective of Their Role in Nutrition and Health. Rijeka: InTech; 2012. p255-272.

[127] Deep G., Oberlies NH., Kroll DJ., Agarwal R. Isosilybin B and isosilybin A inhibit growth, induce G1 arrest and cause apoptosis in human prostate cancer LNCaP and 22Rv1 cells. Carcinogenesis. 2007; 28(7) 1533-1542.

[128] Svobodová A., Zdařilová A., Walterová D., Vostálová J. Flavonolignans from Silybummarianummoderate UVA-induced oxidative damage to HaCaT keratinocytes. Journal of DermatologicalScience. 2007; 48(3) 213-224.

[129] Zermeño-Ayala, P. DegreeThesis. Evaluación del efecto quimiopreventivo de la silimarina sobre el daño genotóxico hepático producido por el consumo subcrónico de etanol. Institute of Health Sciences, Autonomous University of Hidalgo State, México. 2011

Chapter 8

Inflammatory Environmental, Oxidative Stress in Tumoral Progression

César Esquivel-Chirino, Jaime Esquivel-Soto,
José Antonio Morales-González,
Delina Montes Sánchez, Jose Luis Ventura-Gallegos,
Luis Enrique Hernández-Mora and
Alejandro Zentella-Dehesa

Additional information is available at the end of the chapter

http://dx.doi.org/10.5772/51789

1. Introduction

The incidence and prevalence of cancer has been increasing in such as degree that it has become the second or third leading cause of death worldwide, depending on ethnicity or country in question and is consequently a major public health, cancer is a leading cause of death in many countries, accounting for 7.6 million deaths (around 13% of all deaths) in 2008. Deaths from cancer worldwide are projected to continue rising, with an estimated 13.1 million deaths in 2030. About 30% of cancer deaths are due to the five leading behavioral and dietary risks: high body mass index, low fruit and vegetable intake, lack of physical activity, tobacco use, alcohol use [1-8].

Cancer is a generic term for a large group of diseases that can affect any part of the body, cancer cells are significantly influenced by the surrounding stromal tissues for the initiation, proliferation, and distant colony formation. When a tumour successfully spreads to other parts of the body and grows, invading and destroying other healthy tissues, this process is referred to as metastasis, and the result is a serious condition that is very difficult to treat, because the progression to metastases is the leading cause of death associated to cancer. Metastatic cells in this process must interact with the endothelium in three stages of tumor progression. In recent years, the interaction between these cell populations has been seen as part of a complex microenvironment tumor-associated. Mantovani et al. have even postulated that this tumor microenvironment inflammatory plays an important role in tumorigenesis and tumor progression [18]. Tumor and normal surrounding cells such as endothelial cells, soluble factors derived from this two cell populations and extracellular matrix [9-12], compose the tumor microenvironment.

There are other factors in the development of cancer such as genetic, environmental, as well as the role of oxidative stress and free radicals in response to damage caused by chemicals, radiation, cellular aging, ischemic lesions and cells immune system [13]. Substances such as antioxidants can protect cells against free radical damage. The damage caused by free radicals can lead to cancer. Antioxidants interact with free radicals to stabilize them so that, being able to avoid some of the damage that free radicals can cause. It is important to analyze the role of antioxidants as an alternative that contributes to cancer treatment and to promote their use and consumption in cancer prevention

2. Tumoral progression

Tumors often become more aggressive in their behavior in more aggressive and their characteristics, although the time course may be quite variable, this phenomenon has been termed tumor progression by Foulds [15].

In the early stages of the tumor progression, there is a detachment of cancer cells from the primary tumor, followed by tumor cell adhesion to endothelial cells of venules in the target organs. After the extravasations occurs extracellular matrix invasion by tumor cells, these cells of primary lesions enter the lymphatics or the bloodstream depending on their anatomical location. In the circulation, many tumor cells are eradicated by physical forces exerted on them to pass through the microvasculature of secondary organs, and immunological mechanisms of action of host defense. Furthermore, once inside the target tissue tumor cells must find favorable conditions for survival and proliferation [16-18]. The biological characteristic that define tumor progression have been extensively described, although the underlying mechanisms are still not completely defined, however there are two theories have been proposed to explain how tumor cells invade secondary sites where metastasis occurs are the following [18-20]. The first is similar to the inflammatory process by cell adhesion and migration, while the second involves the aggregation of circulating tumor cells, and that these cells blocked blood vessels. In this theory in which a cell stably adhered frequently starts a homotypic aggregation, capturing other circulating tumor cells, followed by the formation of multicellular aggregates, these aggregates once grow and emerge from the primary tumor site, is carried out which triggers tumor progression in metastasis, where it requires a coordinated interaction of tumor cells and vascular endothelial cells, play a critical role in most of the events that characterize tumor progression and metastasis [21-23], so it is important to mention the general aspects of endothelial biology:

2.1. Endothelial Biology

The endothelium is the thin layer of cells that lines the interior surface of blood vessels and lymphatic vessels, forming an interface between circulating blood or lymph in the lumen and the rest of the vessel wall. The cells that form the endothelium are called endothelial cells, these cells have very distinct and unique functions that are paramount to vascular biology. These functions include fluid filtration, formation of new blood vessels in the angiogenesis, neutrophil recruitment. The endothelium acts as a semi-selective barrier between the vessel lumen and surrounding tissue, controlling the passage of materials and the transit of white blood cells, hormones into and out of the bloodstream. Excessive or prolonged increases in permeability of the endothelial monolayer, as in cases of chronic inflammation, may lead to tissue edema. It is also important in controlling blood pressure, blood coagulation, vascular tone, degradation of lipoproteins an in the secretion of growth factors and cytokines [24-25]. In recent decades, it has become clear that the endothelium of venules and smaller capillaries, and lymphatic vessels play a central role in the process of tumor growth, dissemination of metastatic cells, which is accompanied by the development of a characteristic tumor vasculature and tumors formed by endothelial cells [26].

There are two phenotypes endothelial (constitutive and activated);

2.1.1. *The constitutive phenotype of endothelial cells*

Quiescent, resting endothelial cells in the adult form a highly heterogeneous cell population that varies not only in different organs but also in different vessel calibers within an organ. Endothelium in the normal adult male, although being metabolically active, considered quiescent because the turnover of these cells is very low and this called: constitutive phenotype Fig (1).

In this condition, the apical membrane of endothelial cells exhibits a very low amount of intercellular adhesion molecules, so that no adhesion of cellular blood components to the vessel walls [27].

2.1.2. *The activated phenotype of endothelial cells*

Endothelial cell activation is associated with a number of distinct phenotype changes that, much like differentiation processes of the constitutive phenotype of endothelial cells, serve their need to adapt to functional requirements. [28] The cytokine-induced phenotype of endothelial cells during inflammation has been characterized most extensively in the last few years. When endothelial cells are activated by these cytokines are functional disorders involving immediate responses, for example, some pathological conditions such as sepsis, are associated with endothelial conversion to a phenotype activated [29-30]. The activated phenotype characterized by activation of constitutive nitric oxide synthase (NOS), also accompanied by changes such as increased expression of cell adhesion molecules (CAMs) and E-selectin (CD62E), ICAM-1(CD54), VCAM-1(CD106), P-selectin (CD62P) Fig (1). These changes allow the endothelium to participate in pathological conditions including inflammation, coagulation, cell proliferation, metastasis, tumor angiogenesis. All these cellular interactions are regulated by temporal and spatial presentation of various cell adhesion molecules and chemotactical molecules displaying appropriate specificity and affinity for proper development and functioning of the organism [31-32]. Has been postulated that this phenotype or variants of it, are involved in the processes of metastasis [33].

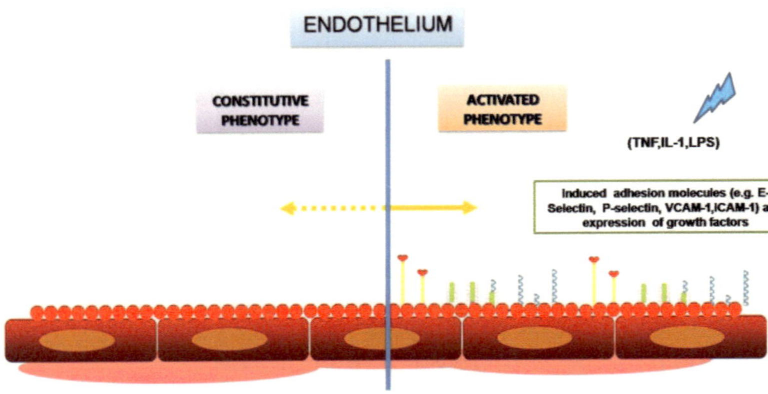

Figure 1. There are two phenotypes endothelial constitutive and activated.

2.2. Metastasis

Metastasis is the result of cancer cell adaptation to a tissue microenvironment at a distance from the primary tumor, is a complex process involving multiple steps: first, when cancer cells break away from the primary tumor, they invade the host stroma, intravasate into lymphatic or blood vessels, spread to the capillary bed of distant organs, where they invade into new surrounding tissues and proliferate to form secondary tumors [34-35]. When cancer is detected at an early stage, before it has spread, it can often be treated successfully by conventional cancer therapies such as surgery, chemotherapy, local irradiation, metastatic diffusion of cancer cells remains the most important clinical problem, because when cancer is detected after known to have metastasized, treatment are much less successful [36]. The metastatic capacity of tumor cells correlates with their ability to exit from the blood circulation, to colonize distant organs, and to grow in distant organs. Metastasis is a complex process that includes local infiltration of tumor cells into the adjacent tissue, transendothelial migration of cancer cells into vessels known as intravasation, survival in the circulatory system, extravasation and subsequent proliferation in competent organs leading to colonization [36-38]. Initially, tumor cell aggregates detachment from the primary tumor, next the cells actively infiltrate the surrounding stroma and enter into the circulatory system, traveling to distinct sites to establish the secondary tumor growth. In the bloodstream, a very small number of tumor cells survive to reach the target organ, indicating that metastasis formation must be regarded as a very ineffective event. Millions of carcinoma cells enter into the circulatory system, but the majority of them die during transportation, and only 1-5% of viable cells are successful in formation of secondary deposits in distinct sites [37-40]. Fig (2).

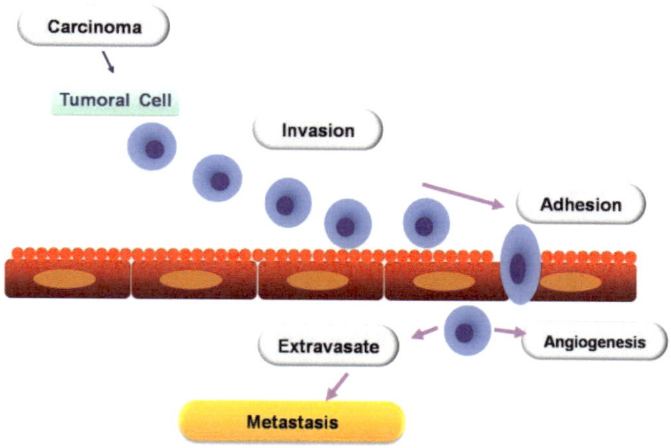

Figure 2. Steps in the Metastasis process.

Metastasis is facilitated by cell-cell interactions between tumor cells and the endothelium in distant tissues and determines the spread. Metastatic cells must act with the endothelium in three different stages of tumor progression: initially during the formation of blood vessels that enable tumor growth (vascularization), during the migration process that allows the passage from tissue into the bloodstream (intravasation), and finally during the process allowing extravasation into the target tissue [41-43]. Metastatic cancer cells require properties that allow them not only to adapt to a foreign microenvironment but also to subvert it in a way that is conducive to their continued proliferation and survival [36-38]. In addition, direct tumor cell interactions with platelets, fibroblasts and monocytes/macrophages, polymorphonuclear cells, soluble components, cytokines, chemokines, proteins of the extracellular matrix, growth factors, and other molecules secreted by host cells, significantly contribute to cancer cell adhesion, extravasation, and the establishment of metastatic lesions [44-47].

2.3. Cellular interactions in the inflammatory reaction and spread tumor

In the early stages of inflammation, neutrophils are cells that migrate to the site of inflammation under the influence of growth factors, cytokines and chemokines, which are produced by macrophages and mast cells residing in the tissue [48]. The process of cell extravasation from the bloodstream can be divided into four stages:

1. bearing
2. activation by stimulation chemoattractant
3. adherence
4. transendothelial migration.

If the inflammatory response is not regulated, the cellular response will become chronic and will be dominated by lymphocytes, plasma cells, macrophages metastasis, which is favored by the microenvironment of the organ target. The installation of tumor cells in blood vessels of the organ target to invade, is related to phenotypic changes in the endothelium allowing vascular extravasation of blood circulation of leukocytes in the inflammatory reaction and, as hypothesized current of tumor cells with metastatic capacity. The phenomenon of extravasation in response to a tumor cell interaction cell endothelial or not allowing the passage of cells whether there are appropriate conditions for the invasion with varied morphology [53-55].

Within the process of inflammation, a phenomenon is well-studied cell migration, which is the entrance of polymorphonuclear neutrophils and the vascular system. This involves a sequential mechanism of recognition, contact formation, and migration mediated by adhesion molecules such as (ICAM-1, VCAM-1, E and P Selectins, Integrins) it has been demonstrated that some of these adhesion molecules, such as E-selectin are not only involved in inflammation, but also in tumor metastasis and play a significant role in cancer progression and metastasis, in some cases of colon cancer. [56-58]. The expression of ICAM-1 has also proven to be a marker associated with an invasive phenotype [59].

Hanahanan et al. suggest that diversity of cancer cell genotypes is a manifestation of six basics alterations in cell physiology that together indicate development of malignant growth Fig (3), these alterations are shared in common by most all types of tumors [19]. In recent years, it has been demonstrated that metastatic dissemination can be influenced by inflammatory-reparative processes [46]. The interaction between these cell populations has been seen as part of a complex inflammatory microenvironment tumor-associated. Mantovani et al. have even postulated that this tumor microenvironment inflammatory plays an important role in tumorigenesis and tumor progression [60].

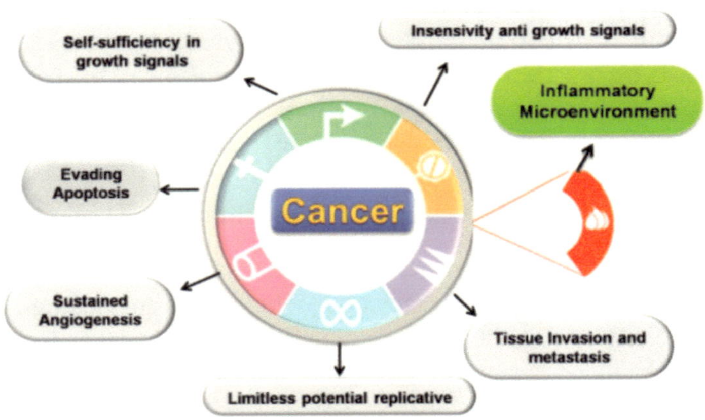

Figure 3. Capabilities of cancer and inflammation [19,60].

2.4. Inflammatory Microenvironment

The tumor microenvironment is composed of stromal fibroblasts, myofibroblasts, myoepithelial cells, macrophages, endothelial cells, leucocytes, and extracellular matrix (ECM) and soluble factors derived from tumor cells. This inflammatory environment surrounding a tumor promotes the breaking of the basal membrane, a process required for the invasion and migration of metastatic cells [60]. Tumor cells are also capable of produce cytokines and chemokines that facilitate evasion of the system immune and help to establishment and development of metastasis (Fig. 4). The increase of tumor-associated macrophages (TAMs) is associated with poor prognosis through various mechanisms: a) release by macrophage IL-10 and prostaglandin E2 which suppress antitumor response, b) easy to release angiogenic factors as VEGF, EGF, endothelin-2 and plasminogen activator promote tumor growth, c) to facilitate cell invasion metastasis by releasing matrix metalloproteinases and induce TNF production and vasodilatation enzyme nitric oxide synthase [61-62].

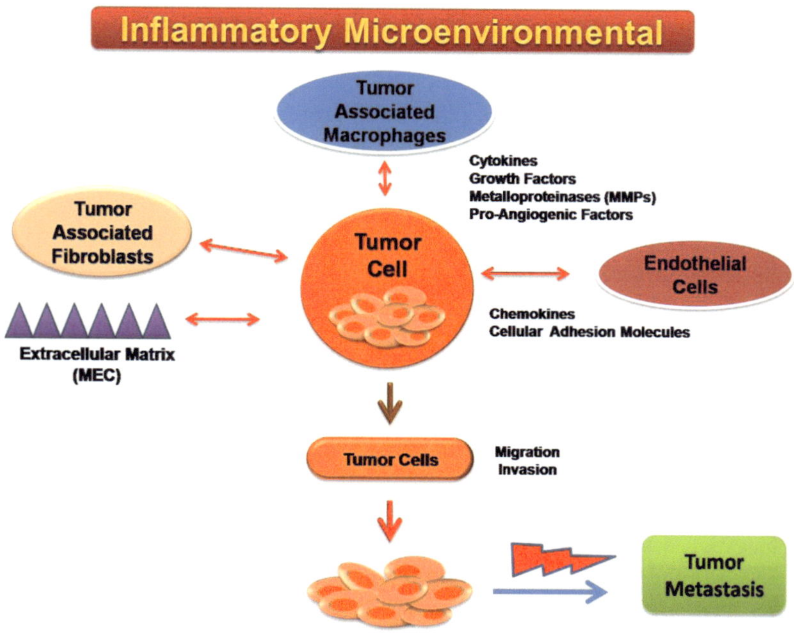

Figure 4. The tumor microenvironment and its role in promoting tumor growth

Cells grow within defined environmental sites and are subject to microenvironmental control. Outside of their sites, normal cells lack appropriate survival signals. During tumor development and progression, malignant cells escape the local tissue control and escape death. Diverse chemoattractant factors promote the recruitment and infiltration of these cells to the tumor microenvironment where they suppress the antitumor immunity or promote tumor angiogenesis and vasculogenesis.

TNF is expressed in low amounts by other cells such as fibroblasts, smooth muscle cells and tumor cells, his target cell are primary endothelial cells, inducing their activation by changing expression levels of some membrane proteins, primarily as adhesion molecules E-selectin, ICAM-1 and VCAM-1 whose expression and synthesis, are regulated by the transcription factor kB nuclear. Activated endothelial phenotype induced by TNF and characteristics of the inflammatory response, have served to comparing the endothelial phenotype has been observed, is produced in response to contact with soluble tumor factors [63-64].

The nuclear factor kappa B (NF-kB) is a transcription factor paramount in regulation of inflammatory response genes Early involved in cell-cell interaction, communication intercellular recruitment or transmigration, amplification of signals pathogenic and acceleration of tumorigenesis [52-53,65].

The study of tumor cells to modify their microenvironment has been a growing area of interest, which has identified the secretion of pro-inflammatory cytokines such as TNF, IL-6) and chemokines such as IL-8, it is interesting to note that these products are known modulators of endothelial function [53].

In recent years, it has been found that tumor cells secrete soluble factors, which modify the endothelial constitutive phenotype, and that exposure to these factors increase to a greater or less extent the capacity to adhere endothelial human tumor cells. It has been recognized that these soluble factors released by tumor cells or non-tumor cells surrounding the tumor play an important role in tumor progression [66].

Our group has shown that breast cancer cells, lymphomas, with high metastatic potentially induces a change in human endothelial cells (HUVECs) that is characterized by promoting a pro-adhesive endothelial phenotype, the expression of intercellular adhesion molecules (ICAM-1, VCAM-1 and E-Selectin) and the activation of NF-kB [66]. These studies include soluble factor leukemias (EUHE, Eusebia), cervical cancer (HeLa) and mammary gland cancer (MCF-7, ZR), oral cancer. In all cases, we have used primary cultures of human endothelial cells (HUVECs), which have generated an *in vitro* model to study the tumor microenvironment. This model is based on induction of a pro-adhesive endothelial phenotype that is associated with expression of adhesion molecules E-selectin, ICAM-1, VCAM dependent activation of NF-kB. These effects are considered essential in the process of adhesion and extravasation during the inflammatory reaction.

Moreover, we have analyzed the biochemical composition of the soluble factors derived from tumor cells. Proteins such as 27, cytokines, chemokines and growth factors associated with the inflammatory reaction; (IL-1 ra, IL-1 beta, IL-2, IL-4, IL-5, IL-6, IL-7, IL-9, IL-10, IL-12, IL-13, IL-15, IL-16, TNF, IFN-gamma, IP-10,RANTES), (IL-8, Eotaxina, MCP-1, MIP-1 beta, MIP-1 alpha), (G-CSF, GM-CSF, PDGF, FGF, VEGF). The molecules identified most significant expression in breast carcinomas, were (VEGF, GM-CSF, IL-1ra, IL6, IL8, IP10, RANTES), which play an important role in endothelial cell activation [52-53,65-66].

VEGF:

An unexpected finding was the abundant presence of the expression of vascular endothelial growth factor (VEGF) in breast carcinoma lines. Whereas in normal cells the expression of VEGF is dependent on a condition of hypoxia, surprisingly VEGF production by tumor cell lines and occurred at a concentration of oxygen partial indicating 20% indicative of impaired VEGF expression. In cancer constitutive secretion of VEGF independent of a condition that favors hypoxic tumor angiogenesis and growth in a clinical setting and is a marker of poor prognosis

[67]. In cancer, constitutive secretion of VEGF-independent hypoxic condition favors tumor angiogenesis and growth in a clinical setting and is a marker of poor prognosis [68].

GM-CSF

The involvement of GM-CSF (Granulocyte Macrophage Colony Stimulating Factor) in cancer is complex, since it seems to require the presence of other cytokines such as IL-4 and IL-6, and there are reports antagonistic to their involvement in tumorigenesis. For example, expression of GM-CSF and interleukin-12, inhibits the immune response and being expressed on B16 murine melanoma cells, decreasing completely tumorigenic capacity of syngenic mice C57B/6. In contrast to the line described breast epithelial MCF7-A benign growth factors secreted able to induce expression of IL-6 and GM-CSF in the tumor line R2T1AS breast cancer that is associated with a higher rate of growth and a higher tumorigenic capacity in vivo. Moreover, continued exposure of GM-CSF plus IL-4 of mesenchymal cells from human bone marrow, resulted in an increase in the morphological transformation and increased rate of growth both *in vitro* and *in vitro* mesenchymal cells, indicating the induction of a transformed phenotype. Whereas previous reports indicate that the involvement of GM-CSF requires other cytokines [69-70].

IL-1ra

It is reported that the soluble form of IL-1 receptor could serve as a functional antagonist of IL-1, the presence of this protein is controversial, since IL-1 by trapping an anti-inflammatory effect would reduce the adhesion of cells added by tumor cells and their extravasation, given this scenario could be proposed that tumor cells not only produce pro-inflammatory factors through the release of soluble factors tumor, but it can regulate the inflammatory process [71].

IL-6

The IL-6 is recognized as a classic inducer of the states of chronic inflammation and that promote the activation of vascular endothelium. It has also been reported that IL-6 is produced by tumors that develop metastases to the liver, as the case of colon and mammary gland cancer. The activity of this cytokine in the soluble factors tumor could be further enhanced by the presence of other co-factors secreted by cells [72-73].

IL-8

Expression of IL-8 in colorectal cancer favors an increase in tumorigenesis and metastasis, this increase is due to IL-8 is associated with expression of MMP-2 and MMP-9, the activity of these metalloproteases have been identified in physiological tissue remodeling processes like normal healing, but also participate in tissue remodeling associated with pathological processes including invasion [74]. Also in melanomas have been identified as IL-8 acts as an angiogenic factor and also promotes mitosis, therefore the IL-8 has been postulated as a potential therapeutic target. In particular, we have sought to interfere with IL-8 with the purpose of reducing tumor growth, an alternative that has developed is the use of an anti-IL-8 in nude mice with liver cancer. The results show that administration of neutralizing anti-IL8 significantly decreased tumor growth, even more interestingly this decrease is associated with decreased expression of MMP-2 and MMP-9. Something similar is observed using the same experimental treatment of melanoma with a decrease in angiogenesis [75]. In our results, we found increases in IL-8 in the soluble factors soluble breast cancer [66]. This indicates that IL-8 could be used as a marker associated with tumor progression, regardless of tumor type. In patients where they identified the production of IL-8, using neutralizing antibodies that interfere with their signaling could serve as a therapeutic alternative.

IP10

IP10 is a protein induced by interferon, has been reported that this protein inhibits proliferation and metastatic tumors, that expression of IP10 in patients with stages II and III colorectal cancer correlate with the development of metastasis and a poor prognosis. The detection of IP10 could be used as a prognostic marker in stage II and III colorectal cancer patients. This has led to propose that IP10 might be used as a prognostic marker in stage II and III colorectal cancer patients [76-77].

RANTES

Regulated upon activation normal T-cell expressed, and secreted is a chemokine that belongs to the CC class, which distinguishes it from IL8, which belongs to the CXC class. RANTES expression in tumor cells has been associated with tumorigenesis and is consistent with our finding of RANTES the products secreted by breast carcinomas. The presence of RANTES in the tumor microenvironment may be chemoattractant to tumor cells. From this point of view is interesting that the tumor-associated endothelial cells, when stimulated with MIF1-alpha can release RANTES [78]. Study that evaluated the expression of RANTES and its receptor CCR5 in 60 patients with metastatic gastric cancer, were identified elevated expression levels, where it is concluded that RANTES and its receptor may contribute to gastric cancer metastasis by promoting responses TH1 and TH2. By comparing a variety of biological markers in a group of biopsies of mammary gland cancer, RANTES was the only marker present in all biopsies [79].

The reported findings strengthen the idea that soluble factors of tumor microenvironment may be relevant in the final stages of the metastatic spread and that these effects may be mediated by cytokines, chemokines, and growth factors present in the soluble factors secreted by tumor cell lines. These elements found in high concentrations are known to be capable of inducing the activated phenotype of endothelial cells to a variety of physiological and pathological cellular responses.

3. Oxidative stress and free radicals: role in cancer development

During the inflammatory process macrophages and endothelial cells, generate a large amount of growth factors, cytokines and reactive oxygen species (ROS) and nitrogen (RNS) that can cause DNA damage. If macrophages and remain on the endothelium may allow the tissue damage continues chronic inflammation predisposes to malignancy [56,80].

3.1. Introduction

In different pathological process the cell injury is induced by free radicals, is an important mechanism of cell damage in many pathological conditions, such as chemical and radiation injury, ischemia injury, cellular aging and some immune system cells such as the phagocytes [82-84]. The free radicals are an example of instability in a biological system; namely, are chemical substances that they have an unpaired electron in its final orbit, this causes that their energy to be unstable and for they become stable they need for molecules which are adjacent to it either organic or inorganic for example: lipids, proteins, carbohydrates and nucleic acids essentials compounds in the cells mainly in the membrane and the core.

These types of chemical species may be either.

1. Oxygen derived (reactive oxygen species ROS)
2. Nitrogen derived (reactive nitrogen species RNS)

Reactive oxygen species (ROS) and reactive nitrogen (RNS) (Table 1) are created in the some cells such as the hepatocytes and in different normal physiologic processes, including, oxidative respiration, growth, regeneration, apoptosis and microbial killing by phagocytes [83]. The generation of this species chemical types, is normal in a normal cells; however, when these start to produce in excess and the antioxidant system is deficient, oxidative stress occurs. This causes damage cells: hepatocytes, kupffer cells and endothelial cells, through induction of inflammation, ischemia, fibrosis, apoptosis, necrosis or other atypical transformation in the cell structure and function.

Type of radical	Activating enzyme	Physiological process
Nitrogen derived (nitric oxide)	Nitric oxide synthase	Smooth muscle (control or vascular tone) and other cGMP- depended functions. (glycogenolysis, apoptosis, conductance regulator ion channels, vasodilatation and increased blood flow)
Oxygen derived (ROS)	NADPH oxidase	Oxidation-reduction reactions within the cell, muscle relaxation, control of erythropoietin production, signal transduction from various receptors, enhancement of immunological functions and oxidative

Table 1. *Important physiological process involved with the free radicals.ROS, Reactive oxygen species.*

3.2. Reactive oxygen species

Reactive oxygen species are produced in normal condition them in a living cell during cellular respiration, energy production and various events of growth and cell death, these are degrade by the defensive systems. Therefore, the cells self-regulate their production and degradation of ROS is found transiently in the cell without causing any damage to the cellular level, and for that reason the cell maintains an equilibrium constant but as this production increases oxidative stress is generated, this relates whit different pathological process such as damage in the cell structure and function, degenerative process and cancer, also influence within the inflammation and the immune response, as these are generated by macrophages and neutrophils as mediators for the destruction of pathogens and dead tissue. The generation of free radicals can be made by different pathways [85-86]. (Table 2)

Redox reactions in metabolic processes.	During cellular respiration O_2 is reduced by four electrons to the transport of H_2 for generating two molecules of water through an oxidative enzyme which results is the formation of superoxide anion (electron), hydrogen peroxide (two electron) and hydroxyl ions (three electrons).
Absorption of radiant energy	Electromagnetic radiation (x rays), gamma rays, infrared, UV, microwave etc. These to hydrolyze the water and generate hydroxyl ions and hydrogen
Inflammation and immune response.	This response induced by activated leukocytes, that is caused by a protein complex located at the plasma membrane that employs NADPH oxidase and some intracellular oxidase and this generated a superoxide anion.
Metabolism of drugs	Most chemicals do not show biological activity in its native form these have to become toxic reactive metabolites to act on their target molecules. This is made by oxidases. These metabolites induce the formation of free radicals, such as acetaminophen.
Metals	The transmission metals donate or accept electrons intracellular free during the reactions thus causing the formation of free radicals
Nitric oxide	Nitric oxide is generated by different cells are an example of this are the endothelial cells giving rise to reactive species of nitrogen, such as the peroxynitrite anion ($ONOO^-$) also NO_2 and $NO_3^{\cdot}$.

Table 2. The generation of free radicals can be made by different pathways.

3.3. Free radical and carcinogenesis

Free radicals are atoms or groups of atoms that in their atomic structure present one or more unpaired electrons in the outer orbit. These free radicals steal electrons from other molecules in effort to heal themselves, ultimately creating new free radicals in the process by stealing electrons. Free radicals are formed from a number of causes such as cigarette smoke, pollution, exposure to sunlight all cause the formation of free radicals. Other factors include normal daily processes like food digestion and breathing.

When increased production of reactive species and have a deficiency in the antioxidant system they cause significant neoplastic changes. In some diseases, such as Bloom syndrome develops lymphomas, leukemias and carcinomas, in anemia are implicated the production of these and alterations of antioxidant defense mechanisms at the systemic level [82-83]. Some epidemiological information indicates that tumor incidence is lower in populations where the diet is rich in antioxidants like fruits and vegetables [84].

Tumor cells have a high activity of free radical formation in contrast to healthy cells. It is known that tumor cells not only produce oxygen peroxide (H_2O_2) but also decreases the production of antioxidants such as glutathione peroxidase and SOD.

Some pathways by which cancer cells have high amounts of ROS is for multiple factors:

Increasing the metabolic activity of a neoplastic cell, lead to the increased energy requirements (ATP) produced in the mitochondria this in order to enhance the growth and proliferation.

1. The progression of cancer, primarily because of the damage they cause in to the genetic material of a normal cell. Has been shown that the oxidation of guanine to 8-oxo-dG (oxidation product DNA) induces errors in their replication. By dependent of DNA polymerase that generates the nitrogenous base pairing, not complementary, to permit the establishment of hydrogen bonds with (A) adenine and (T) thymine [85].

2. Oncogenic transformation is conditioned by the presence of mutated genes or oncogenes that control essential cellular functions in which the redox state within or outside the cellular microenvironment is very important ROS are potential carcinogens because they facilitate mutagenesis, tumor growth, and metastasis and all those process that has been showed [86].

Cancer is a multifactorial disease where endogenous and exogenous factors are involved but the roles played by free radicals in this disease are very important because, these produce damage in de DNA structure and this produces an important negative effects [86].

4. Antioxidants and Chemoprevention in cancer

Antioxidants are substances that prevent damage to cells caused by free radicals, it can cause damage to DNA, leading to the possible development of cancer [87]. Antioxidants search for these free radicals and lend them an, this stabilizes the molecule, thus preventing damage to other cells. Antioxidants also turn free radicals into waste by products, and they eventually are eliminated from the body. The inability of our body to neutralize free radicals we are exposed daily forces us to rely on foods with antioxidant properties capable of neutralizing them [88].

4.1. Flavonoids

Flavonoids are found in numerous plants and vegetables, with a wide distribution through the plant kingdom. This class compounds numbers more than 4000 members and can be divided into five subcategories: flavones, monomeric flavanols, flavanones, flavonols and anthocyanidines. Are natural compounds chemically derivate from bezo-y-pirone (phenyl-chromone) or flavone. They are considered important constituents of the human diet. It has been reported that they exert multiple biological effects due to their antioxidant and free radical-scavenging abilities [89].

Flavanoids possess a lot of pharmacological and therapeutically properties; antioxidant, antitumor, antiangiogenic (vascular protective), anti-inflammatory, antiallergic, antihepatotoxic, anticancerigenic, antimutagenic, anticancer effects, antiosteoporotic, and antiviral properties [90]. Many studies emphasize take adequate diets that are active allies against cancer. These diets are based on enzymes and antioxidant substances in certain foods that are rich in components that collect above [91].

They also have the ability to repair previous damage to cells, examples of antioxidants include (beta-carotene, lycopene, vitamins C, E, and A), and other substances. Nutrients such as; green tea, flavonoids, vitamins C, E, and Beta-carotene in the carcinogen process, has been showed that have function in the elimination of carcinogenic factors, inhibition of pre-carcinogens and reparation of DNA damage. The mechanisms are diverse and range from inhibition to an active reaction of the immune system in general. This has caused the use of multiple antioxidant micronutrients as preventive agents [90]. Several experimental data have demonstrated the antiproliferative and anti-carcinogenic and the role of chemopreventive agent of flavonoids [91-92].

Currently investigations are performed to determine the mechanisms by which act flavonoids, because it has been observed that their effects are greater at high doses, which gives them inducing side effects, so it is important to moderate their consumption by a balanced diet.

5. Conclusions

It is important to analyze the role of tumor-associated inflammatory microenvironment and has been identified that plays an important role in tumor progression. This microenvironment is composed of molecules that play an important role in inflammatory processes and chronic, and favor the invasion and metastasis process that triggers the death of many people with any cancer.

The installation of tumor cells in blood vessels of the target organ to invade, is related to phenotypic changes in the endothelium allowing vascular extravasation of blood circulation of leukocytes in the inflammatory reaction and, as hypothesized current of tumor cells with metastatic capacity. The phenomenon of extravasation in response to a cell interactions between tumor cells and endothelial cells or not allowing the passage of cells whether there are appropriate conditions for the invasion.

Understanding the molecular basis of these interactions between metastatic cells and endothelial cells, will enable us to design strategies to interfere with this inter-cellular communication. It is important to recognize the tumor-associated inflammatory microenvironment and what is the contribution to tumor progression. The importance of these factors on endothelial activation being evaluated by reconstituting the mixture with cytokines, chemokines and growth factors recombinant depleted mixtures of tumor soluble factors of each of these proteins by specific monoclonal antibodies.

Is important mention that during the inflammatory process macrophages, fibroblasts and endothelial cells generate a large amount of growth factors, cytokines, chemokines and reactive oxygen species (ROS) and nitrogen (RNS) that can cause DNA damage. These process allow the tissue damage continues chronic inflammation predisposes to malignancy. Therefore, it is important to note that people with chronic degenerative diseases, which clearly show chronic inflammatory processes, they may promote or contribute to present or develop a tumor lesion.

The use of antioxidants consumed in a balanced diet can be used as an element in the diet that can become a preventive or contributing to diminish the appearance of a tumor lesion.

Author details

César Esquivel-Chirino[1,4,5,6*], Jaime Esquivel-Soto[1,6], José Antonio Morales-González[2],
Delina Montes Sánchez[3,4,5], Jose Luis Ventura-Gallegos[4,5],
Luis Enrique Hernández-Mora[1] and Alejandro Zentella-Dehesa[4,5]

*Address all correspondence to: cesquivelch@gmail.com

1 Facultad de Odontología, Universidad Nacional Autónoma de México, México

2 Instituto de Ciencias de la Salud, Universidad Autónoma del Estado de Hidalgo (UAEH), México

3 Programa de Genómica Funcional de Procariotes, Centro de Ciencias Genómicas, Universidad Nacional Autónoma de México, Campus Morelos, México

4 Departamento de Medicina y Toxicología Ambiental, Instituto de Investigaciones Biomédicas. Universidad Nacional Autónoma de México, México

5 Departamento de Bioquímica. (INNCMSZ) Instituto Nacional de Ciencias Médicas y Nutrición "Salvador Zubirán" México D.F., México

6 Facultad de Odontología, Universidad Intercontinental, México

References

[1] Ferlay, J., Bray, P., Pisani, P., & Parkin, D. M. (2004). GLOBOCAN 2002: Cancer incidence, mortality and prevalence worldwide. *IARC Cancer Base Version 2.0. Lyon: IARC Press* [5].

[2] World Health Organization. (2008). World cancer report 2008. *Lyon (France): IARC.*

[3] World Health Organization. (2007). Ten statistical highlights in global public health. *World health statistics 2007. Geneva: WHO.*

[4] Jemal, A., Ward, E., & Thun, M. (2005). Cancer statistics. *In: DeVita VJ, Hellman S, Rosenberg S, editors. Cancer principles and practice of oncology. 7th ed. Baltimore (MD): Lipppincott Williams & Wilkins*, 226-240.

[5] International Agency for Cancer Research (IARC). (2012). CANCER Mondial. available from: http://www-dep.iarc.fr/Accessed on June 22

[6] Jemal, A., Bray, F., Center, M. M., Ferlay, J., Ward, E., & Forman, D. (2011). Global cancer statistics. *CA Cancer J Clin*, 61, 69-90.

[7] Kanavos, P. (2006). The rising burden of cancer in the developing world. *Ann Oncol*, 17(8), 15-23.

[8] Kolonel, L., Wilkens, L., Schottenfeld, D., & Fraumeni, J. F. Jr. (2006). Cancer epidemiology and prevention. *3rd ed Oxford: Oxford University Press*, 189-201.

[9] Bacac, M., & Stamenkovic, I. (2008). Metastatic cancer cell. *Annu Rev Pathol*, 3, 221-47.

[10] Jackson, S. P., & Bartek, J. (2009). The DNA-damage response in human biology and disease. *Nature*, 461, 1071-1078.

[11] Pantel, K., Brakenhoff, R. H., & Brandt, B. (2008). Detection, clinical relevance and specific biological properties of disseminating tumor cells. *Nat Rev Cancer*, 8, 329-340.

[12] Hanahan, D., & Weinberg, R. (2000). The hallmarks of cancer. *Cell*, 100, 57-70.

[13] Oldham Hikman, Elizabeth. (2004). Intrinsic oxidative stress in cancer cells a biological basis for therapeutic selectivity" Cancer". *Cancer chemother pharmacol*, 53, 209-19.

[14] Steinmetz, K. A., & Potter, J. D. (1991). Vegetables, fruit, and cancer. *I. Epidemiology. Cancer Causes Control.*, 2, 325-357.

[15] Foulds, L. (1954). The experimental study of tumor progression: a review. *Cancer Res*, 14, 327-339.

[16] Ann, F., Chambers, Alan. C., & Groom, Iand. Mc Donald. (2002). Dissemination and growth cancer cells in metastatic sites. *Nature Reviews*, 2.

[17] Alby, L., & Auerbach, R. (1984). Differential adhesion of tumor cells to capillary endothelial cells in vitro. *Proc Natl Acad Sci USA*, 81, 5739-43.

[18] Balkwill, F., & Mantovani, A. (2001). Inflammation and cancer: back to Virchow? *Lancet*, 357, 539-45.

[19] Hanahan, D., & Weinberg, R. A. (2011). Hallmarks of cancer: the next generation. *Cell.*, 44, 646-674.

[20] Chaffer, C.L, & Weinberg, R. A. (2011). A perspective on cancer cell metastasis. Science. Mar 25; , 331(6024), 1559-64.

[21] Nicolson, G. L. (1993). Cancer progression and growth: relationship of paracrine and autocrine growth mechanisms to organ preference of metastasis. *Exp Cell Res*, 204, 171-80.

[22] Kopfstein, L., & Christofori, G. (2006). Metastasis: cell-autonomous mechanisms versus contributions by the tumor microenvironment. *Cell Mol Life Sci.*, 63(4), 449-68.

[23] Calorini, Lido, & Bianchini, Francesca. (2010). Environmental control of invasiveness and metastatic dissemination of tumor cells: the role of tumor cell-host cell interactions. *Cell Communication and Signaling*, 8, 24.

[24] Tang, D. G., & Conti, C. J. (2004). Endothelial cell development, vasculogenesis, angiogenesis, and tumor neovascularization: an update. *Semin Thromb Hemost.*, 30(1), 109-17.

[25] Aird, W. C. (2009). Cell Tissue Res. Molecular heterogeneity of tumor endothelium. Epub 2008 Aug 23., 335(1), 271-81.

[26] Risau, W. (1995). Differentiation of endothelium. *FASEB J*, 9, 926-33.

[27] Ribatti, Domenico, Nico, Beatrice, Vacca, Angelo, Roncali, Luisa, & Dammacco, Franco. (2002). Journal of Hematotherapy & Stem. *Cell Research.*, 11(1), 81-90.

[28] Augustin, H. G., Kozian, D. H., & Johnson, R. C. (1994). Differentiation of endothelial cells: Analysis of the constitutive and activated endothelial cell phenotypes. *Bioessays*, 16, 901-906.

[29] Geng, J. G. (2003). Interaction of vascular endothelial cells with leukocytes, platelets and cancer cells in inflammation, thrombosis and cancer growth and metastasis. *Acta Pharmacol Sin.*, 24(12), 1297-300.

[30] Pober, J.S. (2002). Arthritis Res. 4(3), 109-16, *Epub May 9. Endothelial activation: intracellular signaling pathways.*

[31] Riscoe, D. M., Cotran, R. S., & Pober, J. S. (1992). Effects of tumor necrosis factor, lipopolysaccharide, and IL-4 on the expression of vascular cell adhesion molecule-1 in vivo. *Correlation with CD3+ T cell infiltration. J Immunol.*, 149, 2954-2960.

[32] Joan, M., & Cook-Mills, Tracy. L. Deem. (2005). Active participation of endothelial cells in inflammation. *J Leukoc Biol.*, 77(4), 487-495.

[33] Wagner, M., Bjerkvig, R., Wiig, H., Melero-Martin, J. M., Lin, R. Z., Klagsbrun, M., & Dudley, A. C. (2012). Inflamed tumor-associated adipose tissue is a depot for macrophages that stimulate tumor growth and angiogenesis. Angiogenesis. May 22.

[34] Bendas, G, & Borsig, L. (2012). Cancer cell adhesion and metastasis: selectins, integrins, and the inhibitory potential of heparins. *Int J Cell Biol*, 676-731.

[35] Chambers, A. F., Groom, A. C., & Mac Donald, I. C. (2002). Dissemination and growth of cancer cells in metastatic sites. *Nature Reviews Cancer*, 2(8), 563-572.

[36] Calorini, Lido, & Bianchini, Francesca. (2010). Environmental control of invasiveness and metastatic dissemination of tumor cells: the role of tumor cell-host cell interactions. Cell Communication and Signaling, , 8, 24.

[37] Rubin, H. (2008). Contact interactions between cells that suppress neoplastic development: can they also explain metastatic dormancy? *Adv Cancer Res*, 100, 159-202.

[38] Aguirre-Ghiso, J. A. (2007). Models, mechanisms and clinical evidence for cancer dormancy. *Nat Rev Cancer*, 7, 834-846.

[39] Nicolson, G. L. (1993). Cancer progression and growth: relationship of paracrine and autocrine growth mechanisms to organ preference of metastasis. *Exp Cell Res*, 204, 171-80.

[40] Gasic, G. J. (1986). Role of plasma, platelets and endothelial cells in tumor metastasis. *Cancer Metastasis Rev*, 3, 99-116.

[41] De Visser, K. E., Korets, L. V., & Coussens, L. M. (2005). De novo carcinogenesis promoted by chronic inflammation is B lymphocyte dependent. *Cancer Cell*, 7, 411-423.

[42] Di Carlo, E., Forni, G., Lollini, P., Colombo, M. P., Modesti, A., & Musiani, P. (2001). The intriguing role of polymorphonuclear neutrophils in antitumor reactions. *Blood*, 97, 339-345.

[43] De Larco, J. E., Wuertz, B. R., & Furcht, L. T. (2004). The potential role of neutrophils in promoting the metastatic phenotype of tumors releasing interleukin-8. *Clin Cancer Res*, 10, 4895-4900.

[44] Silzle, T., Randolph, G. J., Kreutz, M., & Kunz-Schughart, L. A. (2004). The fibroblast: sentinel cell and local immune modulator in tumor tissue. *Int J Cancer*, 108, 173-180.

[45] Kalluri, R., & Zeisberg, M. (2006). Fibroblasts in cancer. *Nat Rev Cancer*, 6, 392-401.

[46] Nicolson, G. L. (1993). Cancer progression and growth: relationship of paracrine and autocrine growth mechanisms to organ preference of metastasis. *Exp Cell Res*, 204, 171-80.

[47] Mehlen, P., & Puisieux, A. (2006). Metastasis: a question of life or death Nat. *Rev. Cancer*, 6, 449-458.

[48] Fidler, I. J. (2005). Cancer biology is the foundation for therapy. *Cancer Biol. Ther.*, 4, 1036-1039.

[49] Almog, N. (2010). Molecular mechanisms underlying tumor dormancy. *Cancer Lett*, 294, 139-146.

[50] Klein, C. A. (2009). Parallel progression of primary tumours and metastases. *Nat. Rev. Cancer*, 9, 302-312.

[51] Lu, H., Ouyang, W., & Huang, C. (2006). Inflammation, a key event in cancer development. *Mol Cancer Res*, 4, 221-233.

[52] Lopez-Bojorquez, L. N. (2004). Regulation of NF-kappaB transcription factor. *A molecular mediator in inflammatory process. Rev Invest Clin.*, 56, 83-92.

[53] Lopez-Bojorquez, L. N., Arechavaleta-Velasco, F., Vadillo-Ortega, F., Montes-Sanchez, D., Ventura-Gallegos, J. L., & Zentella-Dehesa, A. (2004). NF-kappaB translocation and endothelial cell activation is potentiated by macrophage released signals co-secreted with TNF-alpha and IL-1beta. *Inflamm Res.*, 53, 567-575.

[54] Smid, M., Wang, Y., Zhang, Y., Sieuwerts, A. M., Yu, J., Klijn, J. G., Foekens, J. A., & Martens, J. W. (2008). Subtypes of breast cancer show preferential site of relapse. *Cancer Res*, 68, 3108-3114.

[55] Wu, J. M., et al. (2008). Heterogeneity of breast cancer metastases: comparison of therapeutic target expression and promoter methylation between primary tumors and their multifocal metastases. *Clin Cancer Res*, 14, 1938-1946.

[56] O'Hanlon, D. M, Fitzsimons, H, Lynch, J, Tormey, S, Malone, C, & Given, H. F. (2002). Soluble adhesion molecules (E-selectin, ICAM-1 and VCAM-1) in breast carcinoma. *Eur J Cancer.*, 38, 2252-2257.

[57] Kim, I., Moon, S. O., Kim, S. H., Kim, H. J., Koh, Y. S., & Koh, G. Y. (2001). Vascular endothelial growth factor expression of intercellular adhesion molecule 1 (ICAM-1), vascular cell adhesion molecule 1 (VCAM-1), and E-selectin through nuclear factor-kappa B activation in endothelial cells. *J Biol Chem.*, 276, 7614-7620.

[58] Lieder, A. M., Prior, T. G., Wood, K. J., & Werner, J. A. (2005). The relevance of adhesion molecules in the classification of 72 squamous cell carcinoma of the head and neck. *Anticancer Res*, 25, 4141-4147.

[59] Okegawa, T, Li, Y, Pong, R. C., & Hsieh, J. T. (2002). Cell adhesion proteins as tumor suppressors. *J Urol.*, 167, 1836-1843.

[60] Colotta, F., Allavena, P., Sica, A., Garlanda, C., & Mantovani, A. (2009). Cancer-related inflammation, the seventh hallmark of cancer: links to genetic instability. *Carcinogenesis.*, 30, 1073-81.

[61] Kawaguchi, T. (2005). Cancer metastasis: characterization and identification of the behavior of metastatic tumor cells and the cell adhesion molecules, including carbohydrates. *Curr Drug Targets Cardiovasc Haematol Disord*, 39-64.

[62] Thorne, R. F., Legg, J. W., & Isacke, C. M. (2004). The role of the CD44 transmembrane and cytoplasmic domains in coordinating adhesive and signalling events. *J Cell Sci.*, 117, 373-380.

[63] Li, A., Li, H., Jin, G., & Xiu, R. (2003). A proteomic study on cell cycle progression of endothelium exposed to tumor conditioned medium and the possible role of cyclin D1/E. *Clin Hemorheol Microcirc.*, 29, 383-390.

[64] Watts, M. E., Parkins, C. S., & Chaplin, D. J. (2002). Influence of hypoxia and tumour-conditioned medium on endothelial cell adhesion molecule expression in vitro. *Anticancer Res.*, 22, 953-958.

[65] Estrada-Bernal, A., Mendoza-Milla, C., Ventura-Gallegos, J. L., Lopez-Bojorquez, L. N., Miranda-Peralta, E., Arechavaleta-Velasco, F., Vadillo-Ortega, F., Sanchez-Sanchez, L., & Zentella-Dehesa, A. (2003). NF-kappaB dependent activation of human endothelial cells treated with soluble products derived from human lymphomas. *Cancer Lett*, 191, 239-48.

[66] Montes-Sanchez, D., Ventura, J. L., Mitre, I., Frias, S., Michan, L., Espejel-Nunez, A., Vadillo-Ortega, F., & Zentella, A. (2009). Glycosylated VCAM-1 isoforms revealed in 2D western blots of HUVECs treated with tumoral soluble factors of breast cancer cells. *BMC Chem Biol*, 9, 7.

[67] Baldewijns , M. M., van Vlodrop, I. J., Vermeulen, P. B., Soetekouw, P. M., van Engeland, M., & de Bruïne, A. P. (2010). VHL and HIF signalling in renal cell carcinogenesis. *J Pathol.*, 221(2), 125-38.

[68] Kamada, H., Tsutsumi, Y., Kihira, T., Tsunoda, S., Yamamoto, Y., & Mayumi, T. (2000). In vitro remodeling of tumor vascular endothelial cells using conditioned medium from various tumor cells and their sensitivity to TNF-alpha. *Biochemical and Biophysical Research Communications*, 268, 809-813.

[69] Edeline, J., Vigneau, C., Patard, J. J., & Rioux-Leclercq, N. (2010). Signalling pathways in renal-cell carcinoma: from the molecular biology to the future therapy]. *Bull Cancer*, 97, 5-15.

[70] Kulbe, H., Chakravarty, P., Leinster, D. A., Charles, K. A., Kwong, J., Thompson, R. G., Gallagher, W. M., Galletta, L., Salako, M. A., Smyth, J. F., Hagemann, T., Brennan, D. J., Bowtell, D. D., & Balkwill, F. R. (2011). A dynamic inflammatory cytokine network in the human ovarian cancer microenvironment. *Cancer Res.*

[71] Friberg, E., Orsini, N., Mantzoros, C. S., & Wolk, A. (2007). Diabetes mellitus and risk of endometrial cancer: a meta-analysis. *Diabetologia.*, 50(7), 1365-74.

[72] Mc Lean, M. H., Murray, G. I., Stewart, K. N., Norrie, G., Mayer, C., Hold, G. L., Thomson, J., & El -Omar, E. M. (2011). The inflammatory microenvironment in colorectal neoplasia. PLoS One. Jan 7 , 6(1), e15366.

[73] Rajkumar, T., Shirley, S., Raja, U. M., & Ramakrishnan, S. A. Identification and validation of genes involved in gastric tumorigenesis. *Cancer Cell Int.*, 10, 45.

[74] Huang, S., Mills, L., Mian, B., Tellez, C., Mc Carty, M., Yang, X. D., Gudas, J. M., & Bar-Eli, M. (2002). Fully humanized neutralizing antibodies to interleukin-8 (ABX-IL8) inhibit angiogenesis, tumor growth, and metastasis of human melanoma. *Am J Pathol.*, 161(1), 125-34.

[75] Mc Conkey, D. J., & Bar-Eli, M. (2003). Fully human anti-interleukin 8 antibody inhibits tumor growth in orthotopic bladder cancer xenografts via down-regulation of matrix metalloproteases and nuclear factor-kappaB. *Clin Cancer Res.*, 9(8), 3167-75.

[76] Jiang, Z., Xu, Y., & Cai, S. (2010). CXCL10 expression and prognostic significance in stage II and III colorectal cancer. *Mol Biol Rep.*, 37(6), 3029-36.

[77] Utoguchi, N. H., Makimoto, Y., Wakai, Y., Tsutsumi, S., Nakagawa, , & Mayumi, T. (1996). Effect of tumour cell-conditioned medium on endothelial macromolecular permeability and its correlation with collagen. *British Journal of Cancer*, 73, 24-28.

[78] Cao, Z., Xu, X., Luo, X., Li, L., Huang, B., Li, X., Tao, D., Hu, J., & Gong, J. J. (2011). Huazhong Role of RANTES and its receptor in gastric cancer metastasis. Univ Sci Technolog Med Sci. Epub Jun, 31(3), 342-7.

[79] The high level of RANTES in the ectopic milieu recruits macrophages and induces their tolerance in progression of endometriosis. (2010). *J Mol Endocrinol*, 45, 291-299.

[80] Lu, H., Ouyang, W., & Huang, C. (2006). Inflammation, a key event in cancer development. *Mol Cancer Res*, 4, 221-233.

[81] Yan, B., Wang, H., Rabbani, Z. N., Zhao, Y., Li, W., Yuan, Y., Li, F., Dewhirst, M. W., & Li, C. Y. (2006). Tumor necrosis factor-alpha is a potent endogenous mutagen that promotes cellular transformation. *Cancer Res*, 66, 11565-11570.

[82] Kumar et.al. (2010). Robbins and Cortan structural and functional pathology. *ed. the eighth edition. Elsevier sounders Barcelona Spain.*, 19-20.

[83] Hikman, Elizabeth Oldham. (2004). Intrinsic oxidative stress in cancer cells a biological basis for therapeutic selectivity" Cancer". *Cancer chemother pharmacol*, 53, 209-19.

[84] Cerutti, P. (1985). Pro-oxidant states and tumor promotion. *Science*, 227, 375-80.

[85] Migliori, L., et al. (1991). Genetic and environmental factors in cancer an neurodegenerative disease. *Mut Res*, 202(512), 135-153.

[86] Kouchakgjian, M., et al. (1991). MR structural studies of the ionizing radiation adduct 7-hydro-8oxodeoxyguanosine (8-oxo-7H-dG) opposites deoxyadenisine in a D A duplex 8-oxo-7H-7dG(syn)-dA(anti)aligment a lesion site. *Biochem*, 30, 1403.

[87] Elejalde Guerra, J. I. (2001). Oxidative Stress, diseases and antioxidants treatments. *An Med Int (Madrid-Spain)*, 18, 326-335.

[88] Aherne, S. A., & y O'Brien, N. M. (2002). Dietary flavonols: chemistry, food content, and metabolism. *Nutrition*, 18, 75-81.

[89] Jovanovic, S. V., Steenken, S., Simic, M. G. y., & Hara, Y. (1998). Antioxidant properties of flavonoids: reduction potentials and electron transfer reactions of flavonoid radicals. En: Rice Evans C, Parker L (eds.): Flavonoids in health and disease. *Marcel Dekker, Nueva York*, 137-161.

[90] Letan, A. (1966). The relation of structure to antioxidant activity of quercitin and some of its derivates. *J Food Sci*, 31, 518-523.

[91] Stahl, W., Ale-Agha, N. Y., & Polidori, M. C. (2002). Non-antioxidant properties of carotenoids. *Biol Chem*, 383, 553-558.

[92] Stacvric, B. (1994). Quercitin in our diet: From potent mutagen to probable anticarcinogen. *Clinical Biochemistry*, 27, 245-248.

[93] Da Silva, J., Herrmann, S. M., Peres, W., Possa, Marroni. N., Gonzalez Gallego, J. Y., & Erdtmann, B. (2002). Evaluation of the genotoxic effect of rutin and quercetin by comet assay and micronucleus test. *Food Chem Toxicol*, 40, 941-947.

Chapter 9

Oxidative Stress in Diabetes Mellitus and the Role Of Vitamins with Antioxidant Actions

Maria-Luisa Lazo-de-la-Vega-Monroy and
Cristina Fernández-Mejía

Additional information is available at the end of the chapter

http://dx.doi.org/10.5772/51788

1. Introduction

Diabetes mellitus is a group of metabolic diseases characterized by hyperglycemia resulting from defects of insulin action, insulin secretion or both [1]. Diabetes has taken place as one of the most important diseases worldwide, reaching epidemic proportions. Global estimates predict that the proportion of adult population with diabetes will increase 69% for the year 2030 [2].

Hyperglycemia in the course of diabetes usually leads to the development of microvascular complications, and diabetic patients are more prone to accelerated atherosclerotic macrovascular disease. These complications account for premature mortality and most of the social and economical burden in the long term of diabetes [3].

Increasing evidence suggests that oxidative stress plays a role in the pathogenesis of diabetes mellitus and its complications [4]. Hyperglycemia increases oxidative stress, which contributes to the impairment of the main processes that fail during diabetes, insulin action and insulin secretion. In addition, antioxidant mechanisms are diminished in diabetic patients, which may further augment oxidative stress [5, 6]. Several studies have addressed the possible participation of dietary antioxidants, such as vitamins, in ameliorating the diabetic state and retarding the development of diabetes complications [7, 8].

The aim of this chapter is to revise the current knowledge of the role of oxidative stress in the pathogenesis of diabetes mellitus and its complications, and to discuss the existing evidence of the effects of vitamins as antioxidant therapy for this disease.

2. Oxidative stress

At the beginning of life, the organisms obtained their energy (ATP) by anoxygenic photosinthesis, for which oxygen was toxic. Most of the metabolic pathways were developed during this anaerobic stage of life, in which oxygen came later. Cyanobacteria started producing oxygen from photosynthesis, which raised the atmospheric oxygen, and favored those organisms which have evolved into eukaryotic cells with mitochondria, able to use oxygen for a more efficient energy production [9].

Whenever a cell's internal environment is perturbed by infections, disease, toxins or nutritional imbalance, mitochondria diverts electron flow away from itself, forming reactive oxygen species (ROS) and reactive nitrogen species (RNS), thus lowering oxygen consumption. This "oxidative shielding" acts as a defense mechanism for either decreasing cellular uptake of toxic pathogens or chemicals from the environment, or to kill the cell by apoptosis and thus avoid the spreading to neighboring cells [9]. Therefore, ROS formation is a physiological response to stress.

The term "oxidative stress" has been used to define a state in which ROS and RNS reach excessive levels, either by excess production or insufficient removal. Being highly reactive molecules, the pathological consequence of ROS and RNS excess is damage to proteins, lipids and DNA [10]. Consistent with the primary role of ROS and RNS formation, this oxidative stress damage may lead to physiological dysfunction, cell death, pathologies such as diabetes and cancer, and aging of the organism [11].

2.1. ROS and RNS production

ROS and RNS are highly reactive molecules, which can be free radicals such as superoxide ($^{\bullet}O_2^{\,-}$), hydroxyl ($^{\bullet}OH$), peroxyl ($^{\bullet}RO_2$), hydroperoxyl ($^{\bullet}HRO_2^{\,-}$), nitric oxide ($^{\bullet}NO$) and nitrogen dioxide ($^{\bullet}NO_2^{\,-}$), or nonradicals such as hydrogen peroxide (H_2O_2), hydrochlorous acid (HOCl), peroxynitrite ($ONOO^-$), nitrous oxide (HNO_2), and alkyl peroxynitrates (RONOO). Most of the studies regarding diabetes and its complications have addressed the role of superoxide ($^{\bullet}O_2^{\,-}$), nitric oxide ($^{\bullet}NO$), and peroxynitrite ($ONOO^-$) in this disease. There are basically two pathways for $^{\bullet}O_2^{\,-}$ production: NADPH oxidases and mitochondrial function, while $^{\bullet}NO$ and $ONOO^-$ are produced by the Nitric Oxide Synthase pathway [10].

2.1.1. NADPH oxidases

Oxidases are enzymes which catalyze redox reactions involving molecular oxygen (O_2). Superoxide is generated by oxidases via one-electron reduction of oxygen and the oxidation of their substrates. Several oxidases exist in the body, such as xantine oxidase, glucose oxidase, monoamine oxidase, cytochrome P450 oxidase, and NADPH oxidases.

NADP in the cell exists in its reduced (NADPH) and oxidized (NADP+) forms. NADPH supplies reducing power in reactions for biosynthesis, and it also serves as electron donor substrate for the NADPH oxidase. This enzyme is a membrane-bound electron transport complex which pumps electrons from NADPH in the cytosol across biological membranes and into intracellular and extracellular compartments, such as nucleus, endoplasmic reticulum, endosome, phagosome, mitochondria and extracellular space. It is the only enzyme whose primary function is generating superoxide and/or hydrogen peroxide, mainly for preventing the transfer of pathogens and for cellular bactericidal function[12, 13].

2.1.2. Mitochondrial electron transport chain

Mitochondrion is the site of eukaryotic oxidative metabolism. It contains the enzymes needed for converting pyruvate into Acetyl-CoA, the citric acid cycle (also known as the Krebs cycle) and for fatty acid oxidation. Additionally, it performs the electron transport and oxidative phosphorylation. Substrate (amino acid, fatty acid and carbohydrate) oxidation in the citric acid cycle release electrons, which are transferred to the coenzymes NAD+ and FAD to form NADH and FADH2. These electrons then pass into the mitochondrial electron-transport chain, a system of linked electron carrier proteins comprised by Complexes I, II, III and IV. Complex I, III, and IV drive the exit of protons from the mitochondrial matrix, producing a proton gradient across the inner mitochondrial membrane. The free energy stored in this electrochemical gradient drives the condensation of ADP with inorganic phosphate in order to form ATP by oxidative phosphorylation. Along this electron transport, molecular oxygen is the final electron acceptor, which will be then reduced to H_2O [14, 15]. However, between 0.4 and 4% of all oxygen consumed will be converted into superoxide anion [16]. There is also a normal threshold for protonic potential above which electron transfer is inhibited at complex III, causing the electrons to go back to complex II where there are transferred to molecular oxygen prematurely and not to complex IV as it naturally occurs. Therefore, the endproduct of this transfer is superoxide [17].

Mitochondria play an important role in the maintenance of cellular redox status, acting as a redox sink and limiting NADPH oxidase activity. However, when the proton potential threshold is surpassed, mitochondria is also a significant source of ROS, which may further stimulate NADPH oxidases, creating a vicious cycle of ROS production [18]. When mitochondria cannot further extract oxygen, cell and tissue oxygen levels rise, decreasing the tissue extraction of oxygen from the blood. This results in tissue vascularity reduction, which may be associated with peripheral vascular disease and, in time, chronic tissue hypoxia and ischemia [9].

2.1.3. NO and RNS production

Nitric oxide •NO is produced by the enzyme nitric oxide synthase (NOS), of which there are three isoforms: neural (nNOS or NOS-I) expressed in neurons, inducible (iNOS or NOS-II) expressed in smooth muscle of bold vessels, hepatocites, macrophagues and neuroendocirne tissue, and endothelial (eNOS or NOS-III) expressed constitutively in endothelial cells. iNOS and eNOS can be stimulated by the redox state in the cell, cytokines, hormones and nutrients [19, 20]. NOS catalyze the oxidation of the terminal guanidine nitrogen of the L–arginine, in presence of oxygen and NADPH, to yield L-citruline and •NO [21].

Once produced and released, $^{\bullet}NO$ can diffuse freely through membranes or act on different cellular targets. $^{\bullet}NO$ participates as mediator of several physiological effects such as vasorelaxation, macrophague activation, gene expression and apoptosis. Usually, $^{\bullet}NO$ is considered as a vasculoprotective molecule. However, one of its multiple effects is also protein nitrosilation at the thiol groups and RNS generation such as peroxynitrite ($ONOO^-$), as $^{\bullet}NO$ easily reacts with $^{\bullet}O_2^{\ -}$. Therefore, the amount of $^{\bullet}O_2^{\ -}$ determines whether $^{\bullet}NO$ acts as a protective or harmful molecule [10, 22].

2.2. Antioxidant defenses in the organism

As a small part the oxygen consumed for aerobic processes will be converted into superoxide anion [16], which will have to be scavenged or converted into less reactive (and harmful) molecules. The main enzymes that regulate this process are Superoxide dismutase (SOD), Glutathione Peroxidase (GSH-Px) and Catalase (Figure 1). When ROS overproduction or chronic hyperglycemia occurs, the activity of these enzymes is insufficient, leading to more ROS and RNS formation and activation oxidative stress pathways.

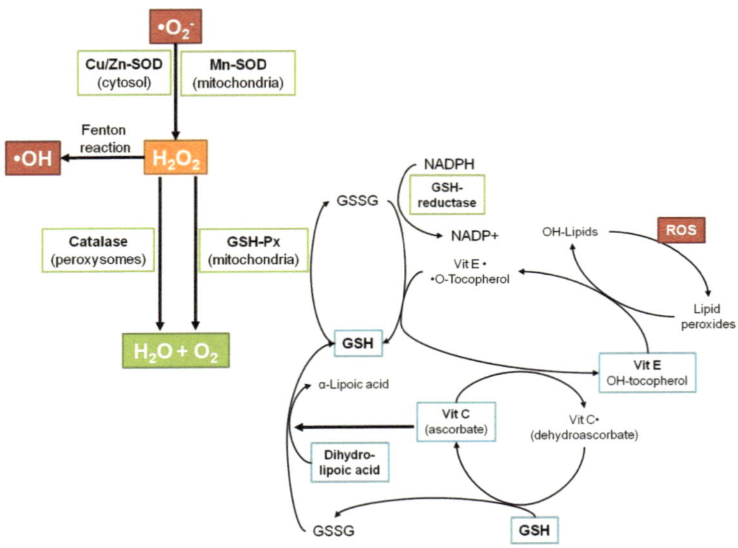

Figure 1. Antioxidant defenses in the organism.

SOD is considered a first-line defense against ROS. This enzyme is present in nearly all cells, and converts $^{\bullet}O_2^{\ -}$ into H_2O_2. Mitochondrial and bacterial SOD contain Mn, while cytosolic SOD is a dimer containing Cu and Zn. As the H_2O_2 may still react with other ROS, it needs to be degraded by either one of the other two antioxidant enzymes, GSH-Px or catalase [10, 12].

GSH peroxidase is located in the mitochondria. It catalyzes degradation of H_2O_2 by reduction, where two gluthathione (GSH) molecules are oxidized to glutathione disulfide (GSSG).

http://dx.doi.org/10.5772/51788

Regeneration of GSH by GSH-reductase, requires NADPH, which is oxidized to NADP+. Catalase, on the other hand, is localized primarily in peroxisomes, and so it detoxifies the H_2O_2 that diffuses from the mitochondria to the cytosol, converting it into water and molecular oxygen [10, 12].

There are also nonenzymatic antioxidant mechanisms, which mostly help regenerate GSSG back into GSH. Antioxidant vitamins such as A, C, E and alpha-lipoic acid are among these mechanisms. Although all these antioxidant defenses work together to eliminate H_2O_2 (and thus superoxide) from the cell, in the presence of reduced transition metals (Cu, Fe), H_2O_2 can be transformed into $^{\bullet}OH$, which is a highly reactive ROS, by the Fenton reaction [10, 23].

2.3. Metabolic and signaling pathways involved in oxidative stress in diabetes

There are several molecular pathways involved in ROS formation and ROS induced damage. Here we will review the ones that have been related to oxidative stress in diabetes. Not surprisingly, most of them are related to glucose and/or lipid metabolism.

2.3.1. Glucose oxidation and GAPDH

In order to generate energy, glucose needs to be first oxidized inside the cells by glycolysis. In this process, once glucose enters the cells, it is phosphorylated to form glucose-6-phosphate, a reaction mediated by hexocinases. Glucose-6-P is then converted to Fructose-6-P by phosphoglucoisomerase, which can undergo two fates: the pentose phosphate pathway, where reduction of $NADP^+$ to NADPH occurs, or to continue glycolysis to yield Gliceraldehyde-3-P. Glyceraldehyde-3-P dehydrogenase (GAPDH) phosphorylates this product and glycolysis is further completed until its end product pyruvate, which enters the Krebs cycle and mitochondrial metabolism (Figure 2).

It has been proposed that hyperglycemia-induced mitochondrial superoxide production activates damaging pathways by inhibiting glyderaldehyde-3-phosphate dehydrogenase (GAPDH) [4, 24], an enzyme that normally translocates in and out of the nucleus [25, 26]. ROS inhibit glyderaldehyde-3-phosphate dehydrogenase through a mechanism involving the activation of enzyme poly-ADP-ribose polymerase-1 (PARP-1). This enzyme is involved in DNA repair and apoptotic pathways. ROS cause strand breaks in nuclear DNA which activates PARP-1. PARP-1 activation results in inhibition of glyderaldehyde-3-phosphate dehydrogenase by poly-ADP-ribosylation [27]. This results in increased levels of all the glycolytic intermediates upstream of GAPDH. Accumulation of glyceraldehyde 3-phosphate activates two major pathways involved in hyperglycemia-complications: a) It activates the AGE pathway deriving glyceraldehyde phosphate and dihydroxyacetone phosphate to the nonenzymatic synthesis of methylglyoxal. b) Increased glyceraldehyde 3-phosphate favors diacylglycerol production which activates PKC pathway. Further upstream, levels of the glycolytic metabolite fructose 6-phosphate increase, which then increases flux through the hexosamine pathway, where fructose 6-phosphate is converted by the enzyme glutamine-fructose-6-phosphate amidotransferase (GFAT) to UDP–N-Acetylglucosamine. Finally, inhibition of GAPDH favors the accumulation of the first glycolytic metabolite, glucose. This increases its flux through the polyol pathway, consuming NADPH in the process [24].

2.3.2. The polyol pathway

The family of aldo-keto reductase enzymes catalyzes the reduction of a wide variety of carbonyl compounds to their respective alcohols. These reactions utilize nicotinic acid adenine dinucleotide phosphate (NADPH). Aldo-keto reductase has a low affinity (high Km) for glucose, and at the normal glucose concentrations, metabolism of glucose by this pathway is a very small percentage of total glucose metabolism. However, in a hyperglycemic environment, increased intracellular glucose results in its increased enzymatic conversion to the polyalcohol sorbitol, with concomitant decreases in NADPH [4] (Figure 2). Since NADPH is a cofactor required to regenerate reduced glutathione, an antioxidant mechanism, and this compound is an important scavenger of reactive oxygen species (ROS), this could induce or exacerbate intracellular oxidative stress [24]. Moreover, sorbitol is oxidated to fructose by sorbitol dehydrogenase, which can lead to PKC activation via the increased NADH/NAD+ ratio [4]. Although this mechanism does not produce ROS in a direct way, it takes part in the redox imbalance causing oxidative stress.

2.3.3. Hexosamine pathway

When glucose levels are within normal range, a relatively low amount of fructose-6-P is drived away from glycolysis. If intracellular glucose rises, excess fructose-6-phosphate is diverted from glycolysis to provide substrate for the rate-limiting enzyme of this pathway, GFAT. This enzyme converts fructose 6-phosphate to glucosamine 6-phosphate, which is then converted to UDP-NAcetylglucosamine, which is essential for making the glycosyl chains of proteins and lipids. Specific O-Glucosamine-N-Acetyl transferases use this metabolite for post-translational modification of specific serine and threonine residues on cytoplasmic and nuclear proteins [24, 28].

2.3.4. Diacylglycerol formation and PKC activation

The Protein Kinase C (PKC) family comprises at least eleven isoforms of serine/threonine kinases, which participate in signaling pathways activated by phosphatidyl serine, Calcium and Diacylglycerol (DAG). DAG levels are elevated chronically in the hyperglycemic or diabetic environment due to an increase in the glycolytic intermediate dihydroxyacetone phosphate (figure 2). This intermediate is reduced to glycerol-3-phosphate, which, conjugated with fatty acids, increases de novo synthesis of DAG [29]. Evidence suggests that the enhanced activity of PKC isoforms could arise from inhibition of the glycolytic enzyme glyceraldehide-3-phosphate dehydrogenase by increased ROS intracellular levels [4, 24]. Other studies suggest that enhanced activity of PKC isoforms could also result from the interaction between AGEs and their extracellular receptors [30]. PKC isoforms constitute a wide range of cellular signals, including activation of NADPH oxidase and NF-κB, resulting in excessive ROS production. They also increase vascular permeability, stabilize vascular endothelial growth factor (VEGF) mRNA expression and increase leukocyte-endothelium interaction [11].

2.3.5. *Glyceraldehyde autoxidation*

Accumulation of glyceraldehyde 3-phosphate, besides activating the AGE formation and the PKC pathway, it can oxidate itself. This autoxidation generates H_2O_2, which further contributes to oxidative stress [31].

2.3.6. *Advanced glycation end-products (AGEs)*

Intracellular hyperglycaemia is the primary initiating event in the formation of both intracellular and extracellular AGEs [32]. AGEs can arise from intracellular auto-oxidation of glucose to glyoxal, decomposition of the Amadori product (glucose-derived 1-amino-1-deoxyfructose lysine adducts) to 3-deoxyglucosone (perhaps accelerated by an amadoriase), and nonenzymatic phosphate elimination from glyceraldehyde phosphate and dihydroxyacetone phosphate to form methylglyoxal. These reactive intracellular dicarbonyl glyoxal, methylglyoxal and 3-deoxyglucosone react with amino groups of intracellular and extracellular proteins to form AGEs [4]. Intracellular production of AGE precursors can damage cells by three general mechanisms: 1) Intracellular proteins modified by AGEs have altered function, 2) Extracellular matrix components modified by AGE precursors interact abnormally with other matrix components and with matrix receptors (integrins) that are expressed on the surface of cells, and 3) Plasma proteins modified by AGE precursors bind to AGE receptors (such as RAGE and AGE-R1,2 and 3) on cells such as macrophages, vascular endothelial cells and vascular smooth muscle cells. AGE receptors binding induces the production of ROS, which in turn activates PKC. It also activates NF-κB and NADPH oxidase, and disturbs MAPK signaling [31].

2.3.7. *Stress-sensitive signaling pathways*

In addition to direct damage of biomolecules in the cells, oxidative stress is also involved in activation of several stress-sensitive signaling pathways, which can result in inflammation, cytokine release, and even apoptosis. Among these pathways we find the transcription factor NF-κB, which together with PARP acts as a transcriptional coactivator of inflammation molecules such as iNOS, intracellular adhesion molecule-1 (ICAM-I), and histocompatibility complex class II [33]. p38 MAPK pathway and c-Jun Nterminal kinase (JNK) (also known as stress-activated protein kinase (SAPK) participate in cellular responses to stress due to osmotic shock, cytokines and UV light, playing a role in cellular proliferation, apoptosis, and inflammatory responses [33]. Jak/STAT is another important signaling pathway, which initiates and mediates cellular responses to cytokines such as interferons and interleukins [33].

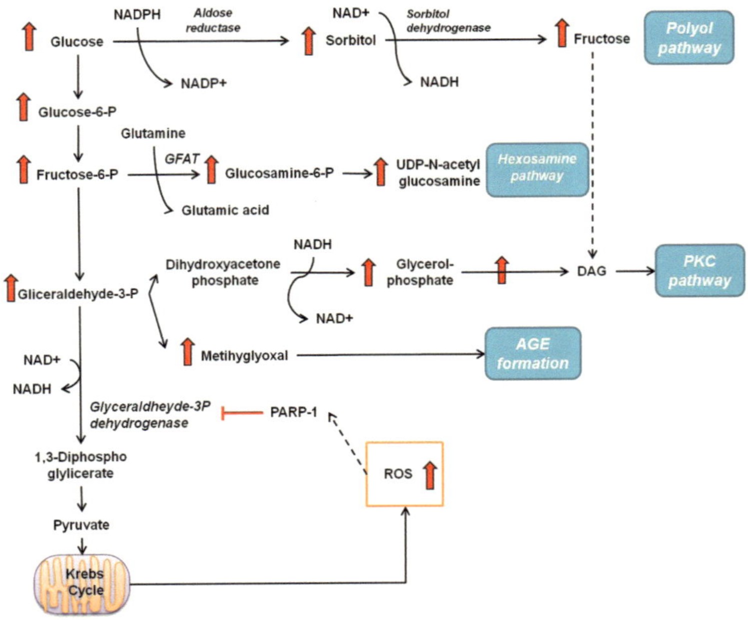

Figure 2. Oxidative stress-related pathways derived from glucose metabolism.

2.4. ROS induced damages

Being highly reactive species, ROS may modify and damage nucleic acids, proteins, lipids and carbohydrates, finally leading to cell damage. Among the motifs that can react with ROS we have the metal ligand from metalloproteases and Fe from oxihemoglobin. $^{\bullet}O_2^{\ -}$ can also modify and inhibit catalases, while $^{\bullet}OH$ can bind to the histidine residue from SOS causing its inhibition. ROS react mostly with insaturated and sulfur containing molecules, thus, proteins with high contents of tryptophan, tyrosine, phenylalanine, histidine, methionine and cysteine can suffer ROS modifications. Finally, ROS may also break peptidic bonds after oxidation of proline residues by $^{\bullet}O_2^{\ -}$ or $^{\bullet}OH$ [31].

ROS and RNS may also modify fatty acids, lipoproteins, and phospholipids, a process termed lipid peroxidation, where $^{\bullet}OH$ and $^{\bullet}O_2^{\ -}$ form hydroperoxide lipids. Hydroperoxyde products cause severe damage to plasma membranes, or they can diffuse to other cells in the organisms and cause vascular permeability and inflammation by binding to (oxidized low-density lipoprotein) LOX receptors, and apoptosis [31].

H_2O_2 in cells can function as a signaling molecule leading to cellular proliferation or can result in cell death. At low concentrations, H_2O_2 serves as a second messenger to activate NF-κB and various kinases (p38 MAPK, ERK, PI3K, Akt, JAK2, STAT). H_2O_2 at slightly higher concentrations can induce the release of cytochrome *c* and apoptosis-inducing factor (AIF) from mitochondria into the cytosol where they trigger the activation of caspase, leading to cell death by apoptosis [12].

3. Diabetes mellitus

Diabetes mellitus is a group of metabolic diseases characterized by hyperglycemia, caused by a defect on insulin production, insulin action or both [1]. There are two main types of diabetes: type 1 and type 2 diabetes.

Type 1 diabetes is due to an autoimmune destruction of the insulin producing pancreatic beta-cells, which usually leads to absolute insulin deficiency. Patients with type 1 diabetes require insulin for survival. This type of diabetes accounts for 5-10% of the total cases of diabetes worldwide. Type 2 diabetes represents approximately 90% of the total diabetes cases, and it is characterized by impairment in insulin action and/or abnormal insulin secretion [1].

The origins of type 2 diabetes are multifactorial. Obesity, age, ethnic origin and familiar history of diabetes are among the factors that contribute to its development. Even though a strong genetic component has been recognized, genotype only establishes the conditions for the individual to be more or less prone to environmental effects and lifestyle factors [34].

Type 2 diabetes develops when insulin secretion or insulin action fails. The impairment of insulin actions is known as insulin resistance, presented as a suppression or retard in metabolic responses of the muscle, liver and adipose tissue to insulin action. This failure is located at the signaling pathways held after insulin binding to its specific receptor [35]. Chronic insulin resistance leads to hyperglycemia.

When the beta cells cannot secrete enough insulin in response to the metabolic demand caused by insulin resistance, frank diabetes type 2 occurs. This failure in the beta cell may be due to an acquired secretory dysfunction and/or a decrease in beta-cell mass [36]. All type 2 diabetic patients have some defect in the ability of beta cells to produce or secrete insulin [37].

3.1. Insulin action and insulin resistance

Once secreted to the portal circulation, insulin is transported to peripheral tissues, on which it will exert mainly anabolic actions [38]. Insulin starts its action by binding to insulin receptor, a transmembrane protein belonging to protein tyrosine kinase activity receptors superfamily, which can autophosphorylate. This initiates a series of events involving protein and membrane lipid phosphorylation, coupling proteins and cytoskeleton activity [39] [40]. The three main signaling pathways activated in response to insulin receptor phosphorylation are 1) PI3K 2)MAPK, and 3) Cb1. These pathways act in a concerted way to translate the signal of insulin receptor into biological actions in target organs, such as glucose transport by transporting GLUT4 vesicles to the membrane, protein, lipid and glycogen synthesis, mitosis and gene expression [40] (Figure 3).

As protein phosphorylation activates these signaling pathways, dephosphorylation inhibits them. Different phosphatases such as protein-tyrosine phosphatase 1B (PTP1B), Phosphatase and tensin homolog (PTEN), SH2-containing tyrosine- protein phosphatase (SHO2), and suppressor of cytokine signaling 3 (SOCS-3) dephosphorylate and shut down insulin signaling [35]. Any alteration in the insulin pathway, being inefficient phosphorylation or increment in phosphatase acticity, causes impairment in insulin action. This is the molecular mechanism leading to insulin resistance.

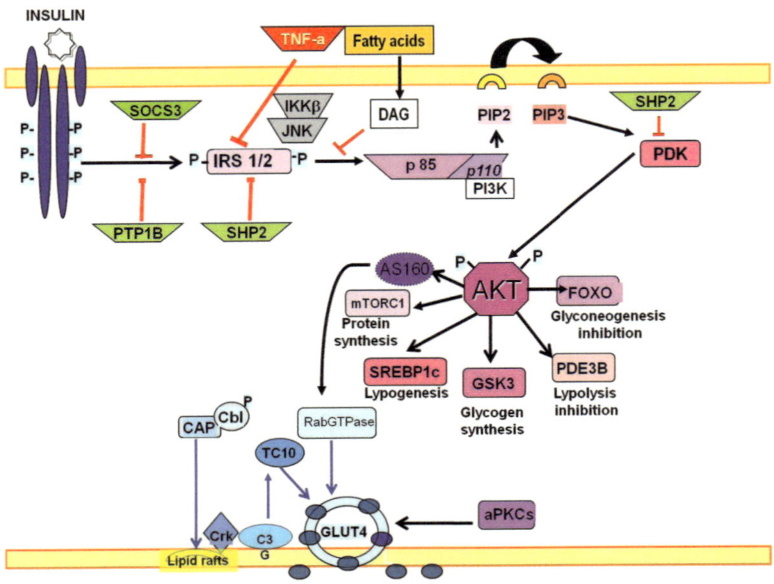

Figure 3. Molecular mechanisms of insulin signaling.

3.2. Insulin secretion

Beta-cells in the endocrine pancreas are responsible for secreting insulin in response to rises in blood nutrient levels during the postprandial state. Glucose is the most important nutrient for insulin secretion. The process by which glucose promotes insulin secretion requires glucose sensing and metabolism by the beta-cell, a process called glucose-stimulated insulin secretion (Figure 4). In the first phase of insulin secretion, glucose enters the cell by glucose transporters (GLUT2 in rodents, GLUT1 in humans). Glucose is then phosphorylated to form glucose-6-phosphate by glucokinase [41]. The generation of ATP by glycolysis, the Krebs cycle and the respiratory chain closes the ATP-sensitive K+ channel (KATP) [42], allowing sodium (Na+) entry without balance. These two events depolarize the membrane and open voltage-dependent T-type calcium (Ca2+) and sodium (Na+) channels. Na+ and Ca2+ entry further depolarizes the membrane and voltage-dependent calcium channels open. This activation increases intracellular Ca2+ ([Ca2+]i) [43], which leads to fusion of insulin-containing secretory granules with the plasma membrane and the first phase insulin secretion [44, 45].

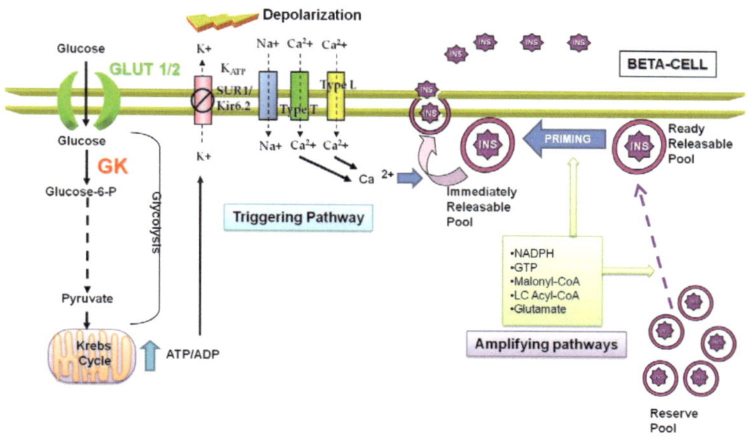

Figure 4. Mechanisms of biphasic glucose-stimulated insulin secretion.

Besides increasing ATP/ADP ratio, glucose metabolism in the beta cell can generate a series of metabolic coupling signals that can initiate and sustain a second insulin secretion phase. Some of these coupling factors participate in mitocondrial metabolism and anaplerosis, constituting cycles involving NADPH, pyruvate, malate, citrate, isocitrate, Acyl-CoA and glutamate [46]. Diverse signaling pathways can also contribute to glucose-induced insulin secretion such as CaMKII [47-49], PKA [50, 51], PKC [51, 52] y PKG [53, 54]. Most secretagogues and potentiators of insulin secretion, such as nutrients, hormones and neurotransmitters, use these pathways to modulate insulin secretion.

4. Oxidative stress in diabetes mellitus

Hyperglycemia and free fatty acid intake are among the causes for oxidative stress conditions [23]. Hence, it may not be surprising that diabetic subjects tend to have more oxidative cell and organism environments than healthy subjects, i.e. an increase in ROS generation [5, 55, 56]. Moreover, diabetic patients present a decrease in antioxidant defenses. The antioxidant enzyme levels are affected by diabetes, which further increase oxidative stress [5, 6].

Oxidative stress has been proposed as a major participant in the patophysiology of diabetic complications [27]. Nevertheless, regarding diabetes onset and development, oxidative stress has also shown to affect the two major mechanisms failing during diabetes: insulin resistance and insulin secretion.

4.1. Oxidative stress processes in insulin resistance

ROS and RNS affect the insulin signaling cascade [5]. As with other ROS effects, low doses play a physiological role in insulin signaling. After insulin stimulation of its receptor in adipocytes, H_2O_2 is produced via NADPH oxidase, which by inhibits PTP1B catalytic activity, thus increasing tyrosine phosphorylation [57].

However, oxidative stress caused by hyperglycemia in diabetes may impair insulin signaling, leading to insulin resistance. Although no mechanisms have been completely established, several responses to ROS excess in the insulin signaling have been proposed.

Disturbs in cellular redistribution of insulin signaling components may alter the insulin cascade, a process mediated by NF-κB [58]. A decrease in GLUT4 gene transcription and increase in GLUT1 (insulin independent glucose transporter) has also been observed, as well as increases in phosphorylation of IRS protein in an insulin receptor-independent fashion (perhaps by the stress kinases). Altogether, hyperglycemia and insulin resistance may also lead to altered mitochondrial function, and insulin action impairment by cytokines in response to metabolic stress [59, 60]. An increase in the hexosamine pathway has also been linked to insulin resistance. Moreover, it has been proposed that this pathway acts as a cellular sensor for the glucose excess. From that point of view, insulin resistance may be a protective mechanism from the glucose excess entrance [28].

4.2. Oxidative stress processes in insulin secretion

Pancreatic beta-cells are especially sensitive to ROS and RNS, because their natural enzymatic antioxidant defenses are lower compared to other tissues such as liver. Moreover, they lack the ability to adapt their low enzyme activity levels in response to stress such as high glucose or high oxygen [61]. Glucose enters to the beta-cell in an insulin independent fashion, because besides providing energy, glucose sensing in the beta-cell is crucial for insulin secretion. It has been suggested that hyperglycemia can generate chronic oxidative stress by the glucose oxidation pathway [62], leading to an excess in mitochondrial superoxide production, which further activates uncoupling protein-2 (UCP-2). This protein lowers ATP/ADP relationship through proton leak in the beta-cell, which reduces insulin secretion [63].

ROS also increase the stress signaling pathways in the beta cells, such as NF-κB activity, which potentially leading to beta-cell apoptosis [64], and the JNK pathway which has been related to suppression of insulin gene expression, possibly by reduction of PDX-1 DNA binding activity, a major regulator of insulin expression [65]. It has also been shown that the activation of the hexosamine pathway in beta-cells leads to suppression of PDX-1 binding to the insulin and other genes involved in insulin secretion, perhaps contributing to the beta-cell dysfunction present in diabetes mellitus [66].

As in other cell types, NO in beta-cells has physiologic roles. NO may regulate glucokinase activity by s-nitrosilation [67] in the beta-cell, and possibly increase insulin secretion. However, NO excess and concomitant NRS may cause apoptosis through caspase-3 activation and decrease in ATP levels [68].

Besides ROS hyperproduction, excess mitochondrial metabolism resulting form hyperglycemia in the beta-cell may also alter mitochondrial shape, volume and behavior, uncoupling K-ATP channels from mitochondrial activity and thus altering glucose-induced insulin secretion [69].

5. Diabetic complications

Hyperglycemia, is the responsible of the development of diabetes complications as well. Hyperglycemia damage is produced in cells in which glucose uptake is independent of insulin, which, similarly to what happens in beta-cells, explains that the cause of the complications resides inside the cells [4]. Prolonged exposure to high glucose levels, genetic determinants of susceptibility and accelerating factors such as hypertension and dyslipidemia participate in the development of diabetic complications. Moreover, the development and progression of damage is proportional to hyperglycemia, which makes the lowering of glucose levels the most important goal for preventing complications and treating diabetes.

The main tissues affected by diabetes complications at the microvasculature levels are retina, renal glomerulus, and peripheral nerves. Diabetes is also associated with accelerated atherosclerotic disease affecting arteries that supply the heart, brain, and lower extremities. In addition, diabetic cardiomyopathy is a major diabetic complication [24].

5.1. Oxidative stress in diabetic complications

Oxidative stress plays a pivotal role in the development of diabetes complications, both at the microvascular and macrovascular levels. Results derived from two decades of diabetes complications investigation point towards mitochondrial superoxide overproduction as the main cause of metabolic abnormalities of diabetes. Thus, all of the above reviewed pathways are involved in microvasculature and macrovasculature hyperglycemic damage [24].

5.2. Microvascular complications

Diabetic retinopathy: Diabetic retinopathy appears in most patients after 10 to 15 years after diabetes onset. Background retinopathy presents small hemorrhages in the middle layers of the retina, appearing as "dots". Lipid deposition occurs at the margins of the hemorraghe, and microaneurisms (small vascular dilatations) and edema may appear. Proliferative retinopathy occurs when new blood vessels on the surface of the retina cause vitreous hemorrhage, and eventually, blindness. As the cells of the retina contain high amounts of aldo-ketoreductase, they have high susceptibility to increase the polyol pathway in the presence of excess glucose, with concomitant decreases in NADPH [4]. Sorbitol produced in this process increases osmotic stress, which has been linked to microaneurysm formation, thickening of the basement membranes and loss of pericytes. It is also thought that retina cells are damaged by glycoproteins, particularly form AGEs. Additionally, ROS by themselves may damage the cells. Importantly, VEGF, growth hormone and TGF-beta increases during diabetes may be the cause of proliferation of blood vessels [70].

Diabetic nephropaty: this complication causes glomerular basement membrane thickness, microaneurism formation, and mesangial nodule formation, all which are reflected in proteinuria and, in the end, renal insufficiency. The mechanisms for injury also involve the increased polyol pathway and AGE formation. AGE binding to its receptors has been proven to play a role as well in renal damage, fibrosis and inflammation associated with diabetic nephropaty. This actions of AGE also potentiate oxidative stress, while synergizing with rennin-angiotensin system activation, which leads to a vicious cycle causing kidney failure. As mentioned, diabetic patients, and particularly those with nephropaty, have lowered antioxidant defenses. Moreover, AGE receptors are significantly increased [71].

Diabetic neuropathy: Diabetic neuropathy is defined as the presence of symptoms and/or signs of peripheral nerve dysfunction in diabetic patients after exclusion of other causes. Peripheral neuropathy in diabetes may manifest in several different forms, including sensory, focal/multifocal, and autonomic neuropathies. [72]. Mechanisms of nerve injury are less known but likely related also to the polyol pathway, AGE formation and ROS themselves [70]. Oxidized proteins and lipoproteins also interact with receptors in the membrane of neurons, initiating inflammatory signaling mechanisms which further produce ROS, damaging cellular components and leading to neuronal injury [73].

5.3. Macrovascular complications

The central pathological mechanism in macrovascular complications is atherosclerotic disease. Atherosclerosis occurs as a result of chronic inflammation and injury to the arterial wall in the peripheral or coronary vascular system. This damages cause accumulation of oxidized lipids from LDL particles in the endothelial wall of arteries, whose rupture leads to acute vascular infarction. Additionally, platelet adhesion and hypercoagulability also occurs in type 2 diabetes, increasing the risk of vascular occlusion [70]. It has been proposed that increased superoxide production is the central and major mediator of endothelial tissue damage, causing direct inactivation of two antiatherosclerotic enzymes, endothelial nitric oxide synthase and prostacyclin synthase and that the activation of oxidative stress pathways is involved in the pathogenesis of complications [24].

Endothelial cells also contain high amounts of aldo-keto reductase, and are thus prone to increased polyol pathway activation. Moreover, a large body of evidence supports hypothesis that hyperglycemia or diabetes leads to vascular diacylglycerol accumulation and subsequent PKC activation, causing a variety of cardiovascular defects [29]. PKC activation has been associated with vascular alterations such as increases in permeability, contractility, extracellular matrix synthesis, cell growth and apoptosis, angiogenesis, leukocyte adhesion, and cytokine activation and inhibition [29]. Hyperglycemia-induced activation of PKC has also been implicated in the overexpression of the fibrinolytic inhibitor, plasminogen activator inhibitor-1 (PAI-1) [74]. In smooth muscle PKC hyperactivity is associated with decreased NO production [75] and has been shown to inhibit insulin-stimulated expression of eNOs in endothelial cells.

In arterial endothelial cells O-glucosamine-acylation participates in vascular complications interfering with the action of Akt/PKB, a critical insulin signaling protein, on eNOS [76]. GFAT activity is associated with increased transcription of transforming growth factor (TGF) alpha and beta and PAI-1, factors involved in the proliferation of vascular smooth-muscle and endothelial cells. This effect appears to be mediated by O-glucosamine-acylation of the transcription factor, Sp1 [77]. Increased TGF-beta and PAI-1 are associated with capillary and vascular occlusion by mechanisms associated with collagen and fibronectin expression causing capillary occlusion, in the case of TGF-beta, and decreased fibrinolysis in the case of PAI-1. O-GlcNAcylation impairs cardiomyocyte calcium cycling decreasing sarcoplasmic reticulum calcium ATPase 2a (Serca 2a) [78-80].

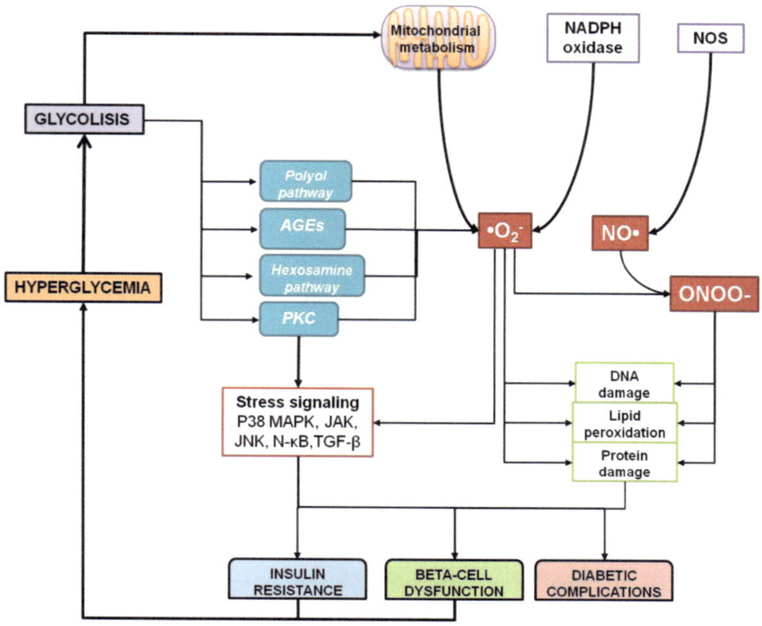

Figure 5. Oxidative stress pathways in diabetes mellitus.

6. Antioxidant vitamins and diabetes mellitus

As mentioned above, vitamins C, E, and A constitute the non enzymatic defense against oxidative stress, by regenerating endogenous antioxidants (Figure 1). Vitamin C has a role in scavenging ROS and RNS by becoming oxidated itself. The oxidized products of vitamin C, ascorbic radical and dehydroascorbic radical are regenerated by glutathione, NADH or NADPH. In addition, vitamin C can reduce the oxidized forms of vitamin E and gluthatione [81]. Vitamin E is a fat-soluble vitamin which may interact with lipid hydroperoxides and scavenge them. It also participates, together with vitamin C, in gluthatione regeneration by interaction with lipoic acid [23]. Vitamin A has a plethora of cellular actions. Besides modulating gene expression, cell growth and differentiation, this vitamin may also act as antioxidant, although the mechanisms of action in this role are not fully deciphered. The antioxidant potential of carotenoids (vitamin A) depends on their distinct membrane-lipid interactions, while some carotenoids can decrease lipid peroxidation, others can stimulate it [82].

Since oxidative stress is present during the progression of diabetes and its complications, amelioration of oxidative status, mainly by increasing antioxidant non-enzymatic defenses, has been largely proposed and studied. Several clinical observational trials have particularly studied the correlation between vitamin E status in plasma and/or diet, and markers of oxidation, inflammation, type 2 diabetes incidence, and diabetic complications. Although inverse association has been found for vitamin E in some studies [7, 83, 84], the association found in other study disappeared after adjustment for cardiovascular risk factors such as obesity, smoking, and hypertension [85], or have observed no beneficial effect at all [7, 8]. Such contrasting results have also been reported for studies looking association of vitamin A and C consumption and amelioration of diabetes status and/or complications [7, 8, 81, 86].

On the other hand, in interventional trials with vitamin supplementation, the effects of vitamins E, C and A, alone or in diverse combinations, have yielded barely any promising result. There appears to be no beneficial effect of vitamin supplementation on diabetes or macrovascular complications [7, 8, 81]. Some of these studies have even evidenced associations between vitamin supplementation and an increased incidence of stroke [7]. Likewise, supplementation with antioxidant vitamins can even block beneficial ROS production during exercise, inhibiting the health-promoting effects of exercise in humans [87].

Paradoxically, in spite of the solid evidence of increased oxidative stress in diabetes, and the well established actions of vitamins as antioxidants, the association studies between antioxidant vitamin status and its beneficial effects in diabetes has no consistent results at all. What is more, interventional studies have failed in demonstrating a favorable effect of vitamin supplementation, discouraging its use as antioxidant therapy for diabetes.

Several reasons have been suggested for these contradictory observations. First, as vitamins may be easily oxidized, a vitamin may have antioxidant or oxidant properties, depending on the presence of other vitamins and the oxidative state in the cells i.e., if the oxidized form of a vitamin is not correctly reversed into the reduced form. Additionally, some vitamins may also activate oxidative stress pathways and further increase the oxidative stress, such as the activation of PKC by retinoids [88].

Vitamin doses may also be part of the problem, as the effect of vitamins depends on dietary concentrations and/or supplement intake. The wide variety of doses reached with diet and supplements, and the lack of an established "pharmacological" dose of vitamins, makes it difficult to ascertain the true net effect of vitamin status or supplementation needed to generate beneficial effects. As well, the required dose for antioxidant effects versus the required for the vitamin's role in the body may differ, which, together with vitamin's bioavailability and its interaction with other vitamins, are caveats for assessing and finding vitamins' effects, if any [7, 88].

Finally, the antioxidant effects of vitamins may not be sufficient to scavenge the great amount of ROS present in diabetes. Certainly, glucose levels have been correlated to the presence and severity of the complications. However once hyperglycemia has established, the incidence of complications after tight glycemic control remains the same. This effect has been termed glycemic memory, and is the cause for accumulative damage rendering diabetic complications. Considering that hyperglycemia is the main cause of oxidative stress in diabetes, in a similar way, the chronic undesirable effects that occur by ROS production may generate a vicious cycle difficult to break, in which ROS damage exacerbates the diabetic state, increasing glucose levels, which will further induce more oxidative imbalance [24].

7. Conclusions

Diabetes mellitus has reached epidemic proportions in the last decade, becoming one of the most important diseases worldwide. Several studies indicate oxidative stress is present in the dysfunction of insulin action and secretion that occur during diabetes, as well as in the development of diabetic complications. Nevertheless, oxidative stress is not the primary cause of diabetes, but rather a consequence of nutrient excess, given that oxidative stress is a natural response to stress, in this case, to glucose and/or lipid overload.

Vitamins such as E, C and A with antioxidant properties constitute the physiological non-enzymatic defense against oxidative stress. However, the evidence in favor of the use of vitamin supplementation as antioxidant therapy remains uncertain. Although some beneficial effects have been proven in observational studies, the results of interventional trials are still ineffective. Perhaps more studies on the physiopathology of oxidative stress and the role of vitamins in it, as well as standardizing vitamin dosage and assessing their undesirable effects are needed in order to determine a clear participation of vitamin supplementation in amelioration of the oxidative balance. More studies addressing the possibility of targeting directly at the enzymes and mechanisms involved in ROS production and not by antioxidants are needed as well.

Given that it is mostly dietary vitamin intake which has shown an association with ameliorating the diabetic state, and that oxidative stress is a response to excess of nutrients, it seems that attending the cause of excessive ROS production represents the best therapeutic option. Thus, adequate dietary interventions that reduce hyperglycemia, and increases in oxygen consumption (i.e. improve mitochondrial function) by exercise remain the primary choices for diabetes treatment and prevention of its complications.

Acknowledgements

This work was supported by grants from CONACyT and from the Dirección General de Asuntos del Personal Académico, UNAM.

Author details

Maria-Luisa Lazo-de-la-Vega-Monroy* and Cristina Fernández-Mejía

Unidad de Genética de la Nutrición, Departamento de Medicina Genómica y Toxicología Ambiental, Instituto de Investigaciones Biomédicas, Universidad Nacional Autónoma de México/ Instituto Nacional de Pediatría, México

References

[1] ADA. (2009). Diagnosis and classification of diabetes mellitus. *Diabetes care*, 32(1), 62-7.

[2] Shaw, J. E., Sicree, R. A., & Zimmet, P. Z. (2010). Global estimates of the prevalence of diabetes for 2010 and 2030. *Diabetes Res Clin Pract*, 87(1), 4-14.

[3] King, H., Aubert, R. E., & Herman, W. H. (1998). Global burden of diabetes, 1995-2025: prevalence, numerical estimates, and projections. *Diabetes Care*, 21(9), 1414-1431.

[4] Brownlee, M. (2001). Biochemistry and molecular cell biology of diabetic complications. *Nature*, 414(6865), 813-820.

[5] Rains, J. L., & Jain, S. K. (2011). Oxidative stress, insulin signaling, and diabetes. *Free Radic Biol Med*, 50(5), 567-575.

[6] Maritim, A. C., Sanders, R. A., & Watkins, J. B. (2003). Diabetes, oxidative stress, and antioxidants: A review. *Journal of Biochemical and Molecular Toxicology*, 17(1), 24.

[7] Sheikh-Ali, M., Chehade, J. M., & Mooradian, A. D. (2011). The antioxidant paradox in diabetes mellitus. *Am J Ther*, 18(3), 266-278.

[8] Cuerda, C., Luengo, L. M., Valero, Vidal. A., Burgos, R., Calvo, F. L., et al. (2011). Antioxidants and diabetes mellitus: review of the evidence]. *Nutr Hosp*, 26(1), 68-78.

[9] Naviaux, R. K. (2012). Oxidative Shielding or Oxidative Stress? *J Pharmacol Exp Ther.*

[10] Johansen, J. S., Harris, A. K., Rychly, D. J., & Ergul, A. (2005). Oxidative stress and the use of antioxidants in diabetes: linking basic science to clinical practice. *Cardiovasc Diabetol*, 4(1), 5.

[11] Ceriello, A. (2006). Oxidative stress and diabetes-associated complications. *Endocr Pract*, 12(1), 60-62.

[12] Fisher, A. B., Zhang, Q. Â. Â., Geoffrey, J. L., & Steven, D.S. (2006). Nadph and nadph oxidase. *Encyclopedia of Respiratory Medicine.*, Oxford, Academic Press, 77.

[13] Drummond, G. R., Selemidis, S., Griendling, K. K., & Sobey, C. G. (2011). Combating oxidative stress in vascular disease: NADPH oxidases as therapeutic targets. *Nat Rev Drug Discov*, 10(6), 453-471.

[14] Maechler, P., & Wollheim, C. B. (2001). Mitochondrial function in normal and diabetic beta-cells. *Nature*, 414(6865), 807-812.

[15] Giacco, F., & Brownlee, M. (2010). Oxidative stress and diabetic complications. *Circ Res*, 107(9), 1058-1070.

[16] Boveris, A. (1984). Determination of the production of superoxide radicals and hydrogen peroxide in mitochondria. *Methods Enzymol*, 105, 429-435.

[17] Korshunov, S. S., Skulachev, V. P., & Starkov, A. A. (1997). High protonic potential actuates a mechanism of production of reactive oxygen species in mitochondria. *FEBS Lett*, 416(1), 15-18.

[18] Dikalov, S. Cross talk between mitochondria and NADPH oxidases. *Free Radical Biology and Medicine*, 51(7), 1289.

[19] Schmidt, H. H., Lohmann, S. M., & Walter, U. (1993). The nitric oxide and cGMP signal transduction system: regulation and mechanism of action. *Biochim Biophys Acta*, 1178(2), 153-175.

[20] Hobbs, A. J., & Ignarro, L. J. (1996). Nitric oxide-cyclic GMP signal transduction system. *Methods Enzymol*, 269, 134-148.

[21] Murad, F. (1999). Cellular signaling with nitric oxide and cyclic GMP. *Braz J Med Biol Res*, 32(11), 1317-1327.

[22] Mc Donald, L. J., & Murad, F. (1995). Nitric oxide and cGMP signaling. *Adv Pharmacol*, 34, 263-275.

[23] Evans, J. L., Goldfine, I. D., Maddux, B. A., & Grodsky, G. M. (2002). Oxidative stress and stress-activated signaling pathways: a unifying hypothesis of type 2 diabetes. *Endocr Rev*, 23(5), 599-622.

[24] Giacco, F., & Brownlee, M. Oxidative stress and diabetic complications. *Circ Res*, 107(9), 1058-1070.

[25] Mazzola, J. L., & Sirover, M. A. (2003). Subcellular localization of human glyceraldehyde-3-phosphate dehydrogenase is independent of its glycolytic function. *Biochim Biophys Acta*, 1622(1), 50-56.

[26] Tristan, C., Shahani, N., Sedlak, T. W., & Sawa, A. The diverse functions of GAPDH: views from different subcellular compartments. *Cell Signal*, 23(2), 317-323.

[27] Brownlee, M. (2005). The pathobiology of diabetic complications: a unifying mechanism. *Diabetes*, 54(6), 1615-1625.

[28] Buse, M. G. (2006). Hexosamines, insulin resistance, and the complications of diabetes: current status. *Am J Physiol Endocrinol Metab*, 290(1), 1-8.

[29] Geraldes, P., & King, G. L. Activation of protein kinase C isoforms and its impact on diabetic complications. *Circ Res*, 106(8), 1319-1331.

[30] Scivittaro, V., Ganz, M. B., & Weiss, M. F. (2000). AGEs induce oxidative stress and activate protein kinase C-beta(II) in neonatal mesangial cells. *Am J Physiol Renal Physiol*, 278(4), 676-683.

[31] Camacho-Ruiz, A., & Esteban-Méndex, M. Diabetes y radicales libres. *In: Morales-González JA, Madrigal-Santillán EO, Nava-Chapa G, Durante-Montiel I, Jongitud-Falcón A, Esquivel-Soto J. 2010Editors. Diabetes. 2nd ed. Pachuca, Hidalgo, México.: Universidad Autónoma del Estado de Hidalgo.*

[32] Degenhardt, T. P., Thorpe, S. R., & Baynes, J. W. (1998). Chemical modification of proteins by methylglyoxal. *Cell Mol Biol (Noisy-le-grand)*, 44(7), 1139-1145.

[33] Gomperts, B. D., Kramer, I. M., & Tatham, P. E. R. (2003). Signal Transduction. London, UK, Elsevier Academic Press.

[34] Permutt, M. A., Wasson, J., & Cox, N. (2005). Genetic epidemiology of diabetes. J Clin Invest , 115(6), 1431-9.

[35] Zick, Y. (2004). Uncoupling insulin signalling by serine/threonine phosphorylation: a molecular basis for insulin resistance. *Biochem Soc Trans*, 32(5), 812-816.

[36] Weir, G. C., Laybutt, D. R., Kaneto, H., Bonner-Weir, S., & Sharma, A. (2001). Beta-cell adaptation and decompensation during the progression of diabetes. *Diabetes*, 50(1), 154-159.

[37] Leahy, J. L., Hirsch, I. B., Peterson, K. A., & Schneider, D. (2010). Targeting beta-cell function early in the course of therapy for type 2 diabetes mellitus. *J Clin Endocrinol Metab*, 95(9), 4206-4216.

[38] Lazo-de-la-Vega-Monroy, M. L., & Fernandez-Mejia, C. Bases moleculares de la diabetes tipo 2. *In: Morales-González JA, Madrigal-Santillán EO, Nava-Chapa G, Durante-Montiel I, Jongitud-Falcón A, Esquivel-Soto J. 2010Editors. Diabetes. 2nd ed. Pachuca, Hidalgo, México:Universidad Autónoma del Estado de Hidalgo.*

[39] Bjornholm, M., & Zierath, J. R. (2005). Insulin signal transduction in human skeletal muscle: identifying the defects in Type II diabetes. *Biochem Soc Trans*, 33(2), 354-357.

[40] Withers, D. J., Gutierrez, J. S., Towery, H., Burks, D. J., Ren, J. M., Previs, S., et al. (1998). Disruption of IRS-2 causes type 2 diabetes in mice. *Nature*, 391(6670), 900-904.

[41] Matschinsky, F. M. (1996). Banting Lecture 1995. A lesson in metabolic regulation inspired by the glucokinase glucose sensor paradigm. *Diabetes*, 45(2), 223-241.

[42] Aguilar-Bryan, L., Clement, J. P., Gonzalez, G., Kunjilwar, K., Babenko, A., & Bryan, J. (1998). Toward understanding the assembly and structure of KATP channels. *Physiol Rev*, 78(1), 227-245.

[43] Hiriart, M., & Aguilar-Bryan, L. (2008). Channel regulation of glucose sensing in the pancreatic beta-cell. *Am J Physiol Endocrinol Metab*, 295(6), 1298-1306.

[44] Rorsman, P., & Renstrom, E. (2003). Insulin granule dynamics in pancreatic beta cells. *Diabetologia*, 46(8), 1029-1045.

[45] Straub, S. G., & Sharp, G. W. (2002). Glucose-stimulated signaling pathways in biphasic insulin secretion. *Diabetes Metab Res Rev*, 18(6), 451-463.

[46] Jitrapakdee, S., Wutthisathapornchai, A., Wallace, J. C., & Mac Donald., M.J. (2010). Regulation of insulin secretion: role of mitochondrial signalling. *Diabetologia*, 53(6), 1019-1032.

[47] Krueger, K. A., Bhatt, H., Landt, M., & Easom, R. A. (1997). Calcium-stimulated phosphorylation of MAP-2 in pancreatic betaTC3-cells is mediated by Ca2+/calmodulin-dependent kinase II. *J Biol Chem*, 272(43), 27464-27469.

[48] Nielander, H. B., Onofri, F., Valtorta, F., Schiavo, G., Montecucco, C., Greengard, P., et al. (1995). Phosphorylation of VAMP/synaptobrevin in synaptic vesicles by endogenous protein kinases. *J Neurochem*, 65(4), 1712-1720.

[49] Easom, R. A. (1999). CaM kinase II: a protein kinase with extraordinary talents germane to insulin exocytosis. *Diabetes*, 48(4), 675-684.

[50] Sharp, G. W. (1979). The adenylate cyclase-cyclic AMP system in islets of Langerhans and its role in the control of insulin release. *Diabetologia*, 16(5), 287-296.

[51] Jones, P. M., & Persaud, S. J. (1998). Protein kinases, protein phosphorylation, and the regulation of insulin secretion from pancreatic beta-cells. *Endocr Rev*, 19(4), 429-461.

[52] Doyle, M. E., & Egan, J. M. (2003). Pharmacological agents that directly modulate insulin secretion. *Pharmacol Rev*, 55(1), 105-131.

[53] Laychock, S.G., Modica, M.E., & Cavanaugh, C.T. (1991). L-arginine stimulates cyclic guanosine 3',5'-monophosphate formation in rat islets of Langerhans and RINm5F insulinoma cells: evidence for L-arginine:nitric oxide synthase.(6), *Endocrinology*, 129(6), 3043-3052.

[54] Russell, M. A., & Morgan, N. (2010). Expression and functional roles of guanylate cyclase isoforms in BRIN-BD11 beta-cells. *Islets*, 2(6), 23-31.

[55] Guzik, T. J., Mussa, S., Gastaldi, D., Sadowski, J., Ratnatunga, C., Pillai, R., et al. (2002). Mechanisms of increased vascular superoxide production in human diabetes mellitus: role of NAD(P)H oxidase and endothelial nitric oxide synthase. *Circulation*, 105(14), 1656-1662.

[56] Ceriello, A., Mercuri, F., Quagliaro, L., Assaloni, R., Motz, E., Tonutti, L., et al. (2001). Detection of nitrotyrosine in the diabetic plasma: evidence of oxidative stress. *Diabetologia*, 44(7), 834-838.

[57] Mahadev, K., Motoshima, H., Wu, X., Ruddy, J. M., Arnold, R. S., Cheng, G., et al. (2004). The NAD(P)H oxidase homolog Nox4 modulates insulin-stimulated genera-

tion of H2O2 and plays an integral role in insulin signal transduction. *Mol Cell Biol*, 24(5), 1844-1854.

[58] Ogihara, T., Asano, T., Katagiri, H., Sakoda, H., Anai, M., Shojima, N., et al. (2004). Oxidative stress induces insulin resistance by activating the nuclear factor-kappa B pathway and disrupting normal subcellular distribution of phosphatidylinositol 3-kinase. *Diabetologia*, 47(5), 794-805.

[59] Eriksson, J. W. (2007). Metabolic stress in insulin's target cells leads to ROS accumulation- a hypothetical common pathway causing insulin resistance. *FEBS Lett*, 581(19), 3734-3742.

[60] Bloch-Damti, A., & Bashan, N. (2005). Proposed mechanisms for the induction of insulin resistance by oxidative stress. *Antioxid Redox Signal*, 7(11-12), 1553-1567.

[61] Tiedge, M., Lortz, S., Drinkgern, J., & Lenzen, S. (1997). Relation between antioxidant enzyme gene expression and antioxidative defense status of insulin-producing cells. *Diabetes*, 46(11), 1733-1742.

[62] Robertson, R. P., Harmon, J., Tran, P. O., Tanaka, Y., & Takahashi, H. (2003). Glucose toxicity in beta-cells: type 2 diabetes, good radicals gone bad, and the glutathione connection. *Diabetes*, 52(3), 581-587.

[63] Brownlee, M. (2003). A radical explanation for glucose-induced beta cell dysfunction. *J Clin Invest*, 112(12), 1788-90.

[64] Rhodes, C. J. (2005). Type 2 diabetes-a matter of beta-cell life and death? *Science*, 307(5708), 380-384.

[65] Kaneto, H., Matsuoka, T. A., Nakatani, Y., Kawamori, D., Matsuhisa, M., & Yamasaki, Y. (2005). Oxidative stress and the JNK pathway in diabetes. *Curr Diabetes Rev*, 1(1), 65-72.

[66] Kaneto, H., Xu, G., Song, K. H., Suzuma, K., Bonner-Weir, S., Sharma, A., et al. (2001). Activation of the hexosamine pathway leads to deterioration of pancreatic beta-cell function through the induction of oxidative stress. *J Biol Chem*, 276(33), 31099-31104.

[67] Rizzo, M. A., & Piston, D. W. (2003). Regulation of beta cell glucokinase by S-nitrosylation and association with nitric oxide synthase. *J Cell Biol*, 161(2), 243-248.

[68] Tejedo, J., Bernabe, J. C., Ramirez, R., Sobrino, F., & Bedoya, F. J. (1999). NO induces a cGMP-independent release of cytochrome c from mitochondria which precedes caspase 3 activation in insulin producing RINm5F cells. *FEBS Lett*, 459(2), 238-243.

[69] Drews, G., Krippeit-Drews, P., & Dufer, M. (2010). Oxidative stress and beta-cell dysfunction. *Pflugers Arch*, 460(4), 703-718.

[70] Fowler, M. J. (2008). Microvascular and Macrovascular Complications of Diabetes. *Clinical Diabetes*, 26(2), 77-782.

[71] Thomas, M. C. (2011). Advanced glycation end products. *Contrib Nephrol*, 170, 66-74.

[72] ADA. (2012). Standards of Medical Care in Diabetes. *Diabetes Care*, 170(1), 11-63.

[73] Vincent, A. M., Callaghan, B. C., Smith, A. L., & Feldman, E.L. Diabetic neuropathy: cellular mechanisms as therapeutic targets. *Nat Rev Neurol*, 7(10), 573.

[74] Feener, E. P., Xia, P., Inoguchi, T., Shiba, T., Kunisaki, M., & King, G. L. (1996). Role of protein kinase C in glucose- and angiotensin II-induced plasminogen activator inhibitor expression. *Contrib Nephrol*, 118, 180-187.

[75] Ganz, M. B., & Seftel, A. (2000). Glucose-induced changes in protein kinase C and nitric oxide are prevented by vitamin E. *Am J Physiol Endocrinol Metab*, 278(1), 146-152.

[76] Akimoto, Y., Kreppel, L. K., Hirano, H., & Hart, G. W. (2001). Hyperglycemia and the O-GlcNAc transferase in rat aortic smooth muscle cells: elevated expression and altered patterns of O-GlcNAcylation. *Arch Biochem Biophys*, 389(2), 166-175.

[77] Du, X. L., Edelstein, D., Rossetti, L., Fantus, I. G., Goldberg, H., Ziyadeh, F., et al. (2000). Hyperglycemia-induced mitochondrial superoxide overproduction activates the hexosamine pathway and induces plasminogen activator inhibitor-1 expression by increasing Sp1 glycosylation. *Proc Natl Acad Sci USA*, 97(22), 12222-12226.

[78] Clark, R. J., Mc Donough, P. M., Swanson, E., Trost, S. U., Suzuki, M., Fukuda, M., et al. (2003). Diabetes and the accompanying hyperglycemia impairs cardiomyocyte calcium cycling through increased nuclear O-GlcNAcylation. *J Biol Chem*, 278(45), 44230-44237.

[79] Pang, Y., Bounelis, P., Chatham, J. C., & Marchase, R. B. (2004). Hexosamine pathway is responsible for inhibition by diabetes of phenylephrine-induced inotropy. *Diabetes*, 53(4), 1074-1081.

[80] Liu, J., Pang, Y., Chang, T., Bounelis, P., Chatham, J. C., & Marchase, R. B. (2006). Increased hexosamine biosynthesis and protein O-GlcNAc levels associated with myocardial protection against calcium paradox and ischemia. *J Mol Cell Cardiol*, 40(2), 303-312.

[81] Garcia-Bailo, B., El -Sohemy, A., Haddad, P. S., Arora, P., Benzaied, F., Karmali, M., et al. (2011). Vitamins D, C, and E in the prevention of type 2 diabetes mellitus: modulation of inflammation and oxidative stress. *Biologics*, 5, 7-19.

[82] Mc Nulty, H., Jacob, R. F., & Mason, R. P. (2008). Biologic activity of carotenoids related to distinct membrane physicochemical interactions. *Am J Cardiol*, 101(10A), 20-29.

[83] Salonen, J. T., Nyyssonen, K., Tuomainen, T. P., Maenpaa, P. H., Korpela, H., Kaplan, G. A., et al. (1995). Increased risk of non-insulin dependent diabetes mellitus at low plasma vitamin E concentrations: a four year follow up study in men. *BMJ*, 311(7013), 1124-1127.

[84] Mayer-Davis, E. J., Costacou, T., King, I., Zaccaro, D. J., & Bell, R. A. (2002). Plasma and dietary vitamin E in relation to incidence of type 2 diabetes: The Insulin Resistance and Atherosclerosis Study (IRAS). *Diabetes Care*, 25(12), 2172-2177.

[85] Reunanen, A., Knekt, P., Aaran, R. K., & Aromaa, A. (1998). Serum antioxidants and risk of non-insulin dependent diabetes mellitus. *Eur J Clin Nutr*, 52(2), 89-93.

[86] Sinclair, A. J., Taylor, P. B., Lunec, J., Girling, A. J., & Barnett, A. H. (1994). Low plasma ascorbate levels in patients with type 2 diabetes mellitus consuming adequate dietary vitamin C. *Diabet Med*, 11(9), 893-898.

[87] Ristow, M., Zarse, K., Oberbach, A., Kloting, N., Birringer, M., Kiehntopf, M., et al. (2009). Antioxidants prevent health-promoting effects of physical exercise in humans. *Proc Natl Acad Sci USA*, 106(21), 8665-8670.

[88] Chertow, B. (2004). Advances in diabetes for the millennium: vitamins and oxidant stress in diabetes and its complications. *MedGenMed*, 6(3), 4.

Chapter 10

Oxidative Stress and Antioxidant Therapy in Chronic Kidney and Cardiovascular Disease

David M. Small and Glenda C. Gobe

Additional information is available at the end of the chapter

http://dx.doi.org/10.5772/51923

1. Introduction

Chronic kidney disease (CKD) and cardiovascular disease (CVD) have major impacts upon the health of populations worldwide, especially in Western societies. The progression of CKD or CVD independently exerts synergistic deleterious effects on the other, for example, patients with CKD are more likely to die of CVD than to develop renal failure. This overlap between CKD and CVD, in part, relates to common etiologies such as diabetes mellitus and hypertension, but important dynamic and bidirectional interactions between the cardiovascular system and kidneys may also explain the occurrence of concurrent organ dysfunction [1]. Cardio-renal syndrome (or reno-cardiac syndrome, the prefix depending on the primary failing organ) is becoming increasingly recognised [2]. Conventional treatment targeted at either syndrome generally reduces the onset or progression of the other [3]. Even though our understanding of various factors and steps involved in the pathogenesis of CKD and CVD and their obvious links has improved, a complete picture of the mechanisms involved is still unclear. Oxidative stress has been identified as one unifying mechanism in the pathogenesis of CKD and CVD [4]. This current chapter gives a brief review of recent literature on the relationship between CKD, CVD and oxidative stress and indicates how, by applying knowledge of the molecular controls of oxidative stress, this information may help improve targeted therapy with antioxidants for these diseases.

2. Pathogenesis of chronic kidney and cardiovascular disease – The links

It is, in fact, very difficult to separate these chronic diseases, because one is a complication of the other in many situations. The development and progression of CKD are closely linked with hypertension and dyslipidemia, both causes of renal failure. Diabetic nephropathy is arguably the leading cause of renal failure. CKD, hypertension and diabetes mellitus all involve endothelial dysfunction, a change well known in the development of atherosclerosis and CVD that includes coronary artery disease, heart failure, stroke and peripheral arterial disease [5]. Vascular calcification occurs in progressive atherosclerosis and CVD, but it is also an important part of vascular injury in end-stage renal disease (ESRD), where patients need renal replacement therapy to survive. It is paradoxical that approximately 50% of individuals with ESRD die from a cardiovascular cause [6]. Thus, CKD and CVD patients have closely-linked diseases with increasing morbidity and mortality. Prevention and treatment of these diseases are major aims in health systems worldwide.

The initiating causes of CKD are highly variable, with previously-mentioned hypertension and diabetes being two of the key ones [7]. Epidemiological studies reveal other strong risk factors for CKD, such as a previous episode of acute kidney damage, exposure to nephrotoxins, obesity, smoking, and increasing age [8, 9]. However, no matter the cause, the progressive structural changes that occur in the kidney are characteristically unifying [10]. The characteristics of CKD are tubulointerstitial inflammation and fibrosis, tubular atrophy, glomerulosclerosis, renal vasculopathy, and presence of granulation tissue. Alterations in the glomerulus include mesangial cell expansion and contraction of the glomerular tuft, followed by a proliferation of connective tissue which leads to significant damage at this first point of the filtration barrier. Structural changes that occur in the kidney produce a vicious cycle of cause and effect, thereby enhancing kidney damage and giving CKD its progressive nature. Whilst early pathological changes in the kidney can occur without clinical presentations, due to the high adaptability of the kidney [10], once the adaptive threshold is reached, the progression of CKD is rapid and the development of ESRD imminent. Vascular pathology exacerbates development of CKD, and it is perhaps here that the links with CVD are closest. Hypertension induces intimal and medial hypertrophy of the intrarenal arteries, leading to hypertensive nephropathy. This is followed by outer cortical glomerulosclerosis with local tubular atrophy and interstitial fibrosis. Compensatory hypertrophy of the inner-cortical glomeruli results, leading to hyperfiltration injury and global glomerulosclerosis. Note, however, that although glomerulopathy is an important characteristic of CKD, the incidence of tubulointerstitial fibrosis has the best correlation with CKD development [11]. As such, kidney tubular cells and renal fibroblasts may be the founding cell types in the progressive development of CKD.

The main clinical manifestation of CKD is a loss of glomerular filtration rate (GFR), allowing for staging of CKD with progressively decreasing (estimated) GFR. CKD staging was facilitated by the National Kidney Foundation (NKF) Kidney Disease Outcomes Quality Initiative (KDOQI) and the Kidney Disease - Improving Global Outcomes (KDIGO), an outcome that highlighted the condition and facilitated its increased diagnosis [12]. The first two stages have normal, or slightly reduced kidney function but some indication of structural deficit in two samples at least 90 days apart. Stages 3-5 are considered the most concerning, with Stage 3 now being sub-classified into Stages 3a and b because of their diagnostic importance. It is thought that stages 2 and 3 should be targeted with prophylactic therapies, such

http://dx.doi.org/10.5772/51923

as lipid lowering drugs or RAS modifiers [13], to minimize the progression of CKD. Table 1 summarises GFR classification and staging for CKD.

Stage	GFR*	Description
1	90mL/Min	Normal renal function but abnormal urine findings, or structural abnormalities, or a genetic trait indicating kidney disease
2	60-89mL/min	Mildly reduced renal function, and other findings (as for stage 1) indicate kidney disease
3A 3B	45-59mL/min 30-44mL/min	Moderately reduced kidney function
4	15-29mL/min	Severely reduced kidney function
5	<15mL/min or on dialysis	Very severe, or end-stage kidney failure (sometimes called established renal failure)

* Measured using the MDRD formula (MDRD= Modification of Diet in Renal Disease). All GFR values are normalized to an average surface area (size) of $1.73m^2$

Table 1. Classification and description of the different stages of CKD

Similar to CKD, the initiating causes for CVD are complex. Although exposure to cardiovascular risk factors such as hypertension, dyslipidemia and diabetes mellitus contributes to CVD, obesity, lack of physical exercise, smoking, genetics, and even depression, also play a role [14]. Common themes for causality are oxidative stress and inflammation, be they local or systemic. The prevalence of CVD also has a strong positive correlation with age, with more than 80% of cases of coronary artery disease and 75% of cases of congestive heart failure observed in geriatric patients [14]. Intrinsic cardiac aging, defined as the development of structural and functional alterations during aging, may render the heart more vulnerable to various stressors, and this ultimately favours the development of CVD. In the early stages of CVD, left ventricular hypertrophy and myocardial fibrosis may be seen in many patients [15]. The processes involved in their development, particularly in association with CKD, can be attributed to hypervolaemia, systemic arterial resistance, elevated blood pressure, large vessel compliance, and activation of pathways related to the parathyroid hormone–vitamin D–phosphate axis. Left ventricular hypertrophy and myocardial fibrosis also predispose to an increase in electric excitability and ventricular arrhythmias [16].

Heart failure resulting from CVD may be staged in a system similar to CKD. In its 2001 guidelines, the American College of Cardiology (ACC) and the American Heart Association working groups introduced four stages of heart failure [17]: Stage A with patients at high risk for developing heart failure in the future but no functional or structural heart disorder; Stage B with a structural heart disorder but no symptoms at any stage; Stage C with previous or current symptoms of heart failure in the context of an underlying structural heart problem, but managed with medical treatment; and Stage D with advanced disease requiring hospital-based support, a heart transplant or palliative care. The ACC staging system is useful in that Stage A may be considered pre-heart failure where intervention with treatment may prevent progression to overt symptoms.

The links between CKD and CVD are so close that it is often difficult to tease out individual causes and mechanisms, given their chronic nature. However, children with CKD present as a particular population without pre-existing symptomatic cardiac disease. This population could also receive significant benefit from preventing and treating CKD and thereby minimising the forthcoming development of CVD which is a major cause of death in children with advanced CKD. Left ventricular hypertrophy and dysfunction, and early markers of atherosclerosis such as increased intimal-medial thickness and stiffness of the carotid artery, and coronary artery calcification, may develop in children with CKD. Early CKD, before needing dialysis, is the optimal time to identify and modify risk factors and intervene in an effort to avert risk of premature cardiac disease and death in these children [18]. These observations have sparked added interest in the mechanisms of the chronic diseases, and in ways to target these mechanisms with additional therapies, such as antioxidants.

2.1. Inflammation and chronic kidney and cardiovascular disease

The circulating nature of many inflammatory mediators such as cytokines, and inflammatory or immune cells, indicates that the immune system can act as a mediator of kidney-heart cross-talk and may be involved in the reciprocal dysfunction that is encountered commonly in the cardio-renal syndromes. Chronic inflammation may follow acute inflammation, but in many chronic diseases like CKD and CVD, it is likely that it begins as a low-grade response with no initial manifestation of an acute reaction. There are many links with visceral obesity and with increased secretion of inflammatory mediators seen in visceral fat [15]. Proinflammatory cytokines are produced by adipocytes, and also cells in the adipose stroma. The links with oxidative stress as an endogenous driver of the chronic diseases become immediately obvious when one admits the close association between oxidative stress and inflammation. The characteristics of dyslipidaemia (elevated serum triglycerides, elevated low-density lipoprotein cholesterol, and/or low high-density lipoprotein cholesterol) are also often seen in obese patients and these are all recognized as risk factors for atherosclerosis. The links between obesity, inflammation, dyslipidaemia, CKD and CVD also occur through yet another syndrome, metabolic syndrome. An improved understanding of the precise molecular mechanisms by which chronic inflammation modifies disease is required before the full implications of its presence, including links with persistent oxidative stress as a cause of chronic disease can be realized.

3. Oxidative stress and chronic kidney and cardiovascular disease

3.1. Understanding oxidative stress

Oxidative stress has been implicated in various pathological systems that are prevalent in both CKD and CVD, most importantly inflammation and fibrosis. Chronic inflammation is induced by biological (eg. infections, autoimmune disease), chemical (eg. drugs, environ-

http://dx.doi.org/10.5772/51923

mental toxins), and physical factors (eg. lack of physical activity) [19]. The inflammatory cells are then a source of free radicals in the forms of reactive oxygen and nitrogen species, although reactive oxygen species (ROS) are considered the most common. The highly reactive ROS are capable of damaging various structures and functional pathways in cells. In consequence, the presence of inflammatory cells is stimulated by cell damage caused by ROS, creating a cycle of chronic damage that is difficult to break. Oxidative stress arises from alterations in the oxidation-reduction balance of cells. Normally, ROS are countered by endogenous natural defences known as antioxidants, and it is the imbalance between ROS and antioxidants which favours greater relative levels of ROS, thereby giving rise to a state of oxidative stress [20-22]. The simple oxidant "imbalance" theory has now grown to incorporate the various crucial pathways and cell metabolism that are also controlled by the interplay between oxidants and antioxidants [23-27]. The rationale for antioxidant therapies lies in restoring imbalances in the redox environment of cells.

The main ROS are superoxide ($O_2^{\bullet-}$), the hydroxyl radical ($OH^{\bullet}$) and hydrogen peroxide (H_2O_2). Mitochondria are considered the major source of ROS, however other contributing sites of ROS generation include the endoplasmic reticulum, peroxisomes and lysosomes [28-30]. Estimated levels of ROS within mitochondria are 5-10 fold higher than cystolic and nuclear compartments in cells [31] due to the presence of the electron transport chain (ETC) within the mitochondrial inner membrane. 1-3% of inspired molecular oxygen (O_2) is converted to the most common of the ROS, $O_2^{\bullet-}$ [32, 33], a powerful precursor of H_2O_2. Although cellular H_2O_2 is stable in this form, it has the potential to interact with a variety of substrates to cause damage, especially in the presence of the ferrous iron (Fe^{2+}), which leads to cleavage and formation of the most reactive and damaging of the ROS, the $OH^{\bullet}$ [34]. In healthy metabolic cells, the production of the potentially harmful H_2O_2 is countered by the catalizing actions of mitochondrial or cystolic catalase (CAT) or thiol peroxidases into water and O_2. The ETC consists of 5 multi-enzyme complexes responsible for maintaining the mitochondrial membrane potential and ATP generation. Each of these complexes presents a site of ROS generation, however complexes I and III have been identified as primary sites of $O_2^{\bullet-}$ generation [35-38]. ROS generation from mitochondrial complexes increases with age in mice [39]. In humans, Granata and colleagues [40] have demonstrated that patients with CKD and haemodialysis patients display impaired mitochondrial respiration.

Agreement on the role of oxidative stress in the pathogenesis of chronic disease is, however, not complete. Oxidants are involved in highly conserved basic physiological processes and are effectors of their downstream pathways [41, 42]. The specific mechanisms for "oxidative stress" are difficult to define because of the rapidity of oxidant signalling [31]. For example, protein tyrosine phosphatases are major targets for oxidant signalling since they contain the amino acid residue cysteine that is highly susceptible to oxidative modification [43]. Meng and colleagues [25] demonstrated the oxidation of the SH2 domain of the platelet-derived growth factor (PDGF) receptor, which contains protein tyrosine phosphatases, in response to PDGF binding. This may indicate the induction of free radicals in response to receptor activation by a cognate ligand in a process that is similar to phosphorylation cascades of intracellular signalling.

3.2. Endogenous antioxidants – Metabolism or disease modifiers

The production of ROS is usually in balance with the availability and cellular localisation of antioxidant enzymes such as superoxide dismutase (SOD), CAT and glutathione peroxidase (Gpx). *In vivo* studies have found accumulated oxidative damage occurs from decreased levels of these enzymes rather than increased ROS production [44, 45]. However, adequate levels of both are likely to be vital for normal cell function. Mitochondria possess their own pool of antioxidants to counteract their generation of ROS. Mitochondrial manganese-SOD (Mn-SOD) converts $O_2^{\bullet -}$ to H_2O_2 which is then decomposed to harmless H_2O and O_2 by CAT and Gpx [46]. Copper/zinc-SOD (Cu/Zn-SOD) has been implicated in stabilizing $O_2^{\bullet -}$ within other cellular compartments, especially peroxisomes, and must be considered in maintenance of the redox state of the whole cell [47, 48]. Limited antioxidant actions of Cu/Zn-SOD may also occur within the inter-membrane space [49]. There is no evidence to indicate that glutathione synthesis occurs within mitochondria, however the mitochondria have their own distinct pool of glutathione required for the formation of Gpx [50].

Among the various endogenous defences against ROS, glutathione homeostasis is critical for a cellular redox environment. Glutathione-linked enzymatic defences of this family include Gpx, glutathione-S-transferase (GST), glutaredoxins (Grx), thioredoxins (Trx), and peroxiredoxins (Prx) [51]. Many of these proteins are known to interact with each other, forming redox networks that have come under investigation for their contribution to dysfunctional oxidant pathways. Mitochondrial-specific isoforms of these proteins also exist and include Grx2, Grx5, Trx2 and Prx3 [52-54], which may be more critical for cell survival compared to their cystolic counterparts [50]. Mitochondrial dysfunction, resulting in depleted ATP synthesis, has the potential to reduce the redox control of glutathione since the rate of glutathione synthesis is ATP-dependent [55]. Intracellular synthesis of glutathione from amino acid derivatives (glycine, glutamic acid and cysteine) accounts for the majority of cellular glutathione compared with extracellular glutathione uptake [56]. Antioxidant networks in which there is interplay, crosstalk and synergism to efficiently and specifically scavenge ROS, may also exist. If this is the case, these antioxidant networks could be harnessed to develop poly-therapeutic antioxidant supplements to combat oxidant-related pathologies, like CKD and CVD.

3.3. Oxidative stress and transcriptional control

The role of oxidative stress in upstream transcriptional gene regulation is becoming increasingly recognised. Not only does this provide insight into the physiological role of oxidative stress, but presents regulatory systems that are possibly prone to deregulation. Furthermore, these sites present targets for pharmacological intervention. Peroxisome proliferator-activated receptors (PPARs) are members of the nuclear hormone receptor superfamily of ligand-dependant transcription factors which have been shown to alter during CKD and CVD [57-59]. They have important roles in the transcriptional regulation of cell differentiation, lipid metabolism, glucose homeostasis, cell cycle progression, and inflammation. There are three PPAR isoforms – α, β/δ and γ. Peroxisome proliferator gamma coactivator (PGCα), in association with PPARγ activation, leads to a variety of cellular protective responses including mitochondrial biogenesis [57]. PPARγ regulation in chronic disease is increasingly recognised, with oxidative stress as the unifying initiating feature. Omega-3 polyunsaturated fatty acids (PUFA) reduce inflammation in kidney tubular epithelial cells by upregulating PPARγ [60]. PPARγ activation by pioglitazone reduced cyclo-oxygenase 2 (COX2) expression in smooth muscle cells from hypertensive rats, and upregulated endogenous antioxidants Mn- and Cu/Zn-SOD [61].

Recently, the protective responses of the nuclear factor E2-related factor 2/Kelch-like ECH-associated protein 1 (Nrf2/Keap1)/antioxidant response element (ARE) were noted [62]. Nrf2 is a nuclear transcription factor that is suppressed in the cytoplasm by the physical binding of Keap1 preventing its translocation into the nucleus. Nrf2 is activated by a loss of Keap1 binding by alterations in cellular redox status, such as increased ROS, by-products of oxidative damage, and reduced antioxidant capacity, thereby promoting its transcriptional response at the ARE [63]. The ARE is a vital component of the promoter regions of genes encoding detoxifying, antioxidant, and glutathione-regulatory enzymes such as quinone-reductase, glutathione-peroxidases, glutathione-reductase, thioredoxins and thioredoxin-reductase, peroxiredoxins, gamma-glutamyl cysteine, heme-oxygenase-1 (HO-1), CAT, SOD metallothionein and ferritin [64-67]. Important to note is that by-products of oxidative damage such a 4-hydroxynoneal and J-isoprostanes act as endogenous activators of Nrf2 [68, 69]. Thus, NRF2/Keap1 and the ARE play a crucial role in cellular defence against ROS. Recent pharmacological protocols have allowed the modulation of this pathway to enhance the capabilities of cells to combat oxidative stress and inflammation [70].

3.4. CKD and CVD are unified by oxidative stress

Chronic diseases of the kidney possess various commonalities to chronic disease of the cardiovascular system which can be linked through pathways controlled by oxidative stress, as shown in Figure 1. Vascular, cellular and biochemical factors all contribute. Increased serum uric acid levels (hyperuricaemia) can arise from increased purine metabolism, increasing age and decreased renal excretion, and have harmful systemic effects. Hyperuricaemia is associated with an increased risk for development and progression of CKD. Hyperuricemia is also a risk factor associated with coronary artery disease [71], left ventricular hypertrophy [72], atrial fibrillation [73], myocardial infarction [74] and ischemic stroke [75]. A 20.6% prevalence of hyperuricemia was found in a cross-sectional study of 18,020 CKD patients [76], and a positive correlation was found between serum uric acid and serum creatinine with impaired renal function [77]. Retention of uremic toxins promotes inflammation and oxidative stress, by priming the acute inflammatory polymorphonuclear lymphocytes, activating interleukin (IL)-1β and IL-8 [78] and stimulating the innate immune response through CD8+ cells [79]. Additionally, uric acid synthesis can promote oxidative stress directly through the activity of xanthine oxidoreductase. This enzyme is synthesized as xanthine dehydrogenase, which can be converted to xanthine oxidase by calcium-dependant proteolysis [80] or modification of cysteine residues [81]. In doing so, the enzyme loses its capacity to bind NADH by alterations in its catalytic site and, instead, transfers electrons from O_2, thereby generating O_2^- [82]. However, the role of uric acid in many conditions associated with oxidative stress is not clear and there are experimental and clinical data showing that uric acid also has a role *in vivo* as an anti-oxidant [83].

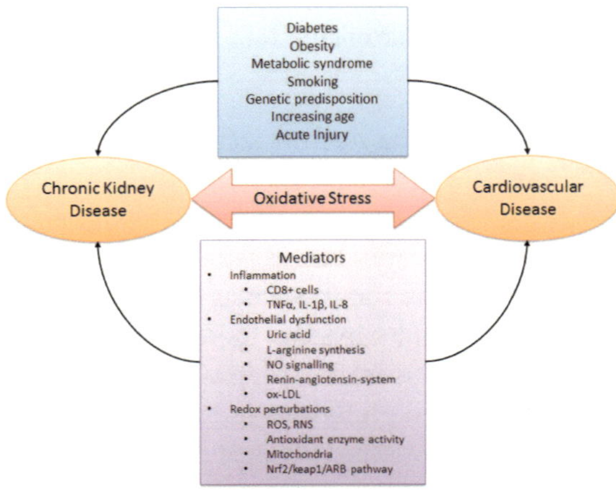

Figure 1. Chronic kidney disease and cardiovascular disease are unified by oxidative stress. Mutual risk factors influence the development and progression of CKD and CVD and can either be modifiable (diabetes, obesity, metabolic syndrome, smoking) or non-modifiable (genetic predisposition, increasing age, acute injury). Oxidative stress has been implicated in the majority of initiating factors. The progression of CKD to CVD, or vice versa, is mediated through: (1) inflammation and the release of pro-inflammatory cytokines such as tumor necrosis factor-α (TNFα), interleukin-1β (IL-1β) and IL-8 from activated lymphocytes; (2) endothelial dysfunction due to increased retention of uremic toxins, and decreased L-arginine synthesis which causes alterations in nitric oxide (NO) signalling - dyslipidaemia and associated pro-oxidative/inflammatory state lead to increased oxidised-low density lipoproteins (ox-LDL), a major component in the pathogenesis of atherosclerosis; (3) redox perturbations that ultimately underlie oxidative stress due to an imbalance between the production of reactive oxygen species (ROS)/reactive nitrogen species (RNS) and endogenous antioxidants, leading to mitochondrial dysfunction and alterations in redox sensitive pathways such as Nrf2/keap1/ARB.

The kidney is a vital source of L-arginine which is a precursor for nitric oxide (NO). A reduction in renal mass can therefore reduce the production of L-arginine and NO activity. NO is vital for regular vascular endothelial cell function, and decreased amounts have the potential to manifest into CVD [84]. Additionally, oxidized low density lipoprotein (ox-LDL), a by-product of oxidative damage in human blood, plays a pivotal role in the pathogenesis of atherosclerosis [85]. There is also a possible link between CVD and CKD that is regulated by oxidative stress through a functional mitochondrial angiotensin system [86]. Angiotensin type II receptors were co-localised with angiotensin on the inner mitochondrial membrane of human mononuclear cells and mouse renal tubular cells. This system was found to modulate mitochondrial NO production and respiration.

4. Antioxidant therapies in chronic kidney and cardiovascular disease

The current state of antioxidant therapies for CKD and CVD is one of promise, but not without controversy. *In vitro* studies commonly identify agents that are able to detoxify harmful

http://dx.doi.org/10.5772/51923

oxidants. However, these studies are criticised for their isolated, non-holistic, nature [87, 88]. It is largely the positive pre-clinical results from *in vivo* studies, usually in rodents, which drive progress for applicability in chronic human disease, but even these show considerable discrepancies in translation into patients. Despite the well-documented dysregulated endogenous oxidant/antioxidant profile in chronic degenerative disorders such as CVD and CKD, there is still evidence that certain antioxidants have no effect [89-92]. It may first be important to identify patients having an altered oxidative stress profile, since this population provides an ideal "intention to treat" cohort. The following trials of antioxidants need then to be rigorous, identifying not only any positive patient outcomes, but also the underlying mechanism, and of course any deleterious outcome. Various approaches have been taken to reduce oxidative stress in models of CKD and accelerated CVD, ranging from reducing oxidant intake in food stuffs [93, 94] to targeted polypharmaceutical compounds. The benefit of rigorous review of outcome from antioxidant therapies in either CKD or CVD is that the primary and secondary outcomes related to both can be measured. In the following section, some antioxidants used for CKD or CVD are reviewed, as shown in Figure 2.

4.1. N-acetylcysteine – An antioxidant with promise

N-acetyl cysteine (NAC) acts as an essential precursor to many endogenous antioxidants involved in the decomposition of peroxides [95]. NAC attenuates oxidative stress from various underlying causes by replenishing intracellular glutathione stores. Glutathione is synthesized in the body by three amino acids by the catalysing of intracellular enzymes gamma-glutamylcysteine synthetase and glutathione synthetase. L-glutamic acid and glycine are two precursors of glutathione that are biologically and readily available. However, the limiting precursor to glutathione biosynthesis and the third amino acid, L-cysteine, is not readily available in a human diet. Although the primary basis for NAC supplementation is to replenish cellular cysteine levels to maintain intracellular glutathione and thus redox control, the sulfhydral-thiol group of L-cysteine is also able to exert direct antioxidant effects by scavenging free radicals, and NAC may also exert its protective effects against 2,3,5-tris(glutathion-S-yl)-hydroquinone toxicity. This was demonstrated in isolated renal tubular epithelial cells, in part by the activation of extracellular signal regulated protein kinase (ERK) 1/2 [96].

The results of NAC supplementation in kidney disease have been variable and largely dependent on the type and cause of kidney injury and also the timing of treatment. In cultured human proximal tubular epithelial cells, NAC reduced lipid peroxidation and maintained the mitochondrial membrane potential, thereby preventing apoptosis following H_2O_2 administration [97]. Although NAC had no significant effect on markers of oxidative stress and inflammation in rats following unilateral ureteral obstruction [98], it reduced kidney malondialdehyde (MDA) levels in a diabetic mouse model [99]. The treatment of CKD patients with NAC with the aim of improving renal function and preventing ESKD has been largely disappointing, with no evidence of reduction in proteinuria [100, 101]. However, NAC seems to exert the greatest antioxidant and anti-inflammatory properties when used against the greatest injury, such as in ESKD patients receiving either haemodialysis or peritoneal dialysis. In those cases, NAC reduced serum 8-isoprostane and the inflammatory cytokine IL-6 [102, 103]. A recent systemic review on antioxidant therapy in hemodialysis patients highlighted NAC as the most efficacious agent in decreasing oxidative stress [104].

The effect of NAC on cardiovascular pathologies is less well investigated than CKD. Crespo *et al*., (2011) demonstrated *in vivo* that, although long-term NAC supplementation improved cardiac function, it did not delay progression to cardiomyopathy [105]. Endothelial dysfunction caused by uremic toxins such as indoxyl sulphate induced ROS-dependent expression of the pro-inflammatory and pro-oxidant nuclear factor-κB (NF-κB), which was ameliorated by NAC pre-treatment [106].

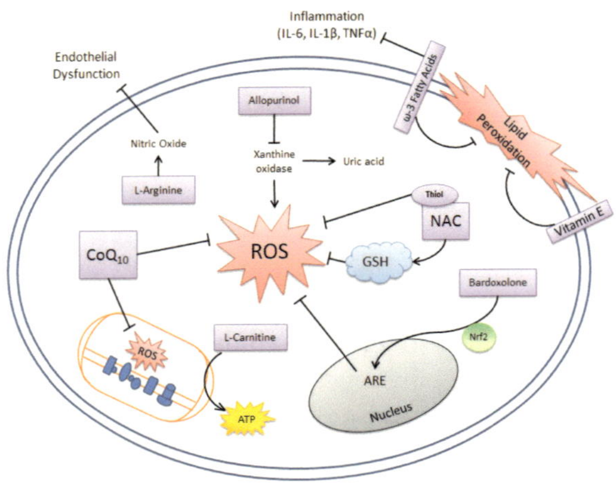

Figure 2. Cellular sites for antioxidant therapy targets in CKD and CVD. Inflammation, lipid peroxidation and reactive oxygen species (ROS) from mitochondrial, cytoplasmic and extracellular sources contribute to oxidative stress. Vitamin E incorporates into the phospholipid bilayer halting lipid peroxidation chain reactions. Omega (ω)-3 fatty acids displace arachadonic acid in the cell membrane and thus reduce arachadonic acid-derived ROS, but also significantly reduce inflammation and subsequent fibrosis. The cysteine residue of N-acetyl-cysteine (NAC) is a precursor for glutathione (GSH) synthesis, and the thiol group is able to scavenge ROS directly. Bardoxolone exerts transcriptional control by promoting nuclear translocation of Nrf2, facilitating antioxidant response element (ARE) binding that upregulates endogenous antioxidant enzyme activity. Allopurinol inhibits xanthine oxidase-derived ROS and the damaging effects of hyperuricemia. Coenzyme Q_{10} (CoQ_{10}) enhances the efficacy of electron transport in the mitochondria, thereby reducing mitochondrial-derived ROS – it is also able to directly scavenge ROS. L-carnitine enhances mitochondrial fatty acid synthesis and subsequent ATP production and thereby maintains cell health. L-arginine is a precursor for nitric oxide which restores endothelial function.

4.2. Vitamin E – An established antioxidant with controversial outcomes

Vitamin E, or α-tocopherol, is a lipid-soluble antioxidant that incorporates into the plasma membrane of cells, thereby scavenging free radicals, mainly the peroxyl radical, and halting lipid peroxidation chain reactions [107]. A benefit of α-tocopherol is its ability to restore its antioxidant capacity from its oxidized form following free radical scavenging, and incorporate back into the plasma membrane. Vitamin C (ascorbic acid) is able to directly reduce α-tocopherol [108-110], and intracellular glutathione and lipoic acid can restore α-tocopherol indirectly by restoring vitamin C [111]. This is a prime example of a cellular antioxidant network prone to dysregulation. Administration of α-tocopherol to kidney proximal tubular cells in culture decreased cisplatin-induced ROS and increased cell viability [112]. The beneficial effects of α-tocopherol are not limited to its antioxidant properties, and recently attention has focused on its blood oxygenising and endogenous cell signalling functions [113]. Vitamin E foodstuffs primarily consist of α-tocotreinol, an isoform of α-tocopherol which has higher antioxidant efficacy in biological membranes. Despite this, the uptake and distribution of α-tocotreinol is far less than α-tocopherol. Therefore, the basis of vitamin E supplementation is to enhance α-tocopherol levels in cell plasma membranes to prevent lipid peroxidation and resultant oxidative stress. One drawback of α-tocopherol is that it takes several days of pre-treatment to exhibit antioxidant effects [114].

Vitamin E therapy has been extensively researched for renal and cardiovascular benefits in human disease populations. Nevertheless, confounding reports mean there is a lack of consensus as to whether vitamin E therapy induces an overall benefit. It is known that patients with CKD stage 4 display the largest decrease in serum α-tocopherol levels following a progressive decline from stage 1 indicating an increased need for α-tocopherol in the CKD population [115]. Interestingly, within the same cohort of patients, a positive correlation of serum α-tocopherol levels and GFR was found [115]. A large scale trial concluded that vitamin E supplementation to cardiovascular high-risk patients over 4.5 years induced no benefit to cardiovascular outcome [92]. The results from the Selenium and Vitamin E Cancer Prevention Trial (SELECT) are of greater concern. They suggest that vitamin E supplementation significantly increases the risk of prostate cancer for young healthy men [116]. Most studies finding beneficial outcomes of α-tocopherol supplementation have largely focused on the ESKD dialysis populations compared to healthy controls and found a reduced risk of CVD, decreased oxidative stress and increased erythrocyte antioxidants SOD, Gpx and CAT [117-119]. The use of α-tocopherol in CKD patients is not without controversy. Miller and colleagues (2005) concluded that high-dose (≥400 IU/day) vitamin E supplementation may increase all cause mortality which may be due to α-tocopherol displacing gamma-(γ)-tocopherol and delta-(δ)-tocopherol in the body [120]. However, this study was highly criticized owing to a bias in data analysis and numerous methodological flaws [121-130]. The apparent lack of clarity surrounding vitamin E supplementation and associated renal and cardiovascular outcomes appears to stem largely from differences in trial design and failure to specify the form of tocopherol used.

4.3. Coenzyme Q_{10} - Maintaining mitochondrial health

The heart and kidneys contain the highest endogenous levels of co-enzymes (Co)Q_9 and CoQ_{10} compared to all other organs [131, 132]. This is likely due to the respective reliance on aerobic metabolism and high density of mitochondria in the intrinsic functioning cells from these organs. It is imperative that endogenous CoQ_{10} levels are maintained to ensure mitochondrial health, and this forms the rationale for CoQ_{10} therapy. CoQ_{10} is a fundamental lipid-soluble component of all cell membranes including those enclosing subcellular compartments. The physiological roles of CoQ_{10} act mostly within the mitochondria where it has three well-characterised functions: (1) the transfer of electrons from complexes I and II to complex III along the ETC of the inner mitochondrial membrane and subsequent membrane polarisation and ATP generation [133, 134]; (2) the pro-oxidant generation of $O_2^{\bullet-}$ and H_2O_2 [135, 136]; and (3) the anti-oxidant quenching of free radicals [137]. The continual oxidation-reduction cycle, and existence of CoQ_{10} in three different redox states, explains its actions as an important cellular redox modulator through its pro-oxidant and antioxidant actions. The fully oxidised form of CoQ_{10}, or ubiquinone, is able to accept electrons, primarily from NADH, to become fully reduced (ubiquinol - CoQ_{10}-H_2). The reduced form of CoQ_{10} is able to give up electrons, thereby scavenging free radicals. The intermediate of ubiquinone and ubiquinol is the univalently-reduced ubisemiquinone (CoQ_{10}-H^+) which acts as a pro-oxidant to form $O_2^{\bullet-}$ and, subsequently, H_2O_2.

The major antioxidant role of CoQ_{10} is in preventing lipid peroxidation directly, and by interactions with α-tocopherol [138]. Ubiquinol is able to donate a hydrogen atom and thus quench peroxyl radicals, preventing lipid peroxidation chain reactions. CoQ_{10} and α-tocopherol co-operate as antioxidants through the actions of CoQ_{10}-H_2 restoring α-tocopheroxyl back to α-tocopherol [109, 139]. However, the reactivity of α-tocopherol with peroxy radicals far exceeds that of ubiquinol with peroxyl radicals, suggesting that, *in vivo*, ubiquinols do not act as antioxidants but regenerate the antioxidant properties of α-tocopherols [140]. This is in accordance with *in vivo* studies investigating the effects of CoQ_{10} supplementation which have primarily found a limited antioxidant capacity. CoQ_{10}, acting as a pro-oxidant in all biological membranes including the Golgi, endosome/lysosome systems, as well as mitochondria, has led to much criticism regarding the claimed antioxidant power of CoQ_{10} supplementation in humans [141]. Nonetheless, many *in vitro* studies demonstrate antioxidant properties of CoQ_{10} in single cells, and benefits of CoQ_{10} supplementation in humans are attributed to its ability to maintain efficient mitochondrial energy metabolism and thus prevent mitochondrial dysfunction, rather than act as a direct cellular antioxidant. CoQ_{10} supplementation *in vivo* reduced protein oxidation in skeletal muscle of rats but had no effect on mitochondrial H_2O_2 production in the kidney [142]. However, Ishikawa and colleagues (2011) demonstrated a decrease in kidney $O_2^{\bullet-}$ levels in hemi-nephrectomised rats on a CoQ_{10} supplemented diet, and increased renal function compared with rats on a control diet [143]. Recently, CoQ_{10} supplementation improved left ventricular diastolic dysfunction and remodelling and reduced oxidative stress in a mouse model of type 2 diabetes [144]. CoQ_{10} supplementation in CVD patients also receiving statin therapy is becoming increasingly popular due to the CoQ_{10}-inhibitory actions of statins. CoQ_{10} levels decrease with age, but there are no studies measuring endogenous CoQ_{10} levels in CKD or CVD patients and this could prove vital in the identification of population where CoQ_{10} therapy may have beneficial outcomes.

4.4. Omega-3 poly-unsaturated fatty acids – Inflammation and oxidative stress

Inflammation and fibrosis are causes, as well as consequences, of oxidative stress [145, 146]. Direct targeting of inflammatory and fibrotic pathways with more specific modifying compounds presents a way to indirectly decrease oxidative stress in chronic pathologies. Long chain omega-3 PUFA, including docosahexanoic acid (DHA) and eicosapentanoic acid (EPA), have been investigated in a large range of *in vitro* and *in vivo* models and found to possess anti-inflammatory properties. Recently, omega-3 fatty acid treatment of peripheral blood mononuclear cells from pre-dialysis CKD patients reduced the inflammatory markers IL-6, IL-1β, tumor necrosis factor (TNF)-α and C-reactive protein to levels observed in healthy subjects [147]. Although the beneficial effects of EPA/DHA are attributed to their anti-inflammatory properties, they are also known to enhance endogenous antioxidant defence systems such as γ-glutamyl-cysteinyl ligase and glutathione reductase [148]. DHA and EPA incorporate into the phospholipid bilayer of cells where they displace arachidonic acid. Arachidonic acid can generate ROS through the COX2 and xanthine oxidase inflammatory pathways. DHA/EPA administration to renal epithelial cells and macrophages suppresses this pro-oxidant pathway [149]. Furthermore, chemoattractants derived from EPA are less potent that those derived from arachidonic acid [150, 151]. Recently, *in vitro* studies determined that EPA and DHA attenuated TNF-α-stimulated monocyte chemoattractant protein (MCP)-1 gene expression by interacting with ERK and NF-κB in rat mesangial cells [152]. Earlier evidence had shown that EPA and DHA inhibit NF-κB expression by stimulating PPARs in human kidney-2 cells *in vitro* [60]. *In vivo* studies have now confirmed an improvement in kidney function and structure using EPA/DHA supplementation, with reduced oxidative stress, inflammation and tubulointerstitial fibrosis through the reversal of inflammatory and oxidant pathways [153, 154]. Recently, a highly beneficial outcome of fish oil supplementation was found with heart failure patients with co-morbid diabetes [155]. Clinical studies have found fish oil treatment modulates lipid levels [156, 157], and has antithrombotic [158, 159] and anti-hypertensive effects due to its vascular and endothelial actions [160].

4.5. Allopurinol – A xanthine oxidase inhibitor

Allopurinol treatment aims is to inhibit xanthine oxidase to decrease serum uric acid and its associated toxic effects. Allopurinol and its metabolite, oxypurinol, act as competitive substrates for xanthine oxidase. They enhance urinary urate excretion and block uric acid reabsorption by urate transporters in the proximal tubule, thereby facilitating enhanced uric acid excretion [161-163]. Allopurinol treatment of diabetic mice attenuated hyperuricaemia, albuminuria, and tubulointerstitial injury [164]. Allopurinol may also have antioxidant activities in addition to its enzyme inhibitory activities, by scavenging $OH^{\bullet}$ as well as chlorine dioxide and HOCl [165, 166]. Although later *in vivo* studies revealed that rat serum obtained after oral administration of allopurinol did not contain allopurinol levels sufficient to scavenge free radicals [167], inhibition of xanthine oxidase-dependent production of $NO^{\bullet}$ and ROS provides allopurinol an indirect mechanism for decreasing oxidative stress in hyperuricaemic CKD patients. Interventional studies of use of allopurinol in renal disease have shown improved uric acid levels, GFR, cardiovascular outcomes and delayed CKD progression. A prospective randomised trial of 113 patients with GFR <60 ml/min/1.73m^3 given allopurinol 100mg/d for 2 years found an increase in GFR of 1.31 ml/min/1.73m^3 compared to the controls which decreased, and a 71% decreased risk of CVD [168]. Interestingly, Kanbay and colleagues (2007) found that allopurinol at 300mg/d over 3 months improved GFR, uric acid and C-reactive protein levels but made no change to proteinuria [169]. Allopurinol given to ESKD patients on hemodialysis reduced the risk of CVD by decreasing serum low density lipoproteins, triglycerides and uric acid [170]. Large, long-term interventional studies investigating kidney function in the CKD, and CVD, populations are needed to fully determine if allopurinol is cardio- and reno-protective via anti-oxidant mechanisms.

4.6. Bardoxolone methyl - Targeting the Nrf/Keap1/ARE pathway

A different approach has been investigated by modulating pathways that respond to oxidative stress, rather than targeting ROS by directly increasing endogenous antioxidants. The Nrf2/keap1/ARE pathway presents an exciting target to enhance the oxidant detoxifying capabilities of cells. Bardoxolone methyl [2-cyano-3,12-dioxooleana-1,9(11)-dien-28-oic acid (CDDO-Me)] is a potent activator of the Nrf2/keap1/ARE pathway and currently shows promise for halting the progressive decline of GFR in type 2 diabetic CKD patients [171, 172]. Bardoxolone methyl is a triterperoid derived from natural plant products that has undergone oleanolic acid-based modification [173]. Its mechanism of action is largely unknown, however, it induces an overall antioxidative protective effect with anti-inflammatory and cytoprotective characteristics [174, 175]. Bardoxolone methyl administered to mice ameliorated ischemia-reperfusion induced acute kidney injury by Nrf2-dependant expression of HO-1 and PPARγ [176]. Its mechanism may also reside in regulating mitochondrial biogenesis given the involvement of PPARγ. A large international study evaluating the full scale of bardoxolone methyl's effects on CKD progression is in progress, the results of which could determine if bardoxolone methyl should become a standard treatment in renal disease patients. Concurrent benefits to CVD will undoubtedly also be measured.

4.7. L-Carnitine – Improving cardiovascular health in dialysis

Carnitine is an essential cofactor required for the transformation of free fatty acids into acylcarnitine and its subsequent transport into the mitochondria for β-oxidation [177]. This underlies its importance in the production of ATP for cellular energy. Acylcarnitine is also essential for the removal of toxic fat metabolism by-products. Carnitine is obtained primarily from food stuffs, however it can be synthesised endogenously from the amino acid L-lysine and methionine [177]. L-carnitine supplementation primarily benefits ESRD patients on hemodialysis and their associated cardiovascular complications, especially anemia. This is primarily due to the well-described decrease in serum free carnitine in maintenance hemodialysis patients compared to non-dialysis CKD and healthy patients [178]. L-carnitine supplementation offsets renal anemia, lipid abnormalities and cardiac dysfunction in hemodialysis patients [179]. Left ventricular hypertrophy regressed in hemodialysis patients receiving 10mg/kg of L-carnitine immediately following hemodialysis for a 12 month period. [180]. Other measures of cardiac morbidity such as reduced left ventricular ejection fraction and increased left ventricular mass also significantly improved following low dose L-carnitine supplementation [181]. Benefits to the peripheral vasculature have also been demonstrated by L-carnitine through a mechanism thought to involve an associated decrease in homocysteine levels [182]. Interestingly, oxidative stress is a major characteristic of hemodialysis patients [183].

As well as the physiological role of L-carnitine in mitochondrial fatty acid synthesis, oxidant reducing capabilities have also been demonstrated and may underlie the health benefits of L-carnitine therapy in CKD and CVD. L-carnitine infusions significantly improved blood urea nitrogen (BUN) and creatinine levels in a 5/6 nephrectomy model of CKD with a concomitant increase in plasma SOD, Gpx, CAT and GSH, and decrease in the oxidative stress marker malondialdehyde [184]. Ye et al., (2010) suggest that L-carnitine attenuates renal tubular cell oxidant injury and subsequent apoptosis by reducing mitochondrial-derived ROS [97]. They suggest that this anti-apoptotic mechanism may also explain the demonstrated reduction in morbidity from cardiomyopathies in L-carnitine supplemented hemodialysis patients.

4.8. L-Arginine - Maintaining endothelial function

The premise of L-arginine supplementation is to maintain NO signalling and thereby maintain vascular endothelial cell function. L-arginine is a physiological precursor to NO and its availability and transport determine the rate of NO biosynthesis. CKD patients most often present with atherosclerosis, thromboembolitic complications, and endothelial dysfunction, primarily due to altered endothelium-dependant relaxation factors [185]. It is believed that the impaired NO synthesis, common in CKD individuals, contributes significantly to their disease pathogenesis [186]. L-arginine synthesis occurs in the liver and kidney, with the kidney functioning to maintain homeostatic plasma levels since the liver processes NO from the diet [187]. The addition of L-aspartic acid or L-glutamic acid with L-citrulline and arginirosuccinic acid synthase as the rate determining enzyme forms L-arginine [188]. The proximal tubular cells account for the majority of kidney NO synthesis [189, 190], thus kidney damage and atrophy, a primary corollary of CKD, results in decreased synthesis of L-arginine. The majority of research demonstrates decreased levels of NO production in CKD and CVD patients [191-193]. However, some research suggests NO activity increases [194, 195]. These disparate findings highlight the need to measure L-arginine levels in patients before commencing L-arginine supplementation. Rajapaske *et al.* (2012) demonstrated impaired kidney L-arginine transport and a contributing factor to hypertension in rats, irrespective of an underlying renal disease [196]. During a state of oxidative stress, L-arginine supplementation was shown to decrease MDA, myeloperoxidase and xanthine oxidase and increase glutathionine in both heart and kidney tissue from rats [197]. As such, L-arginine supplementation represents an approach to restoring a dysregulation of NO signalling and subsequent endothelial dysfunction in both chronic kidney and heart diseases.

4.9. Combination antioxidants

Compounds commonly used to alleviate oxidative stress exhibit different antioxidant actions, and so there exists the potential for different antioxidants to work together to improve whole cell and organ function through a targeted polypharmaceutical approach to decrease oxidative stress. However, most clinical studies investigating the effects of combination antioxidants have demonstrated confounding results. Mosca *et al.*, (2002) demonstrated that daily intake of NAC 100mg, L-carntine 100mg, selenomethionine 0.05mg, α-tocopherol 10mg, CoQ_{10} 100mg and α-lipoic acid 100mg successfully increased plasma CAT, Gpx and total antioxidant capacity whilst decreasing lipid peroxides and ROS generation by lymphocyte mitochondria [198]. However, this trial only included healthy participants and cannot be extrapolated to the CKD and CVD populations.

In a murine model of diabetic nephropathy, a major cause of CKD with associated CVD, the beneficial effects of NAC, L-ascorbic acid (vitamin C) and α-tocopherol were demonstrated [199]. Daily supplementation for 8 weeks decreased lipid peroxidation, BUN, serum creatinine and blood glucose, mainly due to a reduction in the inflammatory response induced by hyperglycemia. In comparison, a prospective trial investigating oral supplementation of mixed tocopherols and α-lipoic acid in stage 3 and 4 CKD patients has revealed disappointing results. Over 2 months, supplementation did not reduce biomarkers of oxidative stress (F_2-isoprostanes and protein thiol concentration) or inflammation (CRP and IL-6). The short period of time (2 months) of the intervention may explain this result and longer trials need to be carried out. The inclusion of vitamin E in these interventions has polarized discussion on the outcomes, because of its negligible benefits when cardiovascular outcomes were measured [91, 92, 200] and also because of contraindications, discussed previously. Despite this, long-term treatment in with the antioxidants vitamin C, vitamin E, CoQ_{10} and selenium has been shown to reduce multiple cardiovascular risk factors [201]. Recently, multiple antioxidants in combination with L-arginine have shown promise in animal models of CKD and associated CVD. Korish (2010) has demonstrated in a 5/6 nephrectomy CKD model that L-arginine improved the effects of L-carnitine, catechin and vitamins E and C on blood pressure, dyslipidemia, inflammation and kidney function [84].

5. Conclusion

CKD is a progressive disease with increasing incidence, having very little success in current conventional therapies once CKD reaches stage 4. Stages 2 and 3 are best to target to slow or stop further development of the disease. There is an almost inseparable connection between CKD and CVD, with many patients with CKD dying of the cardiovascular complications before renal failure reaches its fullest extent. Oxidative stress and inflammation are closely interrelated with development of CKD and CVD, and involve a spiralling cycle that leads to progressive patient deterioration. Given the complex nature of oxidative stress and its molecular pathways, antioxidants may need to be given as a polypharmacotherapy to target each aberrant pathway, with the aim of reducing the burden of these chronic diseases. It is vital for the progression of antioxidant therapy research in CKD and CVD that measures of oxidative stress are compared with pathophysiological outcome in the diseases, especially in connection with antioxidant therapies that may be delivered with or without more conventional CKD therapies.

Author details

David M. Small and Glenda C. Gobe*

*Address all correspondence to: g.gobe@uq.edu.au

Centre for Kidney Disease Research, School of Medicine, The University of Queensland, Brisbane, Australia

References

[1] Rosner MH, Ronco C, Okusa MD. The role of inflammation in the cardio-renal syndrome: a focus on cytokines and inflammatory mediators. Semin Nephrol. 2012 Jan; 32(1):70-8.

[2] Ronco C, McCullough P, Anker SD, Anand I, Aspromonte N, Bagshaw SM, et al. Cardio-renal syndromes: report from the consensus conference of the acute dialysis quality initiative. Eur Heart J. 2010 Mar;31(6):703-11.

[3] Leung FP, Yung LM, Laher I, Yao X, Chen ZY, Huang Y. Exercise, vascular wall and cardiovascular diseases: an update (Part 1). Sports Med. 2008;38(12):1009-24.

[4] Bongartz LG, Cramer MJ, Doevendans PA, Joles JA, Braam B. The severe cardiorenal syndrome: 'Guyton revisited'. Eur Heart J. 2005 Jan;26(1):11-7.

[5] Sallam N, Fisher A, Golbidi S, Laher I. Weight and inflammation are the major determinants of vascular dysfunction in the aortae of db/db mice. Naunyn Schmiedebergs Arch Pharmacol. 2011 May;383(5):483-92.

[6] Schiffrin EL, Lipman ML, Mann JF. Chronic kidney disease: effects on the cardiovascular system. Circulation. 2007 Jul 3;116(1):85-97.

[7] McDonald SP, Chang S, Excell L, editors. ANZDATA Registry Report. Adelaide 2007.

[8] Tanner RM, Brown TM, Muntner P. Epidemiology of obesity, the metabolic syndrome, and chronic kidney disease. Curr Hypertens Rep. 2012 Apr;14(2):152-9.

[9] Graf J, Ryan C, Green F. An overview of chronic kidney disease in Australia, 2009. Canberra: Australian Inst Health Welfare 2009.

[10] Tesch GH. Review: Serum and urine biomarkers of kidney disease: A pathophysiological perspective. Nephrol (Carlton). 2010 Sep;15(6):609-16.

[11] Rodriguez-Iturbe B, Johnson RJ, Herrera-Acosta J. Tubulointerstitial damage and progression of renal failure. Kidney Int Suppl. 2005 Dec(99):S82-6.

[12] Fassett RG, Venuthurupalli SK, Gobe GC, Coombes JS, Cooper MA, Hoy WE. Biomarkers in chronic kidney disease: a review. Kidney Int. 2011 Oct;80(8):806-21.

[13] Choudhury D, Luna-Salazar C. Preventive health care in chronic kidney disease and end-stage renal disease. Nat Clin Pract Nephrol. 2008 Apr;4(4):194-206.

[14] Dutta D, Calvani R, Bernabei R, Leeuwenburgh C, Marzetti E. Contribution of impaired mitochondrial autophagy to cardiac aging: mechanisms and therapeutic opportunities. Circ Res. 2012 Apr 13;110(8):1125-38.

[15] Manabe I. Chronic inflammation links cardiovascular, metabolic and renal diseases. Circ J. 2011;75(12):2739-48.

[16] Glassock RJ, Pecoits-Filho R, Barberato SH. Left ventricular mass in chronic kidney disease and ESRD. Clin J Am Soc Nephrol: CJASN. 2009 Dec;4 Suppl 1:S79-91.

[17] Hunt SA, Abraham WT, Chin MH, Feldman AM, Francis GS, Ganiats TG, et al. ACC/AHA 2005 Guideline Update for the Diagnosis and Management of Chronic Heart Failure in the Adult: a report of the American College of Cardiology/American Heart Association Task Force on Practice Guidelines (Writing Committee to Update the 2001 Guidelines for the Evaluation and Management of Heart Failure): developed in collaboration with the American College of Chest Physicians and the International Society for Heart and Lung Transplantation: endorsed by the Heart Rhythm Society. Circulation. 2005 Sep 20;112(12):e154-235.

[18] Mitsnefes MM. Cardiovascular disease in children with chronic kidney disease. J Am Soc Nephrol: JASN. 2012 Apr;23(4):578-85.

[19] Whaley-Connell A, Pavey BS, Chaudhary K, Saab G, Sowers JR. Renin-angiotensin-aldosterone system intervention in the cardiometabolic syndrome and cardio-renal protection. Ther Adv Cardiovasc Dis. 2007 Oct;1(1):27-35.

[20] Gomes P, Simao S, Silva E, Pinto V, Amaral JS, Afonso J, et al. Aging increases oxidative stress and renal expression of oxidant and antioxidant enzymes that are associated with an increased trend in systolic blood pressure. Oxid Med Cell Longev. 2009 Jul-Aug;2(3):138-45.

[21] Pias EK, Aw TY. Apoptosis in mitotic competent undifferentiated cells is induced by cellular redox imbalance independent of reactive oxygen species production. FASEB J. 2002 Jun;16(8):781-90.

[22] Zhuang S, Yan Y, Daubert RA, Han J, Schnellmann RG. ERK promotes hydrogen peroxide-induced apoptosis through caspase-3 activation and inhibition of Akt in renal epithelial cells. Am J Physiol Renal Physiol. 2007 Jan;292(1):F440-7.

[23] Blanchetot C, Tertoolen LG, den Hertog J. Regulation of receptor protein-tyrosine phosphatase alpha by oxidative stress. EMBO J. 2002 Feb 15;21(4):493-503.

[24] Jones DP. Redefining oxidative stress. Antioxid Redox Signal. 2006 Sep-Oct;8(9-10): 1865-79.

[25] Meng TC, Fukada T, Tonks NK. Reversible oxidation and inactivation of protein tyrosine phosphatases in vivo. Mol Cell. 2002 Feb;9(2):387-99.

[26] Rao RK, Clayton LW. Regulation of protein phosphatase 2A by hydrogen peroxide and glutathionylation. Biochem Biophys Res Commun. 2002 Apr 26;293(1):610-6.

[27] Tavakoli S, Asmis R. Reactive Oxygen Species and Thiol Redox Signaling in the Macrophage Biology of Atherosclerosis. Antioxid Redox Signal. 2012 Jun 11.

[28] Madesh M, Hajnoczky G. VDAC-dependent permeabilization of the outer mitochondrial membrane by superoxide induces rapid and massive cytochrome c release. J Cell Biol. 2001 Dec 10;155(6):1003-15.

[29] Soubannier V, McBride HM. Positioning mitochondrial plasticity within cellular signaling cascades. Biochim Biophys Acta. 2009 Jan;1793(1):154-70.

[30] Vay L, Hernandez-SanMiguel E, Lobaton CD, Moreno A, Montero M, Alvarez J. Mitochondrial free [Ca2+] levels and the permeability transition. Cell Calcium. 2009 Mar;45(3):243-50.

[31] Cadenas E, Davies KJ. Mitochondrial free radical generation, oxidative stress, and aging. Free Radic Biol Med. 2000 Aug;29(3-4):222-30.

[32] Boveris A, Chance B. The mitochondrial generation of hydrogen peroxide. General properties and effect of hyperbaric oxygen. Biochem J. 1973 Jul;134(3):707-16.

[33] Nohl H, Hegner D. Do mitochondria produce oxygen radicals in vivo? Eur J Biochem. 1978 Jan 16;82(2):563-7.

[34] Lipinski B. Is it oxidative stress or free radical stress and why does it matter? Oxid Antioxid Med Sci. 2012 March;1(1):5-9.

[35] Cadenas E, Boveris A, Ragan CI, Stoppani AO. Production of superoxide radicals and hydrogen peroxide by NADH-ubiquinone reductase and ubiquinol-cytochrome c reductase from beef-heart mitochondria. Arch Biochem Biophys. 1977 Apr 30;180(2):248-57.

[36] Turrens JF, Boveris A. Generation of superoxide anion by the NADH dehydrogenase of bovine heart mitochondria. Biochem J. 1980 Nov 1;191(2):421-7.

[37] Turrens JF, Alexandre A, Lehninger AL. Ubisemiquinone is the electron donor for superoxide formation by complex III of heart mitochondria. Arch Biochem Biophys. 1985 Mar;237(2):408-14.

[38] Turrens JF. Mitochondrial formation of reactive oxygen species. J Physiol. 2003 Oct 15;552(Pt 2):335-44.

[39] Choksi KB, Nuss JE, Boylston WH, Rabek JP, Papaconstantinou J. Age-related increases in oxidatively damaged proteins of mouse kidney mitochondrial electron transport chain complexes. Free Radic Biol Med. 2007 Nov 15;43(10):1423-38.

[40] Granata S, Zaza G, Simone S, Villani G, Latorre D, Pontrelli P, et al. Mitochondrial dysregulation and oxidative stress in patients with chronic kidney disease. BMC Genomics. 2009;10:388.

[41] Nemoto S, Takeda K, Yu ZX, Ferrans VJ, Finkel T. Role for mitochondrial oxidants as regulators of cellular metabolism. Mol Cell Biol. 2000 Oct;20(19):7311-8.

[42] Werner E, Werb Z. Integrins engage mitochondrial function for signal transduction by a mechanism dependent on Rho GTPases. J Cell Biol. 2002 Jul 22;158(2):357-68.

[43] Cooper CE, Patel RP, Brookes PS, Darley-Usmar VM. Nanotransducers in cellular redox signaling: modification of thiols by reactive oxygen and nitrogen species. Trends Biochem Sci. 2002 Oct;27(10):489-92.

[44] Kokoszka JE, Coskun P, Esposito LA, Wallace DC. Increased mitochondrial oxidative stress in the Sod2 (+/-) mouse results in the age-related decline of mitochondrial function culminating in increased apoptosis. Proc Natl Acad Sci U S A. 2001 Feb 27;98(5): 2278-83.

[45] Meng Q, Wong YT, Chen J, Ruan R. Age-related changes in mitochondrial function and antioxidative enzyme activity in fischer 344 rats. Mech Ageing Dev. 2007 Mar; 128(3):286-92.

[46] Raha S, McEachern GE, Myint AT, Robinson BH. Superoxides from mitochondrial complex III: the role of manganese superoxide dismutase. Free Radic Biol Med. 2000 Jul 15;29(2):170-80.

[47] Angermuller S, Islinger M, Volkl A. Peroxisomes and reactive oxygen species, a lasting challenge. Histochem Cell Biol. 2009 Apr;131(4):459-63.

[48] Islinger M, Li KW, Seitz J, Volkl A, Luers GH. Hitchhiking of Cu/Zn superoxide dismutase to peroxisomes-evidence for a natural piggyback import mechanism in mammals. Traffic. 2009 Nov;10(11):1711-21.

[49] Sturtz LA, Diekert K, Jensen LT, Lill R, Culotta VC. A fraction of yeast Cu,Zn-superoxide dismutase and its metallochaperone, CCS, localize to the intermembrane space of mitochondria. A physiological role for SOD1 in guarding against mitochondrial oxidative damage. J Biol Chem. 2001 Oct 12;276(41):38084-9.

[50] Soderdahl T, Enoksson M, Lundberg M, Holmgren A, Ottersen OP, Orrenius S, et al. Visualization of the compartmentalization of glutathione and protein-glutathione mixed disulfides in cultured cells. FASEB J. 2003 Jan;17(1):124-6.

[51] Godoy JR, Oesteritz S, Hanschmann EM, Ockenga W, Ackermann W, Lillig CH. Segment-specific overexpression of redoxins after renal ischemia and reperfusion: protective roles of glutaredoxin 2, peroxiredoxin 3, and peroxiredoxin 6. Free Radic Biol Med. 2011 Jul 15;51(2):552-61.

[52] Lillig CH, Holmgren A. Thioredoxin and related molecules--from biology to health and disease. Antioxid Redox Signal. 2007 Jan;9(1):25-47.

[53] Lonn ME, Hudemann C, Berndt C, Cherkasov V, Capani F, Holmgren A, et al. Expression pattern of human glutaredoxin 2 isoforms: identification and characterization of two testis/cancer cell-specific isoforms. Antioxid Redox Signal. 2008 Mar; 10(3):547-57.

[54] Hanschmann EM, Lonn ME, Schutte LD, Funke M, Godoy JR, Eitner S, et al. Both thioredoxin 2 and glutaredoxin 2 contribute to the reduction of the mitochondrial 2-Cys peroxiredoxin Prx3. J Biol Chem. 2010 Dec 24;285(52):40699-705.

[55] Lash LH, Putt DA, Matherly LH. Protection of NRK-52E cells, a rat renal proximal tubular cell line, from chemical-induced apoptosis by overexpression of a mitochondrial glutathione transporter. J Pharmacol Exp Ther. 2002 Nov;303(2):476-86.

[56] Visarius TM, Putt DA, Schare JM, Pegouske DM, Lash LH. Pathways of glutathione metabolism and transport in isolated proximal tubular cells from rat kidney. Biochem Pharmacol. 1996 Jul 26;52(2):259-72.

[57] Funk JA, Odejinmi S, Schnellmann RG. SRT1720 induces mitochondrial biogenesis and rescues mitochondrial function after oxidant injury in renal proximal tubule cells. J Pharmacol Exp Therapeut. 2010 May;333(2):593-601.

[58] Lepenies J, Hewison M, Stewart PM, Quinkler M. Renal PPARgamma mRNA expression increases with impairment of renal function in patients with chronic kidney disease. Nephrology. 2010 Oct;15(7):683-91.

[59] Sakamoto A, Hongo M, Saito K, Nagai R, Ishizaka N. Reduction of renal lipid content and proteinuria by a PPAR-gamma agonist in a rat model of angiotensin II-induced hypertension. Eur J Pharmacol. 2012 May 5;682(1-3):131-6.

[60] Li H, Ruan XZ, Powis SH, Fernando R, Mon WY, Wheeler DC, et al. EPA and DHA reduce LPS-induced inflammation responses in HK-2 cells: evidence for a PPAR-gamma-dependent mechanism. Kidney Int. 2005 Mar;67(3):867-74.

[61] Martin A, Perez-Giron JV, Hernanz R, Palacios R, Briones AM, Fortuno A, et al. Peroxisome proliferator-activated receptor-gamma activation reduces cyclooxygenase-2 expression in vascular smooth muscle cells from hypertensive rats by interfering with oxidative stress. J Hypertens. 2012 Feb;30(2):315-26.

[62] Hybertson BM, Gao B, Bose SK, McCord JM. Oxidative stress in health and disease: the therapeutic potential of Nrf2 activation. Mol Aspects Med. 2011 Aug;32(4-6): 234-46.

[63] Wilmes A, Crean D, Aydin S, Pfaller W, Jennings P, Leonard MO. Identification and dissection of the Nrf2 mediated oxidative stress pathway in human renal proximal tubule toxicity. Toxicol In Vitro. 2011 Apr;25(3):613-22.

[64] Nelson SK, Bose SK, Grunwald GK, Myhill P, McCord JM. The induction of human superoxide dismutase and catalase in vivo: a fundamentally new approach to antioxidant therapy. Free Radic Biol Med. 2006 Jan 15;40(2):341-7.

[65] Prestera T, Talalay P, Alam J, Ahn YI, Lee PJ, Choi AM. Parallel induction of heme oxygenase-1 and chemoprotective phase 2 enzymes by electrophiles and antioxidants: regulation by upstream antioxidant-responsive elements (ARE). Mol Med. 1995 Nov;1(7):827-37.

[66] Li Y, Jaiswal AK. Regulation of human NAD(P)H:quinone oxidoreductase gene. Role of AP1 binding site contained within human antioxidant response element. J Biol Chem. 1992 Jul 25;267(21):15097-104.

[67] Okuda A, Imagawa M, Maeda Y, Sakai M, Muramatsu M. Structural and functional analysis of an enhancer GPEI having a phorbol 12-O-tetradecanoate 13-acetate responsive element-like sequence found in the rat glutathione transferase P gene. J Biol Chem. 1989 Oct 5;264(28):16919-26.

[68] Itoh K, Mochizuki M, Ishii Y, Ishii T, Shibata T, Kawamoto Y, et al. Transcription factor Nrf2 regulates inflammation by mediating the effect of 15-deoxy-Delta(12,14)-prostaglandin j(2). Mol Cell Biol. 2004 Jan;24(1):36-45.

[69] Levonen AL, Landar A, Ramachandran A, Ceaser EK, Dickinson DA, Zanoni G, et al. Cellular mechanisms of redox cell signalling: role of cysteine modification in controlling antioxidant defences in response to electrophilic lipid oxidation products. Biochem J. 2004 Mar 1;378(Pt 2):373-82.

[70] Rojas-Rivera J, Ortiz A, Egido J. Antioxidants in kidney diseases: the impact of bardoxolone methyl. Int J Nephrol. 2012;2012:321714.

[71] Brand FN, McGee DL, Kannel WB, Stokes J, 3rd, Castelli WP. Hyperuricemia as a risk factor of coronary heart disease: The Framingham Study. Am J Epidemiol. 1985 Jan;121(1):11-8.

[72] Mitsuhashi H, Tamura K, Yamauchi J, Ozawa M, Yanagi M, Dejima T, et al. Effect of losartan on ambulatory short-term blood pressure variability and cardiovascular remodeling in hypertensive patients on hemodialysis. Atherosclerosis. 2009 Nov; 207(1):186-90.

[73] Letsas KP, Korantzopoulos P, Filippatos GS, Mihas CC, Markou V, Gavrielatos G, et al. Uric acid elevation in atrial fibrillation. Hellenic J Cardiol. 2010 May-Jun;51(3): 209-13.

[74] Car S, Trkulja V. Higher serum uric acid on admission is associated with higher short-term mortality and poorer long-term survival after myocardial infarction: retrospective prognostic study. Croat Med J. 2009 Dec;50(6):559-66.

[75] Chen JH, Chuang SY, Chen HJ, Yeh WT, Pan WH. Serum uric acid level as an independent risk factor for all-cause, cardiovascular, and ischemic stroke mortality: a Chinese cohort study. Arthritis Rheum. 2009 Feb 15;61(2):225-32.

[76] Shan Y, Zhang Q, Liu Z, Hu X, Liu D. Prevalence and risk factors associated with chronic kidney disease in adults over 40 years: a population study from Central China. Nephrology (Carlton). 2010 Apr;15(3):354-61.

[77] Chen YC, Su CT, Wang ST, Lee HD, Lin SY. A preliminary investigation of the association between serum uric acid and impaired renal function. Chang Gung Med J. 2009 Jan-Feb;32(1):66-71.

[78] Martinon F, Petrilli V, Mayor A, Tardivel A, Tschopp J. Gout-associated uric acid crystals activate the NALP3 inflammasome. Nature. 2006 Mar 9;440(7081):237-41.

[79] Sakamaki I, Inai K, Tsutani Y, Ueda T, Tsutani H. Binding of monosodium urate crystals with idiotype protein efficiently promote dendritic cells to induce cytotoxic T cells. Cancer Sci. 2008 Nov;99(11):2268-73.

[80] Amaya Y, Yamazaki K, Sato M, Noda K, Nishino T. Proteolytic conversion of xanthine dehydrogenase from the NAD-dependent type to the O2-dependent type. Amino acid sequence of rat liver xanthine dehydrogenase and identification of the cleavage sites of the enzyme protein during irreversible conversion by trypsin. J Biol Chem. 1990 Aug 25;265(24):14170-5.

[81] Nishino T, Okamoto K, Kawaguchi Y, Hori H, Matsumura T, Eger BT, et al. Mechanism of the conversion of xanthine dehydrogenase to xanthine oxidase: identification of the two cysteine disulfide bonds and crystal structure of a non-convertible rat liver xanthine dehydrogenase mutant. J Biol Chem. 2005 Jul 1;280(26):24888-94.

[82] Maia L, Duarte RO, Ponces-Freire A, Moura JJ, Mira L. NADH oxidase activity of rat and human liver xanthine oxidoreductase: potential role in superoxide production. J Biol Inorg Chem. 2007 Aug;12(6):777-87.

[83] Miller NJ, RiceEvans CA. Spectrophotometric determination of antioxidant activity. Redox Report. 1996 Jun;2(3):161-71.

[84] Korish AA. Multiple antioxidants and L-arginine modulate inflammation and dyslipidemia in chronic renal failure rats. Ren Fail. 2010 Jan;32(2):203-13.

[85] Ehara H, Yamamoto-Honda R, Kitazato H, Takahashi Y, Kawazu S, Akanuma Y, et al. ApoE isoforms, treatment of diabetes and the risk of coronary heart disease. World J Diabetes. 2012 Mar 15;3(3):54-9.

[86] Abadir PM, Foster DB, Crow M, Cooke CA, Rucker JJ, Jain A, et al. Identification and characterization of a functional mitochondrial angiotensin system. Proc Nat Acad Sci USA. 2011 Sep 6;108(36):14849-54.

[87] Halliwell B. The wanderings of a free radical. Free Radic Biol Med. 2009 Mar 1;46(5):531-42.

[88] Halliwell B, Whiteman M. Measuring reactive species and oxidative damage in vivo and in cell culture: how should you do it and what do the results mean? Br J Pharmacol. 2004 May;142(2):231-55.

[89] Golbidi S, Ebadi SA, Laher I. Antioxidants in the treatment of diabetes. Curr Diabetes Rev. 2011 Mar;7(2):106-25.

[90] Ramos LF, Kane J, McMonagle E, Le P, Wu P, Shintani A, et al. Effects of combination tocopherols and alpha lipoic acid therapy on oxidative stress and inflammatory biomarkers in chronic kidney disease. J Ren Nutr. 2011 May;21(3):211-8.

[91] Yusuf S, Dagenais G, Pogue J, Bosch J, Sleight P. Vitamin E supplementation and cardiovascular events in high-risk patients. The Heart Outcomes Prevention Evaluation Study Investigators. N Engl J Med. 2000 Jan 20;342(3):154-60.

[92] Mann JF, Lonn EM, Yi Q, Gerstein HC, Hoogwerf BJ, Pogue J, et al. Effects of vitamin E on cardiovascular outcomes in people with mild-to-moderate renal insufficiency: results of the HOPE study. Kidney Int. 2004 Apr;65(4):1375-80.

[93] Harcourt BE, Sourris KC, Coughlan MT, Walker KZ, Dougherty SL, Andrikopoulos S, et al. Targeted reduction of advanced glycation improves renal function in obesity. Kidney Int. 2011 Jul;80(2):190-8.

[94] Vlassara H, Torreggiani M, Post JB, Zheng F, Uribarri J, Striker GE. Role of oxidants/inflammation in declining renal function in chronic kidney disease and normal aging. Kidney Int Suppl. 2009 Dec(114):S3-11.

[95] Zafarullah M, Li WQ, Sylvester J, Ahmad M. Molecular mechanisms of N-acetylcysteine actions. Cell Mol Life Sci. 2003 Jan;60(1):6-20.

[96] Zhang F, Lau SS, Monks TJ. The cytoprotective effect of N-acetyl-L-cysteine against ROS-induced cytotoxicity is independent of its ability to enhance glutathione synthesis. Toxicol Sci. 2011 Mar;120(1):87-97.

[97] Ye J, Li J, Yu Y, Wei Q, Deng W, Yu L. L-carnitine attenuates oxidant injury in HK-2 cells via ROS-mitochondria pathway. Regul Pept. 2010 Apr 9;161(1-3):58-66.

[98] Pat B, Yang T, Kong C, Watters D, Johnson DW, Gobe G. Activation of ERK in renal fibrosis after unilateral ureteral obstruction: modulation by antioxidants. Kidney Int. 2005 Mar;67(3):931-43.

[99] Ribeiro G, Roehrs M, Bairros A, Moro A, Charao M, Araujo F, et al. N-acetylcysteine on oxidative damage in diabetic rats. Drug Chem Toxicol. 2011 Aug 16.

[100] Moist L, Sontrop JM, Gallo K, Mainra R, Cutler M, Freeman D, et al. Effect of N-acetylcysteine on serum creatinine and kidney function: results of a randomized controlled trial. Am J Kidney Dis. 2010 Oct;56(4):643-50.

[101] Renke M, Tylicki L, Rutkowski P, Larczynski W, Aleksandrowicz E, Lysiak-Szydlowska W, et al. The effect of N-acetylcysteine on proteinuria and markers of tubular injury in non-diabetic patients with chronic kidney disease. A placebo-controlled, randomized, open, cross-over study. Kidney Blood Press Res. 2008;31(6):404-10.

[102] Hsu SP, Chiang CK, Yang SY, Chien CT. N-acetylcysteine for the management of anemia and oxidative stress in hemodialysis patients. Nephron Clin Pract. 2010;116(3):c207-16.

[103] Nascimento MM, Suliman ME, Silva M, Chinaglia T, Marchioro J, Hayashi SY, et al. Effect of oral N-acetylcysteine treatment on plasma inflammatory and oxidative stress markers in peritoneal dialysis patients: a placebo-controlled study. Perit Dial Int. 2010 May-Jun;30(3):336-42.

[104] Coombes JS, Fassett RG. Antioxidant therapy in hemodialysis patients: a systematic review. Kidney Int. 2012 Feb;81(3):233-46.

[105] Crespo MJ, Cruz N, Altieri PI, Escobales N. Chronic treatment with N-acetylcysteine improves cardiac function but does not prevent progression of cardiomyopathy in Syrian cardiomyopathic hamsters. J Cardiovasc Pharmacol Ther. 2011 Jun;16(2): 197-204.

[106] Tumur Z, Shimizu H, Enomoto A, Miyazaki H, Niwa T. Indoxyl sulfate upregulates expression of ICAM-1 and MCP-1 by oxidative stress-induced NF-kappaB activation. Am J Nephrol. 2010;31(5):435-41.

[107] Serbinova E, Kagan V, Han D, Packer L. Free radical recycling and intramembrane mobility in the antioxidant properties of alpha-tocopherol and alpha-tocotrienol. Free Radic Biol Med. 1991;10(5):263-75.

[108] Fujisawa S, Ishihara M, Atsumi T, Kadoma Y. A quantitative approach to the free radical interaction between alpha-tocopherol or ascorbate and flavonoids. In Vivo. 2006 Jul-Aug;20(4):445-52.

[109] Kagan VE, Serbinova EA, Packer L. Recycling and antioxidant activity of tocopherol homologs of differing hydrocarbon chain lengths in liver microsomes. Arch Biochem Biophys. 1990 Nov 1;282(2):221-5.

[110] Kagan VE, Serbinova EA, Forte T, Scita G, Packer L. Recycling of vitamin E in human low density lipoproteins. J Lipid Res. 1992 Mar;33(3):385-97.

[111] Guo Q, Packer L. Ascorbate-dependent recycling of the vitamin E homologue Trolox by dihydrolipoate and glutathione in murine skin homogenates. Free Radic Biol Med. 2000 Aug;29(3-4):368-74.

[112] Schaaf GJ, Maas RF, de Groene EM, Fink-Gremmels J. Management of oxidative stress by heme oxygenase-1 in cisplatin-induced toxicity in renal tubular cells. Free Radic Res. 2002 Aug;36(8):835-43.

[113] Sen CK, Khanna S, Roy S, Packer L. Molecular basis of vitamin E action. Tocotrienol potently inhibits glutamate-induced pp60(c-Src) kinase activation and death of HT4 neuronal cells. J Biol Chem. 2000 Apr 28;275(17):13049-55.

[114] Machlin LJ, Gabriel E. Kinetics of tissue alpha-tocopherol uptake and depletion following administration of high levels of vitamin E. Ann N Y Acad Sci. 1982;393:48-60.

[115] Karamouzis I, Sarafidis PA, Karamouzis M, Iliadis S, Haidich AB, Sioulis A, et al. Increase in oxidative stress but not in antioxidant capacity with advancing stages of chronic kidney disease. Am J Nephrol. 2008;28(3):397-404.

[116] Klein EA, Thompson IM, Jr., Tangen CM, Crowley JJ, Lucia MS, Goodman PJ, et al. Vitamin E and the risk of prostate cancer: the Selenium and Vitamin E Cancer Prevention Trial (SELECT). JAMA. 2011 Oct 12;306(14):1549-56.

[117] Boaz M, Smetana S, Weinstein T, Matas Z, Gafter U, Iaina A, et al. Secondary prevention with antioxidants of cardiovascular disease in endstage renal disease (SPACE): randomised placebo-controlled trial. Lancet. 2000 Oct 7;356(9237):1213-8.

[118] Giray B, Kan E, Bali M, Hincal F, Basaran N. The effect of vitamin E supplementation on antioxidant enzyme activities and lipid peroxidation levels in hemodialysis patients. Clin Chim Acta. 2003 Dec;338(1-2):91-8.

[119] Islam KN, O'Byrne D, Devaraj S, Palmer B, Grundy SM, Jialal I. Alpha-tocopherol supplementation decreases the oxidative susceptibility of LDL in renal failure patients on dialysis therapy. Atherosclerosis. 2000 May;150(1):217-24.

[120] Huang HY, Appel LJ. Supplementation of diets with alpha-tocopherol reduces serum concentrations of gamma- and delta-tocopherol in humans. J Nutr. 2003 Oct;133(10): 3137-40.

[121] Baggott JE. High-dosage vitamin E supplementation and all-cause mortality. Ann Intern Med. 2005 Jul 19;143(2):155-6; author reply 6-8.

[122] Blatt DH, Pryor WA. High-dosage vitamin E supplementation and all-cause mortality. Ann Intern Med. 2005 Jul 19;143(2):150-1; author reply 6-8.

[123] Carter T. High-dosage vitamin E supplementation and all-cause mortality. Ann Intern Med. 2005 Jul 19;143(2):155; author reply 6-8.

[124] DeZee KJ, Shimeall W, Douglas K, Jackson JL. High-dosage vitamin E supplementation and all-cause mortality. Ann Intern Med. 2005 Jul 19;143(2):153-4; author reply 6-8.

[125] Hemila H. High-dosage vitamin E supplementation and all-cause mortality. Ann Intern Med. 2005 Jul 19;143(2):151-2; author reply 6-8.

[126] Krishnan K, Campbell S, Stone WL. High-dosage vitamin E supplementation and all-cause mortality. Ann Intern Med. 2005 Jul 19;143(2):151; author reply 6-8.

[127] Lim WS, Liscic R, Xiong C, Morris JC. High-dosage vitamin E supplementation and all-cause mortality. Ann Intern Med. 2005 Jul 19;143(2):152; author reply 6-8.

[128] Marras C, Lang AE, Oakes D, McDermott MP, Kieburtz K, Shoulson I, et al. High-dosage vitamin E supplementation and all-cause mortality. Ann Intern Med. 2005 Jul 19;143(2):152-3; author reply 6-8.

[129] Meydani SN, Lau J, Dallal GE, Meydani M. High-dosage vitamin E supplementation and all-cause mortality. Ann Intern Med. 2005 Jul 19;143(2):153; author reply 6-8.

[130] Possolo AM. High-dosage vitamin E supplementation and all-cause mortality. Ann Intern Med. 2005 Jul 19;143(2):154; author reply 6-8.

[131] Lass A, Forster MJ, Sohal RS. Effects of coenzyme Q10 and alpha-tocopherol administration on their tissue levels in the mouse: elevation of mitochondrial alpha-tocopherol by coenzyme Q10. Free Radic Biol Med. 1999 Jun;26(11-12):1375-82.

[132] Lass A, Sohal RS. Effect of coenzyme Q(10) and alpha-tocopherol content of mitochondria on the production of superoxide anion radicals. FASEB J. 2000 Jan;14(1):87-94.

[133] Merker MP, Audi SH, Lindemer BJ, Krenz GS, Bongard RD. Role of mitochondrial electron transport complex I in coenzyme Q1 reduction by intact pulmonary arterial endothelial cells and the effect of hyperoxia. Am J Physiol Lung Cell Mol Physiol. 2007 Sep;293(3):L809-19.

[134] Ohnishi T, Ohnishi ST, Shinzawa-Ito K, Yoshikawa S. Functional role of coenzyme Q in the energy coupling of NADH-CoQ oxidoreductase (Complex I): stabilization of the semiquinone state with the application of inside-positive membrane potential to proteoliposomes. Biofactors. 2008;32(1-4):13-22.

[135] James AM, Smith RA, Murphy MP. Antioxidant and prooxidant properties of mitochondrial Coenzyme Q. Arch Biochem Biophys. 2004 Mar 1;423(1):47-56.

[136] Linnane AW, Kios M, Vitetta L. Coenzyme Q(10) - its role as a prooxidant in the formation of superoxide anion/hydrogen peroxide and the regulation of the metabolome. Mitochondrion. 2007 Jun;7 Suppl:S51-61.

[137] Nohl H, Gille L, Kozlov AV. Critical aspects of the antioxidant function of coenzyme Q in biomembranes. Biofactors. 1999;9(2-4):155-61.

[138] Frei B, Kim MC, Ames BN. Ubiquinol-10 is an effective lipid-soluble antioxidant at physiological concentrations. Proc Natl Acad Sci U S A. 1990 Jun;87(12):4879-83.

[139] Stoyanovsky DA, Osipov AN, Quinn PJ, Kagan VE. Ubiquinone-dependent recycling of vitamin E radicals by superoxide. Arch Biochem Biophys. 1995 Nov 10;323(2):343-51.

[140] Lass A, Sohal RS. Electron transport-linked ubiquinone-dependent recycling of alpha-tocopherol inhibits autooxidation of mitochondrial membranes. Arch Biochem Biophys. 1998 Apr 15;352(2):229-36.

[141] Linnane AW, Kios M, Vitetta L. Healthy aging: regulation of the metabolome by cellular redox modulation and prooxidant signaling systems: the essential roles of superoxide anion and hydrogen peroxide. Biogerontology. 2007 Oct;8(5):445-67.

[142] Kwong LK, Kamzalov S, Rebrin I, Bayne AC, Jana CK, Morris P, et al. Effects of coenzyme Q(10) administration on its tissue concentrations, mitochondrial oxidant generation, and oxidative stress in the rat. Free Radic Biol Med. 2002 Sep 1;33(5):627-38.

[143] Ishikawa A, Kawarazaki H, Ando K, Fujita M, Fujita T, Homma Y. Renal preservation effect of ubiquinol, the reduced form of coenzyme Q10. Clin Exp Nephrol. 2011 Feb;15(1):30-3.

[144] Huynh K, Kiriazis H, Du XJ, Love JE, Jandeleit-Dahm KA, Forbes JM, et al. Coenzyme Q10 attenuates diastolic dysfunction, cardiomyocyte hypertrophy and cardiac fibrosis in the db/db mouse model of type 2 diabetes. Diabetologia. 2012 May;55(5): 1544-53.

[145] Dendooven A, Ishola DA, Jr., Nguyen TQ, Van der Giezen DM, Kok RJ, Goldschmeding R, et al. Oxidative stress in obstructive nephropathy. Int J Exp Pathol. 2011 Jun;92(3):202-10.

[146] Irita J, Okura T, Jotoku M, Nagao T, Enomoto D, Kurata M, et al. Osteopontin deficiency protects against aldosterone-induced inflammation, oxidative stress, and interstitial fibrosis in the kidney. Am J Physiol Renal Physiol. 2011 Jul 6.

[147] Shing CM, Adams MJ, Fassett RG, Coombes JS. Nutritional compounds influence tissue factor expression and inflammation of chronic kidney disease patients in vitro. Nutrition. 2011 Sep;27(9):967-72.

[148] Arab K, Rossary A, Flourie F, Tourneur Y, Steghens JP. Docosahexaenoic acid enhances the antioxidant response of human fibroblasts by upregulating gamma-glutamylcysteinyl ligase and glutathione reductase. Br J Nutr. 2006 Jan;95(1):18-26.

[149] Kim YJ, Chung HY. Antioxidative and anti-inflammatory actions of docosahexaenoic acid and eicosapentaenoic acid in renal epithelial cells and macrophages. J Med Food. 2007 Jun;10(2):225-31.

[150] Mayer K, Meyer S, Reinholz-Muhly M, Maus U, Merfels M, Lohmeyer J, et al. Short-time infusion of fish oil-based lipid emulsions, approved for parenteral nutrition, reduces monocyte proinflammatory cytokine generation and adhesive interaction with endothelium in humans. J Immunol. 2003 Nov 1;171(9):4837-43.

[151] Sperling RI, Benincaso AI, Knoell CT, Larkin JK, Austen KF, Robinson DR. Dietary omega-3 polyunsaturated fatty acids inhibit phosphoinositide formation and chemotaxis in neutrophils. J Clin Invest. 1993 Feb;91(2):651-60.

[152] Diaz Encarnacion MM, Warner GM, Cheng J, Gray CE, Nath KA, Grande JP. n-3 Fatty acids block TNF-alpha-stimulated MCP-1 expression in rat mesangial cells. Am J Physiol Renal Physiol. 2011 May;300(5):F1142-51.

[153] An WS, Kim HJ, Cho KH, Vaziri ND. Omega-3 fatty acid supplementation attenuates oxidative stress, inflammation, and tubulointerstitial fibrosis in the remnant kidney. Am J Physiol Renal Physiol. 2009 Oct;297(4):F895-903.

[154] Peake JM, Gobe GC, Fassett RG, Coombes JS. The effects of dietary fish oil on inflammation, fibrosis and oxidative stress associated with obstructive renal injury in rats. Mol Nutr Food Res. 2011 Mar;55(3):400-10.

[155] Kazemian P, Kazemi-Bajestani SM, Alherbish A, Steed J, Oudit GY. The use of omega-3 poly-unsaturated fatty acids in heart failure: a preferential role in patients with diabetes. Cardiovasc Drugs Ther. 2012 May 30.

[156] Bouzidi N, Mekki K, Boukaddoum A, Dida N, Kaddous A, Bouchenak M. Effects of omega-3 polyunsaturated fatty-acid supplementation on redox status in chronic renal failure patients with dyslipidemia. J Ren Nutr. 2010 Sep;20(5):321-8.

[157] Harris WS. n-3 fatty acids and serum lipoproteins: human studies. Am J Clin Nutr. 1997 May;65(5 Suppl):1645S-54S.

[158] Cohen MG, Rossi JS, Garbarino J, Bowling R, Motsinger-Reif AA, Schuler C, et al. Insights into the inhibition of platelet activation by omega-3 polyunsaturated fatty acids: Beyond aspirin and clopidogrel. Thromb Res. 2011 May 26.

[159] Woodman RJ, Mori TA, Burke V, Puddey IB, Watts GF, Beilin LJ. Effects of purified eicosapentaenoic and docosahexaenoic acids on glycemic control, blood pressure, and serum lipids in type 2 diabetic patients with treated hypertension. Am J Clin Nutr. 2002 Nov;76(5):1007-15.

[160] Matsumoto T, Nakayama N, Ishida K, Kobayashi T, Kamata K. Eicosapentaenoic acid improves imbalance between vasodilator and vasoconstrictor actions of endothelium-derived factors in mesenteric arteries from rats at chronic stage of type 2 diabetes. J Pharmacol Exp Ther. 2009 Apr;329(1):324-34.

[161] El-Sheikh AA, van den Heuvel JJ, Koenderink JB, Russel FG. Effect of hypouricaemic and hyperuricaemic drugs on the renal urate efflux transporter, multidrug resistance protein 4. Br J Pharmacol. 2008 Dec;155(7):1066-75.

[162] Riegersperger M, Covic A, Goldsmith D. Allopurinol, uric acid, and oxidative stress in cardiorenal disease. Int Urol Nephrol. 2011 Jun;43(2):441-9.

[163] Sanders SA, Eisenthal R, Harrison R. NADH oxidase activity of human xanthine oxidoreductase--generation of superoxide anion. Eur J Biochem. 1997 May 1;245(3): 541-8.

[164] Kosugi T, Nakayama T, Heinig M, Zhang L, Yuzawa Y, Sanchez-Lozada LG, et al. Effect of lowering uric acid on renal disease in the type 2 diabetic db/db mice. Am J Physiol Renal Physiol. 2009 Aug;297(2):F481-8.

[165] Das DK, Engelman RM, Clement R, Otani H, Prasad MR, Rao PS. Role of xanthine oxidase inhibitor as free radical scavenger: a novel mechanism of action of allopurinol and oxypurinol in myocardial salvage. Biochem Biophys Res Commun. 1987 Oct 14;148(1):314-9.

[166] Moorhouse PC, Grootveld M, Halliwell B, Quinlan JG, Gutteridge JM. Allopurinol and oxypurinol are hydroxyl radical scavengers. FEBS Lett. 1987 Mar 9;213(1):23-8.

[167] Klein AS, Joh JW, Rangan U, Wang D, Bulkley GB. Allopurinol: discrimination of antioxidant from enzyme inhibitory activities. Free Radic Biol Med. 1996;21(5):713-7.

[168] Goicoechea M, de Vinuesa SG, Verdalles U, Ruiz-Caro C, Ampuero J, Rincon A, et al. Effect of allopurinol in chronic kidney disease progression and cardiovascular risk. Clin J Am Soc Nephrol. 2010 Aug;5(8):1388-93.

[169] Kanbay M, Ozkara A, Selcoki Y, Isik B, Turgut F, Bavbek N, et al. Effect of treatment of hyperuricemia with allopurinol on blood pressure, creatinine clearence, and proteinuria in patients with normal renal functions. Int Urol Nephrol. 2007;39(4):1227-33.

[170] Shelmadine B, Bowden RG, Wilson RL, Beavers D, Hartman J. The effects of lowering uric acid levels using allopurinol on markers of metabolic syndrome in end-stage renal disease patients: a pilot study. Anadolu Kardiyol Derg. 2009 Oct;9(5):385-9.

[171] Pergola PE, Krauth M, Huff JW, Ferguson DA, Ruiz S, Meyer CJ, et al. Effect of bardoxolone methyl on kidney function in patients with T2D and Stage 3b-4 CKD. Am J Nephrol. 2011;33(5):469-76.

[172] Pergola PE, Raskin P, Toto RD, Meyer CJ, Huff JW, Grossman EB, et al. Bardoxolone methyl and kidney function in CKD with type 2 diabetes. N Engl J Med. 2011 Jul 28;365(4):327-36.

[173] Honda T, Yoshizawa H, Sundararajan C, David E, Lajoie MJ, Favaloro FG, Jr., et al. Tricyclic compounds containing nonenolizable cyano enones. A novel class of highly potent anti-inflammatory and cytoprotective agents. J Med Chem. 2011 Mar 24;54(6): 1762-78.

[174] Eskiocak U, Kim SB, Roig AI, Kitten E, Batten K, Cornelius C, et al. CDDO-Me protects against space radiation-induced transformation of human colon epithelial cells. Radiat Res. 2010 Jul;174(1):27-36.

[175] Nagaraj S, Youn JI, Weber H, Iclozan C, Lu L, Cotter MJ, et al. Anti-inflammatory triterpenoid blocks immune suppressive function of MDSCs and improves immune response in cancer. Clin Cancer Res. 2010 Mar 15;16(6):1812-23.

[176] Wu QQ, Wang Y, Senitko M, Meyer C, Wigley WC, Ferguson DA, et al. Bardoxolone methyl (BARD) ameliorates ischemic AKI and increases expression of protective genes Nrf2, PPARgamma, and HO-1. Am J Physiol Renal Physiol. 2011 May; 300(5):F1180-92.

[177] Kelly GS. L-Carnitine: therapeutic applications of a conditionally-essential amino acid. Altern Med Rev. 1998 Oct;3(5):345-60.

[178] Fouque D, Holt S, Guebre-Egziabher F, Nakamura K, Vianey-Saban C, Hadj-Aissa A, et al. Relationship between serum carnitine, acylcarnitines, and renal function in patients with chronic renal disease. J Ren Nutr. 2006 Apr;16(2):125-31.

[179] Eknoyan G, Latos DL, Lindberg J. Practice recommendations for the use of L-carnitine in dialysis-related carnitine disorder. National Kidney Foundation Carnitine Consensus Conference. Am J Kidney Dis. 2003 Apr;41(4):868-76.

[180] Sakurabayashi T, Miyazaki S, Yuasa Y, Sakai S, Suzuki M, Takahashi S, et al. L-carnitine supplementation decreases the left ventricular mass in patients undergoing hemodialysis. Circ J. 2008 Jun;72(6):926-31.

[181] Matsumoto Y, Sato M, Ohashi H, Araki H, Tadokoro M, Osumi Y, et al. Effects of L-carnitine supplementation on cardiac morbidity in hemodialyzed patients. Am J Nephrol. 2000 May-Jun;20(3):201-7.

[182] Signorelli SS, Fatuzzo P, Rapisarda F, Neri S, Ferrante M, Oliveri CG, et al. Propionyl-L-carnitine therapy: effects on endothelin-1 and homocysteine levels in patients with peripheral arterial disease and end-stage renal disease. Kidney Blood Pressure Res. 2006;29(2):100-7.

[183] Zhou Q, Wu S, Jiang J, Tian J, Chen J, Yu X, et al. Accumulation of circulating advanced oxidation protein products is an independent risk factor for ischemic heart disease in maintenance hemodialysis patients. Nephrology. 2012 Jun 28.

[184] Sener G, Paskaloglu K, Satiroglu H, Alican I, Kacmaz A, Sakarcan A. L-carnitine ameliorates oxidative damage due to chronic renal failure in rats. J Cardiovasc Pharmacol. 2004 May;43(5):698-705.

[185] Pecoits-Filho R, Lindholm B, Stenvinkel P. The malnutrition, inflammation, and atherosclerosis (MIA) syndrome -- the heart of the matter. Nephrol Dial Transplant. 2002;17 Suppl 11:28-31.

[186] Brunini TM, da SCD, Siqueira MA, Moss MB, Santos SF, Mendes-Ribeiro AC. Uremia, atherothrombosis and malnutrition: the role of L-arginine-nitric oxide pathway. Cardiovasc Hematol Disord Drug Targets. 2006 Jun;6(2):133-40.

[187] Reyes AA, Karl IE, Klahr S. Role of arginine in health and in renal disease. Am J Physiol. 1994 Sep;267(3 Pt 2):F331-46.

[188] Morris SM, Jr. Enzymes of arginine metabolism. J Nutr. 2004 Oct;134(10 Suppl): 2743S-7S; discussion 65S-67S.

[189] Stuehr DJ. Enzymes of the L-arginine to nitric oxide pathway. J Nutr. 2004 Oct;134(10 Suppl):2748S-51S; discussion 65S-67S.

[190] Morel F, Hus-Citharel A, Levillain O. Biochemical heterogeneity of arginine metabolism along kidney proximal tubules. Kidney Int. 1996 Jun;49(6):1608-10.

[191] Mendes RAC, Brunini TM, Ellory JC, Mann GE. Abnormalities in L-arginine transport and nitric oxide biosynthesis in chronic renal and heart failure. Cardiovasc Res. 2001 Mar;49(4):697-712.

[192] Brunini TM, Roberts NB, Yaqoob MM, Ellory JC, Mann GE, Mendes RAC. Activation of L-arginine transport in undialysed chronic renal failure and continuous ambulatory peritoneal dialysis patients. Clin Exp Pharmacol Physiol. 2006 Jan-Feb;33(1-2): 114-8.

[193] Mendes RAC, Hanssen H, Kiessling K, Roberts NB, Mann GE, Ellory JC. Transport of L-arginine and the nitric oxide inhibitor NG-monomethyl-L-arginine in human erythrocytes in chronic renal failure. Clin Sci (Lond). 1997 Jul;93(1):57-64.

[194] Noris M, Benigni A, Boccardo P, Aiello S, Gaspari F, Todeschini M, et al. Enhanced nitric oxide synthesis in uremia: implications for platelet dysfunction and dialysis hypotension. Kidney Int. 1993 Aug;44(2):445-50.

[195] Aiello S, Noris M, Remuzzi G. Nitric oxide synthesis and L-arginine in uremia. Miner Electrolyte Metab. 1997;23(3-6):151-6.

[196] Rajapakse NW, Kuruppu S, Hanchapola I, Venardos K, Mattson DL, Smith AI, et al. Evidence that renal arginine transport is impaired in spontaneously hypertensive rats. Am J Physiol Renal Physiol. 2012 Jun;302(12):F1554-62.

[197] Huang CC, Tsai SC, Lin WT. Potential ergogenic effects of L-arginine against oxidative and inflammatory stress induced by acute exercise in aging rats. Exp Gerontol. 2008 Jun;43(6):571-7.

[198] Mosca L, Marcellini S, Perluigi M, Mastroiacovo P, Moretti S, Famularo G, et al. Modulation of apoptosis and improved redox metabolism with the use of a new antioxidant formula. Biochem Pharmacol. 2002 Apr 1;63(7):1305-14.

[199] Park NY, Park SK, Lim Y. Long-term dietary antioxidant cocktail supplementation effectively reduces renal inflammation in diabetic mice. Br J Nutr. 2011 Nov;106(10): 1514-21.

[200] Dietary supplementation with n-3 polyunsaturated fatty acids and vitamin E after myocardial infarction: results of the GISSI-Prevenzione trial. Gruppo Italiano per lo Studio della Sopravvivenza nell'Infarto miocardico. Lancet. 1999 Aug 7;354(9177): 447-55.

[201] Shargorodsky M, Debby O, Matas Z, Zimlichman R. Effect of long-term treatment with antioxidants (vitamin C, vitamin E, coenzyme Q10 and selenium) on arterial compliance, humoral factors and inflammatory markers in patients with multiple cardiovascular risk factors. Nutr Metab (Lond). 2010;7:55.

Chapter 11

Role of Oxidative Stress in Calcific Aortic Valve Disease: From Bench to Bedside - The Role of a Stem Cell Niche

Nalini Rajamannan

Additional information is available at the end of the chapter

http://dx.doi.org/10.5772/52275

1. Introduction

Calcific Aortic Stenosis is the most common cause of aortic valve disease in developed countries. This condition increases in prevalence with the advancing age of the U.S. population, afflicting 2-3 % by age 65 [1]. Aortic valve replacement is the number one indication for surgical valve replacement in the United States and in Europe. The natural history of severe symptomatic aortic stenosis is associated with 50% mortality within 5 years [2]. Bicuspid aortic valve disease is the most common congenital heart abnormality and it is the most common phenotype of calcific aortic stenosis. The bicuspid aortic valve (BAV) is the most common congenital cardiac anomaly, having a prevalence of 0.9 to 1.37% in the general population [3]. Understanding the cellular mechanisms of tricuspid versus bicuspid aortic valve lesions will provide further understanding the mechanisms of this disease. Currently, there are three fundamental mechanisms defined in the development of aortic valve disease: 1) oxidative stress via traditional cardiovascular risk factors [4-8,6, 7, 9-12], 2) cellular proliferation [13] and 3) osteoblastogenesis in the end stage disease process [14, 15]. Previously, the Wnt/Lrp5 signaling pathway has been identified as a signaling mechanism for cardiovascular calcification [5, 16, 17]. The corollaries necessary to define a tissue stem cell niche: 1) physical architecture of the endothelial cells signaling to the adjacent subendothelial cells: the valve interstitial cell along the valve fibrosa. 2) defining the oxidative-mechanical stress gradient necessary to activate Wnt3a/Lrp5 in this tissue stem niche to induce disease. Recently, the mechanisms of oxidative stress have been identified in the development of calcific aortic valve disease. This chapter will outline the factors important in the role of calcific aortic valve disease.

2. The role of lipids in vascular and valvular disease

The role of lipids in vascular atherosclerosis has been defined in the literature for years. Atherosclerosis is a complex multifactorial process which produces a lesion composed of lipids [18, 19], macrophages [20], and proliferating smooth muscle cells [21] apoptosis [22] and extracellular bone matrix production [23] in the vascular wall [24, 25]. The activation of these cellular processes is regulated by a number of pathways. Integrins provide an important role in the regulation of cellular adhesion in atherosclerosis [26]. Another critical regulator of vascular endothelial biology is nitric oxide (NO) [27, 28]. Cholesterol-rich LDL also has a critical role in the onset and further progression of the atherosclerotic lesion via an inactivation of endothelial nitric oxide synthase (eNOS) [22, 29-31] contributing to an abnormal oxidation state within the vessel. In this inflammatory environment, growth factors and cytokines are secreted to induce vascular smooth cell proliferation and recruitment of macrophage cells [32-37] which are important in the development of the atherosclerotic plaque lesion.

Recently, similar risk factors for calcific aortic valve disease have recently been described including male gender, hypertension, elevated levels of LDL, and smoking [38, 39] which mimic those that promote the development of vascular atherosclerosis. Surgical pathological studies have demonstrated the presence of LDL and atherosclerosis in calcified valves, demonstrating similarities between the genesis of valvular and vascular disease and suggesting a common cellular mechanism [40, 41]. Patients who have the diagnosis of familial hypercholesterolemia develop aggressive peripheral vascular disease, coronary artery disease, as well as aortic valve lesions which calcify with age [10, 42]. Rajamannan et al, have shown that the development of atherosclerosis occurs in the aortic valve in a patient with Familial Hypercholesterolemia with the Low density lipoprotein receptor mutation [10]. The atherosclerosis develops along the aortic surface of the aortic valve and in the lumen of the left circumflex artery [10]. This provides the first index case of atherosclerotic aortic valve disease in this patient population. Studies have confirmed in experimental hypercholesterolemia that both atherosclerosis and osteoblast markers are present in the aortic valves [4, 6, 13]. This background provides the foundation for studying valve calcification in an experimental atherosclerotic *in vivo* model.

3. Aortic valve calcification

The presence of calcification in the aortic valve is responsible for valve stenosis. Severe aortic stenosis can result in symptomatic chest pain, as well as syncope and congestive heart failure in patients with severe aortic valve stenosis. For years, aortic valve stenosis was thought to be a degenerative process. However, the pathologic lesion of calcified aortic valves demonstrate indicate the presence of complex calcification in these tissues. Furthermore, there are a growing number of descriptive studies delineating the presence of bone formation in the aortic valve [15, 43, 44].

Until recently the etiology of valvular heart disease has been thought to be a degenerative process related to the passive accumulation of calcium binding to the surface of the valve leaflet. Recent descriptive studies have demonstrated the critical features of aortic valve calcification, including osteoblast expression, cell proliferation and atherosclerosis [6, 14, 15, 45] and mitral valve degeneration, glycosaminglycan accumulation, proteoglycan expression, and abnormal collagen expression [46-49]. These studies define the biochemical and histological characterization of these valve lesions. We and others, have also shown that specific bone cell phenotypes are present in calcifying valve specimens in human specimens [16, 50]. These data provide the evidence that the aortic valve calcification follows the spectrum of bone formation in calcifying tissues.

4. The role of Lrp5/beta-catenin activation in cardiovascular calcification and osteoblast bone formation: Connection with the bone axis

Bone and cartilage are major tissues in the vertebrate skeletal system, which is primarily composed of three cell types: osteoblasts, chrondrocytes, and osteoclasts. In the developing embryo, osteoblast and chrondrocytes, both differentiate from common mesenchymal progenitors in situ, where as osteoclasts are of hematopoietic origin and brought in later by invading blood vessels. Osteoblast differentiation and maturation lead to bone formation controlled by two distinct mechanisms: intramembranous and endochondral ossification, both starting from mesenchymal condensations.

To date only two osteoblast-specific transcripts have been identified: 1) Cbfa1 and 2) osteocalcin (OC). The transcription factor Cbfa1 [51] has all the attributes of a 'master gene' differentiation factor for the osteoblast lineage and bone matrix gene expression. During embryonic development, Cbfa1 expression precedes osteoblast differentiation and is restricted to mesenchymal cells destined to become osteoblast. In addition to its critical role in osteoblast commitment and differentiation, Cbfa1 appears to control osteoblast activity, i.e., the rate of bone formation by differentiated osteoblasts [51]. We have shown previously that cholesterol upregulates Cbfa1 gene expression in the aortic valve and atorvastatin decreases the gene expression [6] in an animal model. We have also demonstrated that Sox9 and Cbfa1 are expressed in human degenerative valves removed at the time of surgical valve replacement [16]. The regulatory mechanism of osteoblast differentiation from osteoblast progenitor cells into terminally differentiated cells is via a well orchestrated and well studied pathway which involves initial cellular proliferation events and then synthesis of bone matrix proteins, which requires the actions of specific paracrine/hormonal factors and the activation of the canonical Wntpathway [52].

Genes which code for the bone extracellular matrix proteins in osteoblast cells include alkaline phosphatase (AP), osteopontin (OP), osteocalcin (OC), and bone sialoprotein (BSP). This data supports a potential regulatory mechanism that these matrix proteins play a critical role in the development of biomineralization. To date, many of these markers have been shown to be critical in the extracellular mineralization and bone formation that develops in normal osteoblast differentiation (Fig.5). Dr. Spelsberg and Dr. Rajamannan have extensive experience in osteoblast cell biology and will contribute to the translational studies in the aortic valve involving the differentiation and mineralization [53, 54].

A link between lipids and osteoporosis have been studied extensively [55-60]. These groups have shown in *in vitro* and *in vivo* studies that lipids decrease bone formation and increase vascular calcification. Hurska's group from the University of Washington have studied this important hypothesis in the LDLR$^{-/-}$ mice with renal disease [55]. This studied correlated the important understanding of chronic kidney disease with decreased bone formation rates and increase in vascular calcification. This study demonstrates that accelerated vascular calcification found in patients with end stage renal disease may be related to multifactorial mechanisms including traditional atherosclerotic risk factors and elevated serum phosphate levels. Giachelli has also studied extensively the hypothesis of a sodium phosphate abnormality in the vascular smooth muscle cell [61]. Her group has also shown that osteopontin expression by vascular smooth muscle cells may have an inhibitory effect in the development of calcification [62] which further defines the complexity of the matrix synthesis phase of bone formation. Demer's laboratory has also studied extensively the correlation of lipids with vascular calcification and osteoporosis via inhibition of Cbfa1 in osteoblast cells [60, 63]. This paradoxical finding between the calcifying vascular aorta and osteoporosis is an important link in the hypercholesterolemia hypothesis. The development of cardiovascular calcification is a multifactorial process which includes a number of mechanisms. Studies in the different laboratories provide important evidence towards the development of therapies depending on the patient population i.e. end stage renal disease versus treatment of the traditional risk factors for vascular disease.

Our lab (43) and Towler's laboratory (44) have shown that the Lrp5/Wnt/beta-catenin pathway plays an important role in the development of vascular and valvular calcification. Studies have shown that different mutations in Lrp5, an LDL receptor related protein; develop a high bone mass phenotype and an osteoporotic phenotype (45, 46). In the presence of the palmitoylation of Wnt an active beta-catenin accumulates in the cytoplasm, presumably in a signaling capacity, and eventually translocates to the nucleus via binding to nucleoporins [64], where it can interacts with LEF-1/TCFs in an inactive transcription complex [65, 66], The Wnt/Lrp5/frizzled complex turns on downstream components such as Dishevelled (Dvl/Dsh) which leads to repression of the glycogen synthase kinase-3 (GSK3) [67]. Inhibition of GSK3 allows beta-catenin to accumulate in the nucleus, interacting with members of the LEF/TCF class of architectural HMG box of transcription factors including Cbfa1 involved in cell differentiation and osteoblast activation [68, 69, 70-72] and Sox 9, a HMG box transcription factor, is required for chondrocyte cell fate determination and marks early chondrocytic differentiation of mesenchymalprogenitors [73].

To determine a potential signaling pathway for the development of aortic valve disease there are numerous pathways which may be implicated in this disease process [50, 74, 75]. Recent evidence suggests that the Wnt pathway regulates the expression of bone mineral markers in cells responsive to the Wnt pathway. Furthermore the Wnt pathway has been shown to be activated by lipids. Therefore we chose to assess this pathway in our model of experimental hypercholesterolemia to determine how lipids may be regulating Lrp5 in the aortic valve. This background outlines the potential for lipids in the regulation of aortic valve mineralization via the canonical Wnt pathway.

5. Echocardiography and Computerized Tomography (CT) evaluation of the development of calcification and stenosis

Currently the non-invasive "gold standard" for the diagnosis of aortic valve stenosis is 2-Dimensional doppler echocardiography. It is the test of choice to quantify the severity of valve stenosis and pressure differential across the aortic valve. There are a increasing number of studies which have demonstrated the utility of calculating the volume of calcium and the rate of progression of the disease process in the aortic valve [76-80]. Confirmation of hemodynamic valve stenosis by echo will provide the degree of valve stenosis using ultrasound techniques[4]. MicroCT will assess the degree of calcification within the mineralizing tissues.

6. Development of future medical therapies for calcific aortic stenosis

The natural history studies of valvular aortic stenosis as defined by clinical and histopathologic parameters have provided landmark developments towards the understanding of this disease. HMG CoA reductase inhibitors may provide an innovative therapeutic approach by employing both lipid lowering and possibly non-lipid lowering effects to forestall critical stenosis in the aortic valve. Our laboratory has shown that atorvastatin has a number of effects in the aortic valve including: 1) inhibition of foam cell accumulation [6], 2) inhibition of Cbfa1 activation [6], 3) eNOS enzymatic activation [11] and 4) attenuation of Lrp5 receptor activation [81]. Statins have potent LDL lowering effects via inhibition of the rate-limiting step in cholesterol synthesis. There are a number of experimental models which demonstrate the potential for treating the vasculature with statins to inhibit matrix formation [24, 25], cellular proliferation [6] and vascular aneurysm formation [82]. Although valve replacement is the current treatment of choice for severe critical aortic stenosis, future insights into the mechanisms of calcification and its progression may indicate a role for lipid lowering therapy in modifying the rate of progression of stenosis.

There are a growing number of retrospective studies demonstrating that statins may have benefits in slowing the progression of aortic stenosis [83-85]. A recent clinical trial by Cowell et al, demonstrated that high dose atorvastatin did not slow the progression of aortic stenosis in patients [86]. However, the timing of the initiation of the statin therapy was at a later stage of aortic valve disease. A clinical trial in Portugal called RAAVE- Rosuvastatin Affecting Aortic Valve Endothelium demonstrated prospectively that statins slow progression in CAVD in an open label study. In the RAAVE study we found a change in aortic valve area (AVA) in the control group was -0.10 ± 0.09 cm^2 per year versus -0.05 ± 0.12 cm^2 per yearin the

Rosuvastatin group (p=0.041). In addition there was an increase in peak aortic valve velocity was +0.24±0.30 m/sec/yr in the control group as compared to the increase in +0.04±0.38 m/sec/yr in the Rosuvastatin group (p=0.007), indicating that in this prospective hypothesis driven study we found by echocardiography a slowing of progression in the aortic valve disease. SALTIRE initiated atorvastatin in patients who had more advanced aortic stenosis as defined by the mean aortic valve area 1.03 cm^2 as compared to the average aortic valve area in RAAVE of 1.23 cm^2 as the baseline aortic valve area prior to treatment with Rosuvastatin [87]. The investigators of RAAVE hypothesize that the beneficial effect of the statin was secondary to the early initiation of treatment. Furthermore, the SALTIRE investigators recently acknowledged the potential of medical therapies may be found if the treatment of this disease is initiated earlier in the disease process [86]. The studies planned in this application should lead to an important understanding of the molecular and cellular mechanisms of aortic valve disease. Furthermore, the experimental approach will also correlate the development of valve calcium by MicroCT and hemodynamic progression by echocardiography in this important disease process.

The bicuspid aortic valve (BAV) is the most common congenital cardiac anomaly, having a prevalence of 0.9 to 1.37 % in the general population [3]. The natural history of the BAV is progressive stenosis that typically occurs at a faster rate than tricuspid aortic valves requiring earlier surgical intervention in the BAV patients [2, 3]. With the decline of acute rheumatic fever, calcific aortic stenosis has become the most common indication for surgical valve replacement. Despite the high prevalence of aortic stenosis, few studies have investigated the mechanisms responsible for aortic valve disease. The cellular mechanism for the development of this disease is not well known. Previously, we and others have demonstrated that aortic valve calcification is associated with an osteoblast bone-like phenotype [14, 15]. This bone phenotype is regulated by the canonical Wnt pathway in experimental cardiovascular calcification [5, 17]. We have alsoshown that the canonical Wnt/Lrp5 pathway is upregulated in diseased human valves from patients with valvular heart disease [16]. These studies implicate that inhibition of the canonical Wnt pathway provides a therapeutic approach for the treatment of degenerative valvular heart diseases. A recent study [88], discovered that a loss of function mutation in Notch1 was associated with accelerated aortic valve calcification and a number of congenital heart abnormalities. Normal Notch1 receptor functions to inhibit osteoblastogenesis [89, 90]. Evaluation of Notch1 gene and protein expression in human bicuspid calcified valves compared to normal aortic valves removed at the time of surgical valve replacement is shown in Figure 1, Panel A. Notch1 protein expression was decreased in the BAV compared to controls by immunhistochemistry and Western Blot expression Figure 1, Panel B1 and B2, and C. RNA expression by RTPCR indicates a spliced Notch1 receptor in the diseased valves as compared to controls as shown in Figure 1, Panel D. This Notch1 splicing may be the regulatory switch important for the activation of the Wnt pathway and downstream calcification in these diseased valves [5, 17, 90].

Risk factors for the development of calcific aortic valve disease(CAVD) have been elucidated in a number of epidemiologic databases [38]. The risk factors for CAVD are similar to those of vascular atherosclerosis which include: elevated LDL, hypertension, male gender, smok-

http://dx.doi.org/10.5772/52275

ing and increased body mass index [38]. The elucidation of these risk factors have provided the experimental basis for hypercholesterolemia as a method to induce aortic valve disease [4-8]. Furthermore, studies have shown that the eNOS$^{-/-}$ mouse is a novel mouse model which develops anatomic bicuspid aortic valves (BAV) [91]. To understand if eNOS$^{-/-}$ mice with the BAV phenotype, develops accelerated stenosis earlier than tricuspid aortic valves via the Lrp5 pathway activation, eNOS$^{-/-}$ mice were given a cholesterol diet versus cholesterol and atorvastatin. The Visual Sonics mouse echocardiography machine was used to screen for the BAV phenotype. Echocardiography hemodynamics was also performed to determine the timing of stenosis in bicuspid vs. tricuspid aortic valves eNOS-/- mice on different diets.

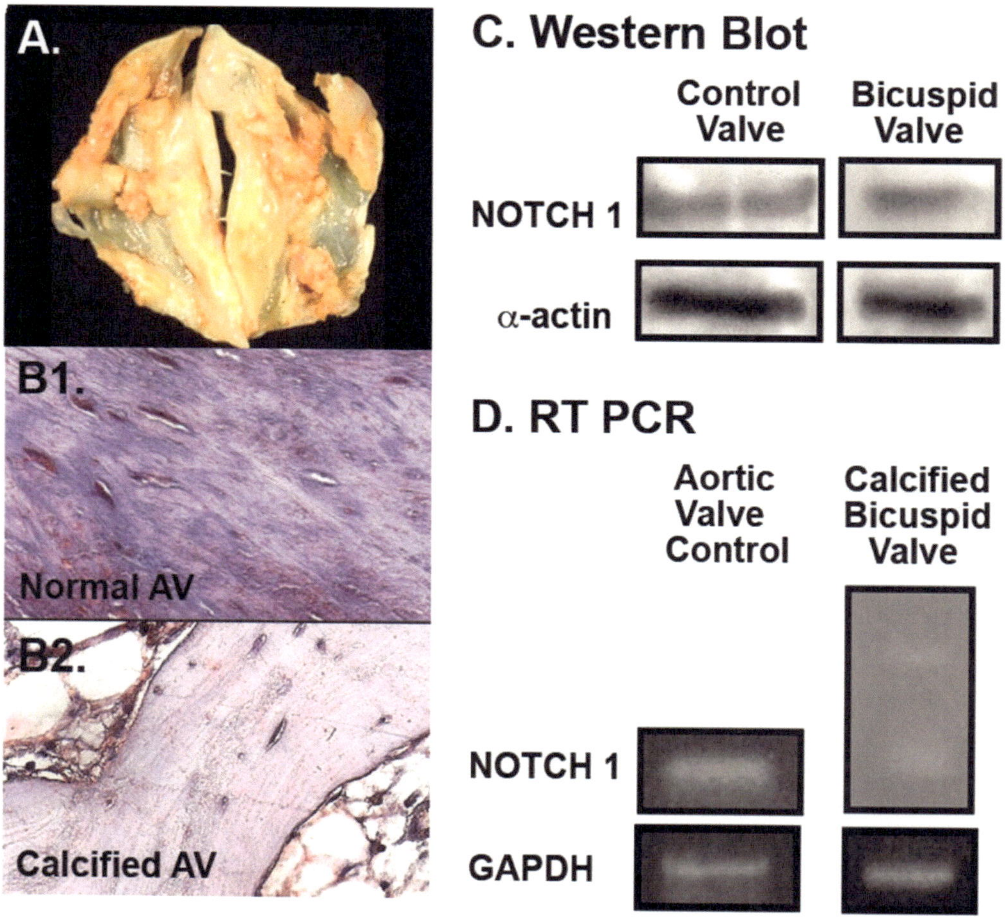

Figure 1. Histology of the aortic valves from human bicuspid calcified valves compared to normal aortic valves removed at the time of surgical valve replacement; **Panel A.** Bicuspid Aortic Valve Removed from patient at the time of surgical valve replacement. **Panel B1.**Notch1 Immunohistochemistry of a Normal Aortic Valve. **Panel B2.**Notch1 Immunohistochemistry of a Bicuspid Aortic Valve. **Panel C.** Notch1 protein expression was decreased in the BAV compared to controls by immunhistochemistry and Western Blot expression. **Panel D.** Notch1 RNA expression was decrease in the BAV as compared to Control aortic Valve.

Figure 2 demonstrates the characterization of the eNOS phenotype as defined by histology, RTPCR and echocardiography. In Figure 2, Panel A is the histology for BAV, Figure 2, Panel B is the semi-quantitative RTPCR from the BAV eNOS$^{-/-}$ mice, and echocardiographic data for the bicuspid vs. tricuspid aortic valves Figure 2, Panel C. We measured Notch1, Wnt3a and downstream markers of the canonical Wnt pathway by protein and RNA expression. Notch1 protein was diminished and the RNA expression demonstrates a similar spliced variant with lipid treatments which was not present with the control and atorvastatin treatment. Cholesterol diets increased the members of the canonical Wnt pathway and Atorvastatin diminished these markers significantly ($p<0.05$).

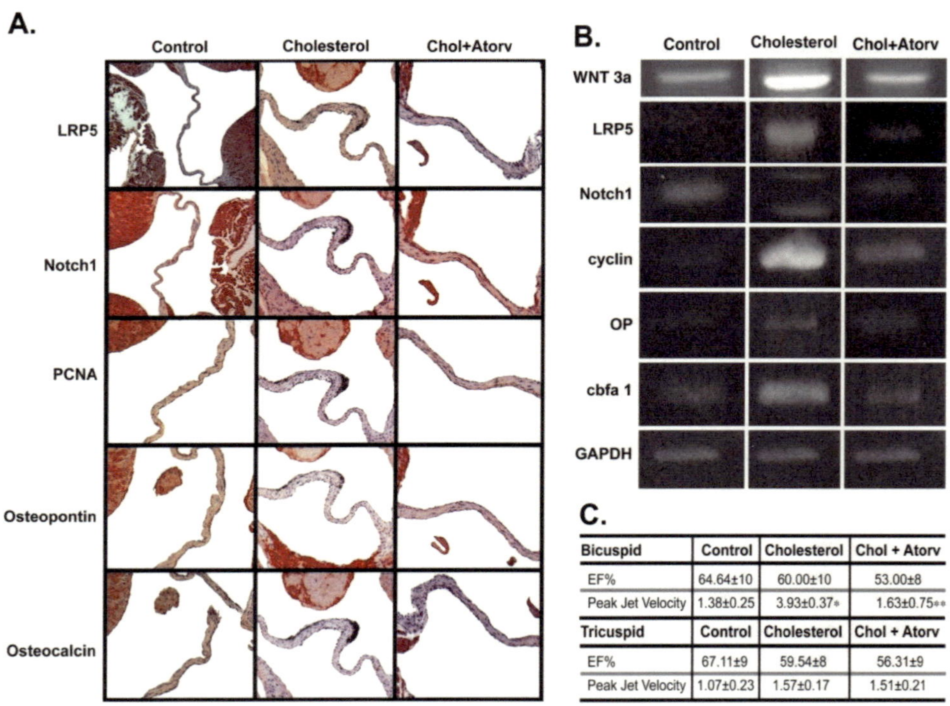

C.

Bicuspid	Control	Cholesterol	Chol + Atorv
EF%	64.64±10	60.00±10	53.00±8
Peak Jet Velocity	1.38±0.25	3.93±0.37*	1.63±0.75**
Tricuspid	**Control**	**Cholesterol**	**Chol + Atorv**
EF%	67.11±9	59.54±8	56.31±9
Peak Jet Velocity	1.07±0.23	1.57±0.17	1.51±0.21

Figure 2. Characterization of the Bicuspid Aortic valve from the eNOS$^{-/-}$ bicuspid aortic valves. **Panel A**. Immunohistochemistry stain for Lrp5, Notch1, Proliferating Cell Nuclear Antigen, Osteopontin and Osteocalcin from the eNOS$^{-/-}$ aortic valves on the control, Left column, control diet; middle column, cholesterol diet; right column, cholesterol diet plus atorvastatin. In each panel, the aortic valve leaflet is in the center. (All frames 20X magnification) **Panel B**. RTPCR for Wnt3a, Lrp5, Notch1, cyclin1, Osteopontin and Cbfa1 from the eNOS$^{-/-}$ aortic valves on the control, A1, Cholesterol A2, and the Cholesterol + Atorvastatin diets A3. **Panel C.** Echocardiographic results of the tricuspid versus bicuspid eNOS$^{-/-}$ null mice.

BAV is a complex model to study the mechanisms of calcification. The importance of cell-cell communication within a stem cell niche is necessary for the development of valvular heart disease. The two corollaries necessary for an adult stem cell niche is to first define the physical architecture of the stem-cell niche and second is to define the gradient of proliferation to differentiation within the stem-cell niche. The endothelial lining cell located along the aortic surface is responsible for the secretion of a growth factors [92]. These cells interact with the subendothelial cells that are resident below the endothelial layer of cells. These cells have been characterized as myofibroblast cells [75, 93, 94].

To test the hypothesis that BAV disease develops secondary to a stem cell niche process, the physical cell-cell communication needed to be established [95]. In the aortic valve the communication for the stem cell niche would be between the aortic valve endothelial cell and the adjacent myofibroblast cell located below the aortic lining endothelial cell. Conditioned media was produced from untreated aortic valve endothelial cells for the microenvironment that activates signaling in the myofibroblast cell. A mitogenic protein (Wnt3a) was isolated from the conditioned media and then tested directly on the responding mesenchymal cell, the cardiac valve myofibroblast [93, 96,95]. This transfer of isolated protein to the adjacent cell was necessary to determine if the cell would proliferate directly in the presence of this protein. This system is appealing because the responding mesenchymal cell is isolated from the anatomic region adjacent and immediately below that of the endothelial cells producing the growth factor activity along the fibrosa surface. Very little is known regarding the characterization of the endothelial cell conditioned media. These experiments test the corollary that the physical architecture described above is necessary for disease development in the aortic valve.

Figure 3 demonstrates the isolation and characterization of the Wnt3a from the conditioned media microenvironment. Figure 3, Panel A, is light microscopy of aortic valve endothelial cells isolated from the aortic surface of the aortic valve. The results of the mitogen assays for fractions eluting from a DEAE- Sephadex column are shown in Figure 3, Panel B. It can be seen that the mitogenic activity appeared as a single peak eluting at approximately 0.25 M NaCl. The material eluting from DEAE- Sephadex was then applied to Sephadex G-100; the results of mitogen assays on fractions eluting from such a gel filtration column are shown in Figure 3, Panel C. It can be seen that under these native, non-denaturing conditions the bulk of the mitogenic activity eluted as a peak corresponding to standard proteins of 30- 40,000 molecular weight. A SDS denaturing protein gel was run on each sample from the eluted proteins and the bulk of activated protein correlated with the protein peak at 46kd as shown in Figure 3, Panel D. The protein size and charge determination is similar to that previously characterized as Wnt3a [97]. This material lost all activity when heated to 100°C for 5 minutes; disulfide bond reduction with dithiothreitol also abolished all mitogenic activity; and treatment with trypsin destroyed all activity, implicating a protein structure.

The second corollary for identifying a stem cell niche is to define the gradient responsible for the proliferation to differentiation process. The main postulate for this corollary stems from the risk factor hypothesis for the development of aortic valve disease. If traditional atherosclerotic risk factors are necessary for the initiation of disease, then these risk factors are responsible for the gradient necessary for the differentiation of myofibroblast cells to become an osteoblast calcifying phenotype [5, 17, 62, 75, 94, 95, 98, 99]. If traditional risk factors are responsible for the development of valvular heart disease, then an oxidative stress mechanism is important for the development of a gradient in this niche.

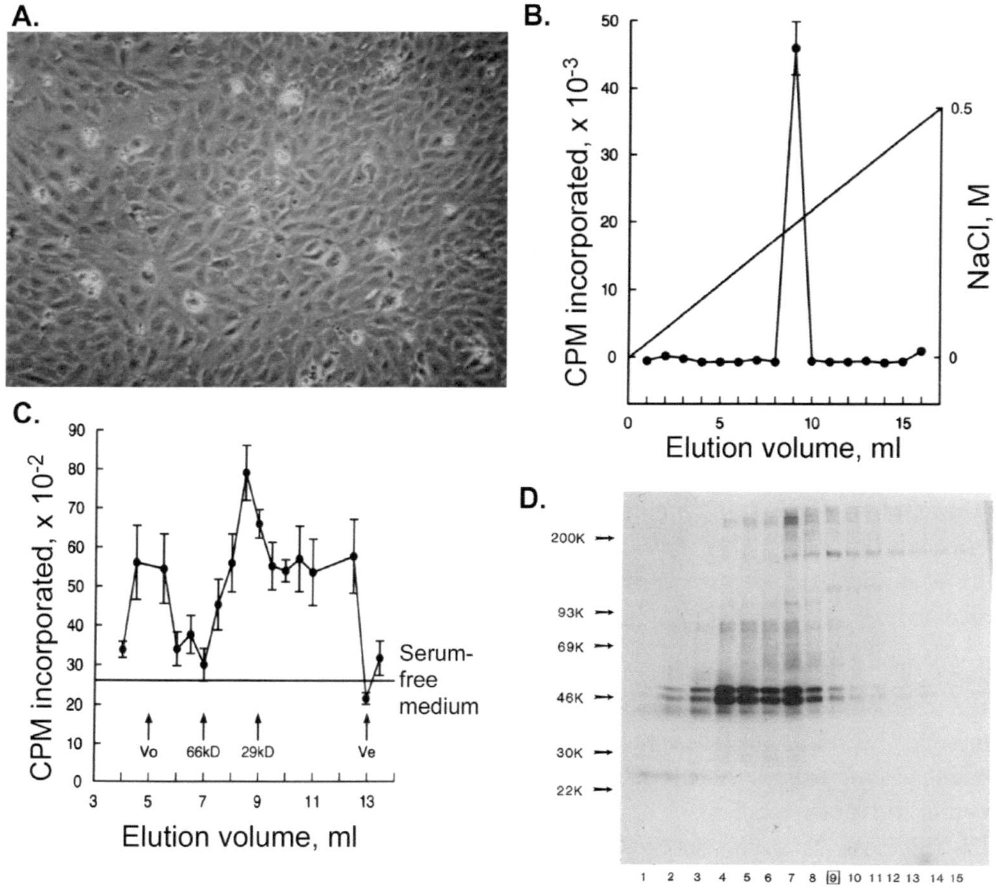

Figure 3. Protein Isolation and Characterization of Aortic Valve Endothelial Cell Conditioned Media; **Panel A**. Light Microscopy for Aortic Valve Endothelial Cells. **Panel B**. Cell Proliferation for fractions eluting from a DEAE- Sephadex column. ($p<0.001$) **Panel C**. Fractions from DEAE- Sephadex to characterize weight with Sephadex G-100. ($p<0.001$) **Panel D.** Southern Blot for Protein Expression of Fractions isolated form the DEAE-Sephadex column.

Nitric oxide is important in terms of the mechanism in adult disease processes and also in the developmental abnormalities such as the bicuspid aortic valve phenotype in the eNOS null mouse. To answer this question of the role of oxidative stress and nitric oxide in the aortic valve, I performed *in vitro* experiments to determine eNOS enzymatic and protein regulation in the presence of lipids and attenuation with Atorvastatin. We have previously published that eNOS is regulated in the aortic valve in an experimental hypercholesterolemia model of valvulardisease [11]. Figure 4, demonstrates the eNOS regulation in the endothelial cells in the presence of lipids with and without Atorvastatin. A number of standard assays were performed to measure eNOS functional activity. Figure 4, Panel A1, tests for eNOS enzymatic activity in the aortic valve endothelial cells (AEC) in the presence of LDL with and without Atorvastatin. ENOS enzymatic activity was decreased in the presence of lipids and Atorvastatin improved functional enzyme activity. Figure 4, Panel A2, shows results for tissue nitrites measured in the endothelial cells providing indirect evidence for the enzyme activity. There was an increase is nitrites with lipid treatments and attenuation with Atorvastatin. This increase in nitrite levels correlates with a decrease in the functional activity of the eNOS enzyme in the aortic valve endothelium.

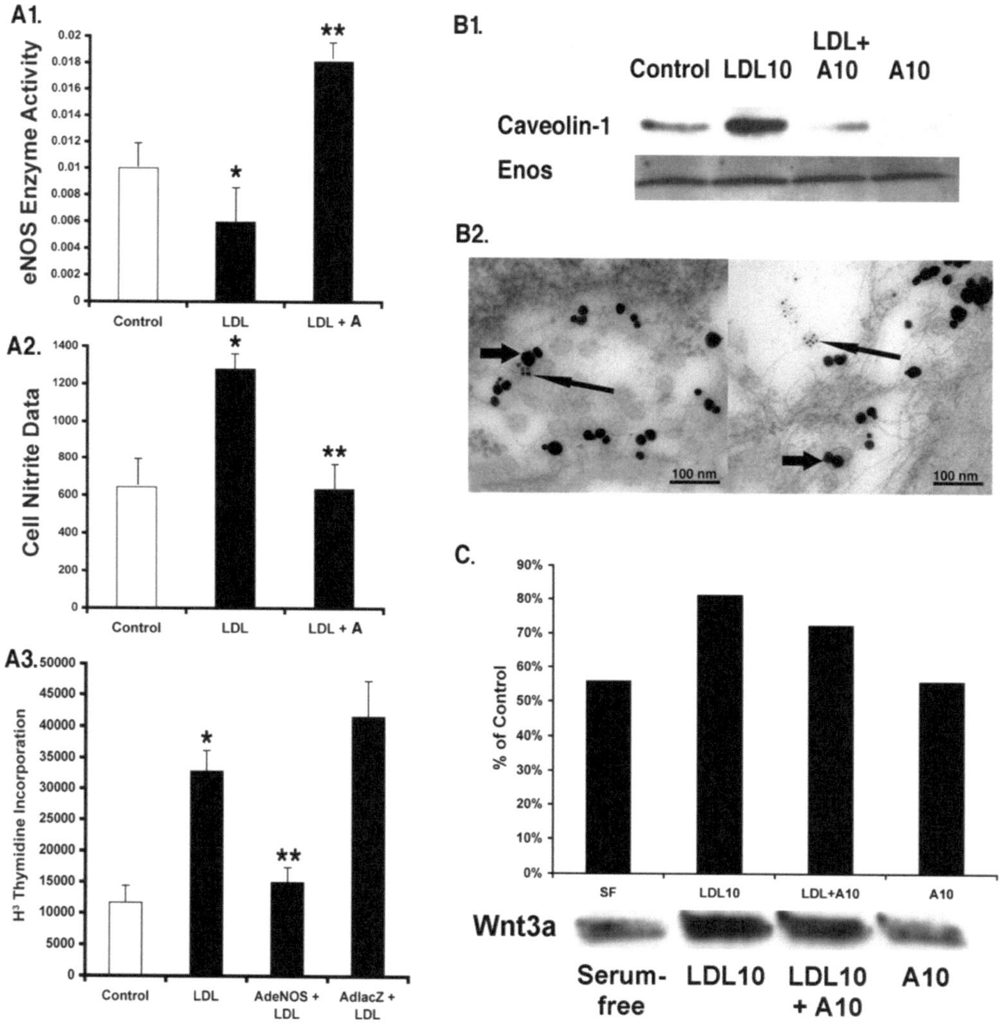

Figure 4. Evidence for eNOS regulation and Wnt3a Secretion from Aortic Valve Endothelial Cells. *p<0.001 for control compared to cholesterol, **p<0.001 for cholesterol compared to cholesterol + Atorvastatin. **Panel A1**. eNOS enzymatic activity in the aortic valve endothelial cells (AEC) in the presence of LDL with and without atorvastatin. **Panel A2.** Cell Nitrite activity in the aortic valve endothelial cells (AEC) in the presence of LDL with and without atorvastatin. **Panel A3.** Thymidine incorporation in LDL treated media, compared to AdeNOS treated myofibroblast cells, versus control LacZ virus. **Panel B1**. Caveolin-1 and eNOS protein expression isolated from the lipid with and without atorvastatin treated cells as shown by Western Blot. **Panel B2**. Electron microscopy immunogold labeling for eNOS and Caveolin-1 localize in the aortic valve endothelial cells in caveolae. **Panel C.** Wnt3a Immunoprecipitate from Conditioned Media treated with LDL with and without Atorvastatin.

The proof of principle experiment to test the importance of eNOS enzymatic activity is an overexpression experiment to determine if eNOS is able to inhibit cell proliferation, an early cellular event in the development of aortic stenosis [13]. Experiments were performed to overexpress eNOS to determine if eNOS overexpression in the aortic valve endothelial cells would regulate cell proliferation. The in vitro myofibroblast cells were directly transduced with an eNOS adenoviral gene construct. Thymidine incorporation was measured to test if overexpressing eNOS can inhibit cellular proliferation. Figure 4, Panel A3, eNOS overexpression inhibits the cell proliferation in the oxidized LDL treated cells induced as compared to the LacZ control treated cells.

A key regulator of eNOS function is caveolin-1 which is expressed in aortic valve endothelial cells [9. Caveolin-1 upregulation in the presence of lipids inactivates eNOS enzymatic function and further promotes oxidative stress [100, 101]. Experiments were performed to localize the expression of Caveolin1 and eNOS in the aortic valve endothelial cell caveolae. A well defined mechanism to inactivate eNOS enzymatic activity is functional binding of eNOS with caveolin1 in the presence of lipids [29, 95, 102]. Figure 4, Panel B1, demonstrates that Caveolin-1 is upregulated in the lipid treated cells and decreases with atorvastatin treatment with no change in the eNOS protein expression as shown by Western Blot. Figure 4, Panel B2, demonstrates the ultrastructural evidence by immunogold labeling for eNOS and Caveolin-1 present in the aortic valve endothelial cells in caveolae, similar to previously reported data [9, 11]. This caveolin1 upregulation is indirect evidence in addition to the direct data of a decrease in the enzyme activity, that caveolin-1 may play a similar role in AEC found in the aortic valve similar to the vascular endothelium.

Experiments were performed to determine if Wnt3a secretion changes in the microenvironment of the aortic valve endothelial cells with and without lipids. Figure 4, Panel C, demonstrates that Wnt3a protein concentration in the conditioned media in the presence of LDL with and without Atorvastatin. There is a significant increase in the protein with the lipids and attenuation of this protein secretion with the Atorvastatin treatments. This experiment tests the effects of lipids regulating the development of a "Wnt3a" gradient in the microenvironment. If LDL increases Wnt3a secretion into the conditioned media or the microenvironment of the diseased aortic valve, this further contributes to the activation of the canonical Wnt pathway in the subendothelial space of the aortic valve.

The final experiment to test the importance of a stem cell niche to activate the cellular osteoblast gene program in the subendothelial layer cells was to test for the gene expression of the Wnt/Lrp5 pathway in the myofibroblast cells. The stem cell niche is a unique model for the development of an oxidative stress communication within the aortic valve endothelium. As shown in Figure 5, oxidative stress contributes to the release of Wnt3a into the subendothelial space to activate Lrp5/Frizzeled receptor complex on the extracellular membrane of the myofibroblast. This trimeric complex then induces glycogen synthase kinase to be phosphorylated. This phosphorylation event causes ▪ -catenin translocation to the nucleus. ▪ -catenin acts as a coactivator of osteoblast specific transcription factor Cbfa1 to induce mesenchymalosteoblastogenesis in the aortic valve myofibroblast cell.

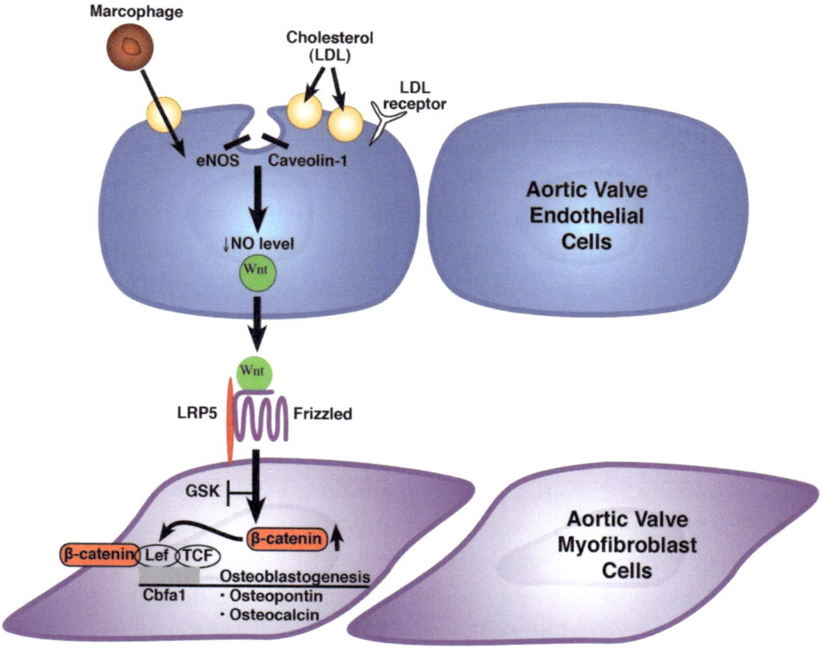

Figure 5. Schematic Modeling for Calcification in the Aortic Valve Stem Cell Niche.

Adult tissues stem cells are a population of functionally undifferentiated cells, capable of (i) homing (ii) proliferation, (iii) producing differentiated progeny, (iv) self-renewing, (v) regeneration, and (vi) reversibility in the use of these options. Within this definition, stem cells are defined by virtue of their functional potential and not by a specific observable characteristic. This data is the first to implicate a cell-cell communication between the aortic valve endothelial cell and the myofibroblast cell to activate the canonical Wnt pathway. Lrp5 is important in normal valve development [103], in this stem cell niche, reactivation of latent Lrp5 expression [5, 16], regulates osteoblastogenesis in these mesenchymal cells. The two corollary requirements necessary for an adult stem cell niche is to first define the physical architecture of the stem-cell niche and second is to define the gradient of proliferation to differentiation within the stem-cell niche. The aortic valve endothelial cell communicates with the myofibroblast cell to activate the myofibroblast to differentiate to form an osteoblast-like phenotype [14]. This concept is similar to the endothelial/mesenchymal transition critical in normal valve development [104]. This data fulfills these main corollaries of the plausibility of a stem cell niche responsible for the development of valvular heart disease. Within a stem cell niche there is a delicate balance between proliferation and differentiation. Cells near the stem-cell zone are more proliferative, and Wnt likely plays a role in directing cell differentiation. Stem cell behavior is determined by the number of its stem cell neighbors, which in the valve is the endothelial cell. This assumption is aimed at simply describing the fact that cytokines, secreted by cells into the micro-environment are capable of activating quiescent stem cells into differentiation [105].

The important inhibitor in this model is Notch1. Notch1 plays a roll in cellular differentiation decisions. In the osteoblast cell, it serves as an inhibitor of osteoblast differentiation [89, 90]. In the aortic valve, it serves to turn off bone formation via the cell-cell crosstalk between the endothelial and the myofibroblast cells. Normal Notch1 receptor functions to maintain normal valve cellular composition and homeostasis. In the presence of lipids, Notch1 is spliced and therefore activates osteoblastogenesis. In turn, the Wnt3a is secreted and binds to Lrp5 and Frizzled on the extracellular membrane to regulate the osteoblast gene program. This developmental disease process follows a parallel signaling pathway that has been observed in the normal embryonic valve development that has been well delineated by previous investigators [104]. A similar cell-cell communication is necessary for the development of valve disease.

This study provides the correlates described in the mathematical modeling by Agur [106]. This mathematical model has demonstrated the principal that the universal properties of the stem cells can be described in a simple discrete model as derived from hemopoietic stem cell behavior [106]. The transition of hemopoietic stem cells from quiescence into differentiation, is governed by their cell-cycling status, by stimulatory hormones secreted by neighboring cells into the micro-environment and by the level of amplification of stem-cell population [105, 107]. The model of Agur, defines the corollaries necessary to identify a stem cell niche, first the physical architecture of the stem cell niche and second the gradient necessary to regulate the niche. In the BAV the gradient is defined by the niche's microenvironment. The initiation of event of oxidative stress inhibits normal endothelial nitric oxide synthase function, activates notch1 splicing which in turn induces Wnt3a secretion to activate bone formation within the valve [5, 17], [99].

The model proposed in the study as described in Figure 5, provides the cellular architecture for the development of this disease process. This model does not take into account other cytokine/growth factor mediated mechanisms that have been shown to also be important in this disease process [108]. However, understanding CAVD from a development disease perspective will provide a foundation for understanding this and other development disease processes. Clinical trials in the field of CAVD are demonstrating variable results [86, 87]. The possible differences in the published trials are secondary to the timing of therapy and the biological targeting of the lipid levels in these patients. Future medical therapies targeting stem cell niche mediated diseases provides a novel model system to test and to translate clinically for patients in the future.

Author details

Nalini Rajamannan

Molecular Biology and Biochemistry, Mayo Clinic School of Medicine, Rochester MN, USA

References

[1] Lindroos, M., Kupari, M., Heikkila, J., & Tilvis, R. Prevalence of aortic valve abnormalities in the elderly: an echocardiographic study of a random population sample. Journal of the American College of Cardiology. (1993). , 21(5), 1220-1225.

[2] Ross, J., Jr , , Braunwald, E., Aortic, stenosis., & Circulation, . (1968). Suppl):, 61-67.

[3] Roberts WC, Ko JM. (2005). Frequency by decades of unicuspid, bicuspid, and tricuspid aortic valves in adults having isolated aortic valve replacement for aortic stenosis, with or without associated aortic regurgitation. *Circulation*, 111(7), 920-925.

[4] Drolet, M. C., Arsenault, M., & Couet, J. (2003). Experimental aortic valve stenosis in rabbits. *Journal of the American College of Cardiology*, 41(7), 1211-1217.

[5] Rajamannan, N. M., Subramaniam, M., Caira, F., Stock, S. R., & Spelsberg, T. C. Atorvastatin inhibits hypercholesterolemia-induced calcification in the aortic valves via the Lrp5 receptor pathway. Circulation(2005). Suppl):I, 229-234.

[6] Rajamannan, N. M., Subramaniam, M., Springett, M., Sebo, T. C., Niekrasz, M., Mc Connell, J. P., Singh, R. J., Stone, N. J., Bonow, R. O., & Spelsberg, T. C. (2002). Atorvastatin inhibits hypercholesterolemia-induced cellular proliferation and bone matrix production in the rabbit aortic valve. *Circulation*, 105(22), 2260-2265.

[7] Weiss, R. M., Ohashi, M., Miller, Young. S. G., & Heistad, D. D. (2006). Calcific aortic valve stenosis in old hypercholesterolemic mice. *Circulation*, 114(19), 2065-2069.

[8] Aikawa, E., Nahrendorf, M., Sosnovik, D., Lok, V. M., Jaffer, F. A., Aikawa, M., & Weissleder, R. (2007). Multimodality molecular imaging identifies proteolytic and osteogenic activities in early aortic valve disease. *Circulation*, 115(3), 377-386.

[9] Rajamannan NM, Springett MJ, Pederson LG, Carmichael SW.Localization of caveolin 1 in aortic valve endothelial cells using antigen retrieval. Journal of Histochemistry&Cytochemistry. (2002). , 50(5), 617-628.

[10] Rajamannan NM, Edwards WD, Spelsberg TC. Hypercholesterolemic aortic-valve disease.New England Journal of Medicine. (2003). , 349(7), 717-718.

[11] Rajamannan, N. M., Subramaniam, M., Stock, S. R., Stone, N. J., Springett, M., Ignatiev, K. I., Mc Connell, J. P., Singh, R. J., Bonow, R. O., & Spelsberg, T. C. Atorvastatin inhibits calcification and enhances nitric oxide synthase production in the hypercholesterolaemic aortic valve. Heart (British Cardiac Society). (2005). , 91(6), 806-810.

[12] Makkena, B., Salti, H., Subramaniam, M., Thennapan, S., Bonow, R. H., Caira, F., Bonow, R. O., Spelsberg, T. C., & Rajamannan, N. M. (2005). Atorvastatin decreases cellular proliferation and bone matrix expression in the hypercholesterolemic mitral valve. *Journal of the American College of Cardiology*, 45(4), 631-633.

[13] Rajamannan, N. M., Sangiorgi, G., Springett, M., Arnold, K., Mohacsi, T., Spagnoli, L. G., Edwards, W. D., Tajik, A. J., & Schwartz, R. S. Experimental hypercholesterolemia

induces apoptosis in the aortic valve. Journal of Heart Valve Disease. (2001). , 10(3), 371-374.

[14] Rajamannan, N. M., Subramaniam, M., Rickard, D., Stock, S. R., Donovan, J., Springett, M., Orszulak, T., Fullerton, D. A., Tajik, A. J., Bonow, R. O., & Spelsberg, T. (2003). Human aortic valve calcification is associated with an osteoblast phenotype. *Circulation*, 107(17), 2181-2184.

[15] Mohler, E. R., 3rd Gannon, F., Reynolds, C., Zimmerman, R., Keane, M. G., & Kaplan, F. S. (2001). Bone formation and inflammation in cardiac valves. *Circulation*, 103(11), 1522-1528.

[16] Caira, F. C., Stock, S. R., Gleason, T. G., Mc Gee, E. C., Huang, J., Bonow, R. O., Spelsberg, T. C., Mc Carthy, P. M., Rahimtoola, S. H., & Rajamannan, N. M. (2006). Human degenerative valve disease is associated with up-regulation of low-density lipoprotein receptor-related protein 5 receptor-mediated bone formation. *Journal of the American College of Cardiology*, 47(8), 1707-1712.

[17] Shao, J. S., Cheng, S. L., Pingsterhaus, J. M., Charlton-Kachigian, N., Loewy, A. P., Towler, D. A., Msx, promotes., cardiovascular, calcification., by, activating., paracrine, Wnt., & signals, . The Journal of clinical investigation. (2005). , 115(5), 1210-1220.

[18] Desai, M. Y., Rodriguez, A., Wasserman, Gerstenblith. G., Agarwal, S., Kennedy, M., Bluemke, D. A., & Lima, J. A. Association of cholesterol subfractions and carotid lipid core measured by MRI. ArteriosclerThrombVasc Biol. (2005). e, 110-111.

[19] Subbaiah, P. V., Gesquiere, L. R., & Wang, K. Regulation of the selective uptake of cholesteryl esters from high density lipoproteins by sphingomyelin. J Lipid Res. (2005). , 46(12), 2699-2705.

[20] Kim, W. J., Chereshnev, I., Gazdoiu, M., Fallon, J. T., Rollins, B. J., Taubman, M. B. M. C. P., deficiency, is., associated, with., reduced, intimal., hyperplasia, after., arterial, injury., & Biochem, . BiochemBiophys Res Commun. (2003). , 310(3), 936-942.

[21] Tanner, F. C., Boehm, M., Akyurek, L. M., San, H., Yang, Z. Y., Tashiro, J., Nabel, G. J., & Nabel, E. G. Differential effects of the cyclin-dependent kinase inhibitors Kip1), p21(Cip1), and p16(Ink4) on vascular smooth muscle cell proliferation. Circulation. (2000). , 27.

[22] Zhang, R., Luo, D., Miao, R., Bai, L., Ge, Q., Sessa, W. C., Min, W., Hsp9, , Akt, phosphorylates. A. S. K., inhibits, A. S., & K1-mediated, apoptosis. (2005). *Oncogene*, 24(24), 3954-3963.

[23] (Bostrom K, Watson KE, Horn S, Wortham C, Herman IM, Demer LL. Bone morphogenetic protein expression in human atherosclerotic lesions. The Journal of clinical investigation. 1993;91(4):1800-1809). , 91(4), 1800-1809.

[24] Aikawa, M., Rabkin, E., Sugiyama, S., Voglic, S. J., Fukumoto, Y., Furukawa, Y., Shiomi, M., Schoen, F. J., Libby, P., An-Co, H. M. G., reductase, A., inhibitor, cerivastatin.,

suppresses, growth., of, macrophages., expressing, matrix., metalloproteinases, , tissue, factor., in, vivo., & in, vitro. (2001). *Circulation*, 103(2), 276-283.

[25] Williams, J. K., Sukhova, G. K., Herrington, D. M., & Libby, P. (1998). Pravastatin has cholesterol-lowering independent effects on the artery wall of atherosclerotic monkeys. *Journal of the American College of Cardiology*, 31(3), 684-691.

[26] Shyy, J. Y., & Chien, S. Role of integrins in endothelial mechanosensing of shear stress. Circ Res. (2002). , 91(9), 769-775.

[27] Laufs, U., & Liao, J. K. Post-transcriptional regulation of endothelial nitric oxide synthase mRNA stability by Rho GTPase. Journal of Biological Chemistry. (1998). , 273(37), 24266-24271.

[28] Venema RC, Sayegh HS, Kent JD, Harrison DG. Identification, characterization, and comparison of the calmodulin-binding domains of the endothelial and inducible nitric oxide synthases.Journal of Biological Chemistry. (1996). , 271(11), 6435-6440.

[29] Blair, A., Shaul, P. W., Yuhanna, I. S., Conrad, P. A., & Smart, E. J. Oxidized low density lipoprotein displaces endothelial nitric-oxide synthase (eNOS) from plasmalemmalcaveolae and impairs eNOS activation. J Biol Chem. (1999). , 274(45), 32512-32519.

[30] Smart EJ, Anderson RG.Alterations in membrane cholesterol that affect structure and function of caveolae. Methods Enzymol. (2002). , 353, 131-139.

[31] Pritchard, K. A., Ackerman, A. W., Ou, J., Curtis, M., Smalley, D. M., Fontana, J. T., Stemerman, M. B., & Sessa, W. C. Native low-density lipoprotein induces endothelial nitric oxide synthase dysfunction: role of heat shock protein 90 and caveolin-1. Free RadicBiol Med. (2002). , 33(1), 52-62.

[32] Banka CL, Black AS, Dyer CA, Curtiss LK. THP-1 cells form foam cells in response to coculture with lipoproteins but not platelets.J Lipid Res. (1991). , 32(1), 35-43.

[33] Curtiss LK, Dyer CA, Banka CL, Black AS.Platelet-mediated foam cell formation in atherosclerosis. Clin Invest Med. (1990). , 13(4), 189-195.

[34] Brand, K., Banka, C. L., Mackman, N., Terkeltaub, R. A., Fan, S. T., Curtiss, L. K., Oxidized, L. D. L., enhances, lipopolysaccharide-induced., tissue, factor., expression, in., human, adherent., & monocytes, Arterioscler. ArteriosclerThromb. (1994). , 14(5), 790-797.

[35] Leibovich, S. J., Polverini, P. J., Shepard, H. M., Wiseman, D. M., Shively, V., & Nuseir, N. (1987). Macrophage-induced angiogenesis is mediated by tumour necrosis factor-alpha. *Nature*, 329(6140), 630-632.

[36] Leibovich, S. J., Chen, J. F., Pinhal-Enfield, G., Belem, P. C., Elson, G., Rosania, A., Ramanathan, M., Montesinos, C., Jacobson, M., Schwarzschild, Fink. J. S., & Cronstein, B. Synergistic up-regulation of vascular endothelial growth factor expression in murine macrophages by adenosine A(2A) receptor agonists and endotoxin. Am J Pathol. (2002). , 160(6), 2231-2244.

[37] Subramanian SV, Polikandriotis JA, Kelm RJ, Jr., David JJ, Orosz CG, Strauch AR.Induction of vascular smooth muscle alpha-actin gene transcription in transforming growth factor beta1-activated myofibroblasts mediated by dynamic interplay between the Pur repressor proteins and Sp1/Smadcoactivators. MolBiol Cell. (2004). , 15(10), 4532-4543.

[38] Stewart, B. F., Siscovick, D., Lind, B. K., Gardin, J. M., Gottdiener, J. S., Smith, V. E., Kitzman, D. W., & Otto, C. M. Clinical factors associated with calcific aortic valve disease. Cardiovascular Health Study. Journal of the American College of Cardiology(1997). , 29(3), 630-634.

[39] Aronow, W. S., Ahn, C., Kronzon, I., & Goldman, . Association of coronary risk factors and use of statins with progression of mild valvular aortic stenosis in older persons. American Journal of Cardiology. (2001). , 88(6), 693-695.

[40] O'Brien, K. D., Reichenbach, D. D., Marcovina, S. M., Kuusisto, J., Alpers, Otto. C. M., Apolipoproteins, B., (a, , accumulate, E., in, the., morphologically, early., lesion, of., 'degenerative', valvular., & aortic, stenosis. ArteriosclerosisThrombosis & Vascular Biology. (1996). , 16(4), 523-532.

[41] Olsson, M., Thyberg, J., & Nilsson, J. Presence of oxidized low density lipoprotein in nonrheumaticstenotic aortic valves. ArterioscIerThrombVasc Biol. (1999). , 19(5), 1218-1222.

[42] Sprecher, D. L., Schaefer, E. J., Kent, K. M., Gregg, R. E., Zech, L. A., Hoeg, J. M., Mc Manus, B., Roberts, W. C., Brewer, H. B., & Jr , . Cardiovascular features of homozygous familial hypercholesterolemia: analysis of 16 patients. American Journal of Cardiology. (1984). , 54(1), 20-30.

[43] O'Brien, K. D., Kuusisto, J., Reichenbach, D. D., Ferguson, M., Giachelli, C., & Alpers, Otto. C. M. (1995). Osteopontin is expressed in human aortic valvular lesions. *Circulation*, 92(8), 2163-2168.

[44] Mohler, E. R., 3rd Adam, L. P., Mc Clelland, P., Graham, L., & Hathaway, D. R. Detection of osteopontin in calcified human aortic valves. ArteriosclerosisThrombosis & Vascular Biology. (1997). , 17(3), 547-552.

[45] O'Brien, K. D., Kuusisto, J., Reichenbach, D. D., Ferguson, M., Giachelli, C., & Alpers, Otto. C. M. Osteopontin is expressed in human aortic valvular lesions. [comment]. Circulation. (1995). , 92(8), 2163-2168.

[46] Whittaker, P., Boughner, D. R., Perkins, D. G., & Canham, P. B. Quantitative structural analysis of collagen in chordae tendineae and its relation to floppy mitral valves and proteoglycan infiltration. British heart journal. (1987). , 57(3), 264-269.

[47] Wooley, C. F., Baker, P. B., Kolibash, A. J., Kilman, J. W., Sparks, E. A., Boudoulas, H., The, floppy., myxomatous, mitral., valve, mitral., valve, prolapse., mitral, regurgitation., & Prog, Cardiovasc. Dis. (1991). , 33(6), 397-433.

[48] Grande-Allen, K. J., Borowski, A. G., Troughton, R. W., Houghtaling, P. L., Dipaola, N. R., Moravec, C. S., Vesely, I., & Griffin, B. P. Apparently normal mitral valves in patients with heart failure demonstrate biochemical and structural derangements: an extracellular matrix and echocardiographic study. [see comment].(2005). Jan 2004., 54-61.

[49] Grande-Allen, K. J., Calabro, A., Gupta, V., Wight, T. N., Hascall, V. C., & Vesely, I. Glycosaminoglycans and proteoglycans in normal mitral valve leaflets and chordae: association with regions of tensile and compressive loading.(2004). Jul., 621-633.

[50] Jian, B., Jones, P. L., Li, Q., Mohler, E. R., 3rd Schoen, F. J., Levy, R. J., Matrix, metalloproteinase., is, associated., with-C, tenascin., in, calcific., & aortic, stenosis. (2001). *The American journal of pathology*, 159(1), 321-327.

[51] Ducy, P., Zhang, R., Geoffroy, V., Ridall, A. L., Karsenty, G., Osf, , Cbfa, , transcriptional, a., activator, of., osteoblast, differentiation. [see., & comment], . Cell. (1997). , 89(5), 747-754.

[52] Aubin, J. E., Liu, F., Malaval, L., & Gupta, A. K. Osteoblast and chondroblast differentiation. Bone(1995). Suppl):, 77S EOF-83S EOF.

[53] Robinson JA, Harris SA, Riggs BL, Spelsberg TC. (1997). Estrogen regulation of human osteoblastic cell proliferation and differentiation. *Endocrinology*, 138(7), 2919-2927.

[54] Spelsberg TC, Harris SA, Riggs BL.Immortalized osteoblast cell systems (new human fetal osteoblast systems). Calcif Tissue Int. (1995). Suppl 1):S, 18-21.

[55] Davies, M. R., Lund, R. J., Mathew, S., & Hruska, K. A. Low turnover osteodystrophy and vascular calcification are amenable to skeletal anabolism in an animal model of chronic kidney disease and the metabolic syndrome. J Am SocNephrol. (2005). , 16(4), 917-928.

[56] Davies MR, Lund RJ, Hruska KA. BMP-7 is an efficacious treatment of vascular calcification in a murine model of atherosclerosis and chronic renal failure. J Am SocNephrol. (2003). , 14(6), 1559-1567.

[57] Parhami, F., Garfinkel, A., & Demer, L. L. Role of lipids in osteoporosis. ArteriosclerosisThrombosis & Vascular Biology. (2000). , 20(11), 2346-2348.

[58] Parhami, F., Mody, N., Gharavi, N., Ballard, A. J., Tintut, Y., & Demer, L. L. Role of the cholesterol biosynthetic pathway in osteoblastic differentiation of marrow stromal cells. Journal of Bone & Mineral Research. (2002). , 17(11), 1997-2003.

[59] Parhami, F., Morrow, A. D., Balucan, J., Leitinger, N., Watson, A. D., Tintut, Y., Berliner, J. A., & Demer, L. L. Lipid oxidation products have opposite effects on calcifying vascular cell and bone cell differentiation. A possible explanation for the paradox of arterial calcification in osteoporotic patients. ArteriosclerThrombVasc Biol. (1997). , 17(4), 680-687.

[60] Parhami, F., Tintut, Y., Beamer, W. G., Gharavi, N., Goodman, W., & Demer, L. L. Atherogenic high-fat diet reduces bone mineralization in mice. J Bone Miner Res. (2001)., 16(1), 182-188.

[61] Jono, S., Mc Kee, Murry., Shioi, A., Nishizawa, Y., Mori, K., Morii, H., & Giachelli, C. M. Phosphate regulation of vascular smooth muscle cell calcification. Circ Res. (2000). E, 10-17.

[62] Wada, T., Mc Kee, Steitz. S., & Giachelli, C. M. Calcification of vascular smooth muscle cell cultures: inhibition by osteopontin. CircRes. (1999)., 84(2), 166-178.

[63] Parhami, F., Mody, N., Gharavi, N., Ballard, A. J., Tintut, Y., & Demer, L. L. Role of the cholesterol biosynthetic pathway in osteoblastic differentiation of marrow stromal cells. J Bone Miner Res. (2002)., 17(11), 1997-2003.

[64] Willert, K., Nusse, R., Beta-catenin, a., key, mediator., of, Wnt., & signaling, . (1998). *Current Opinion in Genetics & Development*, 8(1), 95-102.

[65] Behrens, J., von, Kries. J. P., Kuhl, M., Bruhn, L., Wedlich, D., Grosschedl, R., & Birchmeier, W. (1996). Functional interaction of beta-catenin with the transcription factor LEF-1. *Nature*, 382(6592), 638-642.

[66] Huber, O., Korn, R., Mc Laughlin, J., Ohsugi, M., Herrmann, B. G., & Kemler, R. Nuclear localization of beta-catenin by interaction with transcription factor LEF-1. Mech Dev. (1996)., 59(1), 3-10.

[67] Holmen, S. L., Salic, A., Zylstra, C. R., Kirschner, M. W., Williams, B. O. A., novel, set., of-Frizzled, Wnt., fusion, proteins., identifies, receptor., components, that., activate, beta-catenin-dependent., & signaling, . Journal of Biological Chemistry. (2002)., 277(38), 34727-34735.

[68] Caverzasio, J. [., Wnt, L. R. P., new, a., regulation, osteoblastic., pathway, involved., in, reaching., peak, bone., & masses], . Revue Medicale de la Suisse Romande. (2004)., 124(2), 81-82.

[69] Kahler RA, Westendorf JJ. Lymphoid enhancer factor-1 and beta-catenin inhibit Runx2-dependent transcriptional activation of the osteocalcin promoter.Journal of Biological Chemistry. (2003)., 278(14), 11937-11944.

[70] Smith, E., & Frenkel, B. Glucocorticoids Inhibit the Transcriptional Activity of LEF/TCF in Differentiating Osteoblasts in a Glycogen Synthase Kinase-3{beta}-dependent and-independent Manner. J. Biol. Chem. (2005)., 280(3), 2388-2394.

[71] Wang HY, Malbon CC. Wnt signaling, Ca2+, and cyclic GMP: visualizing Frizzled functions. Science (New York,N.Y. (2003)., 300(5625), 1529-1530.

[72] Gregory, Perry., Reyes, E., Conley, A., Gunn, W. G., & Prockop, D. J. Dkk-1-derived Synthetic Peptides and Lithium Chloride for the Control and Recovery of Adult Stem Cells from Bone Marrow. J. Biol. Chem. (2005)., 280(3), 2309-2323.

[73] Yano, F., Kugimiya, F., Ohba, S., Ikeda, T., Chikuda, H., Ogasawara, T., Ogata, N., Takato, T., Nakamura, K., Kawaguchi, H., & Chung, U. I. The canonical Wnt signaling pathway promotes chondrocyte differentiation in a Sox9-dependent manner. BiochemBiophys Res Commun. (2005). , 333(4), 1300-1308.

[74] Jian, B., Narula, N., Li, Q. Y., Mohler, E. R., & 3rd Levy, R. J. Progression of aortic valve stenosis: TGF-beta1 is present in calcified aortic valve cusps and promotes aortic valve interstitial cell calcification via apoptosis. Ann Thorac Surg. (2003). discussion 465-456., 75(2), 457-465.

[75] Mohler, E. R., 3rd Chawla, M. K., Chang, A. W., Vyavahare, N., Levy, R. J., Graham, L., & Gannon, F. H. Identification and characterization of calcifying valve cells from human and canine aortic valves. Journal of Heart Valve Disease. (1999). , 8(3), 254-260.

[76] Alkadhi, H., Wildermuth, S., Plass, A., Bettex, D., Baumert, B., Leschka, S., Desbiolles, L. M., Marincek, B., Boehm, T., Aortic, Stenosis., Comparative, Evaluation., of, 1., Detector, Row. C. T., & Echocardiography, . (2006). *Radiology*.

[77] Budoff, Takasu. J., Katz, R., Mao, S., Shavelle, D. M., O'Brien, K. D., Blumenthal, R. S., Carr, J. J., Kronmal, R., Reproducibility, of. C. T., measurements, of., aortic, valve., calcification, mitral., annulus, calcification., aortic, wall., calcification, in., the, multiethnic., study, of., & atherosclerosis, Acad. AcadRadiol. (2006). , 13(2), 166-172.

[78] Liu, F., Coursey-Clarke, Grahame., Sciacca, C., Rozenshtein, R. R., Homma, A., Austin, S., & , J. H. Aortic valve calcification as an incidental finding at CT of the elderly: severity and location as predictors of aortic stenosis. AJRAm J Roentgenol. (2006). , 186(2), 342-349.

[79] Boxt LM. CT of valvular heart disease.Int J Cardiovasc Imaging. (2005). , 21(1), 105-113.

[80] Shavelle, D. M., Takasu, J., Budoff, Mao. S., Zhao, X. Q., O'Brien, K. D. H. M. G., Co, A., reductase, inhibitor., (statin, , aortic, valve., & calcium. [comment], . Lancet. (2002). , 359(9312), 1125-1126.

[81] Rajamannan, N. M., Subramaniam, M., Caira, F. C., Stock, S. R., & Spelsberg, T. C. Atorvastatin Inhibits Hypercholesterolemia-Induced Calcification in the Aortic Valves via the Lrp5 Receptor Pathway. Circulation. (2005). In Press.

[82] Steinmetz, E. F., Buckley, C., Shames, M. L., Ennis, T. L., Vanvickle-Chavez, S. J., Mao, D., Goeddel, L. A., Hawkins, C. J., & Thompson, R. W. Treatment with simvastatin suppresses the development of experimental abdominal aortic aneurysms in normal and hypercholesterolemic mice. Ann Surg. (2005). , 241(1), 92-101.

[83] Aronow, W. S., Ahn, C., Kronzon, I., & Goldman, . Association of coronary risk factors and use of statins with progression of mild valvular aortic stenosis in older persons. Am J Cardiol. (2001). , 88(6), 693-695.

[84] Shavelle, D. M., Takasu, J., Budoff, Mao. S., Zhao, X. Q., O'Brien, K. D. H. M. G., Co, A., reductase, inhibitor., (statin, , aortic, valve., & calcium, . (2002). *Lancet*, 359(9312), 1125-1126.

[85] Rosenhek, R., Rader, F., Loho, N., Gabriel, H., Heger, M., Klaar, U., Schemper, M., Binder, T., Maurer, G., & Baumgartner, H. (2004). Statins but not angiotensin-converting enzyme inhibitors delay progression of aortic stenosis. *Circulation*, 110(10), 1291-1295.

[86] Cowell, S. J., Newby, D. E., Prescott, R. J., Bloomfield, P., Reid, J., Northridge, D. B., Boon, N. A. A., randomized, trial., of, intensive., lipid-lowering, therapy., in, calcific., & aortic, stenosis. (2005). *The New England journal of medicine*, 352(23), 2389-2397.

[87] Moura, L. M., Ramos, S. F., Zamorano, J. L., Barros, I. M., Azevedo, L. F., Rocha-Goncalves, F., & Rajamannan, N. M. (2007). Rosuvastatin affecting aortic valve endothelium to slow the progression of aortic stenosis. *Journal of the American College of Cardiology*, 49(5), 554-561.

[88] Garg, V., Muth, A. N., Ransom, J. F., Schluterman, M. K., Barnes, R., King, I. N., Grossfeld, P. D., Srivastava, D., Mutations, in. N. O. T. C. H., cause, aortic., & valve, disease. (2005). *Nature*, 437(7056), 270-274.

[89] Sciaudone, M., Gazzerro, E., Priest, L., Delany, A. M., Canalis, E., Notch, ., impairs, osteoblastic., & cell, differentiation. (2003). *Endocrinology*, 144(12), 5631-5639.

[90] Deregowski, V., Gazzerro, E., Priest, L., Rydziel, S., Canalis, E., Notch, ., overexpression, inhibits., osteoblastogenesis, by., suppressing, Wnt/beta-catenin., but, not., bone, morphogenetic., & protein, signaling. J Biol Chem. (2006). , 281(10), 6203-6210.

[91] Lee TC, Zhao YD, Courtman DW, Stewart DJ. (2000). Abnormal aortic valve development in mice lacking endothelial nitric oxide synthase. *Circulation*, 101(20), 2345-2348.

[92] Rajamannan NM, Helgeson SC, Johnson CM.Anionic growth factor activity from cardiac valve endothelial cells: Partial purification and characterization. Clinical Research. (1988). A.

[93] Johnson CM, Hanson MN, Helgeson SC.Porcine cardiac valvularsubendothelial cells in culture: cell isolation and growth characteristics. J Mol Cell Cardiol. (1987). , 19(12), 1185-1193.

[94] Osman, L., Yacoub, M. H., Latif, N., Amrani, M., & Chester, A. H. Role of human valve interstitial cells in valve calcification and their response to atorvastatin. Circulation(2006). Suppl):I, 547-552.

[95] Rajamannan NM.Oxidative-mechanical stress signals stem cell niche mediated Lrp5 osteogenesis in eNOS(-/-) null mice. Journal of cellular biochemistry. (2012).

[96] Johnson CM, Helgeson SC.Glycoproteins synthesized by cultured cardiac valve endothelial cells: unique absence of fibronectin production. BiochemBiophys Res Commun. (1988). , 153(1), 46-50.

[97] Willert, K., Brown, Danenberg. E., Duncan, A. W., Weissman, I. L., Reya, T., Yates, J. R., & 3rd Nusse, R. (2003). Wnt proteins are lipid-modified and can act as stem cell growth factors. *Nature*, 423(6938), 448-452.

[98] Tintut, Y., Alfonso, Z., Saini, T., Radcliff, K., Watson, K., Bostrom, K., & Demer, L. L. (2003). Multilineage potential of cells from the artery wall. *Circulation*, 108(20), 2505-2510.

[99] Kirton, J. P., Crofts, N. J., George, S. J., Brennan, K., & Canfield, A. E. Wnt/beta-catenin signaling stimulates chondrogenic and inhibits adipogenic differentiation of pericytes: potential relevance to vascular disease? Circ Res. (2007). , 101(6), 581-589.

[100] Garcia-Cardena, G., Martasek, P., Masters, Skidd. P. M., Couet, J., Li, S., Lisanti, M. P., & Sessa, W. C. Dissecting the interaction between nitric oxide synthase (NOS) and caveolin. Functional significance of the noscaveolin binding domain in vivo. J Biol Chem. (1997). , 272(41), 25437-25440.

[101] Garcia-Cardena, G., Oh, P., Liu, J., Schnitzer, J. E., & Sessa, W. C. Targeting of nitric oxide synthase to endothelial cell caveolae via palmitoylation: implications for nitric oxide signaling. ProcNatlAcadSci U S A. (1996). , 93(13), 6448-6453.

[102] Feron, O., Dessy, C., Moniotte, S., Desager, J. P., & Balligand, J. L. Hypercholesterolemia decreases nitric oxide production by promoting the interaction of caveolin and endothelial nitric oxide synthase. J Clin Invest. (1999). , 103(6), 897-905.

[103] Hurlstone, A. F., Haramis, A. P., Wienholds, E., Begthel, H., Korving, J., Van Eeden, F., Cuppen, E., Zivkovic, D., Plasterk, R. H., & Clevers, H. (2003). The Wnt/beta-catenin pathway regulates cardiac valve formation. *Nature*, 425(6958), 633-637.

[104] Paruchuri, S., Yang, J. H., Aikawa, E., Melero-Martin, J. M., Khan, Z. A., Loukogeorgakis, S., Schoen, F. J., & Bischoff, J. Human pulmonary valve progenitor cells exhibit endothelial/mesenchymal plasticity in response to vascular endothelial growth factor-A and transforming growth factor-beta2. Circ Res. (2006). , 99(8), 861-869.

[105] de Haan, G., Dontje, B., & Nijhof, W. Concepts of hemopoietic cell amplification. Synergy, redundancy and pleiotropy of cytokines affecting the regulation of erythropoiesis. Leuk Lymphoma. (1996).

[106] Agur, Z., Daniel, Y., & Ginosar, Y. The universal properties of stem cells as pinpointed by a simple discrete model. J Math Biol. (2002). , 44(1), 79-86.

[107] Veiby OP, Mikhail AA, Snodgrass HR.Growth factors and hematopoietic stem cells. HematolOncolClin North Am. (1997). , 11(6), 1173-1184.

[108] O'Brien KD.Pathogenesis of calcific aortic valve disease: a disease process comes of age (and a good deal more). ArteriosclerThrombVasc Biol. (2006). , 26(8), 1721-1728.

Chapter 12

Menopause Induces Oxidative Stress

Claudia Camelia Calzada Mendoza and
Carlos Alberto Jiménez Zamarripa

Additional information is available at the end of the chapter

http://dx.doi.org/10.5772/52082

1. Introduction

1.1. Menopause: endocrinology and symptoms

Menopause is a physiologic process in women that occurs around 45-55 years old, which is defined as permanent cessation of menstruation by one year in row [1]. The age of menopause depends on multiple factors such as number of ovules from the female at birth, the frequency of loss of these ovules through her life and the number of ovarian follicles required maintaining the menstrual cycle. The diagnosis of menopause is retrospective and is established after a year without menses [2], and their symptoms may have different intensity for each woman [3].

This process is characterized by gradual decrease of estrogen (E) secretion and changes related with sex hormones, so that estradiol levels ranging from 5 to 25 pg/mL, while increasing titers of gonadotrophins, so that the values of follicle stimulating hormone (FSH) between 40 and 250 mU/mL and luteinizing hormone (LH), from 30 to 150 mU/Ml [4, 5].

Irregular uterine bleeding is a characteristic symptom which is due to both depletion and resistance of ovarian receptors to gonadotropins and increased FSH, leading to alterations in the volume and frequency of bleeding (polymenorrhea, hypo-or menorrhagia, oligomenorrhea) [6, 7].

Among symptoms are those related to the genitourinary tract by the common embryological origin of vulva, vagina, bladder, and urethra, consequently alterations as dysuria, urinary urgency and incontinence, epithelial atrophy, decreased production of mucus and vaginal dryness (phenomena that can cause dyspareunia), urethritis, vaginitis or cystitis and local infections [8, 9, 10] (Figure 1).

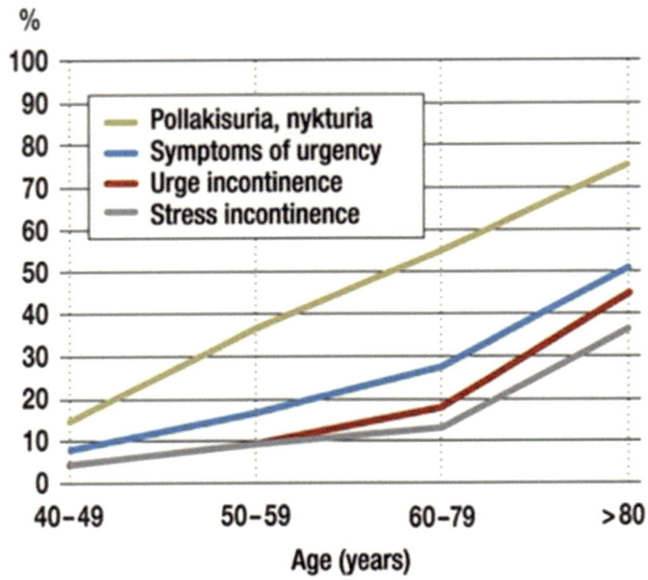

Figure 1. Main genitourinary abnormalities according to age in women [10].

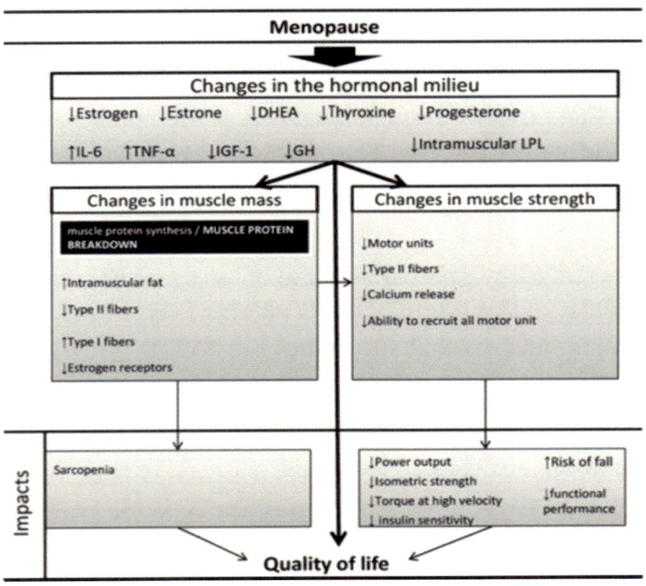

Figure 2. Changes observed in muscle mass and strength after menopause [16].

Hot flushes are one of the main symptoms associated with menopause and occur in more than 75% of menopausal, consisting of intense episodes of heat that begins on chest and spreads to face, sweating, and flushing of face. Hot flushes are associated with headache, anxiety and palpitations, and it usually lasts 2-4 minutes and can vary in frequency, in some women may be daily while others may have one episode per month [11, 12]. The mechanism of hot flushes is not clear, however, it is known that hypothalamus, pituitary gonadotropin releasing hormone and gonadotrophins may be involved in hot flushes [13]. Another frequent symptom is an oral dryness and intense burning sensation that affects mainly the tongue and sometimes lips and gums [14].

On the other hand decreases the content of collagen and elastic fibers of the skin, so that it becomes thinner and brittle losing elasticity and firmness. The epidermis thins, increases water loss and reduces the number of blood vessels, compromising the supply of oxygen and nutrients [15]. Additionally aging is associated with a natural decline in physiological functions, including a loss of muscle mass and strength. Overall, the decline in muscle mass averages 0.4 to 0.8 kg per decade, starting at the age of 20 years, especially around menopause [16] (Figure 2).

Another alteration that occurs is the osteoporosis, which is defined as a skeletal disorder characterized by decreased bone density and an increased risk of fractures [17, 18]. Before reports have confirmed that postmenopausal women have highest incidence of hip fractures [19, 20, 21] (Figure 3).

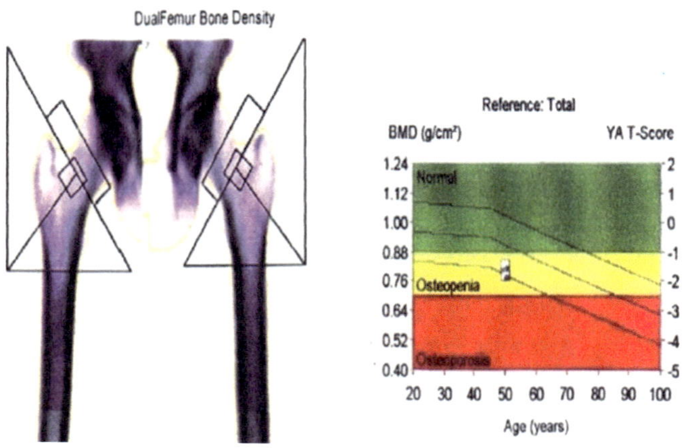

Figure 3. Bone mineral density values by age in women. Bone mineral density decreases around menopause [21].

Menopause is a stage that favors weight gain and development or worsening of obesity, and causes of this problem are many; some are clearly related to hypoestrogenism and other age-dependent, conditioning increased intake and decreased energy expenditure [22, 23] (Figure 4).

During this period there is an abnormal atherogenic lipid profile characterized by increased lipoprotein cholesterol, low density (LDL-C), triglycerides (TG) and small dense LDL particles [24] with reduced HDL-C and elevated serum glucose and insulin, perhaps as a direct result of ovarian failure or indirectly as a result of central redistribution of body fat, and this favors the formation of atheromatous plaques and progression of coronary atherosclerosis and therefore cardiovascular disease incidence increases substantially in postmenopausal women [25, 26]. Other disorders such as obesity and metabolic syndrome also occurs at menopause, suggesting that menopause may be the trigger of the metabolic syndrome at that stage of life [27, 28].

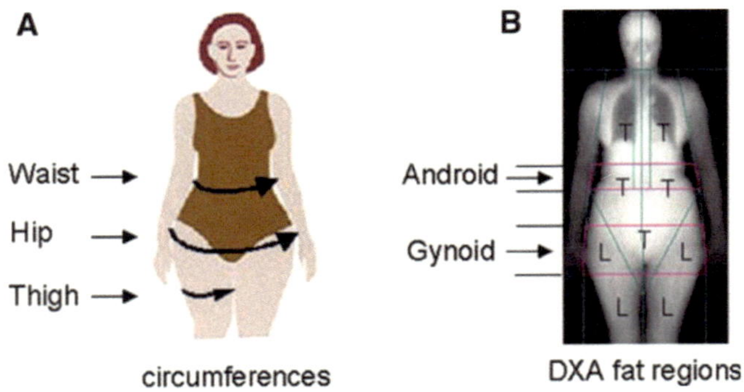

Figure 4. Body fat distribution. Android-type distribution is present in postmenopausal women. DXA= Dual-energy X-ray absorptiometry [23].

Postmenopausal women have higher insulin resistance than premenopausal, which could participate to age, the increase in total body fat, central adiposity, estrogen deficiency, alterations in lipid profile and glucose homeostasis and insulin are more frequent and favor the high cardiovascular morbidity and mortality after menopause. In this sense the transition of menopause is marked by changes in hormonal balance, with increased visceral fat, which are associated with insulin resistance, although it has been found that the change in insulin sensitivity does not alter the lipid profile in early postmenopausal women [24, 26] (Figure 5).

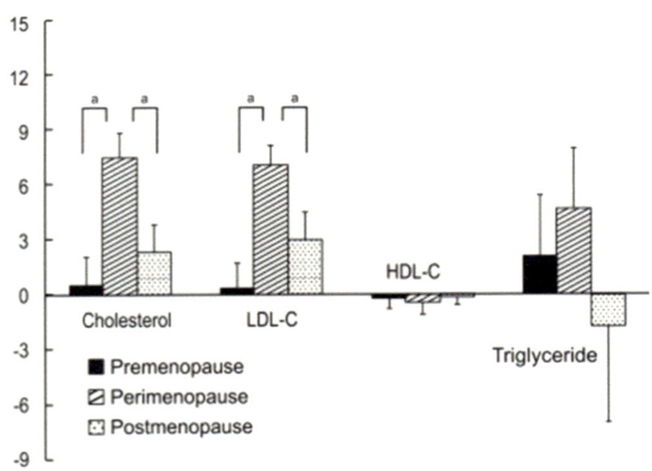

Figure 5. Changes lipid during transition from premenopause to Postmenopause [24].

Depression occurs frequently in postmenopausal women, which is explained by the loss of estrogenic effect in modulating neuronal excitability, synaptic plasticity, neuronal survival induced expression of regenerative responses, regional neurogenesis, regulation of differentiation and neuronal development [29], in the processes of cognition, modulation of mood and other mental states, as well as improving learning and memory [30], regulate the synthesis of tryptophan hydroxylase which is the limiting enzyme in serotonin synthesis so this decline in estrogen at menopause may explain the occurrence of psychological symptoms characteristic of depression (fatigue, irritability, sleep problems, abrupt changes of mood, [31]. With respect depressive symptoms in the Multiethnic Study of Atherosclerosis were analyzed testosterone, estradiol, steroid hormone binding globulin (SHBG) and dehydroepiandrosterone; indicating that in early postmenopausal women, sex hormones were associated with incident depressive symptoms [32].

2. Pro and antioxidants propierties of estrogens

Throughout menopause there are factors that predispose women to the development of oxidative stress, such as estrogen deficiency, as it has been confirmed that they have an antioxidant capacity independently of its binding to receptors, so for example the 17β-estradiol (E2), estriol, estrone, ethinylestradiol and 2-hidroxiestradiol besides reducing neuronal death with antioxidant activity, due to the presence of an intact hydroxyl group on ring A of the molecule [33].

Estrogens are synthesized from different androgen precursors such as androstenedione and testosterone, yielding as products estrone and 17β-estradiol, respectively. The synthesis is catalyzed by aromatase (ARO), the enzyme cytochrome P450 (CYP19) and estrogen synthesizing different tissue-specific manner, and the major estrogen in adipose tissue is estrone, the placenta is estriol and in cells granulosa is 17β-estradiol [34].

The 2-hidroxiestradiol and 2-hydroxyestrone (4-hidroxiestradiol type) (Figure 6) can participate in redox cycling to generate free radicals such as superoxide and chemically reactive estrogen semiquinone/quinone, which can damage DNA and other intracellular constituents.

4-hidroxiestradiol participates in a redox cycle to generate free radicals such as superoxide, and intermediate semiquinone/quinone, these intermediaries may induce cell transformation and initiate tumoral growth [35].

4 - hydroxyestrogens have estrogenic effects and can stimulate the growth of cell lines of breast cancer, with greater intensity than the 4-hydroxyestrone are unstable and can become highly reactive quinone with the formation of semiquinones as intermediary, this reaction produces oxygen free radicals, which can have toxic effects on DNA, such effects include the formation of 8-hydroxy-2-deoxiguanosine a mutagen, resulting from oxidative damage. The toxic effect of 4-hydroxyestrogens probably is prevented under normal conditions intracellular defense mechanisms. Oxygen free radicals can be removed immediately transformed into water by enzymes such as catalase and superoxide dismutase and antioxidant vitamins

such as ascorbic acid and alpha tocopherol, quinone themselves can be inactivated by sulfo compounds, such as glutathione [36].

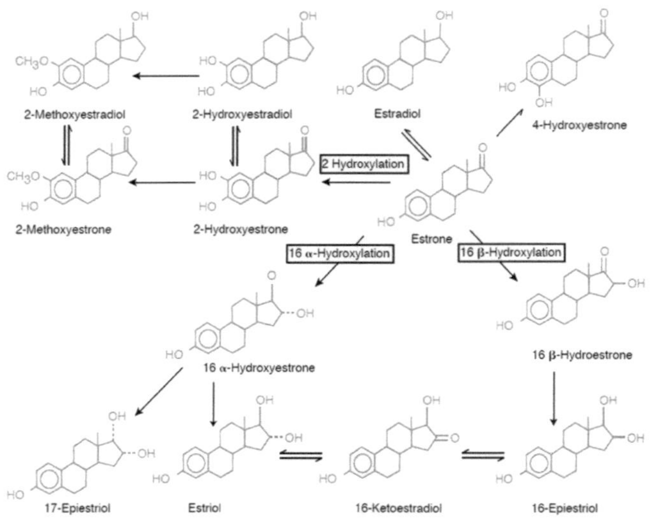

Figure 6. Estrogen metabolism.

Menopause seems to accelerate the development of atherosclerosis and cardiovascular diseases and in order to identify this correlation, was assessed the correlations between intima-media thickness, homocysteine serum levels and oxidative stress both in fertile and postmenopausal women and it was founded that were increased levels of homocysteine, oxidative stress and intima-media tickness (IMT) in postmenopausal women having a positive correlation with IMT, which reinforce the idea that a hyperhomocysteinemia may play a role in the progression of atherosclerosisas a result the lack of estrogens [37].

Vasculo protective effects of estrogen are due in part to the modulation of the balance between nitric oxide (mainly derived from endothelial vasodilator molecule) [38] and superoxide anion (oxygen-free radical highly reactive), promoting the availability of the first such so the lack of protection induces high levels of oxidative stress and low concentrations of NO, these processes are interacting with hypertension, as seen in menopause. In addition, estrogen induces the expression of oxide reductasesthiol / disulfide, such as disulfide isomerase, thioredoxin, thioredoxinreductase and glutaredoxin in the endothelium and inhibits apoptosis mediated by hydrogen peroxide. On the other hand has been described that genetic factors related to dyslipidemia are most important than due to age, for example antioxidant enzymes (SOD, catalase, GR, inflammatory markers CPR, ALT), oxidative stress (O(2)(-), LOO•), hypoxia (HBNO) and all this related to increase vascular resistance, disorders in oxygen supply in tissue and hypoxic competitions of there metabolism may cause, postmenopausal hypertension, hart ischemic disease, impaired hepatic beta-oxidation of fatty acids and hepathosteatosis [39].

17-β-estradiol plays a critical role in neuroprotection through both genomic and non-genomic mechanisms and recently was discovered that a new G-protein-coupled receptor 30 (GPR30) participates in the neuroprotection against oxidative insult, which is agonist G1. E2 attenuated apoptosis induced by H_2O_2 exposure, furthermore, G1 or E2 significantly increased the levels of phosphorylated extracellular signal-regulated kinase 1/2 (p-ERK1/2), Bcl-2 and pro-caspase-3, which is an anti-apoptotic effect [40].

3. Oxidative stress and postmenopause

Actually several oxidative stress biomarkers have been studied in menopause, however, each researcher has used different marker, methodologies and women with dissimilar characteristics (age, ethnic group, postmenopause time), fact does difficult to make a conclusion about the development of oxidative stress during peri, menopause and postmenopause.

Recently has been propose as indicator to γ-glutamyltransferase (GGT) which is an enzyme involved in the transfer of the γ-glutamyl residue from γ-glutamyl peptides to amino acids, H_2O, and other small peptides and can be donated by glutathione [41]. On the other hand, GGT is also involved in the production of glutathione [42], which is limited by cysteine availability. GGT participates in the pathway of extracellular GSH in consequence the biosynthesis of cellular glutathione, the most important cell antioxidant, depends of GGT activity; hence this enzyme may play an important role in the anti-oxidative defense system of the cell [43].

Abdul et al, founded a highly significant reduction in glutathione levels in the post-menopausal-group which could be due to the increase in its free radical scavenging property and increased consumption to counteract the oxidative stress and to inhibit membrane lipid peroxidation which indicates that the increase in serum GGT with enhanced oxidative stress and reduced antioxidant defense system in the post-menopausal women may lead to the speculation that GGT could be considered an index or a oxidative stress marker [43] (Table 1).

Serum level	Premenopausal Group (n=17)	Postmenopausal Group (n=16)	p value
GGT (U/L)	5.96±2.99	9.44±2.89	0.025
GSH (mmole/L)	0.62±0.17	0.47±0.11	0.008
MDA (μmole/L)	1.04±0.06	1.32±0.05	0.035

Table 1. Serum γ-glutamyltransferase, glutathione and malondialdehyde levels in the pre- and postmenopausal women [43].

Supplementary it was found that perimenopausal women have higher total cholesterol values and lower paraoxonase-1 (PON1) activity compared to reference values, 8-oxoG levels were unchanged compared with those of healthy control women, lipoperoxide ranks were significantly increased compared with those of premenopausal women and an indirect correlation between PON1 arylesterase (PON1 A) activity and lipoperoxide levels, between PON1 A activity and atherogenic index, between age and TAS, and between age and 8-oxoG levels. Moreover perimenopausal women had higher total cholesterol levels and PON1 A levels were lower than physiological values (table 2) [44].

Variable	Average±SD or median	Physiological values
TCH	**5.673±0.856 mmol/L**	□5.17 mmol/L
TG	1.424±0.66 mmol/L	□1.9 mmol/L
LDL	3.103±0.649 mmol/L	□3.5 mmol/L
HDL	1.563±0.445 mmol/L	□1.4 mmol/L
Atherogenic index (TCH/HDL)	3.853±1.009	□5.2
PON1 A	**89.628±14.798 U/mL**	100-200 U/mL
Pon1 L	**12.213±2.956 U/mL**	13-20 U/mL
Homocysteine	8.48±2.97 µmol/L	□12 µmol/L
Glycemia	5.43µ0.65 mmol/L	4.2-6.2 mmol/L
Uric acid	246.5 (209.9-296.9) µmol/L	□339µmol/L

Table 2. Data showing departures from normality are expressed as median values with the respective lower and upper quartile. The boldfaced entries indicate values beyond the reference range. PON1 A, paraoxonase with arylesterase activity; PON1 L, paroxonase-1 with lactonase activity.Paroxonase-1 levels in perimenonausal women [44].

Another finding is the lipoperoxide level which was significantly increased in perimenopausal women (Table 3). The levels of the marker of oxidative damage to DNA-8-oxoG were not statistically between pre and perimenopausal. In contrast women in perimenopause had repair ability 4 times higher compared with premenopausal women and significantly increased plasma total antioxidant capacity (TAS) [44] (Table 3).

Variable	Perimenopausal women	Controls (premenopausal)
TAS	1.532±0.095 mmol/L[a]	1.230±0.100 mmol/mL
Lipoperoxides	37.995 (32.035-44.849) nMol/mL[a]	28.096 (23.103-30.850) nmol/mL
8-oxoG	0.464 (0.283-0.957) per 10^6 G	0.503 (0.337-0.674) per 10^6 G
Repair ability	36.919% (30.679%-47.046%)[a]	10.539% (8.665%-11.475%)

Table 3. Data showing departures from normality are given as median values with the respective lower and upper quartile. [a]these values are significantly different (P□0.005) compared with controls. Profile oxidant and antioxidant between premenopausal and perimenopausal women [44].

In another study were determined age, body weight, and superoxide dismutase (SOD), catalase (CAT) and malondialdehyde (MDA) in disease-free women aged 25-65 years and did found that postmenopausal women had the highest oxidative stress and body weight, also superoxide dismutase, catalase and malondialdehyde were correlated significantly with body weight [45].

Pansini demonstrated that the total body fat mass increases significantly in postmenopause in comparison with premenopause, with specific increases in fat deposition at the level of trunk (abdominal and visceral) and arms. Concomitantly, the antioxidant status adjusted for age showed that antioxidant status was retained. Also both antioxidant status and hydroperoxide level increased with trunk fat mass [46].

Also has been carried out protocols that analyze the connection between menopause and periodontal conditions, though to compare serum and gingival crevicular fluid (GCF) total antioxidant capacity (TAOC) and superoxide dismutase (SOD) concentrations in postmenopausal patients with chronic periodontitis (PMCP) with those of pre-menopausal chronic periodontitis patients (CP). The results showed Serum and GCF TAOC and SOD concentrations were significantly lower in menopause and periodontitis, the lowest values were in the PMCP group, whereas the highest values were in premenopausal. While the effect of menopause was more evident in serum antioxidant analysis, the consequence of periodontitis was observed to be more apparent in GCF and a decrease in systemic and local AO defense was observed owing to both menopause and periodontitis [47].

Risk factor	**OR**	**95% CI**	**P[a]**
Menopause (hypoestrogenism)	2.62	1.35-5.11	0.005
Consumption of alcoholic beverages (≥2 glasses/d)	2.49	0.28-22.50	0.417
Smoking (≥2 cigarettes/d)	1.98	0.58-6.82	0.277
Overweight (⍰25 kg/m2)	1.43	0.64-3.18	0.383
Insomnia (AIS score ≥8)	1.13	0.58-2.22	0.715
Age, y	1.04	0.94-1.14	0.466
Physical inactivity (⍰30 min/d of physical activity)	0.85	0.43-1.68	0.632

Table 4. OR, odds ratio; AIS, Athens Insomnia Scale.[a]Logistic regression, R^2=0.106, P=0.036.Risk factors for high lipoperoxide levels, as oxidative stress biomarker, in perimenopausal women [49].

Unsaturated fatty acids have a role in the pathogenesis of atherosclerosis. They are very sensitive to oxidation caused by excess free oxygen radicals and the consequent oxidative status, and it is well known that lipid and lipoprotein metabolism is markedly altered in postmenopausal women as it was demonstrated by Signorelli who founded that the oxidative stress is involved in the pathophysiology of atherosclerosis. Malonaldehyde (MDA), 4-hydroxynenal (4-HNE), oxidized lipoproteins (ox LDL) were higher in postmenopausal while GSH-PX concentrations were significantly higher in fertile women [48].

Similar findings were found in pre and postmenopausal Mexican women by Sánchez. Lipoperoxides, erythrocyte superoxide dismutase and glutathione peroxidase activities, the total antioxidant status, pro-oxidant factors, body mass index were evaluated. The lipoperoxide levels were significantly higher in the postmenopausal group than in the premenopausal group, which concluded that menopause is the main risk factor for oxidative stress [49].

However, there are other contrasting studies, in example in a report was found than postmenopausal women had lower levels of lipid hydroperoxide oxidation, the MDA levels did not differ between pre- and postmenopausal women, no differences in advanced oxidation protein products (AOPP) and nitrite levels were observed between pre- and postmenopausal women. Postmenopausal women also exhibited a higher total radical antioxidant level [50].

Another study included pre-menopausal, peri-menopausal, and post-menopausal women classified according to the Staging of Reproductive Aging Workshop (STRAW) criteria. No significant correlations between E2 levels and OS markers were detected and consequently, estrogen decline during menopausal transition is not a determinant factor for oxidative stress [51].

4. Associated diseases to oxidative stress

There are several evidences that related to oxidative stress with diseases present in postmenopausal women in example depression, osteoporosis, cardiovascular diseases and leg vasoconstriction.

DEPRESSION

The depression is the most frequent symptom in postmenopausal women, even is a major cause of medical consultation. This disorder has cerebral implications, as showed post-mortem studies in patients with depressive disorder pointed a significant decrease of neuronal and glial cells in cortico-limbic regions which can be seen as a consequence of alterations in neuronal plasticity. This could be triggered by an increase of free radicals which in its turn eventually leads to cell death and consequently atrophy of vulnerable neuronal and glial cell population in these regions [52]. In addition elevated levels of MDA adversely affected the efficiency of visual-spatial and auditory-verbal working memories; short-term declarative memory and the delayed recall declarative memory were founded. 1. Higher concentration of plasma MDA in recurrent depressive disorder (rDD) patients is associated with the severity of depressive symptoms 2. Elevated levels of plasma MDA are related to the impairment of visual-spatial and auditory-verbal working memory and short-term and delayed declarative memory [53]. Actually too is known that estrogen protect neurons against oxidative damage excitotoxins, and beta-amyloid-induced toxicity in cell culture, reduces the serum monoamino oxidase levels and might regulate learning and memory. Nitric oxide (NO) is a messenger and in the central nervous system and acts as neurotransmitter/neuromodulator like serotonin, bradykinin, endothelin, acetylcholine and noradrenaline. Estrogen induces activity of constitutive NO synthase, reduces hyperphosphorylated of Tau and stimulates phosphorylated GSK3b [54]; due to in menopause its reduction induces a depressive disorder [55].

OSTEOPOROSIS

Oxidative stress participates in decreasing bone formation and stimulating bone resorption. Furthermore, antioxidant enzymes have been observed to have low protective activity in women with osteoporosis, also has been determined higher urine deoxypyridinoline, total Peroxide (TPx), MDA, nitric oxide, also lower TAS and glutathione reductase, compared with postmenopausal women whitout osteoporosis [56, 57]. Likewise has been studied polymorphism associated with enzymes involved in oxidative balance such as of the glutathione S-reductase (GSR), superoxide dismutase (SOD1 and SOD2), and catalase (CAT), of which polymorphisms from GSR were associated to bone mineral density [58]. Both oxidative stress and associated polymorphisms are useful tool to predict which patients might develop osteoporosis.

CARDIOVASCULAR DISEASES

Oxidative stress biomarkers have been linked with the presence and severity of the CVD, and to the presence and number of risk factors. It is known that young women during their fertile life are at lower risk of cardiovascular events compared with men, being protected by estrogen action and that oxidative stress is generally higher in men than in premenopausal women. However, after menopause the risk of experiencing cardiovascular events rapidly rises in women, in conjunction with a parallel increase in oxidative stress. Moreover, although oxidative stress results are lower in females compared to males during the first decades of life, this difference decreases until the age range which corresponds to the onset of menopause for women [59].

An analyses of relationship among excess iron, oxidative stress, and centralized fat mass in healthy postmenopausal women showed that almost 14% of the variability in oxLDL was accounted for by centralized fat mass AndGynFM ratio (waistþhip=thigh¼AndGynFM), age, and serum iron. Similarly, 16% the variability in 15-isoprostane $F2_{a\alpha}$ (PGF $F2_{a\alpha}$) was accounted for by the AndGynFM ratio, HOMA, and serum iron. Also it was accounted for 33% of the variability in AndGynFM ratio by high-density lipoprotein cholesterol (HDL-C), ferritin, HOMA, oxLDL, and PGF F2aα, all of before suggests that reducing centralized fat mass and maintaining a favorable lipid profile, antioxidant status, and iron status all may be important in protecting postmenopausal women from atherosclerotic CVD [60]. Similar findings has been observed in diabetic postmenopausal women in whom it has been reported higher levels of total cholesterol (TC), triglyceride (TG), low density lipoprotein cholesterol (LDL-C), very low density lipoprotein cholesterol (VLDL-C), catalase (CAT), and malondialdehyde (MDA) and significantly lower levels of HDL-C, reduced glutathione (GSH), glutathione reductase (GR), glutathione peroxidase (GPx), and superoxide dismutase (SOD) [61]. Fact means a cardiovascular risk.

LEG VASOCONSTRICTION

Leg vasoconstriction has been linked to oxidative stress due to the fact that Intravenous administration of a supraphysiological dose of the antioxidant ascorbic acid increased leg blood flow in the postmenopausal women as a result of an increase in leg vascular conductance, but it did not affect leg blood flow in premenopausal controls or mean arterial pressure, also changes in leg blood flow and leg vascular conductance with ascorbic acid were related to high plasma oxidized LDL an low antioxidant status [62] (Table 5, Figure 7).

Variable	Premenopausal	Postmenopausal
Oxidized LDL, U/l	36.9±4.9	55.6±3.3*
TAS, mmol/l	1.4±0.1	1.1±0.1*
ACE, U/l	26.2±3.9	30.3±2.6
Endothelin-1, pg/ml	4.8±0.5	5.3±0.3
Norepinephrine, pg/ml	149±39	343±28*
Epinephrine, pg/ml	20±3	27±2

Table 5. Values are means ±SE, ACE angiotensin-converting enzyme. *P□0.05 vs premenopausal.Serum biomarkers associated to leg vasoconstriction [62].

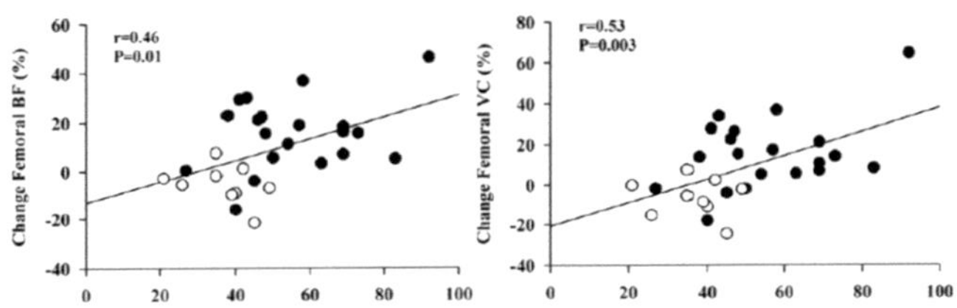

Figure 7. Relationship between plasma oxidized low-density lipoprotein (LDL) and the change in femoral artery BF (top) and vascular conductance VC (bottom) with ascorbic acid in premenopusal (□ and postmenopausal (●) women [62].

In addition, long-term studies indicate that total cholesterol (TC), LDL cholesterol (LDL-C), triglycerides (TG), MDA and common carotid artery wall intima-media thickness (IMT) are higher in women with hormonal depletion over 5 years, reveling a close temporal correlation between plasma oxidative and carotid wall IMT as postmenopause proceeds [63].

Further investigations are needed to examine the roll of oxidative stress as an endogenous bioactive agent related to disease in post-menopausal women. Since oxidative stress is the imbalance between total oxidants and antioxidants in the body, any single oxidant/ antioxidant parameter may not reflect oxidative stress. Further studies are needed to understand the underlying mechanisms of before findings.

5. Hormonal replacement therapy

Hormone replacement therapy (HRT) is defined as treatment that estrogen provides women to improve the characteristic symptoms of menopause [64], especially osteoporosis, dyslipidemias, mood among others, is also important to note that hormone replacement therapy is not without risks.

There are three Hormonal Replacement Therapies (HRT) treatment regimens:

- 1.- Estrogens. They may be natural or synthetic. Estrogens (17β-Estradiol and Estriol) and conjugated equine estrogens, and these are administered orally. Estrogens may administrated by oral, subcutaneous routes, also intravaginal estrogen (tablets, creams, ovules), alone or combined with progestin, are suitable for vaginal symptoms, with no significant increase in endometrial hyperplasia or proliferation.

- 2 - Progestogens. They are administered in combination with estrogen to reduce the risk of endometrial hyperplasia and cancer. Currently most used active ingredients TH are: oral micronized progesterone, medroxyprogesterone and norethisterone. Progestins are mainly used orally, although there are preparations to be administered in combination with estrogen transdermal route [65].

- 3 - Another group of drugs called STEAR (Selective Tissue Estrogenic Activity Regulator) is widely used because it has tissue-specific metabolism, and a main representative is Tibolone, this is a synthetic steroid with weak estrogenic, androgenic and gestagenic activities, which controls vasomotor symptoms, prevents bone demineralization and improves mood [66]. Tibolone improves vaginal symptoms and no significant differences when compared to estrogen, decreases menopausal symptoms, although moderately increases bone density and inhibit bone resorption. In the cardiovascular system there is no evidence of efficacy for the primary or secondary prevention of diseases associated with menopause at this level [67].

6. Effects of hormonal replacement therapy on oxidative stress

As mentioned above, there are different pathologies in the menopause that improve after administrating of hormone replacement therapy, fact that aroused the interest in evaluating their effects on biomarkers of oxidative stress, which has been recognized its participation in illnesses as cancer, atherogenesis, Alzheimer's and aging among others. Below are described the findings on changes in oxidative stress biomarkers after administrating HRT by periods time.

LESS THAN THREE MONTHS

In African American and Caucasian posmenopausal women the HRT reduced plasma levels of free 8-isoprostane after 6 weeks of HT, at the same time nitrite increased, principally in Caucasian women. Both ethnics groups have reduced levels of oxidative stress but the differences were not statistically significant [68]. Even the combined therapy for 3 months had an antioxidant effect in posmenopausal hemodialysis women, who showed reduced levels of MDA although TAC, uric acid and C- reactive protein were not changed [69].

Oxidized low-density lipoprotein (oxLDL)/β2-glycoprotein I (β2GPI) complexes are etiologically important in the development of atherosclerosis. Combined HRT led to a significant increase in TAC and a minor but statistically nonsignificant decrease of oxLDL/β2GPI complexes when compared with the baseline control levels. There was also no significant association between TAC and oxLDL/β2GPI complexes changes related to HRT. This study indicates that, HRT in postmenopausal women leads to an increase in TAC without an equivalent change in serum levels of oxLDL/β2GPI complexes. It is concluded that beneficial effects of HRT could be explained, at least in part, by improving antioxidant status, but may not be directly associated with a change in oxidized lipoprotein production [70] (Figure 8).

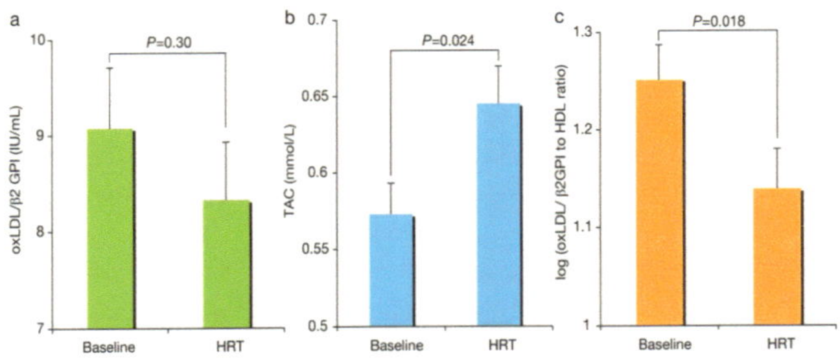

Figure 8. Effects of hormone replacement therapy (HRT) on oxLDL/β2GPI (a), total antioxidant capacity (TAC) (b), oxLDL/β2GPI to HDL-C ratio (c) in the studied subjects. Values are means ± SE. $P < 0.05$ (paired sample t-test). total serum antioxidant capacity (TAC) as ferric reducing ability of plasma (FRAP) and related these to HRT and oxLDL/β2GPI complexes level [70].

SIX MONTHS

The main studies about effect of HRT has been carried out with combination estrogen plus gestagen, and in them has been founded in example that carbonils groups determined by ELISA showed a reduction of serum levels after six months oral or transdermal treat when compared with control group, and there was not difference between oral and transdermal, which indicates that hormonal therapy reduces the of carbonyl protein, a marker of oxidative stress, suggesting potential protective effect [71]. Similar result were founded with the serum level of malondialdehyde, superoxide dismutase and sulfhydryl groups without changes on plasma total homocysteine (tHcy) (used as atherogenic indicator) [72]. Another study with equal number of months of follow-up showed that carbonyls, MDA and oxLDL were reduced, while erythrocyte glutathione (GSH) were increased, and nitrotyrosine (NT) levels were not changed [73, 74] (Table 6).

On the other hand Tibolone treatment leads to a decrease in concentrations of plasma lipid peroxide, increase plasma concentrations of vitamin E and alpha-tocopherol and significant decrease in lipid peroxide concentrations [75, 76].

ONE YEAR

However, the combined treatment by one year significantly reduced the levels of catecholamines, mean blood pressure and LDL cholesterol while it increased levels of nitrite/nitrate, indicating cardiovascular benefit in healthy recent postmenopausal women. Levels of 8-epi PGF2alpha did not change, suggesting no evident relationship between HRT and oxidative stress [77]. Although another study reports that conjugated estrogens alone (EHRT) or conjugated estrogen with medroxyprogesterone acetate can reduce lipoprotein lipase (LPL), hepatic lipase (HL), oxidized apolipoprotein B in LDL [78] also platelet MDA, glutathione-S-transferase (GST) and SOD levels were lower and total thiol (t-SH) content was higher than pre-treatment levels. These results indicate that hormone replacement therapy may affect platelet membrane fatty acid content and oxidant-antioxidant balance in postmenopausal women [79].

	MPA n=25	NETA n=20	Total n=45
Total-C Pre/postreatment (mg/100ml)	204.1±30.0/ 178.6±10.7*	225.7±30.1/ 191.6±21.1*	213.7±31.7/ 184.4±17.2*
HDL-C Pre/postreatment (mg/100ml)	45.7±10.6/ 53.2±7.6*	49.1±11.0/ 56.6±8.2	47.2±10.8/ 54.8±8.0*
LDL-C Pre/postreatment (mg/100ml)	131.7±24.9/ 102.8±13.6*	147.5±25.2/ 112.7±21.2*	138.71±26.08/ 107.2±17.9*
Triglycerides Pre/postreatment (mg/100ml)	133.5±62.0/ 115.2±35.2	138.4±74.4/ 113.5±37.0	135.67±67.1/ 114.4±35.6
MDA Pre/postreatment (mg/100ml)	4.7±0.4/ 4.0±0.4*	4.9±0.4/ 3.6±0.3*	4.82±0.4/ 3.8±0.4*
OxLDL Pre/postreatment (mg/100ml)	54.9±8.5/ 47.9±4.0*	53.1±8.3/ 44.8±4.4*	54.13±8.34/ 46.6±4.4*
PON 1 Pre/postreatment (mg/100ml)	51.5±6.7/ 73.9±11.2*	51.4±9.0/ 67.4±10,0*	51.47±7.7/ 71.0±11.1*

Table 6. Comparinson of parameters before and after HRT in postmenopausal women p[illegible]0.005Serum lipid parameters, MDA, oxLDL and PON1 levels in postmenopausal women before and after HRT [74].

Similarly the effect of DNA damage by oxidative stress has been evaluated. The 8-hydroxy-deoxyguanosine (8-OHdG) is widely used for determination of DNA damage since it is excised from oxidative damaged DNA with endonuclease repair enzymes coded (OGG1). After HT, mean blood 8-OHdG (DNA damage marker) level significantly decreased compared to those before HT, while urinary 8-OHdG level did not show any difference, this without relation with S326C polymorphism [80].

Tibolone acts as an antioxidant upon increase the concentration of reduced sulfhydryl [81], however, the exact mechanism has not been elucidated, but it could participate in a direct mechanism, ie through the structure tibolone and its metabolites as it is similar to the structure of 17β-estradiol, considered as an antioxidant for its phenol ring, which can act neutralizing to free radical [82, 83]. Moreover tibolone reduces the concentration of malondialdehyde compared to those who had no treatment [84, 85].

Although there are reports that indicate the antioxidant effect of HRT, there are also studies that indicate otherwise; in example in another study combined HRT led to decreased plasma total and LDL cholesterol, but did not affect oxidizability and oxidation of LDL. Circulating levels of antioxidant vitamins (beta-carotene, vitamin C, vitamin E/triglycerides) and total antioxidant capacity of plasma and lipid peroxidation, assessed by plasma TBARs, were not different from controls in postmenopausal women receiving HRT, which indicates that combined HRT modifies the blood lipid profile, however it does not appear to influence oxidative status [86]. Additionaly DNA damage, GPx activity and nitrite level as well as a decreased GSH level were observed after oral administrating of estrogens alone or combinated [87].

With respect to hot flushes, they have been associated to smaller level of total antioxidant activity in plasma, without differences in nitrite-nitrate concentrations, and after HRT there is an increase in total antioxidant activity level and nitrite-nitrate concentrations in menopausal women, with and without hot flushes [88].

On the other side estrogen increases vasodilatation and inhibits the response of blood vessels to injury and the development of atherosclerosis, it has been related to hormone's effect on serum lipid concentration, that is reducing MDA and oxLDL levels and increasing activity of paroxonase PON1, which a calcium-dependent enzyme and in serum is exclusively located on HDL. PON is synthesized and secreted by liver and tightly binds to HDL subfractions that also contain apoA-1 and apoJ or clusterin and it has the capacity to protect LDL against oxidation [89, 90, 91].

7. Effect of nutrition and exercise on oxidative stress biomarkers

Adequate nutrition and physical exercise are two factors of health promotion and its effect on oxidative stress has been investigated in postmenopausal women, which has given controversial data. With respect to foods, they contain large amount of antioxidant molecules from there arouse the interest to check if their use can reduce the oxidative stress observed in postmenopause.

For example it was reported that the intake of fresh, greenhouse-grown vegetables for 3-wk did not induced changes in the urine concentrations of 8-isoprostane F2α, hexanoyl lysine, and serum high sensitivity C-reactive protein despite that plasma carotenoids were elevated in overweight postmenopausal women [92]. Something similar was established with a 2-month supplementation period with the Klamath algae extract, which is an extract naturally rich in powerful algal antioxidant molecules (AFA-phycocyanins) and concentrated with Klamath algae's natural neuromodulators (phenylethylamine as well as natural selective MAO-B inhibitors), whose effect was to increase in the plasma levels of carotenoids, tocopherols and retinol, however in this study oxidative stress was not measured [93, 94].

Otherwise is soy milk consumption for four weeks, which did not reduced markers of inflammation and oxidative stress as (tumor necrosis factor alpha [TNF-alpha], interleukin

[IL]-1beta, IL-6) and oxidative stress (superoxide dismutase [SOD], glutathione peroxidase [GPx], cyclooxygenase-2 [COX-2]) [95]. In this sense has also been found that de-alcoholised wine (DAW) with different polyphenol content by one month does not exert a protective activity towards oxidative DNA damage by comet assay, nor modifies significantly the gene expression profile of peripheral lymphocytes, whereas it shows blood-fluidifying actions, expressed as a significant decrease in blood viscosity. However, this effect does not correlate with the dosage of polyphenols from (DAW) [96].

In contrast, the intake for 8 weeks of soya protein diet and soya nut diet decreased MDA and increased the total antioxidant capacity in postmenopausal women with metabolic syndrome [97]. Extra-virgin olive oils (EVOO) is another example of food which reduces DNA damage by oxidative stress when is administrated for 8 weeks in healthy postmenopausal, which was evaluated by the comet assay in peripheral blood lymphocytes [98].

Nevertheless there are also reports of foods that raise oxidative stress as with American ginseng (AG) and wine. The AG supplementation at 500 mg every day for 4 months causes oxidative stress in postmenopausal women, due to reduced total antioxidant capacity, elevated plasma malondialdehyde and urine 8-hydroxydeoxyguanosine concentrations and increased erythrocyte antioxidant enzyme activity such as erythrocyte superoxide dismutase and GSH reductase [99].

Previous studies do not allow a conclusion on the effect of foods rich in antioxidant compounds, because were used different markers and administration time, and still more the age range of the postmenopausal differs considerably. This shows the need for more studies. Before shows the importance of further research with other foods with antioxidant properties known such as:

Vitamin C - Citrus fruits and their juices, berries, dark green vegetables (spinach, asparagus, green peppers, brussel sprouts, broccoli, watercress, other greens), red and yellow peppers, tomatoes and tomato juice, pineapple, cantaloupe, mangos, papaya and guava.

Vitamin E - Vegetable oils such as olive, soybean, corn, cottonseed and safflower, nuts and nut butters, seeds, whole grains, wheat, wheat germ, brown rice, oatmeal, soybeans, sweet potatoes, legumes (beans, lentils, split peas) and dark leafy green vegetables. Selenium - Brazil nuts, brewer's yeast, oatmeal, brown rice, chicken, eggs, dairy products, garlic, molasses, onions, salmon, seafood, tuna, wheat germ, whole grains and most vegetables.

Beta Carotene - Variety of dark orange, red, yellow and green vegetables and fruits such as broccoli, kale, spinach, sweet potatoes, carrots, red and yellow peppers, apricots, cantaloupe and mangos [100].

With respect to the lycopene, the following mechanism of the role of lycopene in chronic diseases has been mentioned by Agarwal and Rao [101] and Waliszewski and Blasco [102].

This highlights the importance of promote healthy lifestyles (balanced diet and moderate intensity exercise) in vulnerable populations, such as menopausal women, in order to prevent aging induced oxidative stress-related diseases.

Whit respect to the exercise has been reported that who practice yoga or tai chi (TC) for a year have same BMI, and even more no effects were shown on erythrocyte superoxide dismutase activity, plasma lipid peroxidation (TBARS) or total homocysteine concentrations, but the activity of erythrocyte glutathione peroxidase - an aerobic training-responsive enzyme - was higher in TC practitioners [103]. Although short term aerobic physical activity program (8 or 15 weeks) exhibited similar results, namely a reduction of serum glucose, LDL-cholesterol (LDL-C), plasma TBARS concentrations, decreasing of HOMA(IR) as well as an increased of total antioxidant status (TAS) of plasma and reduced glutathione (GSH) concentrations in red blood cells (RBC) increased significantly [104, 105].

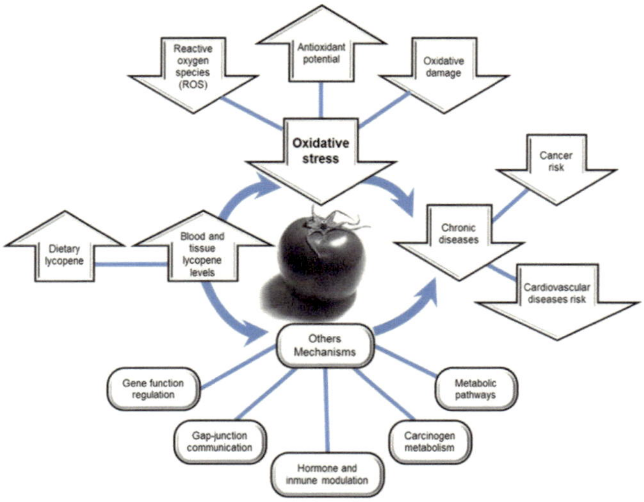

Figure 9. Lycopene and its mechanism in preventing of chronic diseases (Adapted from 101 and 102).

Regular physical training has been shown to upregulate antioxidant enzymatic systems, which may slow down the usual increase of oxidative stress in postmenopausal women, since it has been identified significant negative associations between oxidative stress and indices of physical fitness-activity (malondialdehyde, 8-iso-prostaglandin F2alpha, 8-hydroxy-2'-deoxyguanosine). Conversely, glutathione peroxidase is positively correlated with fitness level, furthermore mean arterial blood pressure (MABP) and cerebrovascular conductance (CVC) are directly associated with 8-hydroxy-2'-deoxyguanosine, nitrotyrosine and nitric oxide (NO) These findings demonstrate that, after menopause, fitness level and regular physical activity mediate against oxidative stress by maintaining antioxidant enzyme efficiency. Furthermore, these results suggest that oxidative stress and NO production modulate MABP and CVC [106].

Contrary to the above has also been reported, that the exercise does not modify the antioxidant status (although this is lower in metabolic healthy obese postmenopausal women than non-metabolic healthy obese postmenopausal women) and worse increases serum levels of thiobarbituric acid-reactive substances [107].

This highlights the importance of promote healthy lifestyles (balanced diet and moderate intensity exercise) in vulnerable populations, such as menopausal women, in order to prevent aging induced oxidative stress-related diseases.

8. Conclusion

The studies presented here were performed with different number of patients, methodologies and biomarkers, but most of them indicate that estrogen depletion induces oxidative stress and hormone replacement therapy seems to reduce it. With respect to the modification of biomarkers of oxidative stress damage by food and exercise needs more research because so far no conclusive data have been obtained.

Author details

Claudia Camelia Calzada Mendoza[1*] and Carlos Alberto Jiménez Zamarripa[2]

*Address all correspondence to: cccalzadam@yahoo.com.mx

1 Section of Post graduate Studies and Research of Escuela Superior de Medicina- Instituto Politécnico Nacional. Street Salvador Díaz Mirón S/N, Colony Casco de Santo Tomás, Delegation Miguel Hidalgo, C.P. 11340, México D.F.

2 Hospital psychiatry "Dr. Samuel Ramírez Moreno"-psychiatric careservices- Secretaria de Salud, highway México-Puebla Km 5.5, Colony Santa Catarina, Tláhuac, C.P. 13100, México D.F.

References

[1] Reddish, S. (2011). Menopausal transition- assessment in general practice. *Australian Family Physician*, 40(5), 266-72.

[2] Burger, H. G., Hale, G. E., Robertson, D. M., & Dennerstein, L. (2007). Review of hormonal changes during the menopausal transition: focus on findings from the Melbourne Women's Midlife Health Project. *Human Reproduction Update*, 13(6), 559-65.

[3] de Cetina , Canto T. (2006). Los síntomas en la menopausia. *Revista de Endocrinología y Nutrición*, 14(3), 141-148.

[4] Guthrie, J. R., Dennerstein, L., Hopper, J. L., & Burger, H. G. (1996). Hot flushes, menstrual status and hormone levels in a population-based sample of midlife women. *Obstetrics and Gynecology*, 88, 437-441.

[5] Zárate, A., Hernández-Valencia, M., Austria, E., Saucedo, R., & Hernánde, M. (2011). Diagnosis of premature menopause measuring circulating anti-Müllerian hormone. *Ginecologia Obstetricia Mexicana*, 79(5), 303-7.

[6] Lockwood, C.J. (2011). Mechanisms of normal and abnormal endometrial bleeding. *Menopause*, 18(4), 408-411.

[7] Goldstein, S.R. (2011). Significance of incidentally thick endometrial echo on transvaginal ultrasound in postmenopausal women. *Menopause*, 18(4), 434-436.

[8] Chapple, C. R., Wein, A. J., Abrams, P., Dmochowski, R. R., Giuliano, F., Kaplan, S. A., Mc Vary, K. T., & Roehrborn, C. G. (2008). Lower urinary tract symptoms revisited: a broader clinical perspective. *European Urology*, 54(3), 563-569.

[9] Labrie, F., Archer, D., Bouchard, C., Fortier, M., Cusan, L., Gomez, J. L., Girard, G., Baron, M., Ayotte, N., Moreau, M., Dubé, R., Côté, I., Labrie, C., Lavoie, L., Berger, L., Gilbert, L., Martel, C., & Balser, J. (2009). Intravaginaldehydroepiandrosterone (Prasterone), a physiological and highly efficient treatment of vaginal atrophy. *Menopause*, 16(5), 907-22.

[10] Goepel, M., Kirschner-Hermanns, R., Welz-Barth, A., Steinwachs, K. C., & Rübben, H. (2010). Urinary incontinence in the elderly: part 3 of a series of articles on incontinence. *Dtsch ArzteblInt*, 107(30), 531-6.

[11] Morrison, L. A., Sievert, L. L., Brown, D. E., Rahberg, N., & Reza, A. (2010). Relationships between menstrual and menopausal attitudes and associated demographic and health characteristics: the Hilo Women's Health Study Women Health,, 50(5), 397-413.

[12] Pachman, D. R., Jones, J. M., & Loprinzi, C. L. (2010). Management of menopause-associated vasomotor symptoms: Current treatment options, challenges and future directions. *International Journal Womens Health*, 9(2), 231-35.

[13] Archer, D. F., Sturdee, D. W., Baber, R., de Villiers, T. J., Pines, A., Freedman, R. R., Gompel, A., Hickey, M., Hunter, M. S., Lobo, R. A., Lumsden, M. A., Mac, Lennan. A. H., Maki, P., Palacios, S., Shah, D., Villaseca, P., & Warren, M. (2011). Menopausal hot flushes and night sweats: where are we now ? *Climacteric*, 14(5), 155-128.

[14] Frutos, R., Rodríguez, S., Miralles-Jorda, L., & Machuca, G. (2002). Oral manifestations and dental treatment in menopause. *Medicine Oral*, 7(1), 26-30.

[15] Safoury, O., Rashid, L., & Ibrahim, M. A. (2010). Study of androgen and estrogen receptors alpha, beta in skin tags. *Indian Journal Dermatology*, 55(1), 20-4.

[16] Maltais, M.L., Desroches, J., & Dionne, I.J. (2009). Changes in muscle mass and strength after menopause. *Journal Musculo skeletand Neuronal Interactions*, 9(4), 186-197.

[17] Sosa, M., Saavedra, P., Jódar, E., Lozano, T. C., Quesada, J. M., Torrijos, A., Pérez, C. R., Nogués, X., Díaz, C. M., Moro, M. J., Gómez, C., Mosquera, J., Alegre, J., Olmos, J.,

Muñoz, T. M., Guañabens, N., Del Pino, J., & Hawkins, F. (2009). Bone mineral density and risk of fractures in aging, obese post-menopausal women with type 2 diabetes. The GIUMO Study. *Aging Clinical Experimental Research*, 21(1), 27-32.

[18] Koroglu, B. K., Kiris, F., Ersoy, I. H., Sutcu, R., Yildiz, M., Aksu, O., Ermis, F., Ersoy, S., & Tamer, M. N. (2011). Relation of leptin, adiponectin and insulin resistance to bone mineral density in type 2 diabetic postmenopausal women. *Endokrynologia Polska*, 62(5), 429-35.

[19] Nelson, H. D., Helfand, M., Woolf, S. H., & Allan, J. D. (2002). Screening for Postmenopausal Osteoporosis: A Review of the Evidence for the U.S. Preventive Services Task Force. *Annals of Internal Medicine*, 137(6), 529-41.

[20] del Rio, B. L., Romera, B. M., Pavia, S. J., Setoain, Q. J., Serra, M. L., Garces, R. P., & Lafuente, N. C. (1992). Bone mineral density in two different socio-economic population groups. *Bone and Mineral*, 18(2), 159-68.

[21] Suresh, S., Kumar, T. S., Saraswathy, P. K., & Pani, S. K. H. (2010). Periodontitis and bone mineral density among pre and post menopausal women: A comparative study. *Journal of Indian Society of Periodontology*, 14(1), 30-34.

[22] Pavón, P., Alameda, H. C., & Olivar, R. J. (2006). Obesidad y menopausia. *Nutrición Hospitalaria*, 21, 633-637.

[23] Zillikens, M. C., Uitterlinden, A. G., van Leeuwen, J. P., Berends, A. L., Henneman, P., van Dijk, K. W., Oostra, B. A., van Duijn, C. M., Pols, H. A., & Rivadeneira, F. (2010). The role of body mass index, insulin, and adiponectin in the relation between fat distribution and bone mineral density. *Calcified Tissue International*, 86(2), 116-25.

[24] Cho, E. J., Min, Y. J., Oh, M. S., Kwon, J. E., Kim, J. E., Lee, W. S., Lee, K. J., Kim, S. W., Kim, T. H., Kim, M. A., Kim, C. J., & Ryu, W. S. (2011). Effects of the transition from premenopause to postmenopause on lipids and lipoproteins: quantification and related parameters. *Korean Journal Internal Medicine*, 26(1), 47-53.

[25] Kallikazaros, I., Tsioufis, C., Zambaras, P., Skiadas, I., Toutouza, M., Tousoulis, D., Stefanadis, C., & Toutouzas, P. (2008). Estrogen-induced improvement in coronary flow responses during atrial pacing in relation to endothelin-1 levels in postmenopausal women without coronary disease. *Vascular Health Risk and Management*, 4(3), 705-14.

[26] Zavala, A. G., & Grover, F. (2007). Perfil lipídico y cambio en sensibilidad a la insulina en posmenopáusicas. *Revista Médica deChile* 135:, 613-619.

[27] Garay, S., & Arellano, S. (2006). Diabetes mellitus (DM), menopausia y reemplazo hormonal. *Revista de Endocrinología y Nutrición*, 14(3), 191-195.

[28] Whitcroft, S., & Herriot, A. (2011). Insulin resistance and management of the menopause: a clinical hypothesis in practice. *Menopause International*, 17(1), 24-8.

[29] Ryan, J., Scali, J., Carrière, I., Peres, K., Rouaud, O., Scarabin, P. Y., Ritchie, K., & Ancelin, M. L. (2011). Estrogen receptor alpha gene variants and major depressive episodes. *Jornal of Affective Disorders*, 136(3), 1222-6.

[30] Parry, B.L. (2010). Optimal management of perimenopausal depression. *International Joornal of Women´s Health*, 9(2), 143-151.

[31] Zender, R., & Olshansky, E. (2009). Women's mental health: depression and anxiety. *The Nursing Clinics of North America*, 44(3), 250-256.

[32] Colangelo, L. A., Craft, L. L., Ouyang, P., Liu, K., Schreiner, P. J., Michos, E. D., & Gapstur, S. M. (2012). Association of sex hormones and sex hormone-binding globulin with depressive symptoms in postmenopausal women: the Multiethnic Study of Atherosclerosis. *Menopause.* Mar 12.PMID: 22415566 [PubMed- as supplied by publisher] PMCID: PMC3376685.

[33] Escalante, G. C., Quesada, M. S., & Zeledón, S. F. (2009). Oxidative Profile of the Menopausal Woman: Estrogens´ Rol in the Prevention and Treatment of Diseases. *Acta Médica Costarricense*, 51(4), 206-2011.

[34] Ghosh, D., Griswold, J., Erman, M., & Pangborn, W. (2009). Structural basis for androgen specificity and oestrogen synthesis in human aromatase. *Nature*, 457(7226), 219-223.

[35] Liehr J.G. (2000). Is Estradiol a Genotoxic Mutagenic Carcinogen?. *Endocrine Reviews*, 21(1), 40-54.

[36] Lippert, T. H., Seeger, H., & Mueck, A. O. (2000). The impact of endogenous estradiol metabolites on carcinogenesis. *Steroids*, 65, 357-369.

[37] Pulvirenti, D., Signorelli, S., Sciacchitano, S., Di Pino, L., Tsami, A., Ignaccolo, L., & Neri, S. (2007). Hyperhomocysteinemia, oxidative stress, endothelial dysfunction in postmenopausal women. *La Clinica Terapeutica*, 158(3), 213-7.

[38] Nascimento, G. R., Barros, Y. V., Wells, A. K., & Khalil, R. A. (2009). Research into Specific Modulators of Vascular Sex Hormone Receptors in the Management of Postmenopausal Cardiovascular Disease. *Current Hypertension Reviews*, 5(4), 283-306.

[39] Ratiani, L., Parkosadze, G., Cheishvili, M., Ormotsadze, G., Sulakvelidze, M., & Sanikidze, T. (2012). Role of estrogens in pathogenesis of age-related disease in women of menopausal age. *Georgian Medical News*, 203, 11-6.

[40] Liu, S. B., Han, J., Zhang, N., Tian, Z., Li, X. B., & Zhao, M. G. (2011). Neuroprotective effects of oestrogen against oxidative toxicity through activation of G-protein-coupled receptor 30 receptor. *Clinical and Experimental Pharmacology and Physiology*, 38(9), 577-85.

[41] Johnson, D. K., Mc Millin, G. A., Bishop, M. L., Fody, E. P., & Schoeff, L. E. (2010). Enzymes. *In: Clinical Chemistry Techniques, Principles, Correlations* 6th ed. Philadelphia: Lippincott Williams and Wilkins., 300.

[42] Vasudevan, D. M., & Sreekumari, S. (2005). Iso-enzymes and clinical enzymology. *In: Vasudevan DM, Sreekumari S, editors. Textbook of Biochemistry (for medical students).* 4th ed. New Delhi: Jaypeebrothers medical publishers (P) Ltd., 57.

[43] Abdul, R. O. F., Al, S. G. A., & Bushra, H. Z. B. H. (2010). Serum γ-glutamyltransferase as Oxidative Stress Marker in Pre-and Postmenopausal Iraqi Women. *Oman Medical Journal*, 25(4), 286-8.

[44] Zitňanová, I., Rakovan, M., Paduchová, Z., Dvořáková, M., Andrezálová, L., Muchová, J., Simko, M., Waczulíková, I., & Duračková, Z. (2011). Menopause Oxidative stress in women with perimenopausal symptoms. *Menopause*, 18(11), 1249-55.

[45] Mittal, P. C., & Kant, R. (2009). Correlation of increased oxidative stress to body weight in disease-free post menopausal women. *Clinical Biochemistry* 42(10-11) 1007-11.

[46] Pansini, F., Cervellati, C., Guariento, A., Stacchini, M. A., Castaldini, C., Bernardi, A., Pascale, G., Bonaccorsi, G., Patella, A., Bagni, B., Mollica, G., & Bergamini, C. M. (2008). Oxidative stress, body fat composition, and endocrine status in pre- and postmenopausal women. *Menopause*, 15(1), 112-8.

[47] Baltacioğlu, E., Akalin, F. A., Alver, A., Balaban, F., Unsal, M., & Karabulut, E. (2006). Total antioxidant capacity and superoxide dismutase activity levels in serum and gingival crevicular fluid in post-menopausal women with chronic periodontitis. *Journal of Clininal Periodontology*, 33(6), 385-92.

[48] Signorelli, S. S., Neri, S., Sciacchitano, S., Pino, L. D., Costa, M. P., Marchese, G., Celotta, G., Cassibba, N., Pennisi, G., & Caschetto, S. (2006). Behaviour of some indicators of oxidative stress in postmenopausal and fertile women. *Maturitas*, 53(1), 77-82.

[49] Sánchez, R.M.A., Zacarías, F.M., Arronte, R.A., Correa, M.E., & Mendoza, Núñez V.M. (2012). Menopause as risk factor for oxidative stress. *Menopause*, 19(3), 361-7.

[50] Victorino, V. J., Panis, C., Campos, F. C., Cayres, R. C., Colado-Simão, A. N., Oliveira, S. R., Herrera, A. C., Cecchini, A. L., & Cecchini, R. (2012). Decreased oxidant profile and increased antioxidant capacity in naturally postmenopausal women. *Studio controversial.* Age (Dordrecht Netherlands) May 28. PMID: 22645022.

[51] Cervellati, C., Pansini, F. S., Bonaccorsi, G., Bergamini, C. M., Patella, A., Casali, F., Fantini, G. F., Pascale, G., Castaldini, C., Ferrazzini, S., Ridolfi, F., Cervellati, G., Cremonini, E., Christodoulou, P., & Bagni, B. (2011). Estradiol levels and oxidative balance in a population of pre-, peri-, and post-menopausal women. *Gynecological Endocrinology*, 27(12), 1028-32.

[52] Michel, T. M., Pülschen, D., & Thome, J. (2012). The role of oxidative stress in depressive disorder. *Current Pharmaceutical Design* Jun 6 PMID: 22681168.

[53] Talarowska, M., Gałecki, P., Maes, M., Bobińska, K., & Kowalczyk, E. (2012). Total antioxidant status correlates with cognitive impairment in patients with recurrent depressive disorder. *Neurochemical Research*, 37(8), 1761-7.

[54] Pinto, A. R., Calzada, M. C. C., Campos, L. M. G., & Guerra, A. C. (2012). Effect of Chronic Administration of Estradiol, Progesterone, and Tibolone on the Expression and Phosphorylation of Glycogen Synthase Kinase-3b and the Microtubule-Associated Protein Tau in the Hippocampus and Cerebellum of Female Rat. *Journal of Neuroscience Research*, 90, 878-886.

[55] Wagner, J. A., Tennen, H., Finan, P. H., White, W. B., Burg, M. M., & Ghuman, N. (2011). Lifetime History of Depression, Type 2 Diabetes, and Endothelial Reactivity to Acute Stress in Postmenopausal Women. *International Jornal of Behavioral Medicine* Oct 2. PMID: 21964983.

[56] Yilmaz, N., & Eren, E. (2009). Homocysteine oxidative stress and relation to bone mineral density in post-menopausal osteoporosis. *Aging Clinical and Experimental Research* 21(4-5)353-7.

[57] Sendur, O. F., Turan, Y., Tastaban, E., & Serter, M. (2009). Antioxidant status in patients with osteoporosis: a controlled study. *Joint Bone Spine*, 76(5), 514-8.

[58] Mlakar, S. J., Osredkar, J., Prezelj, J., & Marc, J. (2012). Antioxidant enzymes GSR, SOD1, SOD2, and CAT gene variants and bone mineral density values in postmenopausal women: a genetic association analysis. *Menopause*, 19(3), 368-76.

[59] Vassalle, C., Mercuri, A., & Maffei, S. (2009). Oxidative status and cardiovascular risk in women: Keeping pink at heart. *World Journal of Cardiology*, 1(1), 26-30.

[60] Crist, B. L., Alekel, D. L., Ritland, L. M., Hanson, L. N., Genschel, U., & Reddy, M. B. (2009). Association of oxidative stress, iron, and centralized fat mass in healthy postmenopausal women. *Journal Women´s Health (Larchmt)*, 18(6), 795-801.

[61] Kumawat, M., Sharma, T. K., Singh, N., Ghalaut, V. S., Vardey, S. K., Sinha, M., & Kaushik, G. G. (2012). Study of changes in antioxidant enzymes status in diabetic post menopausal group of women suffering from cardiovascular complications. *Clinical Laboratory*58(3-4) 203-7.

[62] Moreau, K. L., De Paulis, A. R., Gavin, K. M., & Seals, D. R. (2007). Oxidative stress contributes to chronic leg vasoconstriction in estrogen-deficient postmenopausal women. *Journal of Applied Physiology*, 102, 890-895.

[63] Signorelli, S. S., Neri, S., Sciacchitano, S., Di Pino, L., Costa, M. P., Pennisi, G., Ierna, D., & Caschetto, S. (2001). Duration of menopause and behavior of malondialdehyde, lipids, lipoproteins and carotid wall artery intima-media thickness. *Maturitas*, 39(1), 39-42.

[64] Smith, C. C., Vedder, L. C., Nelson, A. R., Bredemann, T. M., & Mc Mahon, L. L. (2010). Duration of estrogen deprivation, not chronological age, prevents estrogen's ability to enhance hippocampal synaptic physiology. *Proceedings of National Academy of Science of United States of America*, 107(45), 19543-19548.

[65] Griffiths, F., & Convery, B. (1995). Women's use of hormone replacement therapy for relief of menopausal symptoms, for prevention of osteoporosis, and after hysterectomy. *British Journal of General Practice*, 45(396), 355-358.

[66] Modelska, K., & Cummings, S. (2002). Tibolone for postmenopausal women: systematic review of randomized trials. *The Journal of Clinical Endocrinology and Metabolism*, 87(1), 16-23.

[67] Huang, K. E., & Baber, R. (2010). Updated clinical recommendations for the use of tibolone in Asian women Climateric,13:, 317-327.

[68] Ke, R. W., Todd, Pace. D., & Ahokas, R. A. (2003). Effect of short-term hormone therapy on oxidative stress and endothelial function in African American and Caucasian postmenopausal women. *Fertiland Steril*, 79(5), 1118-22.

[69] Chang, S. P., Yang, W. S., Lee, S. K., Min, W. K., Park, J. S., & Kim, S. B. (2003). Effects of hormonal replacement therapy on oxidative stress and total antioxidant capacity in postmenopausal hemodialysis patients. *Renal Failure 2002*, 24(1), 49-57.

[70] Darabi, M., Ani, M., Movahedian, A., Zarean, E., Panjehpour, M., & Rabbani, M. (2010). Effect of hormone replacement therapy on total serum anti-oxidant potential and oxidized LDL/ß2-glycoprotein I complexes in postmenopausal women. *Endocrine Journal*, 57(12), 1029-1034.

[71] Polac, I., Borowiecka, M., Wilamowska, A., & Nowak, P. (2011). Oxidative stress measured by carbonyl groups level in postmenopausal women after oral and transdermal hormone therapy. *Journal ofObstetric andGynaecology Research 2012 Apr 30.* doi:j. , 1447-0756.

[72] Gökkuşu, C., Özbek, Z., & Tata, G. (2012). Hormone replacement therapy: relation to homocysteine and prooxidant-antioxidant status in healthy postmenopausal women Archives of Gynecology and Obstetretics,, 285(3), 733-9.

[73] Telci, A., Cakatay, U., Akhan, S. E., Bilgin, M. E., Turfanda, A., & Sivas, A. (2002). Postmenopausal hormone replacement therapy use decreases oxidative protein damage. *Gynecolicand Obstetric Investigation*, 54(2), 88-93.

[74] Topçuo◉lu, A., Uzun, H., Aydin, S., Kahraman, N., Vehid, S., Zeybek, G., & Topçuo◉lu, D. (2005). The Effect of Hormone Replaceent Therapy on Oxidized Low Density Lipoprotein Levels and Paroxonase Activity in Postmenopausal women. *Tohoku Journal of Experimental Medicine*, 205(1), 79-86.

[75] Vural, P., Akgül, C., & Canbaz, M. (2005). Effects of menopause and tibolone on antioxidants in postmenopausal women. *Annals of Clinical Biochemistry*, 42(3), 220-3.

[76] Bednarek, T. G., Tworowska, U., Jedrychowska, I., Radomska, B., Tupikowski, K., Bidzinska, S. B., & Milewicz, A. (2006). Effects of oestradiol and oestroprogestin on erythrocyte antioxidative enzyme system activity in postmenopausal women. *ClinicalEndocrinolology (Oxf)*, 64(4), 463-8.

[77] Maffei, S., Mercuri, A., Prontera, C., Zucchelli, G. C., & Vassalle, C. (2006). Vasoactive biomarkers and oxidative stress in healthy recently postmenopausal women treated with hormone replacement therapy. *Climacteric x*, 9(6), 452-8.

[78] Castanho, V. S., Gidlund, M., Nakamura, R., & de Faria, E. C. (2011). Post-menopausal hormone therapy reduces autoantibodies to oxidized apolipoprotein B100. *Gynecological Endocrinology*, 27(10), 800-6.

[79] Gökkuşu, C., Tata, G., Ademoğlu, E., & Tamer, S. (2010). The benefits of hormone replacement therapy on plasma and platelet antioxidant status and fatty acid composition in healthy postmenopausal women. *Platelets*, 21(6), 439-44, PMID: 20459351.

[80] Kim, H., Ku, S. Y., Kang, J. W., Kim, H., Kim, Y. D., Kim, S. H., Choi, Y. M., Kim, J. G., & Moon, S. Y. (2011). The 8-hydroxydeoxyguanosine concentrations according to hormone therapy and S326C polymorphism of OGG1 gene in postmenopausal women. *Molecular Genetics and Metabolism*, 104(4), 644-7.

[81] Dlugosz, A., Roszkowska, A., & Zimmer, M. (2009). Oestradiol protects against the harmful effects of fluoride more by increasing thiol group levels than scavenging hydroxyl radicals. *Basic and ClinicalPharmacolology and Toxicology*, 105(6), 366-373.

[82] Wen, Y., Doyle, M., Cooke, T., & Feely, J. (2000). Effect of menopause on low density lipoprotein oxidation: is estrogen an important determinant? *Maturitas*, 34, 233-238.

[83] Subbiah, M. T., Kessel, B., Agrawal, M., Rajan, R., Abplanalp, W., & Rymaszewski, Z. (1993). Antioxidant potential of specific estrogens on lipid peroxidation. *The Journal of Clinical Endocrinology and Metabolism*, 77, 1095-1097.

[84] Kassia, E., Dalamagaa, H., Hroussalasa, G., Kazanisa, K., Merantz, A., & Zacharic, E. J. (2010). Giamarellos-Bourbouli, A. Dionyssiou-Asterioua. Adipocyte factors, high-sensitive C-reactive protein levels and lipoxidative stress products in overweight postmenopausal women with normal and impaired OGTT. *Maturitas*, 67, 72-77.

[85] Arteaga, E., Rojas, A., Villaseca, P., Bianchi, M., Arteaga, A., & Duran, D. (1998). In vitro effect of oestradiol, progesterone, testosterone, and of combined estradiol/progestins on low density lipoprotein (LDL) oxidation in postmenopausal women. *Menopause*, 5(1), 16-23.

[86] Bureau, I., Laporte, F., Favier, M., Faure, H., Fields, M., Favier, A. E., & Roussel, A. M. (2002). No antioxidant effect of combined HRT on LDL oxidizability and oxidative stress biomarkers in treated post-menopausal women. *Journal of the American College of Nutrition*, 21(4), 333-8.

[87] Akcay, T., Dincer, Y., Saygili, E. I., Seyisoğlu, H., & Ertunğalp, E. (2010). Assessment of DNA nucleo base oxidation and antioxidant defense in postmenopausal women under hormone replacement therapy. *Indian Journal of Medical Sciences*, 64(1), 17-25.

[88] Leal, H.M., Abellán, A.J., Carbonell, M.L.F., Díaz, F.J., García, S.F.A., & Martínez, S.J.M. Influence of the presence of hot flashes during menopause on the metabolism

of nitric oxide. Effects of hormonal replacement treatment. *Medicina Clínica (Barc) 200022*, 114(2), 41-5.

[89] Itabe, H. (2003). Oxidized low-density lipopropteins: What is understood and what remains to be clarified. *Biological and Pharmaceutical Bulletin*, 26(1), 1-9.

[90] Mackness, M. I., Mackness, B., Durrington, P. N., Connelly, P. W., & Hegele, R. A. (1996). Paraoxonase: biochemistry, genetics and relationship to plasma lipoproteins: Current Opinionin Lipidology,, 7, 69-76.

[91] Aviram, M., Rosenblat, M., Bisgaier, C. L., Newton, R. S., Primo, P. S. L., & La Du, B. N. (1998). Paraoxonaseinhibitis high-density lipoprotein oxidation and preserves its function. *The Journal of ClinicalInvestigaction*, 101, 1581-1590.

[92] Crane, T. E., Kubota, C., West, J. L., Kroggel, M. A., Wertheim, B. C., & Thomson, C. A. (2011). Increasing the vegetable intake dose is associated with a rise in plasma carotenoids without modifying oxidative stress or inflammation in overweight or obese postmenopausal women. *Journal of Nutrition*, 141(10), 1827-33.

[93] Scoglio, S., Benedetti, S., Canino, C., Santagni, S., Rattighieri, E., Chierchia, E., Canestrari, F., & Genazzani, A. D. (2009). Effect of a 2-month treatment with Klamin, a Klamath algae extract, on the general well-being, antioxidant profile and oxidative status of postmenopausal women. *GynecologicalEndocrinology*, 25(4), 235-40.

[94] Miquel, J., Ramírez, B. A., Ramírez, B. J., & Diaz, A. J. (2006). Menopause: A review on the role of oxygen stress and favorable effects of dietary antioxidants. *Archives of Gerontology and Geriatrics*, 42, 289-306.

[95] Beavers, K. M., Serra, M. C., Beavers, D. P., Cooke, M. B., & Willoughby, D. S. (2009). Soymilk supplementation does not alter plasma markers of inflammation and oxidative stress in postmenopausal women. *Nutrition Research*, 29(9), 616-22.

[96] Giovannelli, L., Pitozzi, V., Luceri, C., Giannini, L., Toti, S., Salvini, S., Sera, F., Souquet, J. M., Cheynier, V., Sofi, F., Mannini, L., Gori, A. M., Abbate, R., Palli, D., & Dolara, P. (2011). Effects of de-alcoholised wines with different polyphenol content on DNA oxidative damage, gene expression of peripheral lymphocytes, and haemorheology: an intervention study in post-menopausal women. *European Journal of Nutrition*, 50(1), 19-29.

[97] Azadbakht, L., Kimiagar, M., Mehrabi, Y., Esmaillzadeh, A., Hu, F. B., & Willett, W. C. (2007). Dietary soya intake alters plasma antioxidant status and lipid peroxidation in postmenopausal women with the metabolic syndrome. *British Journal of Nutrition*, 98(4), 807-13.

[98] Salvini, S., Sera, F., Caruso, D., Giovannelli, L., Visioli, F., Saieva, C., Masala, G., Ceroti, M., Giovacchini, V., Pitozzi, V., Galli, C., Romani, A., Mulinacci, N., Bortolomeazzi, R., Dolara, P., & Palli, D. (2006). Daily consumption of a high-phenol extra-virgin olive oil reduces oxidative DNA damage in postmenopausal women. *British Journal of Nutrition*, 95(4), 742-51.

[99] Dickman, J. R., Koenig, R. T., & Ji, L. L. (2009). American ginseng supplementation induces an oxidative stress in postmenopausal women. *Journal of the American of College of Nutrition*, 28(2), 219-28.

[100] Garcia P.M.C. (2008). Antioxidantes en la dietamediterranea. *Nutrición Clínica en Medicina*, 2(3), 129-140.

[101] Agarwal, S., & Rao, A. V. (2000). Tomato lycopene and its role in human health and chronic diseases. *Canadian Medical Association or its licensors*, 163(6), 739-44.

[102] Waliszewski, K. N., & Blasco, G. (2010). Propiedadesnutraceúticasdellicopeno. *Salud-Publica Mexicana*, 52, 254-265.

[103] Palasuwan, A., Margaritis, I., Soogarun, S., & Rousseau, A. S. (2011). Dietary intakes and antioxidant status in mind-body exercising pre- and postmenopausal women. *The Journal of Nutrition Health and Aging*, 15(7), 577-84.

[104] Karolkiewicz, J., Michalak, E., Pospieszna, B., Deskur-Smielecka, E., Nowak, A., & Pilaczyńska-Szcześniak, Ł. (2009). Response of oxidative stress markers and antioxidant parameters to an 8week aerobic physical activity program in healthy, postmenopausal women. *Archives of Gerontology and Geriatrics* 49(1)e67-71.

[105] Schmitz, K. H., Warren, M., Rundle, A. G., Williams, N. I., Gross, M. D., & Kurzer, M. S. (2008). Exercise effect on oxidative stress is independent of change in estrogen metabolism. *Cancer Epidemiology Biomarkers and Prevention*, 17(1), 220-3.

[106] Pialoux, V., Brown, A. D., Leigh, R., Friedenreich, C. M., & Poulin, M. J. (2009). Effect of cardiorespiratory fitness on vascular regulation and oxidative stress in postmenopausal women. *Hypertension*, 54(5), 1014-20.

[107] Lwow, F., Dunajska, K., Milewicz, A., Jedrzejuk, D., Kik, K., & Szmigiero, L. (2011). Effect of moderate-intensity exercise on oxidative stress indices in metabolically healthy obese and metabolically unhealthy obese phenotypes in postmenopausal women: a pilot study. *Menopause*, 18(6), 646-53.

Chapter 13

Oxidative Stress in Periodontal Disease and Oral Cancer

Mario Nava-Villalba, German González-Pérez,
Maribel Liñan-Fernández and
Torres-Carmona Marco

Additional information is available at the end of the chapter

http://dx.doi.org/10.5772/52492

1. Introduction

The oral cavity is a region interconnected with other systems of the body; it should not be viewed as an isolated area. Diseases that it lays down can have systemic scope and significantly affect the quality of life of individuals who suffer them. Periodontal disease is one of the oral health problems that most often affect the global population, lack of treatment leads to loss of tooth organs and consequently alters the digestion and nutrition, without considering other relevant aspects as phonation, aesthetics and social or emotional impact. The importance of periodontal disease has raised possible bidirectional relationships with systemic diseases such as diabetes, metabolic syndrome and cardiovascular disease. We address herein the role of oxidative stress in the etiopathogeny of periodontal disease. In the same context, another disease that has become relevant in our days is the oral cancer. Epidemiological data show that the incidence of this neoplasm has been increasing in several countries. The impact of oral cancer on patients, who suffer it, is devastating. The role of oxidative stress in the development of this disease and some alternatives for its treatment, are topics addressed in this brief review. These two oral diseases are a sample of the plethora of effects that oxidative stress may have at local and systemic level.

2. Periodontal disease

Periodontitis is the second world health problem since it affects between 10 to 15% of the world population [1]· Although the various states in this disease depend on the degree of destruction and inflammation present, the American Dental Association classifies according to a system development based on the severity of the loss of periodontal insertion. The information obtained in clinical and radiographic examination classifies the patient in four typical cases that are:

- Type I: Gingivitis
- Type II: Mild Periodontitis
- Type III: Moderate Periodontitis
- Type IV: Advanced Periodontitis

There are other classifications of the inflammatory process [2]:

- Ulceronecrotic acute gingivitis
- Acute gingivitis
- Chronic gingivitis
- Marginal periodontitis
- Superficial marginal periodontitis
- Deep marginal periodontitis

Periodontal disease is an inflammatory process involving a set of changes that directly affect tissues that hold the teeth. The etiology plays a role which is essential within the bacterial infection. In fact, within the 300 to 400 species of bacteria located in the oral cavity consider that some of them are exclusive to the periodontal tissues. However in recent years it has been determined that the evolution and spread of the disease will play a decisive role in the host response to bacterial attack. This is reflected in the model of the critical path in the pathogenesis of this disease. Through this one can understand that there are diseases and systemic conditions that have risk factors for periodontal disease, because they are going to modify the host response and favor the development of damage [3].

When it is lost in the inclusion of periodontal fibers, usually after puberty, the cases that are reported before this stage are only 5%. Previously it has reported that there was a ratio of two to one in the frequency of periodontal disease, women being the most affected in this order. Currently known, the presence by gender of this involvement is very similar.

In adults with more than 1 mm of affected dental faces periodontal insertion loss increases with age. An epidemiological report in United States mentions that approximately 80-92% of the population between the ages of 35 and 64 years performed, lost more than 1 mm insertion in 20 to 47% of teeth. From 18 to 22% of the population of 35 to 64 years were more 2 mm deep in the probing of the periodontal bags in 11 to 13% of tooth surfaces. Periodontitis occurs when tissue destruction due to the direct effect of bacterial toxins and removal products, in addition, the effects caused indirectly by the harmful organic defense mechanisms. Microorganisms as *p. gingivalis, a. actinomycetemcomitans* and *Capnocytophaga*sp. produce collagenase (substances similar to trypsin) and phospholipase, among others. Extracellularly there are acid phosphatase and alkaline, lipopolysaccharides, aminopeptidase, epithelium toxin, inhibitor of fibroblasts and a toxin that induces a bone resorption.

Bacteria causes tissue destruction with its deletion, this is a feature of marginal periodontitis products. Destruction of tissues within a radius of 1.5 to 2.5 mm around the plaque has been observed (the so-called *influence radioplate)* in periodontitis. The hydrolysis of the connective tissue associated with the inflammation is due to the reactive oxygen species and the elastase/lysosomic-like enzymes. Collagenase and gelatinase are segregated to the microenvironment. Prostaglandin E, Interleukin 1-/ J and the lipopolysaccharide activates osteoclasts and induce a resorption of alveolar bone. Cellular and humoral components of the immune system, mainly involved in the periodontal immune response are leukocytes, immunoglobulins, complement system and lysozyme. If the immune defenses are working properly, the periodontium is protected from the harmful effect of pathogenic substances secreted by the microorganisms. The immunocompetent host is able to defend itself against microbial attacks that occur every day. Thus prevents *infections,* i.e. the multiplication of microorganisms within the periodontium. We can say that the periodontal inflammation is a local reaction to a tissue injury whose purpose is the destruction of the causal factor, dilution or its encapsulation.

The human immune system can be classified according to their function within the periodontium, follows:

- Secretory system
- Neutrophils, antibodies and complement system
- Leukocytes and macrophages
- Immune regulation system.

The system formed by neutrophils, antibodies and complement is crucial to the immune defense against periodontal infections. When functional defects of neutrophils occur, it increases the frequency of serious marginal periodontitis [4].

3. Oxidative stress

A phenomenon that occurs within the periodontal disease is called oxidative stress. In order to understand the phenomenon of oxidative stress it is important to know what the free radicals (FR) are, where they come from and how to act. A FR is considered that molecule presented an electron unpaired or odd in the orbital external, in its atomic structure giving it a spatial configuration that generates a high instability. In the molecule of oxygen (O_2) know the following FR or also called oxygen reactive species: anion superoxide ($O_2^{\cdot -}$), hydrogen peroxide (H_2O_2), hydroxyl radical ($OH^{\cdot}$) and singlet oxygen ($^1O_2$). The H_2O_2 is not strictly a FR but by its ability to generate the $OH^{\cdot}$ in the presence of metals such as iron, it incorporates it as such. A fundamental characteristic of the reactions of free radicals is that act of chain reactions, where a radical reaction generates another consecutively.

Oxygen (O_2) that this in the air is fundamental to life, many reactions in which participates the O_2 generates reactive oxygen species (ROS), of which some have the chemical character of being free radical (FR), whose biochemical entities in its atomic structure presented an odd or unpaired electron in the outer orbital, giving it a spatial configuration that generates a high instability with an enormous capacity for combined with the diversity of molecules members of the cell structure: carbohydrates, lipids, proteins, nucleic acids, and derivatives of each of them, causing important functional alterations. In this sense, the body has an antioxidant system to counteract the generation of ROS, which maintains a homeostatic balance. However there are pro-oxidant factors that favor the generation of FR, causing an imbalance in favor of the latter, generating so-called oxidative stress (OS) [5].

The tetravalent reduction of oxygen to produce water through the electron transport chain in mitochondria is relatively safe. However, the univalent reduction of oxygen generates ROS. The human organism also has antioxidant system to counteract the generation of ROS, which maintains a homeostatic balance. However, there are pro-oxidant factors that favor the generation of FR, causing an imbalance in favor of the latter, generating OS. The antioxidant enzyme superoxide dismutase (SOD), Glutathione peroxidase (GP), glutathione reductase (GR) and catalase (CAT), as well as proteins carriers of metals (ceruplasmina, transferrin, lactoferrin, etc.), and another micronutrients as vitamins A, C and E, bilirubin, uric acid and selenium, constitute the most important elements of the antioxidant system. Also, between the most important pro-oxidant factors we can highlight the process of aging, ionizing radiation, ultraviolet rays, environmental pollution, cigarette smoke, excess of exercise, intake of alcoholic beverages and inadequate diet [6].

The role of Coenzyme Q_{10} is the mitochondrial energy coupling. It is an essential part of the cellular machinery used to produce ATP that provides the energy for muscle contraction and other vital cellular functions. Most of the ATP production occurs in the inner membrane of the mitochondria, where the Coenzyme Q_{10} is located. The most important function is serving as a suppressor of primary free radicals, located in the membranes in the vicinity of unsaturated lipid chains. There are less established functions that include the oxidation/reduction of the control of the origin and transmission of signals in cells that induce the expression of gender, the control of membrane channels, the structure and solubility in lipids [7].

Free radicals cause damage to periodontal tissues by a variety of different mechanisms including:

- DNA damage
- Lipid peroxidation
- Protein damage
- The oxidation of important enzymes (anti proteases)
- Stimulation and release of pro-inflammatory cytokines

ROS covers other reactive species that are not true radicals, but are however capable of react in intra and extracellular environment: peroxide of hydrogen, hypochlorous acid, oxygen, ozone. The living organism has adapted to an existence under a continuous output of radical free flow. Between the different antioxidant defense mechanism adaptation mechanism is of great importance. Antioxidants are "those substances that when they are present in lower concentrations compared to the substrate of an oxidizable, significantly delay or inhibit the oxidation of the substrate". The various possible mechanisms that antioxidants can offer protection against damage from free radicals are:

- The prevention of the formation of radical free.
- Interception of the radical free to eliminate reactive metabolites and their conversion to less reactive molecules.
- Facilitate the repair of the damage caused by free radicals.
- Create a favourable environment for the effective functioning of other antioxidants.

Antioxidant defense system is very dynamic and responsive to any disturbance that occurs in the body redox balance. Antioxidants can be regulated and neutralize the formation of radical free that can occur due to oxidative stress, such as the factor transcription factors Activator protein 1 and nuclear-kb are redox sensitive. Redox potential is a measure of the affinity of a substance for electrons [8].

The presence of inflammatory infiltrate is a constant feature in periodontal disease. It is known that these cells release lots of free radicals; it is suspected that these metabolites are involved in the pathogenesis of the disease. The presence of a dense inflammatory infiltrate in periodontal disease leads to the suspicion that the relationship of periodontal leukocyte-tissue has a double aspect. The role of these cells in the containment of the gingival bacteria and their products must be analyzed according to a balance with the destruction of tissue due to the release of the products of its action (FR and proteases). In this way, a defensive mechanism, under the interaction of various factors, can be harmful to periodontal tissues, and they are therefore involved in the pathogenesis of inflammatory periodontal disease.

There is numerous evidence pointing to the involvement of FR in periodontal disease. It has been reported in patients with rapidly progressive periodontitis, that the polymorphonuclear neutrophils (PMN) are functionally activated, produce high levels of O_2 and have a high response the luminol-dependent (QL) chemiluminescent. There is an increase of the PMN oxidative response peripherals in patients with localized and generalized juvenile periodontitis, as well as in adult patients with periodontitis (AP). This increase is related to clinical periodontal status and is reversed by therapy.

It has also compared the generation O_2 by the activated PMN in the gingival crevicular fluid (GCF) of patients with AP. The PMN activation with phorbolmyristateacetate causes a marked increase in the release of O_2 in patients with AP, while the antioxidant activity of the gum is similar to the controls. The effect of the PMN in crevicular fluid of patients is dependent on variations in the rate of formation of $O_{2,}$ relative to the intrinsic antioxidant capacity of the gingival tissue.

In gingival epithelial cells in culture studies have shown the PMN may cause lysis of these through the action of the free myeloperoxidase(MPO), a leukocyte enzyme generating radicals. Its activity has been increased in the crevicular fluid of sites with gingivitis and periodontitis with respect to healthy sites.

There is a close relationship between free radical production by leukocytes and activation of proteases. Altogether these actions could have profound effects on the function and integrity of the gingival epithelium.

The above evidence leads to consider that in the inflammatory periodontal disease, the general etiological factors causing the breakup of physiological systems of inhibition of lipid peroxidation, creates a low level of antioxidant protection of periodontal tissues. In these circumstances, the local factors lead to the migration of neutrophils to the gingiva and gingival fluid. The activation of these leukocytes in phagocytosis, causes the release of ROS, which leads to the outbreak of the lipid peroxidation of the soft tissues of the periodontium and activation of protease. This lipid peroxidation is the mechanism that triggers the development of morphofunctionalchangesin periodontium and their vessels, which results in destruction of collagen and bone resorption.

Due to numberless evidences that suggest a participation of the ROS in the pathogenesis of the periodontal disease, it has been raised that the factors that promote a rupture of the antioxidant physiological system, contribute to the development of oxidative mechanisms that initiate the periodontitis. The main cause of lipid peroxidation in the periodontal disease seems to lie in the liberation of ROS by leukocytes in phagocytosis. These concepts emphasize the utility of antioxidants in the prophylaxis and treatment of periodontal disease and therefore justify the search of new antioxidant preparations for this purpose. For example the *p. gingivalis* is a major cause of periodontitis, and their presence is a risk factor for systemic inflammatory syndromes, such as atherosclerosis and cardiac dysfunction. The capacity of the virulence factors such as proteases and LPS to induce inflammation has been studied intensely. In some cases, however, the inflammation occurs regardless of these factors, suggesting the existence of other stimulating immune. It was found that the cell death induced by *p. gingivalis*in the tissues is through the production of ROS [9].

4. Oral Cancer

The oral cancer occupies 2-5 % of all whole body cancers. This percentage places this neoplasia within the ten most common cancers [10]. Although its magnitude is relatively low, its impact on affected patients and their costs in health systems is high.There is a considerable variation in the incidence and mortality rates around the world. The incidence is greater in south of India, Australia, North of America, many European countries, Brazil, certain countries of Africa and some of central Asia [11]. 90% of oral cancer is of epithelial origin and the rest 10% are distributing in adenocarcinomas, sarcomas, lymphoproliferative disorders, metastasis, melanomas and malignant odontogenic tumors. The intraoral main site of oral squamous cell cancer (OSCC) is the posterior lateral border of tongue (Figure 1) and floor of mouth (Figure 2). If the lips are considered within the oral territory, then this site has the highest frequency (Figure 3).Since oral squamous cell carcinoma (OSCC) is the main malignant neoplasia we focus in it.

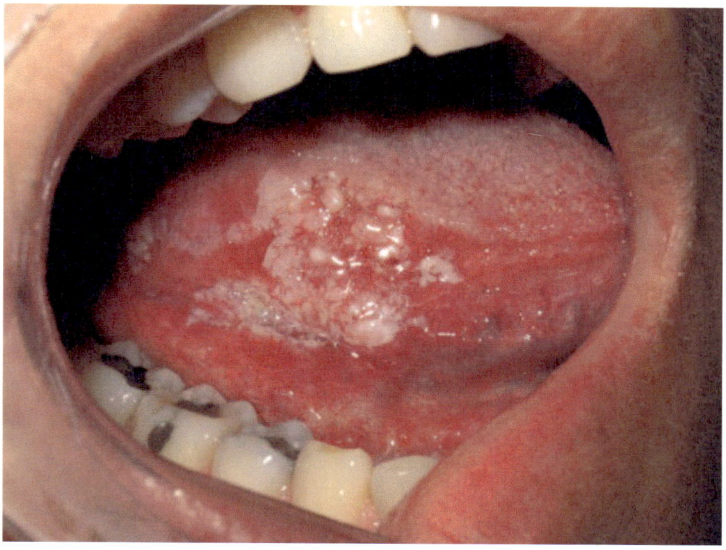

Figure 1. Squamous cell cancer of the posterior lateral border of the tongue in a 28-year-old woman.She smoked a cigarette per day for 15 years.

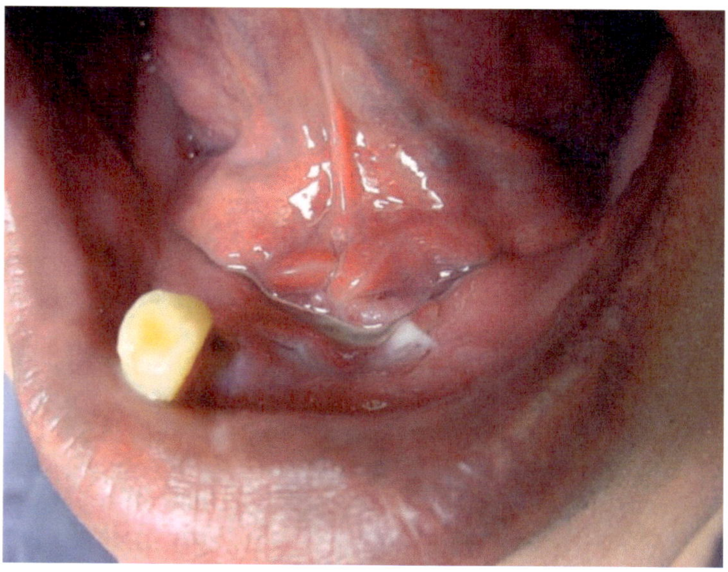

Figure 2. Squamous cell cancer of floor of mouth in a 58-year-old woman.She had a history of poorly controlled diabetes type 2 from 42 years. She also has used ill-fitting dentures since age 50. Note the linear lesion with presence of necrosis in the centre of the fissure.

There are premalignant lesions recognized like: leukoplakia, erythroplakia, oralsubmucosal fibrosis, palatal lesions of reverse cigar smoking, oral liken planus, discoid lupus erythematosus, and hereditary disorders likecongenital dyskeratosis and epidermolysisbullosa, but beyond of a clinical standpoint, diverse carcinogenic molecular mechanisms have been postulated.The main target is the DNA, since mutations that occur in it generates a wide range of deleterious effects in the cell. In a very general overview, the balance between tumor suppressor genes and those genes that induce cell cycle is altered.Allowing cells to escape cell cycle control and developing an unpredictable biological behavior. Subsequently, the cells express molecules that allow them to acquire an invasive phenotype, a phenomenon known as epithelial-mesenchymal transition. Why malignant cells colonize distant sites? Is not yet fully understood, but it is the feature that makes it lethal.

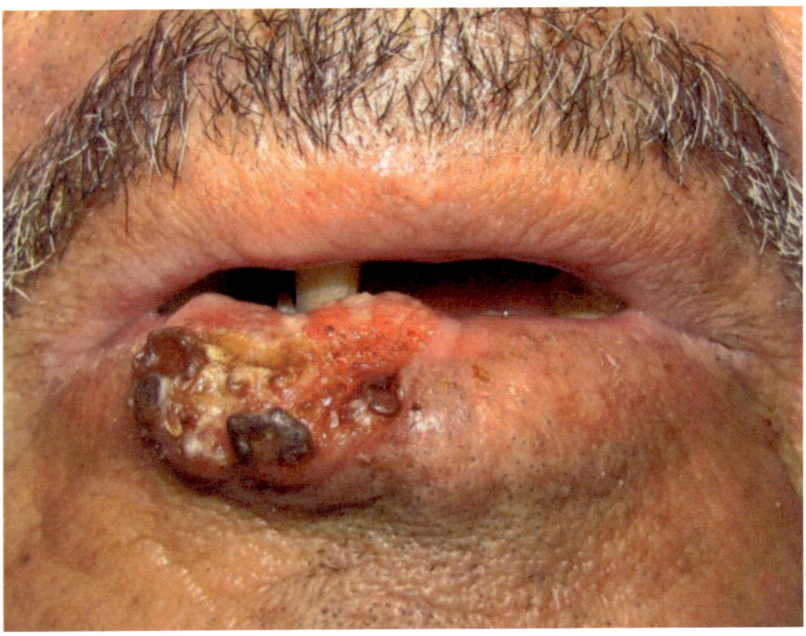

Figure 3. Squamous cell cancer of the lip in a 74-year-old man. He was a farmer and consumed alcohol chronically.

Free radicals are products of the oxidation-reduction systems of the cell and its participation in cellular metabolic functions is essential for cell survival. A classic example is the electron transport chain in mitochondria. However, in whatpathologicalconditions, free radicals can become deleterious? In fact, what are the results of its harmful effects? The involvement of free radicals in cancer development has been studied for 3 decades, and there is sufficient evidence that implicates theirs in the multistage theory of carcinogenesis. They are proposed to cause diverse DNA alterations like: punctual mutations, DNA base oxidations, strand breaks, mutation of tumor suppressor genes and can induce overexpression of proto-oncogenes [12].It should be added that oxidative protein damage participates in facilitating the development of cancer.

Several works explore the levels of oxidative stress in patients with oral cancer [13-15] most of them quantified the products of lipid-peroxidation(mainly malonilaldehyde) and contrast them with the activity of antioxidant enzymes or exogenous antioxidants levels in blood or even saliva. The results agree that there is an imbalance between the high amount of free radicals and insufficient antioxidant system activity.Added to this, some researchers have observed that high levels of lipid-peroxidation combined with low levels of thiols and antioxidant status, correlate with poor survival rate in patients with oral cancer [16].

The OSCC is a multifactorial disease, however, a factor strongly associated, is smoking. 90% of individuals with oral cancer are smokers. It is considered that the smoke from cigarettes have 4000 chemicals, 40 of which have carcinogenic potential. It has been shown that cigarette smoke contains pro-oxidants that are capable of initiating the process of lipid-peroxidation and deplete levels of antioxidants from the diet [17,18].

In contrast, there is epidemiological evidence that demonstrates the protective effect of diet on some populations [19-21].For example in Greece, which has the lowest rates of oral cancer among European countries,its population is exposed to latent risk factors such as alcohol intake and smoking; micronutrients consume such as riboflavin, magnesium and iron correlated inversely with oral cancer [19].

Consequently, several authors have proposed the ingestion of diverse exogenous antioxidants; supporting in those epidemiological studies, where the diet offers protection for the development of cancer, and taking into account that the endogenous antioxidant systems have been overwhelmed by oxidative stress.

For example, vitamin C is one of the most extensively evaluated antioxidants in oral cancer alternative co-therapies. Low or even undetectable levels of vitamin C correlate with the presence of oral cancer [17, 22]; in contrast, is one of the micronutrients that have a consistent inverse correlation in different studies [23].Vitamin C acts as a scavenger of free radicals and impedes the detrimental chain reactions triggered by the free radicals.The l-glutamine is another antioxidant that has shown a beneficial modulating effect in patients with oral cancer in stages III and IV. The l-glutamine is administered in the diet as a complementary therapy; the proposal is that restores glutathione cascade system [15].In addition, other antioxidants such as carotene, vitamin E, thiamine, vitamin B_6, folic acid, niacin and potassium have shown a convincing protective effect [24]. Even more,when them are administered together during the cycles of radiotherapy [25].

Author details

Mario Nava-Villalba, German González-Pérez[2], Maribel Liñan-Fernández[3] and Torres-Carmona Marco[4]

*Address all correspondence to: marionava23@gmail.com

1 Dentistry Department, School of Medicine.Autonomous University of Querétaro.andDentistry Department, Health Science Division,University of the Valley of México, Campus QuerétaroQuerétaro, México

2 Dentistry Department, School of Medicine.AutonomousUniversity of Querétaro, Querétaro, México

3 Dentistry Department, School of Medicine.AutonomousUniversity of Querétaro, Querétaro, México

4 Dentistry Department, School of Medicine.Autonomous University of Querétaro.andGenetics Department,Comprehensive Rehabilitation Center of Querétaro.Querétaro, México

References

[1] (2010). Mendes S.A, A.M Vargas Duarte Ferreira faith, Nogueira GH.Periodontitis in individuals with diabetes treated in the public health system of Belo Horizonte, Brazil., *Revista Brasileira de Epidemiología*, 13(1), 118-25.

[2] Hee-Kyung, L., Sang-Hee, C., Kyu, C. W., Anwar, T. M., Keun-Bae, S., & Seong-Hwa, J. (2009). The effect of intensive oral hygiene care on gingivitis and periodontal destruction in Type 2 diabetic patients., *Yonsei Medical Journal*, 50(4), 529-36.

[3] (2002). Garcia T.B, Saldaña B.A, Soto F.C.Oxidative stress in systemic effects of inflammatory periodontal disease., *Revista Cubana de Investigación Biomédica*, 21(3), 194-196.

[4] (2010). Newman T.C. Clinical Periodontology. McGraw-Hill, 10thEdition,.

[5] Valdez P.A, Mendoza N.V.Relationship of oxidative stress with periodontal disease in older adults with type 2 diabetes mellitus. Revista ADM (2006). , 63(5), 189-94.

[6] (2004). García M.A.Influence of oxidative stress in periodontal disease., *Revista Cubana de Ciencias Médicas*, http://www.cpicmha.sld.cu/hab/10 2_04/hab07204.htm.

[7] Shobha, P., Sunitha, J., & Mayank, H. (2010). Role of coenzyme Q_{10} as an antioxidant and bioenergizer in periodontal diseases., *Indian journal Pharmacology*, 42(6), 334-37.

[8] Gowri, P., Biju, T., & Suchetha, K. (2008). The challenge of antioxidants to free radicals in periodontitis., *Journan Indian Society Periodontology*, 12(3), 79-83.

[9] Kenichi, I., Hiroshi, H., Katsutoshi, I., Tatsuo, A., Mikio, S., Koji, N., et al. (2010). Porphyromonasgingivalis Peptidoglycans induce excessive activation of the innate immune system in silkworm slrvae., *Journal Biological Chemistry*, 285(43), 33338-33347.

[10] Stewart B.W, Kleihues P, editors. World Cancer Report. Lyon:, *WHO International Agency for Research on Cancer; 2003.*

[11] (2009). Petersen P.E. Oral cancer prevention and control- The approach of the World Health Organization., *Oral Oncology*.

[12] Halliwell, B. (2007). Oxidative stress and cancer: have we moved forward?, *Biochemical Journal*, 401(1), 1-11.

[13] Sultan-Beevi, S. S., Hassanal-Rasheed, A. M., & Geetha, A. (2004). Evaluation of oxidative stress and nitric oxide levels in patients with oral cavity cancer., *Japan Journal of Clinical Oncology*, 34(7), 379-385.

[14] Khanna, R., Thapa, P. B., Khanna, H. D., Khanna, S., Khanna, A. K., & Shukla, H. S. (2005). Lipid peroxidation and antioxidant enzyme status in oral carcinoma patients., *Kathmandu University Medical Journal*, 3(4), 334-339.

[15] Das, S., Mahapatra, S. K., Gautam, N., Das, A., & Roy, S. (2007). Oxidative stress in lymphocytes, neutrophils, and serum of oral cavity cancer patients: modulatory array of l-glutamine., *Supportive Care in Cancer*, 15(12), 1399-1405.

[16] (2007). Patel B.P, Rawal U.M, Dave T.K, Rawal R.M, Shukla S.N, Shah P.M, Patel P.S. Lipid peroxidation, total antioxidant status, and total thiol levels predict overall survival in patients with oral squamous cell carcinoma., *Integrative Cancer Therapies*, 6(4), 365-372.

[17] Hedge, N. D., Kumari, S., Hedge, M. N., Bekal, M., & Rajaram, P. (2012). Status of serum vitamin C level and peroxidation in smokers and non-smokers with oral cancer., *Research Journal of Pharmaceutical, Biological and Chemical Sciences*, 3(1), 170-175.

[18] Raghavendra, U. D., Souza, V. D., & Souza, B. (2012). Erythrocyte malonilaldheyde and antioxidant status in oral squamous cell carcinoma patients and tobacco chewers/smokers., *Biomedical Research*, 21(4), 441-444.

[19] Petridou, E., et al. (2002). The role of diet and specific micronutrients in the etiology of oral carcinoma., *Cancer*, 94(11), 2981-2988.

[20] Winn, D. M., Ziegler, R. G., Pickle, L. W., Gridley, G., Blot, W. J., & Hoover, R. N. (1984). Diet in the etiology of oral and pharyngeal cancer among women from the southern United States., *Cancer Research*, 44(3), 1216-1222.

[21] Franceschi S, Favero A, Conti E, Talamini R, Volpe R, Negri E, Barzan L, La Vecchia C. Food groups, oils and butter, and cancer of the oral cavity and pharynx. *British Journal of Cancer* 1999; 80 (3/4) 614-620.

[22] Aravindh, L., Jagathesh, P., Shanmugam, S., Sarkar, S., Kumar, P. M., & Ramasubramanian, S. (2012). Estimation of plasma antioxidants beta carotene, vitamin C and vitamin E levels in patients with OSMF and oral Cancer- Indian population., *International Journal of Biological and Medical Research*, 3(2), 1655-1657.

[23] World Cancer Research Fund & American Institute for CancerResearch. (1997). Food, nutrition and the prevention of cancer.Mouth and pharynx., *Cancers, nutrition, and food. Menasha,WI: World Cancer Research Fund & American Institutefor Cancer Research.*

[24] Negri, E., Franceschi, S., Bosetti, C., Levi, F., Conti, E., Parpinel, M., & La Vecchia, C. (2002). Selected micronutrients and oral and pharyngeal cancer., *International Journal of Cancer*, 86(1), 122-127.

[25] (2009). Shariff A.K, Patil S.R, Shukla P.S, Sontakke A.V, Hendre A.S, Gudur A.K. Effect of oral antioxidant supplementation on lipid peroxidation during radiotherapy in head and neck malignancies., *Indian Journal of Clinical Biochemistry*, 24(3), 307-311.

Section 3

Aging

Chapter 14

Aging, Oxidative Stress and Antioxidants

B. Poljsak and I. Milisav

Additional information is available at the end of the chapter

http://dx.doi.org/10.5772/51609

1. Introduction

Aging is an extremely complex and multifactorial process that proceeds to the gradual deterioration in functions. It usually manifest after maturity and leads to disability and death. Traditionally researchers focused primarily on understanding how physiological functions decline with the increasing age; almost no research was dedicated to investigation of causes or methods of aging intervention. If scientists would discover a drug for healing all major chronic degenerative diseases, the average lifetime would be increased for just 12 years. People would still die from complications connected with the aging process.

Defects formed in human body as a consequence of the aging process start to arise very early in life, probably *in utero*. In the early years, both the fraction of affected cells and the average burden of damage per affected cell are low [1]. The signs of aging start to appear after maturity, when optimal health, strength and appearance are at the peak. After puberty, all physiological functions gradually start to decline (e.g. the maximum lung, heart and kidney capacities are decreased, the secretion of sexual hormones is lowered, arthritic changes, skin wrinkling, etc). The precise biological and cellular mechanisms responsible for the aging are not known, but according to Fontana and Klein [2], "they are likely to involve a constellation of complex and interrelated factors, including [1] oxidative stress–induced protein and DNA damage in conjunction with insufficient DNA damage repair, as well as genetic instability of mitochondrial and nuclear genomes; [2] noninfectious chronic inflammation caused by increased adipokine and cytokine production; [3] alterations in fatty acid metabolism, including excessive free fatty acid release into plasma with subsequent tissue insulin resistance; [4] accumulation of end products of metabolism, such as advanced glycation end products, amyloid, and proteins that interfere with normal cell function; [5] sympathetic nerve system and angiotensin system activation as well as alterations in neuroendocrine systems; and [6] loss of post-mitotic cells, resulting in a decreased number of neurons and muscle cells as well as deterioration in structure and function of cells in all tissues and organs".

In recent years, oxidative stress has been implicated in a wide variety of degenerative processes, diseases and syndromes, including the following: mutagenesis, cell transformation and cancer; heart attacks, strokes, atherosclerosis, and ischemia/reperfusion injury; chronic inflammatory diseases, like rheumatoid arthritis, lupus erythematosus and psoriatic arthritis; acute inflammatory problems; photooxidative stresses to the eye, e.g. cataract; neurological disorders, such as certain forms of familial amyotrophic lateral sclerosis, certain glutathione peroxidase-linked adolescent seizures, Parkinson's and Alzheimer's diseases; and other age-related disorders, perhaps even including factors underlying the aging process itself [3].

2. Aging theories

Scientists estimated that the allelic variation or mutations in up to 7,000 relevant genes might modulate their expression patterns and/or induce senescence in an aging person, even in the absence of aging specific genes [4, 5]. As these are complex processes they may result from different mechanisms and causes. Consequently, there are many theories trying to explain the aging process, each from its own perspective, and none of the theories can explain all details of aging. The aging theories are not mutually exclusive, especially, when oxidative stress is considered [6].

Mild oxidative stress is the result of normal metabolism; the resulting biomolecular damage cannot be totally repaired or removed by cellular degradation systems, like lysosomes, proteasomes, and cytosolic and mitochondrial proteases. About 1% to 4% of the mitochondrially metabolized oxygen is converted to the superoxide ion that can be converted subsequently to hydrogen peroxide, hydroxyl radical and eventually other reactive species, including other peroxides and singlet oxygen that can in turn, generate free radicals capable of damaging structural proteins and DNA [7, 8, 9, 10, 11]. Since extensive research on the relation between polymorphisms likely to accelerate/decelerate the common mechanisms of aging and resistance to the oxidative stress has been neglected in almost all scientific studies, the data do not allow us to conclude that the oxidative theory supports the theory of programmed aging so far [7]. However, the most recent studies support the idea that oxidative stress is a significant marker of senescence in different species. Resistance to oxidative stress is a common trait of long-lived genetic variations in mammals and lower organisms [5, 12]. Theories on aging process can be divided into programmed and stochastic.

2.1. Free radical theory, oxidative stress theory and mitochondrial theory of aging

Denham Harman was first to propose the free radical theory of aging in the 1950s, and extended the idea to implicate mitochondrial production of reactive oxygen species in 1970s, [13]. According to this theory, enhanced and unopposed metabolism-driven oxidative stress has a major role in diverse chronic age-related diseases [13, 14, 7]. Organisms age because of accumulation of free radical damage in the cells. It was subsequently discovered that reactive oxygen species (ROS) generally contribute to the accumulation of oxidative damage to cellular constituents, eventhough some of them are not free radicals, as they do not have an unpaired electron in their outer shells [15, 16]. Consistently, aged mammals contain high quantities of oxidized lipids and proteins as well as damaged/mutated DNA, particularly in the mitochondrial genome [13, 14]. In support of a mitochondrial theory of aging, the mitochondrial DNA damage increases with aging [17, 18]. Thus, a modern version of this tenet is the "oxidative stress theory" of aging, which holds that increases in ROS accompany aging and lead to functional alterations, pathological conditions, and even death [19].

The oxygen consumption, production of ATP by mitochondria and free-radical production are linked processes [20, 21]. Harman first proposed that normal aging results from random deleterious damage to tissues by free radicals [14] and subsequently focused on mitochondria as generators of free radicals [13]. Halliwell and Gutteridge later suggested to rename this free radical theory of aging as the "oxidative damage theory of aging" [22], since aging and diseases are caused not only by free radicals, but also by other reactive oxygen and nitrogen species.

Increases in mitochondrial energy production at the cellular level might have beneficial and/or deleterious effects [23]. Increased regeneration of reducing agents (NADH, NADPH and $FADH_2$) and ATP can improve the recycling of antioxidants and assist the antioxidant defence system. On the other hand, enhanced mitochondrial activity may increase the production of superoxide, thereby aggravating the oxidative stress and further burdening the antioxidant defence system. The mitochondria are the major source of toxic oxidants, which have the potential of reacting with and destroying cell constituents and which accumulate with age. The result of this destructive activity is lowererd energy production and a body that more readily displays signs of age (e.g., wrinkled skin, production of lower energy levels). There is now a considerable evidence that mitochondria are altered in the tissues of aging individuals and that the damage to mitochondrial DNA (mtDNA) increases 1,000-fold with age [24].

The mutation rate of mitochondrial DNA is ten-times higher than that of nuclear DNA. Mitochondrial DNA (mtDNA) is a naked, mostly double-stranded, circular, and is continuously exposed to ROS. It is replicated much faster than nuclear DNA with less proofreading and efficient DNA repair mechanisms [25]. Thus, mtDNA is more vulnerable to attack by ROS. Damaged mitochondria can cause the energy crisis in the cell, leading to senescence and aging of tissue. Accumulation of damage decreases the cell's ability to generate ATP, so that cells, tissues, and individuals function less well. The gradual loss of energy experienced with age is paralleled by a decrease in a number of mitochondria per cell, as well as energy-producing efficiency of remaining mitochondria.

A major effect of mitochondrial dysfunction is an unappropriately high generation of ROS and proton leakage, resulting in lowering of ATP production in relation to electron input from metabolism. Leaked ROS and protons cause damage to a wide range of macromolecules, including enzymes, nucleic acids and membrane lipids within and beyond mitochondria, and thus are consistent with the inflammation theory of aging as being proximal events triggering the production of pro-inflammatory cytokines. The age-related increases in the levels of both oxidative damage and mutational load of mtDNA predicted by the mitochondrial theory of aging have been described in multiple species and organ systems [26]. However, whether this damage affects mitochondrial function or significantly modulates the physiology of aging has remained controversial [27, 28]. As already mentioned, free radicals can damage the mitochondrial inner membrane, creating a positive feedback-loop for increased free-radical creation. Induction of ROS generates mtDNA mutations, in turn leading to a defective respiratory chain. Defective respiratory chain generates even more ROS and generates a vicious cycle. The result is even more damage.

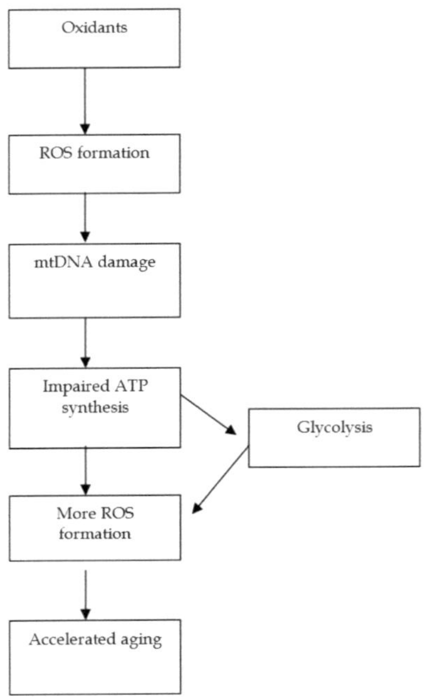

Figure 1. Oxidative stress from endogenous or exogenous sources can trigger the chain reaction, which leads to accelerated aging process of cells and organisms.

On the other hand, the "vicious cycle" theory, which states that free radical damage to mitochondrial DNA leads to mitochondria that produce more superoxide, has been questioned by some scientists since the most damaged mitochondria are degraded by autophagy, whereas the less defective mitochondria (which produce less ATP as well as less superoxide) remain to reproduce themselves [29]. But the efficiency of autophagy to consume malfunctioning mitochondria also declines with age, resulting in more mitochondria producing higher levels of superoxide [30]. Mitochondria of older organisms are fewer in number, larger in size and less efficient (produce less energy and more superoxide).

Free radicals could also be involved in signalling responses, which subsequently stimulate pathways related to cell senescence and death, and in pro-inflammatory gene expression. This inflammatory cascade is more active during aging and has been linked with age-associated pathologies, like cancer, cardiovascular diseases, arthritis, and neurodegenerative diseases [31].

2.2. Other theories of aging

Apart from the free radical theory, the aging is explained by many other theories:

The Telomere shortening hypothesis (also described as "replicative senescence," the "Hayflick phenomenon" or Hayflick limit) is based on the fact that telomeres shorten with each successive cell division. Shortened telomeres activate a mechanism that prevents cell division [32]. The telomere shortening hypothesis cannot explain the aging of the non-dividing cells, e.g. neurons and muscle cells, thus cannot explain the aging process in all the cells of an organism.

The Reproductive-cell cycle theory states that aging is regulated by reproductive hormones, which act in an antagonistic pleiotropic manner through cell cycle signaling. This promotes growth and development early in life in order to achieve reproduction, however later in life, in a futile attempt to maintain reproduction, become dysregulated and drive senescence [32].

The Wear and tear theory of aging is based on the idea that changes associated with aging result from damage by chance that accumulates over time [32]. The wear-and-tear theories describe aging as an accumulation of damage and garbage that eventually overwhelms our ability to function. Similar are Error accumulation and Accumulative waste theories; Error accumulation theory explains aging as the results from chance events that escape proofreading mechanisms of genetic code [32], according to Accumulative waste theory the aging results from build-up of cell waste products in time because of defective repair-removal processes. Terman, [33] believes that the process of aging derives from imperfect clearance of oxidatively damaged, relatively indigestible material, the accumulation of which further hinders cellular catabolic and anabolic functions (e.g. accumulation of lipofuscin in lysosomes). The programmed theories (e.g. aging clock theory) propose a time-switch in our bodies that controls not only our process of development but also triggers our self-destruction. The shortening of telomeres would provide such a clock in rapidly dividing cells. The Autoimmune theory of aging is based on the idea that aging results from an increase in antibodies that attack the body's tissues [32].

Mitohormesis theory of aging is based on the "hormesis effects". It describes beneficial actions resulting from the response of an organism to a low-intensity stressor. It has been known since the 1930s that restricting calories while maintaining adequate amounts of other nutrients can extend the lifespan in laboratory animals. Michael Ristow's group has provided evidence for the theory that this effect is due to increased formation of free radicals within the mitochondria causing a secondary induction of increased antioxidant defense capacity [34]. Finkel et al., [35] stated that the best strategy to enhance endogenous antioxidant levels may actually be oxidative stress itself, based on the classical physiological concept of hormesis (for detailed information on hormesis see paragraph Adaptive responses and hormesis).

Additionally, the Disposable soma theory was proposed [36, 37], which postulated a special class of gene mutations with the following antagonistic pleiotropic effects: these hypothetical mutations save energy for reproduction (positive effect) by partially disabling molecular proofreading and other accuracy promoting devices in somatic cells (negative effect). The

Evolutionary theory of aging is based on life history theory and is constituted of a set of ideas that themselves require further elaboration and validation [38].

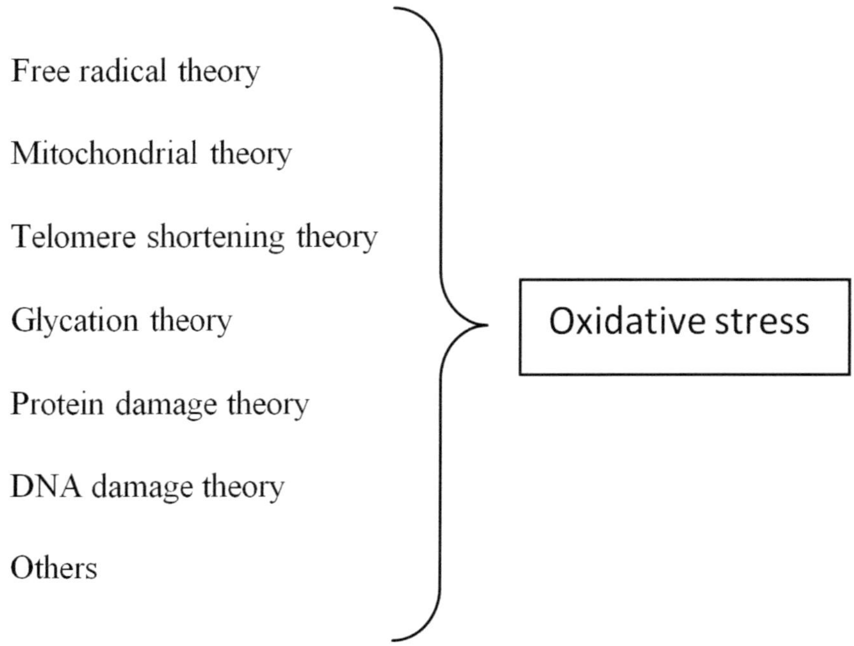

Figure 2. Oxidative stress as the common denominator of majority of aging theories.

Evidence implies that an important theme linking several different kinds of cellular damage is the consequence of exposure to reactive oxygen species [5, 39].

Many of the theories overlap, e.g., ROS can cause DNA damage (free radical theory) and also accelerate telomere shortening (telomere theory), since telomere shortening is accelerated by oxidative stress in vascular endothelial cells [40, 41]. None of the theories explain the aging process, as it may be too complex to be covered by only one theory. Perhaps there is no single mechanism responsible for aging in all living organisms [42]. The definitive mechanisms of aging across species remain equivocal. Diminished capacity for protein synthesis and DNA repair, decline in immune functions, loss of muscle mass and strength, a decrease in bone mineral density as well as a decrease in enzymatic and non-enzymatic antioxidative protections are well established. In essence, aging is progressive accumulation through life of many random molecular defects that build up within the cells and tissues. For this reason, only one "magic bullet" will never be able to prevent or reverse the complex and multicausal process of aging.

3. The Role of Oxidative Stress on the General Aging Process

In order to understand strategies to reduce oxidative stress and aging, it is first important to briefly explain reasons for oxidative stress formation. Oxidative damage is a result of the intrinsic and extrinsic ROS formation factors. The most important endogenous sources of oxidants are mitochondrial electron transport chain and nitric oxide synthase reaction, and the non-mitochondrial soruces: Fenton reaction, reactions involving cytochromes P450 in microsomes, peroxisomal beta - oxidation and respiratory burst of phagocytic cells [6]. Free radical reactions have been implicated also as the consequence of exposure to many environmental pollutants, e.g. cigarette smoke, alcohol, ionizing and UV radiations, pesticides, ozone, etc. Oxidative stress is the direct consequence of an increased generation of free radicals and/or reduced physiological activity of antioxidant defenses against free radicals. The degree of oxidative stress is proportional to the concentration of free radicals, which depends on their formation and quenching.

Causes of increased free-radical production include [43]:

Endogenous

- elevation in O_2 concentration
- increased mitochondrial leakage
- inflammation
- increased respiration
- others

Exogenous

- environment (pollution, pesticides, radiation, etc.)
- smoking
- poor nutrition
- disorders and chronic diseases
- chronic inflammation
- lifestyle
- strenuous excercise
- psychological and emotional stress
- others

Causes of decreased antioxidant defense include:

- reduced activity of endogenous antioxidative enzymes
- reduced biokinetics of antioxidant metabolism
- reduced intake of antioxidants
- reduced bioabsorption of antioxidants
- others

Oxidative stress is caused mainly by:

- mutation or reduced activity of enzymes (catalase, SOD, glutathione peroxidase)
- decreased intake of exogenous antioxidants from food
- increased metal ion intake (e.g., Fe, Cu, Cr)
- easiliy peroxidized amino acids (e.g., lysine)
- increased triplet oxigen ($^3O_2$) concentration
- increased physical activity of an untrained individual
- ROS from ionizing radiation, air pollution, smoking
- chronic inflammation

Excessive generation of free radicals may overwhelm natural cellular antioxidant defenses, leading to oxidation and further functional impairment. There is an oxidative damage potential, as there is a constant free radical formation in small amounts, which escape the cell defense.

The reduction of oxidative stress can be achieved on three levels [44]: i) by lowering exposure to environmental pollutants ii) by increasing the levels of endogenous and exogenous antioxidants in order to scavenge ROS before they can cause any damage; or iii) lowering the generation of oxidative stress by stabilizing mitochondrial energy production and efficiency - reducing the amount of ROS formed per amount of O_2 consumed.

4. Defenses against ROS and strategies to reduce oxidative stress

Generation of ROS and the activity of antioxidant defenses are balanced *in vivo*. In fact, the balance may be slightly tipped in favor of ROS so that there is continuous low-level oxidative damage in the human body.

Besides the endogenous and exogenous antioxidative protection, the second category of defence are repair processes, which remove the damaged biomolecules before they accumulate to cause altered cell metabolism or viability [45].

4.1. Primary Antioxidant Defenses

Superoxide Dismutase (SOD)

SODs are a group of metalloenzymes, which catalyze the conversion of superoxide anion to hydrogen peroxide and dioxygen [46]. This reaction is a source of cellular hydrogen peroxide.

$$2O_2^{\cdot -} + 2H^+ \rightarrow H_2O_2 + O_2 \tag{1}$$

Catalase

Hydrogen peroxide formed by SOD, from other metabolic reactions or from the non-enzymatic reaction of the hydroperoxyl radical, is scavenged by a ubiquitous heme protein catalase. It catalyzes the dismutation of hydrogen peroxide into water and molecular oxygen [47].

$$2\,H_2O_2 \rightarrow O_2 + 2H_2O \tag{2}$$

One antioxidative role of catalases is to lower the risk of hydroxyl radical formation from H_2O_2 via Fenton reaction catalyzed by chromium or ferrous ions.

Glutathione Peroxidase (GPx)

All glutathione peroxidases may catalyze the reduction of H_2O_2 using glutathione (GSH) as a substrate. They can also reduce other peroxides (e.g., lipid peroxides in cell membranes) to alcohols.

$$\mathrm{ROOH} + 2\,\mathrm{GSH} \rightarrow \mathrm{ROH} + \mathrm{GSSG} + H_2O \tag{3}$$

GPx is responsible for detoxification of low H_2O_2 amounts, while in higher H_2O_2 amounts, catalase takes the leading part in cellular detoxification [15].

Glutathione-Related Systems

In addition to enzymatic defenses described above, there is an intracellular non-enzymatic defense system to protect cellular constituents against ROS and for maintaining the redox state. Glutathione (GSH) is the most abundant intracellular thiol-based antioxidant, present in millimolar concentrations in all aerobic cells, eukaryotic and prokaryotic [48]. It is a sulfhydryl buffer, detoxifies compounds through conjugation reactions catalyzed by glutathione S-transferases, directly, as in the case with peroxide in the GPx-catalyzed reaction [47] or with Cr(VI) [49]. GSH is capable of reacting with Cr(VI) to yield Cr(V), Cr(IV), GSH thiyl radicals and Cr(III)-GSH complexes [50, 51]. The ratios of reduced-to-oxidized glutathione (GSH/GSSG) in normal cells are high (> 10 : 1), as the enzyme, glutathione reductase, help to reduce oxidized glutathione in the following reaction:

$$GSSG + NADPH + H^{+} \rightarrow 2GSH + NADP^{+} \tag{4}$$

The NADPH required is from several reactions, the best known from the oxidative phase of pentose phosphate pathway [15]. Both, glutathione reductase and glucose-6-phosphate dehydrogenase are involved in the glutathione recycling system [52].

4.2. Secondary Antioxidant Defenses

Although efficient, the antioxidant enzymes and compounds do not prevent the oxidative damage completely. A series of damage removal and repair enzymes deal with this damage. Many of these essential maintenance and repair systems become deficient in senescent cells, thus a high amount of biological "garbage" is accumulated (e.g., intralysosomal accumulation of lipofuscin) [53, 54]. Age-related oxidative changes are most common in non-proliferating cells, like the neurons and cardiac myocites, as there is no "dilution effect" of damaged structures through cell division [33]. The ability to repair DNA correlates with species-specific lifespan, and is necessary, but not sufficient for longevity [55]. There is an age-related decline in proteasome activity and proteasome content in different tissues (e.g. rat liver, human epidermis); this leads to accumulation of oxidatively modified proteins [56]. Proteasomes are a part of the protein-removal system in eukaryotic cells. Proteasome activity and function may be decreased upon replicative senescence. On the other hand, proteasome activation was shown to enhance the survival during oxidative stress, lifespan extension and maintenance of the juvenile morphology longer in specific cells, e.g. human primary fibroblasts [57]. The total amount of oxidatively modified proteins of an 80-year-old man may be up to 50% [58]. Besides, elevated levels of oxidized proteins, oxidized lipids, advanced DNA oxidation and glycoxidation end products are found in aged organisms [7, 59, 60]. Torres and Perez [61] have shown that proteasome inhibition is a mediator of oxidative stress and ROS production and is affecting mitochondrial function. These authors propose that a progressive decrease in proteasome function during aging can promote mitochondrial damage and ROS accumulation. It is likely that changes in proteasome dynamics could generate a prooxidative conditions that could cause tissue injury during aging, *in vivo* [61].

Numerous studies have reported age-related increases in somatic mutation and other forms of DNA damage, indicating that the capacity for DNA repair is an important determinant of the rate of aging at the cellular and molecular levels [62, 63]. An important player in the immediate cellular response to ROS-induced DNA damage is the enzyme poly(ADP-ribose) polymerase-1 (PARP-1). It recognizes DNA lesions and flags them for repair. Grube and Burkle [64] discovered a strong positive correlation of PARP activity with the lifespan of species: cells from long-lived species had higher levels of PARP activity than cells from short-lived species.

The DNA-repair enzymes, excision-repair enzymes, operate on the basis of damage or mutilation occurring to only one of the two strands of the DNA. The undamaged strand is used as a template to repair the damaged one. The excision repair of oxidized bases involves two

DNA glycosylases, Ogg1p and Ntg2p to remove the damaged bases, like 7,8-dihydro-8-oxo-guanine, 2,6-diamino-4-hydroxy-5-n-methylformamidopyrimidine, thymine glycol, and 5-hydroxycytosine (reviewed in 65). Lipid peroxides or damaged lipids are metabolized by peroxidases or lipases. Overall, antioxidant defenses seems to be approximately balanced with the generation of ROS *in vivo*. There appears to be no great reserve of antioxidant defenses in mammals, but as previously mentioned, some oxygen-derived species perform useful metabolic roles [66]. The production of H_2O_2 by activated phagocytes is the classic example of the deliberate metabolic generation of ROS for organism's advantage [67].

4.3. Exogenous Antioxidant Defenses: Compounds Derived from the Diet

The intake of exogenous antioxidants from fruit and vegetables is important in preventing the oxidative stress and cellular damage. Natural antioxidants like vitamin C and E, carotenoids and polyphenols are generally considered as beneficial components of fruits and vegetables. Their antioxidative properties are often claimed to be responsible for the protective effects of these food components against cardiovascular diseases, certain forms of cancers, photosensitivity diseases and aging [68]. However, many of the reported health claims are based on epidemiological studies in which specific diets were associated with reduced risks for specific forms of cancer and cardiovascular diseases.The identification of the actual ingredient in a specific diet responsible for the beneficial health effect remains an important bottleneck for translating observational epidemiology to the development of functional food ingredients. When ingesting high amounts of synthetic antoxidants, toxic pro-oxidant actions may be important to consider [68].

4.4. Adaptive responses and hormesis

The adaptive response is a phenomenon in which exposure to minimal stress results in increased resistance to higher levels of the same stressor or other stressors. Stressors can induce cell repair mechanisms, temporary adaptation to the same or other stressor, induce autophagy or trigger cell death [69]. The molecular mechanisms of adaptation to stress is the least investigated of the stress responses described above. It may inactivate the activation of apoptosis through caspase-9, i.e. through the intrinsic pathway, one of the main apoptotic pathways [70, 117]. Early stress responses result also in the post-translational activation of pre-existing defenses, as well as activation of signal transduction pathways that initiate late responses, namely the *de novo* synthesis of stress proteins and antioxidant defenses [65]. Hormesis is characterized by dose-response relationships displaying low-dose stimulation and high-dose inhibition [71]. Hormesis is observed also upon the exposure to low dose of a toxin, which may increase cell's tolerance for greater toxicity [35]. Reactive oxygen species (ROS) can be thought of as hormetic compounds. They are beneficial in moderate amounts and harmful in the amounts that cause the oxidative stress. Many studies investigated the induction of adaptive response by oxidative stress [72, 73, 74, 75]. An oxidative stress response is triggered when cells sense an increase of ROS, which may result from exposure of cells to low concentrations of oxidants, increased production of ROS or a decrease in antioxidant defenses. In order to survive, the cells induce the antioxidant defenses and other protective factors, such as stress proteins. Finkel and Holbrook [35] stated that the best strategy to enhance endogenous antioxidant levels may be the oxidative stress itself, based on the classical physiological concept of hormesis.

The enzymatic, non-enzymatic and indirect antioxidant defense systems could be involved in the induction of adaptive response to oxidative stress [76, 77, 78, 79, 80, 81]. It was observed, that a wide variety of stressors, such as pro-oxidants, aldehydes, caloric restriction, irradiation, UV-radiation, osmotic stress, heat shock, hypergravity, etc. can have a life-prolonging effect.The effects of these stresses are linked also to changes in intracellular redox potential, which are transmitted to changes in activity of numerous enzymes and pathways. The main physiological benefit of adaptive response is to protect the cells and organisms from moderate doses of a toxic agent [82, 69]. As such, the stress responses that result in enhanced defense and repair and even cross protection against multiple stressors could have clinical or public-health use.

4.5. Sequestration of metal ions; Fenton-like reactions

Many metal ions are necessary for normal metabolism, however they may represent a health risk when present in higher concentrations. Increased ROS generation has been implicated as a consequence of exposure to high levels of metal ions, like, iron, copper, lead, cobalt, mercury, nickel, chromium, selenium and arsenic, but not to manganese and zinc. The above mentioned transition metal ions are redox active: reduced forms of redox active metal ions participate in already discussed Fenton reaction where hydroxyl radical is generated from hydrogen peroxide [83]. Furthermore, the Haber-Weiss reaction, which involves the oxidized forms of redox active metal ions and superoxide anion, generates the reduced form of metal ion, which can be coupled to Fenton reaction to generate hydroxyl radical [15].

Fenton reaction

$$\mathrm{Metal}^{(n+1)} + H_2O_2 \rightarrow \mathrm{Metal}^{(n+1)+} + HO^{\cdot} + OH^- \quad (5)$$

Haber-Weiss reaction

$$\mathrm{Metal}^{(n+1)+} + 2O_2^{-\cdot} \rightarrow \mathrm{Metal}^{(n+1)} + O_2 \quad (6)$$

Redox cycling is a characteristic of transition metals [84], and Fenton-like production of ROS appear to be involved in iron-, copper-, chromium-, and vanadium-mediated tissue damage [85]. Increases in levels of superoxide anion, hydrogen peroxide or the redox active metal ions are likely to lead to the formation of high levels of hydroxyl radical by the chemical mechanisms listed above. Therefore, the valence state and bioavailability of redox active metal ions contribute significantly to the generation of reactive oxygen species.

- The consequence of formation of free radicals mediated by metals are modifications of DNA bases, enhanced lipid peroxidation, and altered calcium and sulfhydryl homeostasis. Lipid peroxides, formed by the attack of radicals on polyunsaturated fatty acid residues of phospholipids, can further react with redox metals finally producing mutagenic and carcinogenic malondialdehyde, 4-hydroxynonenal and other exocyclic DNA adducts (etheno and/or propano adducts). The unifying factor in determining toxicity and carcinogenicity for all these metals is the abitliy to generate reactive oxygen and nitrogen species. Common mechanisms involving the Fenton reaction, generation of the superoxide radical and the hydroxyl radical are primarily associated with mitochondria, microsomes and peroxisomes. Enzymatic and non-enzymatic antioxidants protect against deleterious metal-mediated free radical attacks to some extent; e.g., vitamin E and melatonin can prevent the majority of metal-mediated (iron, copper, cadmium) damage both in *in vitro* systems and in metal-loaded animals [86, 87].

Iron Chelators

A chelator is a molecule that has the ability to bind to metal ions, e.g. iron molecules, in order to remove heavy metals from the body. According to Halliwell and Gutteridge [22] chelators act by multiple mechanisms; mainly to i) alter the reduction potential or accessibility of metal ions to stop them catalysing $OH^{\cdot}$production (e.g. transferrin or lactoferrin) ii) prevent the escape of the free radical into solution (e.g. albumin). In this case the free radicals are formed at the biding site of the metal ions to chelating agent. Chelators can be man-made or be produced naturally, e.g. plant phenols. Because the iron catalyzes ROS generation, sequestering iron by chelating agents is thought to be an effective approach toward preventing intracellular oxidative damage. Many chelating agents have been used to inhibit iron- or copper-mediated ROS formation, such as ethylenediaminetetraacetic acid (EDTA), diethylenetriaminepenta-acetic acid (DETAPAC), N,n'-Bis- (2-Hydroxybenzyl)ethylenediamine-N,n'-diacetic acid (HBED), 2-3-Dihydroxybenzoate, Desferrioxamine B (DFO), deferasirox (ICL 670), N,N'-bis-(3,4,5-trimethoxybenzyl) ethylenediamine N,N,-diacetic acid dihydrochloride (OR10141), phytic acid, PYSer and others (for details see 22).

Desferrioxamine can react directly with several ROS and is used as iron(III) chelator for prevention and treatment of iron overload in patients who ingested toxic oral doses of iron [22]. Also, the intracellular protein ferritin plays a role in cellular antioxidant defense. It binds nonmetabolized intracellular iron, therefore, aids to regulation of iron availability. In this way it can decrease the availability of iron for participation in Fenton reaction and lipid peroxidations. Body iron burden can be assessed by using a variety of measurements, such as serum ferritin levels and liver iron concentration by liver biopsies [for detailed information see 88, 89, 90].

4.6. Stabilizing mitochondrial ROS production

Oxidative stress and oxidative damage accumulation could be decreased by regulating the electron leakage from electron transport chain and the resultant ROS production [44]. Nutritional and lifestyle modifications may decrease mitochondrial ROS formation, e.g. by caloric restriction (CR), sport activities and healthy eating habits. The anti-aging action of caloric restriction is an example of hormesis [91, 92, 93]. The works of Yu and Lee [94], Koizumi et al. [95] and Chen and Lovry [96] imply that food restriction (energetic stress) increases the overall antioxidant capacity to maintain the optimal status of intracellular environment by balancing ROS in CR thus promotes the metabolic shift to result in more efficient electron transport at the mitochondrial respiratory chain [97]. In this way, the leakage of electrons from the respiratory chain is reduced [98, 99]. There are reports of slower aging by intermittent fasting without the overall reduction of caloric intake [100, 101]. Since it is extremely hard to maintain the long-term CR, the search is on for CR mimetics. These are the agents or strategies that can mimic the beneficial health-promoting and anti-aging effects of CR. Several compounds have been tested for a potential to act as CR mimetic; such as plant-derived polyphenols (e.g., resveratrol, quercetin, butein, piceatannol), insulin-action enhancers (e.g., metformin), or pharmacological agents that inhibit glycolysis (e.g., 2-deoxyglucose) [102].

Mitochondrial uncoupling has been proposed as a mechanism that reduces the production of reactive oxygen species and may account for the paradox between longevity and activity [103]. Moderate and regular exercise enhances health and longevity relative to sedentary lifestyles. Endurance training adaptation results in increased efficiency in ATP synthesis at the expense of potential increase in oxidative stress that is likely to be compensated by enhanced activities of antioxidant enzymes [104] and proteasome [105]. Exercise requires a large flux of energy and a shift in substrate metabolism in mitochondria from state 4 to state 3. This shift may cause an increase in superoxide production [106]. Indeed, a single bout of exercise was found to increase the metabolism and oxidative stress during and immediately after exercise [107, 108, 109]. While a single bout of exercise of sedentary animals is likely to cause increased detrimental oxidative modification of proteins [110], moderate daily exercise appears to be beneficial by reducing the damage in rat skeletal muscle [105]. Organisms exposed to oxidative stress often decrease their rate of metabolism [111, 112]. Metabolic uncoupling may reduce the mitochondrial oxidant production [113]. It may account for the paradox between longevity and activity [103]. Heat is produced when oxygen consumption is uncoupled from ATP generation. When the mitochondria are uncoupled and membrane potential is low animals might produce less free radicals when expending the most energy [114]. Postprandial oxidative stress is characterized by an increased susceptibility of the organism toward oxidative damage after consumption of a meal rich in lipids and/or carbohydrates [115]. The generation of excess superoxide due to abundance of energy substrates after the meal may be a predominate factor resulting in oxidative stress and a decrease in nitric oxide. A mixture of antioxidant compounds is required to provide protection from the oxidative effects of postprandial fats and sugars. No specific antioxidant can be claimed to be the most important, as consumption of food varies enormously in humans. However, a variety of polyphenolic compounds derived from plants appear to be effective dietary antioxidants, especially when consumed with high-fat meals [116].

5. Conclusion and perspectives

In conclusion, excessive production of ROS and reduced antioxidant defence with age significantly contribute to aging. It seems that oxidative damage is the major cause and the most important contributor to human aging. Antioxidant defense seems to be approximatly balanced with the generation of oxygen-derived species in young individuals, however, there is an increase of oxidative stress later in life. Then the approaches to lower the increased ROS formation in our bodies could be implemented by avoiding the exposure to exogenous free radicals, by intake of adequate amounts of antioxidants and/or by stimulating the damage-repair systems of the cells [44 and references within].

Developing natural or pharmacological agents capable of increasing the antioxidative protection and/or modulating the endogenous defense and repair mechanisms may potentially improve health, increase longevity and contribute to treatment of degenerative age-related diseases, such as cardiovascular and neurodegenerative disorders and cancer. The lifestyle changes, e.g. regular physical activity, increased intake of fruits and vegetables, and reduced calorie intake may improve health and increase cellular resistance to stress. Synthetic antioxidant supplements may help to correct the high levels of oxidative stress that cannot be controlled by the sinergy of endogenous antioxidant systems.

Author details

B. Poljsak[1*] and I. Milisav[1,2]

*Address all correspondence to: borut.poljsak@zf.uni-lj.si

1 University of Ljubljana, Laboratory of oxidative stress research, Faculty of Health Sciences, Ljubljana, Slovenia

2 University of Ljubljana, Faculty of Medicine, Institute of Pathophysiology, Ljubljana, Slovenia

References

[1] Kirkwood, B., & Mathers, J. C. (2009). The basic biology of aging. In: Stanner S., Thompson R., Buttriss J. [eds.], *Healthy aging-The role of nutrition and lifestyle*, NY:Wiley-Blackwell.

[2] Fontana, L., & Klein, S. (2007). Aging, Adiposity, and Calorie Restriction. *Jama*, 297, 986-994.

[3] Davies, K. J. (1995). Oxidative stress: the paradox of aerobic life. *Biochem. Soc. Symp.*, 61, 1-31.

[4] Martin, G. M. (1987). Interaction of aging and environmental agents: The gerontological perspective. *Prog. Clin. Bio. Res*, 228, 25-80.

[5] Martin, G. M., Austad, S. N., & Johnson, T. E. (1996). Genetic analysis of ageing: Role of oxidative damage and environmental stress. *Nat. Genet*, 13, 25-34.

[6] Gilca, M., Stoian, I., Atanasiu, V., & Virgolici, B. (2007). The oxidative hypothesis of senescence. *J. Postgrad. Med.*, 53(3), 207-213.

[7] Beckman, K. B., & Ames, B. N. (1998). The free radical theory of aging matures. *Physiol. Rev.*, 78, 547-581.

[8] Casteilla, L., Rigoulet, M., & Penicaud, L. (2001). Mitochondrial ROS metabolism: Modulation by uncoupling proteins. *IUBMB Life*, 52, 181-188.

[9] Hansford, R. G., Hogue, B. A., & Mildaziene, V. (1997). Dependence of H2O2 formation by rat heart mitochondria on substrate availability and donor age. *J. Bioenerg. Biomembr.*, 29, 89-95.

[10] Staniek, K., & Nohl, H. (1999). H(2)O(2) detection from intact mitochondria as a measure for one-electron reduction of dioxygen requires a non-invasive assay system. *Biochim Biophys Acta.*, 1413(2), 70-80.

[11] Speakman, J. R., Selman, C., McLaren, J. S., & Harper, E. J. (2002). Living fast, dying when? The link between aging and energetics. *J. Nutr.*, 132, 1583S-97S.

[12] Mooijaart, S. P., van Heemst, D., Schreuder, J., van Gerwen, S., Beekman, M., & Brandt, B. W. (2004). Variation in the SHC1 gene and longevity in humans. *Exp. Gerontol*, 39, 263-8.

[13] Harman, D. (1972). A biologic clock: the mitochondria? *Journal of the American Geriatrics Society*, 20, 145-147.

[14] Harman, D. (1956). Aging: a theory based on free radical and radiation chemistry. *Journal of Gerontology*, 11, 298-300.

[15] Halliwell, B., & Gutteridge, J. (1999). Free radicals in biology and medicine [3rd edn]. *Oxford: Clarendon Press.*

[16] Reiter, R. J. (1995). Oxygen radical detoxification processes during aging: The functional importance of melatonin. *Aging (Milano)*, 7, 340-51.

[17] Hagen, J. L., Krause, D. J., Baker, D. J., Fu, M. H., Tarnopolsky, M. A., & Hepple, R. T. (2004). Skeletal muscle aging in F344BN F1-hybrid rats: I. mitochondrial dysfunction contributes to the age-associated reduction in CO2max. *J. Gerontol. A. Biol. Sci. Med. Sci.*, 59, 1099-1110.

[18] Hamilton, M. L., Van Remmen, H., Drake, J. A., Yang, H., Guo, Z. M., Kewitt, K., Walter, C. A., & Richardson, A. (2001). Does oxidative damage to DNA increase with age? *Proc. Natl. Acad. Sci. USA*, 98, 10469-10474.

[19] Hagen, T. M. (2003). Oxidative stress, redox imbalance, and the aging process. *Antioxid. Redox Signal*, 5, 503-506.

[20] Sohal, R. (2002). Role of oxidative stress and protein oxidation in the aging process. *Free Radic Biol. Med.*, 33, 37-44.

[21] Sohal, R., Mockett, R., & Orr, W. (2002). Mechanisms of aging: an appraisal of the oxidative stress hypothesis. *Free Radic. Biol. Med.*, 33, 575-86.

[22] Halliwell, B., & Gutteridge, J. (2007). Free radicals in biology and medicine [4th edn]. *Oxford: University Press.*

[23] Ames, B. N. (2004). A Role for Supplements in Optimizing Health: the Metabolic Tune-up. *Archives of Biochemistry and Biophysics*, 423, 227-234.

[24] Arnheim, N., & Cortopassi, G. (1992). Deleterious mitochondrial DNA mutations accumulate in aging human tissues. *Mutat Res.*, 275(3-6), 157-67.

[25] Yang, J. H., Lee, H. C., Lin, K. J., & Wei, Y. H. (1994). A specific 4977- bp deletion of mitochondrial DNA in human aging skin. *Arch. Dermatol. Res*, 286, 386-390.

[26] Golden, T., Morten, K., Johnson, F., Samper, E., & Melov, S. (2006). Mitochondria: A critical role in aging. In: Masoro EJ., Austad S. [eds.], *Handbook of the biology of aging*, Sixth edition. Elsevier.

[27] Jacobs, H. T. (2003). The mitochondrial theory of aging: dead or alive? *Aging cell*, 2, 11.

[28] Pak, J. W., Herbst, A., Bua, E., Gokey, N., McKenzie, D., & Aiken, J. M. (2003). Rebuttal to Jacobs: the mitochondrial theory of aging: alive or dead. *Aging Cell*, 2, 9.

[29] De Grey, A. D. N. J. (2005). Reactive Oxygen Species Production in the Mitochondrial Matrix: Implications for the Mechanism of Mitochondrial Mutation Accumulation. *Rejuvenation Res.*, 8(1), 13-7.

[30] Best, B. Mechanisms of Aging. http://www.benbest.com/lifeext/aging.html, [accessed 10 May 2012].

[31] Chung, H. Y., Sung, B., Jung, K. J., Zou, Y., & Yu, B. P. (2006). The molecular inflammatory process in aging. *Antioxid. Redox. Signal.*, 8, 572-581.

[32] Navratil, V. (2011). Health, Ageing and Entropy. *School and Health 21 Health Literacy Through Education.*, Vyd. 1. Brno : Masarykova Univerzita, 978-8-02105-720-3, 329-336, Brno.

[33] Terman, A. (2001). Garbage catastrophe theory of aging: Imperfect removal of oxidative damage? *Redox. Rep.*, 6, 15-26.

[34] Schulz, T. J., Zarse, K., Voigt, A., Urban, N., Birringer, M., & Ristow, M. (2007). Glucose Restriction Extends Caenorhabditis elegans Lifespan by Inducing Mitochondrial Respiration and Increasing Oxidative Stress. *Cell Metabolism*, 6, 280-293.

[35] Finkel, T., & Holbrook, Nikki J. (2000). Oxidants, oxidative stress and the biology of ageing. *Nature*, 408, 239-247.

[36] Kirkwood, T. B. L., & Holliday, R. (1979). The evolution of ageing and longevity. *Proc. R. Soc. London Ser. B Biol. Sci.*, 205, 531-546.

[37] Kirkwood, T. B. L. (1997). Evolution of ageing. *Nature*, 270, 301-304.

[38] Gavrilov, L. A., & Gavrilova, N. S. (2002). Evolutionary Theories of Aging and Longevity. *The Scientific World JOURNAL*, 2, 339-356.

[39] Von Zglinicki, T., Bürkle, A., & Kirkwood, T. B. (2001). Stress, DNA damage and ageing-an integrative approach. *Exp. Gerontol.*, 36, 1049-1062.

[40] Kurz, D. J., Decary, S., Hong, Y., Trivier, E., Akhmedov, A., & Erusalimsky, J. D. (2004). Chronic oxidative stress compromises telomere integrity and accelerates the onset of senescence in human endothelial cells. *J. Cell Sci*, 117, 2417-2426.

[41] Bayne, S., & Liu, J. P. (2005). Hormones and growth factors regulate telomerase activity in aging and cancer. *Mol. Cell Endocrinol.*, 240, 11-22.

[42] Arking, R. (2006). The biology of aging, observations and principles. *Third edition. New York: Oxford University Press.*

[43] Poljsak, B. (2011). Skin aging, free radicals and antioxidants. *New York: NovaScience Publisher.*

[44] Poljsak, B. (2011). Strategies for reducing or preventing the generation of oxidative stress. *Oxid Med Cell Longev.*, 194586.

[45] Cheeseman, K. H., & Slater, T. F. (1993). An introduction to free radical biochemistry. *Br. Med. Bull*, 49, 481-493.

[46] Hohmann, S. (1997). Yeast Stress Responses. In Hohmann S., Mager WH [eds], *Yeast stress responses*, Austin: HRG. Landes Company.

[47] Santoro, N., & Thiele, D. J. (1997). Oxidative stress responses in the yeast Saccharomyces cerevisiae. In Hohmann S., Mager WH [eds], *Yeast stress responses*, Austin: HRG. Landes Company.

[48] Chesney, J. A., Eaton, J. W., & Mahoney, J. (1996). Bacterial glutathione: a sacrificial defense against chlorine compounds. *J. Bacteriol.*, 178(7), 2131-2135.

[49] Jamnik, P., & Raspor, P. (2003). Stress response of yeast Candida intermedia to Cr(VI). *J. Biochem. Mol. Toxicol.*, 17, 316-23.

[50] Aiyar, J., Berkovits, H. J., Floyd, R. A., & Wetterhahn, K. E. (1990). Reaction of chromium (VI) with hydrogen peroxide in the presence of glutathione: reactive intermediates and resulting DNA damage. *Chem Res Toxicol*, 3(6), 595-603.

[51] Wetterhahn, K. E., & Hamilton, J. W. (1989). Molecular basis of hexavalent chromium carcinogenicity: effect on gene expression. *Sci Total Environ*, 86(1-2), 113-29.

[52] Izawa, S., Inoue, Y., & Kimura, A. (1995). Oxidative stress response in yeast: effect of glutathione on adaptation to hydrogen peroxide stress in Saccharomyces cerevisiae. *Fabs Lett*, 368, 73-76.

[53] Terman, A., & Brunk, U. T. (2006). Oxidative stress, accumulation of biological "garbage," and aging. *Antioxid. Redox Signal*, 8, 197-204.

[54] Brunk, U. T., Jones, C. B., & Sohal, R. S. (1992). A novel hypothesis of lipofuscinogenesis and cellular aging based on interaction between oxidative stress and autophagocitosis. *Mutat. Res*, 275, 395-403.

[55] Cortopassi, G. A., & Wang, E. (1996). There is substantial agreement among interspecies estimates of DNA repair activity. *Mechanisms of Aging and Development*, 91, 211-218.

[56] Grune, T., Reinheckel, T., & Davies, K. J. (1997). Degradation of oxidized proteins in mammalian cells. *Faseb. J.*, 11, 526-34.

[57] Chondrogianni, N., Kapeta, S., Chinou, I., Vassilatou, K., Papassideri, I., & Gonos, E. S. (2010). Anti-ageing and rejuvenating effects of quercetin. *Exp. Gerontol.*, 45(10), 763-71.

[58] Stadtman, E. R. (1992). Protein oxidation and aging. *Science*, 257, 1220-4.

[59] Shringarpure, R., & Davies, K. J. (2002). Protein turnover by the proteasome in aging and disease. *Free Radic. Biol. Med.*, 32, 1084-9.

[60] Sell, D. R., Lane, M. A., Johnson, W. A., Masoro, E. J., Mock, O. B., Reiser, K. M., Fogarty, J. F., Cutler, R. G., Ingram, D. K., Roth, G. S., & Monnier, V. M. (1996). Longevity and the genetic determination of collagen glycoxidation kinetics in mammalian senescence. *Proc. Natl. Acad. Sci. USA*, 93(1), 485-90.

[61] Torres, C. A., & Perez, V. I. (2008). Proteasome modulates mitochondrial function during cellular senescence. *Free Radic. Biol. Med.*, 44(3), 403-14.

[62] Promislow, D. E. (1994). DNA repair and the evolution of longevity: a critical analysis. *J.Theor. Biol.*, 170, 291-300.

[63] Bürkle, A., Beneke, S., Brabeck, C., Leake, A., Meyer, R., Muiras, M. L., & Pfeiffer, R. (2002). Poly(ADP-ribose) polymerase-1, DNA repair and mammalian longevity. *Exp. Gerontol.*, 37(10-11), 1203-5.

[64] Grube, K., & Bürkle, A. (1992). Poly(ADP-ribose) polymerase activity in mononuclear leukocytes of 13 mammalian species correlates with species-specific lifespan. *Proc. Natl. Acad. Sci. USA*, 89, 11759-11763.

[65] Costa, V., & Moradas-Ferreira, P. (2001). Oxidative stress and signal transduction in Saccharomyces cerevisiae: insights into ageing, apoptosis and diseases. *Mol. Aspects Med.*, 22, 217-246.

[66] Stocker, R., & Frei, B. (1991). Endogenous antioxidant defenses in human blood plasma. *In: Oxidative stress: oxidants and antioxidants.*, London: Academic press.

[67] Halliwell, B., & Cross, C. E. (1994). Oxygen-derived species: their role in human disease and environmental stress. *Environ. Health Perspect.*, 102, 5-12.

[68] Rietjens, I., Boersma, M., & de Haan, L. (2001). The pro-oxidant chemistry of the natural antioxidants vitamin C, vitamin E, carotenoids and flavonoids. *Environ Toxicol. Pharmacol*, 11, 321-333.

[69] Milisav, I. (2011). Cellular Stress Responses. In: Wislet-Gendebien S. [Ed.], *Advances in Regenerative Medicine*, 978-9-53307-732-1, InTech, Available from, http://www.intechopen.com/articles/show/title/cellular-stress-responses.

[70] Nipic, D., Pirc, A., Banic, B., Suput, D., & Milisav, I. (2010). Preapoptotic cell stress response of primary hepatocytes. *Hepatology.*, 51(6), 2140-51.

[71] Calabrese, E. J., & Baldwin, L. A. (2002). Hormesis and high-risk groups. *Regul Toxicol Pharmacol.*, 35(3), 414-28.

[72] Feinendegen, L. E., Bond, V. P., Sondhaus, C. A., & Muehlensiepen, H. (1996). Radiation effects induced by low doses in complex tissue and their relation to cellular adaptive responses. *Mutat Res*, 358, 199-205.

[73] Jones, S. A., McArdle, F., Jack, C. I. A., & Jackson, M. J. (1999). Effect of antioxidant supplement on the adaptive response of human skin fibroblasts to UV-induced oxidative stress. *Redox Report*, 4, 291-299.

[74] de Saint-Georges, L. (2004). Low-dose ionizing radiation exposure: Understanding the risk for cellular transformation. *J Biol Regul Homeost Agents*, 18, 96-100.

[75] Shankar, B., Pandey, R., & Sainis, K. (2006). Radiation-induced bystander effects and adaptive response in murine lymphocytes. *Int J Radiat Biol*, 82, 537-548.

[76] Mendez-Alvarez, S., Leisinger, U., & Eggen, R. I. (1999). Adaptive responses in Chlamydomonas reinhardtii. *Int Microbiol*, 2, 15-22.

[77] Chen, Z. H., Yoshida, Y., Saito, Y., Sekine, A., Noguchi, N., & Niki, E. (2006). Induction of adaptive response and enhancement of PC12 cell tolerance by 7-hydroxycholesterol and 15-deoxy-delta(12,14)-prostaglandin J2 through up-regulation of cellular glutathione via different mechanisms. *J Biol Chem*, 281, 14440-14445.

[78] Yan, G., Hua, Z., Du, G., & Chen, J. (2006). Adaptive response of Bacillus sp. F26 to hydrogen peroxide and menadione. *Curr Microbiol*, 52, 238-242.

[79] Tosello, M. E., Biasoli, M. S., Luque, A. G., Magaró, H. M., & Krapp, A. R. (2007). Oxidative stress response involving induction of protective enzymes in Candida dubliniensis. *Med Mycol*, 45, 535-540.

[80] Joksic, G., Pajovic, S. B., Stankovic, M., Pejic, S., Kasapovic, J., Cuttone, G., Calonghi, N., Masotti, L., & Kanazir, D. T. (2000). Chromosome aberrations, micronuclei, and activity of superoxide dismutases in human lymphocytes after irradiation in vitro. *Cell Mol Life Sci*, 57, 842-850.

[81] Bercht, M., Flohr-Beckhaus, C., Osterod, M., Rünger, T. M., Radicella, J. P., & Epe, B. (2007). Is the repair of oxidative DNA base modifications inducible by a preceding DNA damage induction? *DNA Repair*, 6, 367-373.

[82] Crawford, D. R., & Davies, K. J. (1994). Adaptive response and oxidative stress. *Environ Health Perspect.*, 102(10), 25-8.

[83] Nordberg, J., & Arner, E. S. J. (2001). Reactive oxygen species, antioxidants, and the mammalian thioredoxin system. *Free Radic Biol Med*, 31(11), 1287-1312.

[84] Klein, C. B., Frenkel, K., & Costa, M. (1991). The role of oxidative processes in metal carcinogenesis. *Chem. Res. Toxicol.*, 4, 592-604.

[85] Fuch, J., Podda, M., & Zollner, T. (2001). Redox Modulation and Oxidative Stress in Dermatotoxicology. *In: Fuchs, J; Packer, L. [eds]. Environmental stressors in health and disease. NY: Marcel Dekker, Inc.*

[86] Valko, M., Morris, H., & Cronin, M. T. (2005). Metals, toxicity and oxidative stress. *Curr. Med. Chem.*, 12(10), 1161-208.

[87] Valko, M., Leibfritz, D., Moncol, J., Cronin, M. T. D., Mazur, M., & Telser, J. (2007). Free radicals and antioxidants in normal physiological functions and human disease. *The International Journal of Biochemistry & Cell Biology*, 39, 44-84.

[88] Jensen, P. D. (2004). Evaluation of iron overload. *Br J Haematol*, 124(6), 697-71.

[89] Angelucci, E., Brittenham, G. M., McLaren, C. E., et al. (2000). Hepatic iron concentration and total body iron stores in thalassemia major. *N Engl J Med.*, 343(5), 327-331.

[90] Kitazawa, M., Iwasaki, K., & Sakamoto, K. (2006). Iron chelators may help prevent photoaging. *J. Cosmet. Dermatol.*, 5(3), 210-7.

[91] Anderson, R. M., Bitterman, K. J., Wood, J. G., Medvedik, O., & Sinclair, D. A. (2003). Nicotinamide and Pnc 1 govern lifespan extension by calorie restriction in S. *Cerevisiae. Nature*, 432, 181-185.

[92] Iwasaki, K., Gleiser, C. A., Masoro, E. J., McMahan, C. A., Seo, E. J., & Yu, B. P. (1988). The influence of the dietary protein source on longevity and age-related disease processes of Fischer rats. *Journal of gerontology*, 43, B5-B12.

[93] Mattson, M. P. (2003). Energy Metabolism and Lifespan Determination. *Adv. Cell Aging Geronto*, 14, 105-122.

[94] Lee, D. W., & Yu, B. P. (1991). Food restriction as an effective modulator of free radical metabolism in rats. *Korean Biochem J*, 24, 148-154.

[95] Koizumi, A., Weindruch, R., & Walford, R. L. (1987). Influences of dietary restriction and age on liver enzyme activities and lipid peroxidation in mice. *J Nutr*, 117(2), 361-7.

[96] Chen, L. H., & Lowry, S. R. (1989). Cellular antioxidant defense system. *Prog Clin Biol Res*, 287, 247-56.

[97] Sohal, R., & Weindruch, R. (1996). Oxidative stress, caloric restriction, and aging. *Science*, 273, 59-63.

[98] Korshunov, S. S., Skulachev, V. P., & Starkov, A. A. (1997). High protonic potential actuates a mechanism of production of reactive oxygen species in mitochondria. *Febs. Lett.*, 416, 15-18.

[99] Starkov, A. A. (1997). "Mild" uncoupling of mitochondria. *Biosci. Rep.*, 17, 273-279.

[100] Gredilla, R., Sanz, A., Lopez-Torres, M., & Barja, G. (2001). Caloric restriction decreases mitochondrial free radical generation at complex I and lowers oxidative damage to mitochondrial DNA in the rat heart. *Faseb J.*, 15, 1589-1591.

[101] Anson, R. M., Guo, Z., de Cabo, R., Iyun, T., Rios, M., Hagepanos, A., Ingram, D. K., Lane, M. A., & Mattson, M. P. Intermittent fasting dissociates beneficial effects of dietary restriction on glucose metabolism and neuronal resistance to injury from calorie intake. *Proc. Natl. Acad. Sci. U S A., USA, 203*, 100(10), 6216-20.

[102] Ingram, D. K., Zhu, M., & Mamczarz, J. (2006). Calorie restriction mimetics: an emerging research field. *Aging Cell.*, 5, 97-108.

[103] Cámara, Y., Duval, C., Sibille, B., & Villarroya, F. (2007). Activation of mitochondrial-driven apoptosis in skeletal muscle cells is not mediated by reactive oxygen species production. *Int. J. Biochem. Cell Biol.*, 39(1), 146-60.

[104] Hollander, J., Fiebig, R., Gore, M., Bejma, J., Ookawara, T., Ohno, H., & Ji, L. L. (1999). Superoxide dismutase gene expression in skeletal muscle: fiber-specific adaptation to endurance training. *Am. J. Physiol.*, 277, R856-R862.

[105] Radak, Z., Nakamura, A., & Nakamoto, H. (1998). A period of exercise increases the accumulation of reactive carbonyl derivatives in the lungs of rats. *Pfluger Arch: Eur. J. Physiol.*, 435, 439-441.

[106] Barja, G. (1999). Mitochondrial oxygen radical generation and leak: sites of production in states 4 and 3, organ specificity, and relation to aging and longevity. *J Bioenerg Biomembr.*, 31(4), 347-66.

[107] Alessio, H. M., & Goldfarb, A. H. (1988). Lipid peroxidation and scavenger enzymes during exercise. Adaptive response to training. *J Appl Physiol*, 64, 1333-1336.

[108] Ji, L. L. (1993). Antioxidant enzyme response to exercise and aging. *Med Sci Sport Exerc.*, 25, 225-231.

[109] Powers, S. K., & Jackson, M. J. (2008). Exercise-induced oxidative stress: cellular mechanisms and impact on muscle force production. *Physiol Rev.*, 88(4), 1243-76.

[110] Reznick, A. Z., Kagan, V. E., Ramsey, R., Tsuchiya, M., Khwaja, S., Serbinova, E. A., & Packer, L. (1992). Antiradical effects in L-propionyl carnitine protection of the heart against ischemia-reperfusion injury: the possible role of iron chelation. *Arch Biochem Biophys.*, 296(2), 394-401.

[111] Allen, R. G., Farmer, K. J., Newton, R. K., & Sohal, R. S. (1984). Effects of paraquat administration on longevity, oxygen consumption, lipid peroxidation, superoxide dismutase, catalase, glutathione reductase, inorganic peroxides and glutathione in the adult housefly. *Comp Biochem Physiol C.*, 78(2), 283-8.

[112] Allen, R. G., & Sohal, R. S. (1982). Life-lengthening effects of gamma-radiation on the adult housefly, Musca domestica. *Mech Ageing Dev.*, 20(4), 369-75.

[113] Skulachev, V. P. (1996). Role of uncoupled and non-coupled oxidations in maintenance of safely low levels of oxygen and its one-electron reductants. *Q Rev Biophys.*, 169-202.

[114] Speakman, J. R., & Selman, C. (2011). The free-radical damage theory: Accumulating evidence against a simple link of oxidative stress to ageing and lifespan. *Bioessays.*, 33(4), 255-9.

[115] Ursini, F., & Sevanian, A. (2002). Postprandial oxidative stress. *Biol Chem.*, 383(3-4), 599-605.

[116] Sies, H., Stahl, W., & Sevanian, A. (2005). Nutritional, dietary and postprandial oxidative stress. *J Nutr.*, 135(5), 969-72.

[117] Banič, B., Nipič, D., Suput, D., & Milisav, I. (2011). DMSO modulates the pathway of apoptosis triggering. *Cell Mol Biol Lett.*, 16(2), 328-41.

Section 4

Disease and Therapy - A Role for Antioxidants

Chapter 15

Disease and Therapy: A Role for Oxidants

Eva María Molina Trinidad,
Sandra Luz de Ita Gutiérrez,
Ana María Téllez López and Marisela López Orozco

Additional information is available at the end of the chapter

http://dx.doi.org/10.5772/51994

1. Introduction

Oxidative stress is a large increase reduction potential in cell or a decrease in reducing capacity of the cellular redox couples such as glutation. Effects of oxidative stress depend on the magnitude of these changes, if the cell is able to overcome small perturbations and regain its original state. However, severe oxidative stress can cause cell death and even moderate oxidation can trigger apoptosis, whereas if it is too intense can cause necrosis.

A particularly destructive aspect of oxidative stress is the production of reactive oxygen species, which include free radicals and peroxides. Some of the less reactive species (superoxide) can be converted by a redox reaction with transition metals or other compounds quinines redox cycle, more aggressive radical species which can cause extensive damage cellular. Most of these species derived from oxygen are produced at a low level in normal aerobic metabolism and the damage they cause to cells is constantly repaired. However, under the severe levels of oxidative stress that causes necrotic damage produces ATP depletion prevents cell death by apoptosis control.

The antioxidants are substances that may protect your cells against the effects of free radicals. Free radicals are molecules produced when your body breaks down food, or by environmental exposures like tobacco smoke and radiation. Free radicals can damage cells, and may play a role in heart disease, cancer and other diseases.

Antioxidant substances include beta-carotene, lutein, lycopene, selenium, vitamin A; and vitamin C. Antioxidants are found in many foods. These include fruits and vegetables, nuts, grains, and some meats, poultry and fish.

Free radicals damage may lead to cancer. Antioxidants interact with and stabilize free radicals and may prevent some of the damage free radicals might otherwise cause.

Studies in cancer cells *in vitro* and *in vivo* animal's models suggest that the use of free radicals decreases the growth of malignant cells. However, information from recent clinical trials is less clear. In recent years, large-scale, randomized clinical trials reached inconsistent conclusions.

Clinical trials published in the 1990s reached differing conclusions about the effect of antioxidants on cancer. The studies examined the effect of beta-carotene and other antioxidants on cancer in different patient groups. However, beta-carotene appeared to have different effects depending upon the patient population, therefore it is important to personalize treatment, and we must take into account the variability to treatment and individualize or personalize therapy. Studies made by Blot WJ et al., in 1993 for the treatment of cancer published in Chinese Cancer Prevention Study, investigated the effect of a combination of beta-carotene, vitamin E, and selenium on cancer in healthy Chinese men and women at high risk for gastric cancer. The study showed a combination of beta-carotene, vitamin E, and selenium significantly reduced incidence of both gastric cancer and cancer overall.

A 1994 cancer prevention study entitled the Alpha-Tocopherol (vitamin E)/ Beta-Carotene Cancer Prevention Study (ATBC) demonstrated that lung cancer rates of Finnish male smokers increased significantly with beta-carotene and were not affected by vitamin E. Epidemiologic evidence indicates that diets high in carotenoid-rich fruits and vegetables, as well as high serum levels of vitamin E (alpha-tocopherol) and beta carotene are associated with a reduced risk of lung cancer. Another study made by Omenn GS in 1994, the Beta-Carotene and Retinol (vitamin A). Efficacy Trial (CARET) also demonstrated a possible increase in lung cancer associated with antioxidants.

The 1996 Physicians' Health Study I (PHS) found no change in cancer rates associated with beta-carotene and aspirin taken by U.S. male physicians.

The 1999 Women's Health Study (WHS) made by Lee IM, tested effects of vitamin E and beta-carotene in the prevention of cancer and cardiovascular disease among women age 45 years or older. Among apparently healthy women, there was no benefit or harm from beta-carotene supplementation. Investigation of the effect of vitamin E is ongoing.

Three large-scale clinical trials continue to investigate the effect of antioxidants on cancer. The Women's Health Study (WHS) is currently evaluating the effect of vitamin E in the primary prevention of cancer among U.S. female health professionals age 45 and older.

In 2006, the Selenium and Vitamin E Cancer Prevention Trial (SELECT) is taking place in the United States, Puerto Rico, and Canada. SELECT is trying to find out if taking selenium and/or vitamin E supplements can prevent prostate cancer in men age 50 or older. Also the experimental and epidemiologic investigations suggest that alpha-tocopherol (the most prevalent chemical form of vitamin E found in vegetable oils, seeds, grains, nuts, and other foods) and beta-carotene (a plant pigment and major precursor of vitamin A found in many yellow, orange, and dark-green, leafy vegetables and some fruit) might reduce the risk of cancer, particularly lung cancer. The initial findings of the Alpha-Tocopherol, Beta-Carotene Cancer Prevention Study (ATBC Study) indicated, however, that lung cancer incidence was increased among participants who received beta-carotene as a supplement. Similar results were recently reported by the Beta-Carotene and Retinol Efficacy Trial (CARET), which tested a combination of beta-carotene and vitamin A.

The Physicians' Health Study II (PHS II) is a follow up to the earlier clinical trial by the same name. The study is investigating the effects of vitamin E, C, and multivitamins on prostate cancer and total cancer incidence. In another case the supplementation with alpha-tocopherol or beta-carotene does not prevent lung cancer in older men who smoke. Beta-Carotene supplementation at pharmacologic levels may modestly increase lung cancer incidence in cigarette smokers, and this effect may be associated with heavier smoking and higher alcohol intake.

Antioxidants neutralize free radicals as the natural by-product of normal cell processes. Free radicals are molecules with incomplete electron shells which make them more chemically reactive than those with complete electron shells. Exposure to various environmental factors, including tobacco smoke and radiation, can also lead to free radical formation. In humans, the most common form of free radicals is oxygen. When an oxygen molecule (O_2) becomes electrically charged or "radicalized" it tries to steal electrons from other molecules, causing damage to the DNA and other molecules. Over time, such damage may become irreversible and lead to disease including cancer. Antioxidants are often described as "mopping up" free radicals, meaning they neutralize the electrical charge and prevent the free radical from taking electrons from other molecules.

Because of the importance that involves using antioxidants as an alternative in the treatment and prevention of chronic degenerative diseases is useful to express the potential in the use and development of new drugs that include antioxidants.

Free radicals are highly reactive chemical species that possess an unpaired electron. Due to it is reactivity, the radicals react readily with other molecules. When free radicals come into contact with the molecules of the human body such as proteins, lipids, carbohydrates, DNA nucleic acids, react with them. These reactions cause changes in the normal functions of these primary metabolites, which cause severe damage that can cause diseases such as cancer and degenerative diseases like Parkinson´s disease or Alzheimer´s disease and atherosclerosis, coronary heart disease and diabetes [1-4].

When any of these afore mentioned diseases, the patient receive the treatment used to treat the particular disease, however, prevention plays a big role. Oxidation in the body tissues caused by free radicals can be prevented with a daily intake of foods that have antioxidants.

The implications of modern life cause changes in eating habits of people, these results in a lack of antioxidants in the body to cope with free radicals that are in contact. The role of antioxidants is to react with free radicals and thus prevent, to react with the primary metabolites, thus acting as natural shields against diseases like cancer [5, 6].

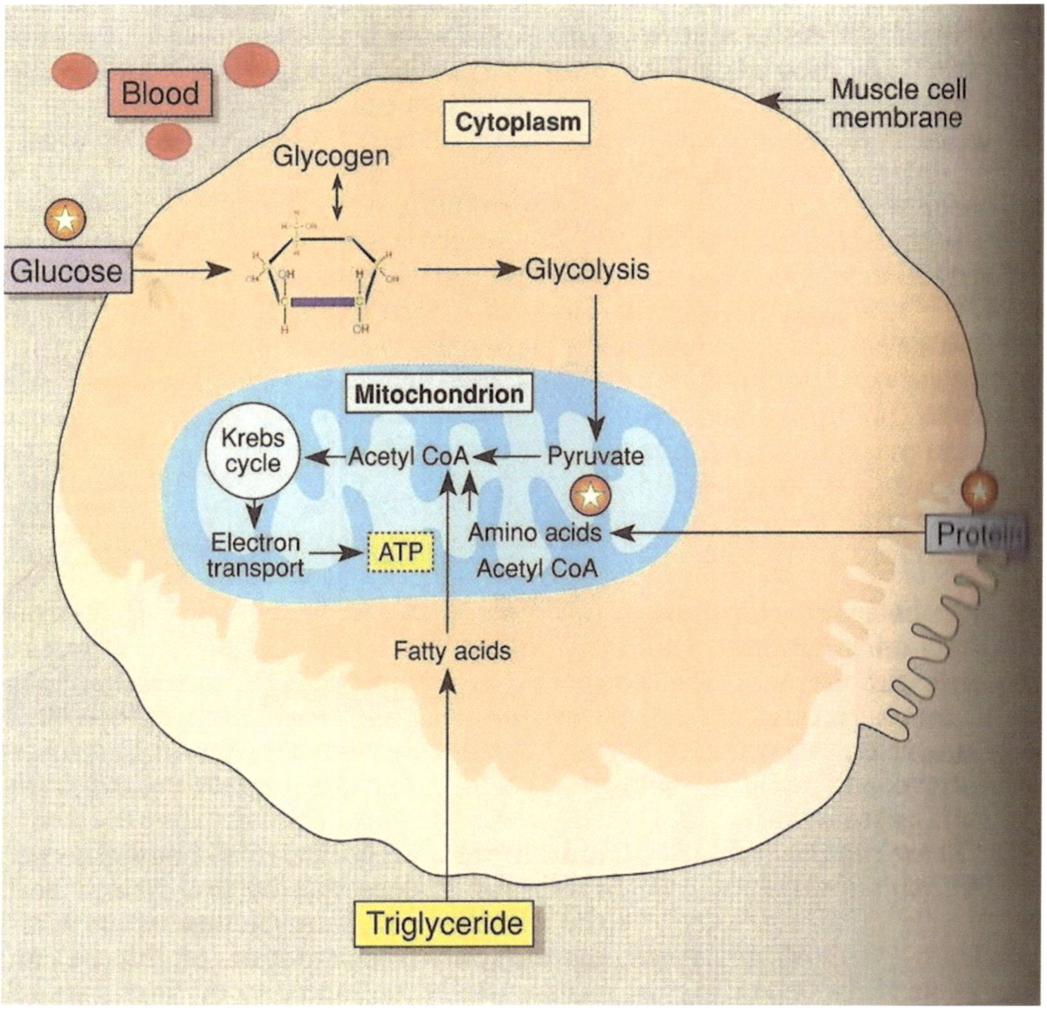

Figure 1. Antioxidants decrement oxidative processes. [Bruce Ames Ph.D., University of California Lecture U.C.T.V. viewed on 08-14-2004].

2. Cancer

2.1. Breast cancer

Currently breast cancer is a disease of high incidence worldwide and causes millions of deaths annually [7]. In the treatment of various cancers have been used drugs that originate from natural products. To get to the application of the drug as a treatment, it requires years of research. The use of treatment leads to the destruction of cancer cells and normal cells in addition, there are numbers side effects resulting from the application of therapies. Prevention of disease is certainly a great alternative to the aggressive use of medications commonly used. The simplest method to prevent cancer and other diseases is undoubtedly add to the diet foods that contain high concentrations of antioxidants, this treatment is easy to perform and causes no adverse side effects. Other organisms containing large amounts of secondary metabolites some of which can act as antioxidants and thereby help prevent cancer and prevent its development (Fruits, vegetables, plants).

Antioxidants can act in two ways:

Blocking cancer, in the initial stage protecting cells against oxidative species and enhancing DNA repair.

Suppressing cancer by inhibiting the progressive stages after formation of pre-neoplastic cells [8].

Studies are underway to help better understand the mechanism of action of antioxidants and test its efficacy against cancer and other diseases. Several studies report that the addition to the diet of foods containing antioxidants may increase the effectiveness of cancer treatment, and help strengthen the body against the side effects associated with treatment [9-11]. The antioxidants found in fruits and vegetables can mention vitamins C and E, carotenoids group and the group of polyphenols. Polyphenols are a group of antioxidant flavonoids to which they belong. There are several types of flavonoids and can be found in foods such as blackberries, blueberries, strawberries, plum, peach, apple, tomato, cherry, broccoli, onion, soya been, legumes like green gram, lupine peas, soy beans, white and horse gram, green leafy spices, citrus fruits, tea, red grapes, chocolate, cocoa and red wine beverages [12].

The following briefly discuss some results of studies using antioxidants from fruits and vegetables for the treatment of breast cancer.

Research conducted in Canada by Hakimuddin and colleagues [13] showed that the polyphenols found in red wine have selective toxicity against MCF-7 cell type of breast cancer; the authors indicate the importance of a diet that incorporates red wine and feeding grapes to serve as a preventive strategy against cancer, which also can be combined with standard therapies.

As mentioned above plums and peaches are fruits that contain phenolic compounds. In a study to test the activity of phenolic species as cancer chemopreventive agents present in extracts of plums and peaches, we found that peaches and plums contain a mixture of phenolic compounds with the ability to inhibit cell lines MCF-7 and MDA -MB435. A very important point to consider is that phenolic acids were isolated chlorogenic and neo-chlorogenic which have great potential for use as chemopreventive agents exerting growth inhibition of the cell line MDA-MB-435 and low toxicity to the normal cell line MCF-10A [14].

Another recent study [15], focused on the action of terpenes located in the skin of the olives suggests that they may serve as natural potential protective against breast cancer. The triterpenes were isolated in significant quantities from the pulp of the olive oil and can act prophylactically and therapeutically.

Currently the investigation for the treatment of breast cancer using apigenin, a flavonoid found in celery. The study conducted at the University of Missouri (United States) [16] was performed in mice that were implanted cell line BT-474, of rapid growth. Mice were also treated with medroxyprogesterone acetate (MPA), which is used in postmenopausal women. Another group of mice was used as a blank. The group of mice treated with MPA was injected apigenin, found that cancerous tumors grew rapidly in mice that were treated with apigenin. Moreover, in mice treated with apigenin was observed a decrease of the tumor when compared with the group of mice used as a blank. Yet unknown mechanism of action of apigenin chemical, however, although the study was conducted in mice, is very promising for future treatment of breast cancer.

2.2. Prostate cancer

Prostate cancer is a very common type of cancer afflicting men; it is now easy detection by prostate specific antigen test (PSA for its initials in English) [17]. Which are still unknown factors that cause this type of cancer, the disease also takes years in some cases to express symptoms, making it necessary for men to undergo regular medical examinations to detect early. One form of treatment of prostate cancer is surgery, whereby the prostate is removed, but this is a procedure which results in urinary incontinence and impotence, which in some cases is permanent.

Prevention through diet prostate cancer has increased because it is recognized as a way to combat this disease [18, 19]. Among the foods that are recommended for the prevention of prostate cancer are generally fruits and vegetables due to its high content of antioxidants. Fruits like pomegranate containing metabolites such as polyphenols and delphinidin urolitina A and B chloride, kaempferol, and punicic acid are considered biologically active against prostate cancer [20, 21].

Other fruit that contains a variety of polyphenolic compounds is strawberry [22] has been found that extracts of strawberry juice cell lines tested against prostate cancer proved effective as antiproliferative agents, is also noteworthy to mention that were tested individually some of the individual components of the extract (cyanidin-3-glucoside, pelargonidin, pelargonidin-3-glucoside, pelargonidin-3-rutinoside, kaempferol, quercetin, kaempferol-3-(6'-coumaroyl) glucoside, 3,4,5 trihydroxyphenyl-acrylic acid-, glucose ester of (E)-*p*-coumaric acid, and ellagic acid) which also showed efficacy individually [23]. These studies confirm the effectiveness of the cutter to inhibit growth of cancer cells.

The apple is considered the quintessential fruit of health, its daily intake is associated with low risk of chronic diseases and cancer, particularly prostate and colon [24-26]. The block contains a variety of compounds polyphenolic that are responsible for their biological activity among these compounds, studies were performed with quercetin which has proven effective as an inhibitor *in vitro* cell growth of prostate cancer [23, 24]. Another study showed that the antioxidant activity of apples is correlated [27] with the total concentration of phenolic compounds present in it clear that this concentration varies according to growing region, and other growth period factors [28-30]. The tomato is another fruit with high antioxidant capacity and owes its activity to lycopene, a carotenoid, which gives the characteristic red color to the fruit [31, 32]. It has been reported that tomato consumption reduces the occurrence of prostate cancer [33-35].

Another study used extracts of potato species Solanum jamesii to test their cytotoxic activity toward antiproliferatva and prostate cancer cells and colon *in vitro*. The extracts were found to inhibit proliferation of cancer cells PC-3 prostate as well as in colon cancer cells LNCaP. Fractions were also tested extract containing anthocyanin and it showed the same activity as the full extract [36].

2.3. Cervical cancer

It is a type of cancer that has one of the top female deaths worldwide [37]. Its main cause is due to Human Papilloma Virus, which is a group of more than 150 types of viruses and is transmitted by sexual contact [38]. To the treatment of cervical cancer, chemotherapy and radiation therapy is performed. As prevention against this type of cancer was recommended not realize sexual contact with infected persons. Another form of prevention is the application of the vaccine that protects against types of HPV high risk of developing cancer. These vaccines Gardasil ® and Cervarix ® were approved by the Federal Drug Administration (FDA) of EU, but these vaccines are only for women, 9 to 26 years of age who are not infected by the virus.

Another recommendation to prevent this cancer is to stimulate the immune system by eating foods rich in antioxidants, because if the body is weakened, the virus is an opportunity to attack and develop cancer [38]. Have also been performed *in vitro* studies to observe foods as antioxidants influence on the growth of cervical cancer cells [39]. One study was carried out with extracts of different types of berries and tested for anti-proliferative activity on HeLa cells (cervical carcinoma). The results show that extracts from blueberry and pomegranate have little effect inhibiting the growth of HeLa cells. The most effective extracts with increasing concentration were: strawberry extract, arctic bramble, lingonberry and cloudberry. It has also been reported [40] that glycoalkaloids present in commercial potatoes inhibit the growth of different types of cancer cell lines, including HeLa cervical cancer cells.

In therapy of cancer selenium doses is 4000 μg in continuous infusion of 1000 μg/9 days, total: 13 mg [41] (Forceville et al, 2007), i.v. bolus 1000 μg in 30 minutes for continuous infusion 1000 μg/d 14 d, total: 15 mg; i.v. bolus 2000 μg in 2 hours continuous infusion 1600 μg/d, 10 d, total 18 mg [42].

3. Diabetes

Diabetes is a metabolic disorder associated with defects in secretion and insulin action [43]. Type 1 diabetes also known as insulin dependent and type 2 diabetes called non-insulin dependent. Both conditions are associated with the formation of free radicals that cause oxidative stress and disease manifestation. Diabetes is associated with health problems such as neuropathy, retinopathy, erectile dysfunction in men, kidney problems, healing and more

[44, 45]. Because diabetes is a disease of oxidative stress, it is expected that the antioxidants in fruits, vegetables and plants to help combat it.

Several studies report that a proper diet that includes antioxidants is important to reduce the risk of diabetes. We have found that various antioxidants present in some foods and plants as coumarins, some terpenes, flavonoids, lignans, phenylpropanoids, tannins and can help people prevent disease and for helping diabetics [46, 47]. These substances exert their activity by inhibiting the action of R-amylase enzyme. Amylase is an enzyme produced in the pancreas and salivary glands; their function is to help the digestion of carbohydrates [48]. Among the flavonoids that can inhibit R-amylase are the quercetin, myricetin, epigallocatechin gallate, and cyanidin. Tannins, present in green and black teas, grapes, wine, raspberry, and strawberry, also seem to be good R-amylase inhibitors. Among fruits and vegetables reported with inhibitory capacity toward the R-amylase in vitro are the red grapes, strawberry, raspberry and, green pepper, broccoli, ginger, and carrot [49-54].

Thanks to these findings, it has been proposed the use of some natural metabolites present in these fruits for the control of hyperglycemia following ingestion of food. The advantage of these natural metabolites is that its use can avoid the side effects that occur when drugs are used for this purpose [55, 56].

Consumption of foods rich in antioxidants can also prevent the complications of this disease has recently been shown that biotin is a vitamin which is part of the B vitamins, which can be found in foods such as biotin find when we eat certain vegetables: cauliflower, peanut butter, mushrooms, yeast, potatoes, mushrooms, almonds, walnuts, soybeans, chickpeas, grapes, strawberries, watermelon, bananas, wheat, flour, pasta, bread, oats, rice, liver, yolk egg, kidney, fish, poultry and offal in general, can help improve metabolism and insulin sensitivity, leading to decreased levels of blood sugar, also sold capsules containing biotin [57, 58].

Resveratrol is a polyphenol present in red wine. According to research Medical Center, University of Texas Southwestern in the U.S. [60], resveratrol administered directly into the brain of diabetic mice, can help control type two diabetes by improving blood sugar levels. What makes the resveratrol is to activate a protein called sirtuin which is expressed in parts of the brain that govern the metabolism of glucose. Much remains to be investigated but it is certainly likely that the intake of red wine under medical supervision can help control diabetes.

Also been studied antioxidants in plants and animals such as the following examples show.

A group of researchers at the University of Jaen in Spain isolated a compound called Cinnamtannin B-1 of the laurel, which has antioxidant properties that can eliminate free radicals that cause diseases such as diabetes. The university has signed an agreement with a pharmaceutical for the distribution of this antioxidant [61].

Lipoic acid, also known as alpha lipoic acid or thioctic acid, is produced in small quantities our bodies, it participates in the metabolism significantly. Can also be found in foods like red meat, yeast and some vegetables such as spinach, broccoli.

http://dx.doi.org/10.5772/51994

In this fatty acid properties are attributed as an antioxidant par excellence also can help re-use of other antioxidants like vitamins C and E, glutathione and coenzyme Q10. Among the many properties that are attributed to reduction of varicose veins, skin moisture, enhances energy levels in the body, cancer protection among others.

Also attributed the reduction in blood glucose levels for type 2 diabetes and help combat the discomforts caused by peripheral neuropathy, and therefore coupled with the effects mentioned above, this antioxidant is ideal for diabetics [62-67].

Currently sold in different forms under different names, but the diabetic patient can take doses of lipoic acid consuming identified through the diet. No indication that lipoic acid has contraindications, although high doses can cause episodes of hypoglycemia [68].

4. Arteriosclerosis

Arteriosclerosis is the hardening of the arteries due to fat accumulation; this may lead to a heart attack that can end life [69]. Atherosclerosis is a preventable disease with a balanced diet and exercise. The diet should include variety of fruits and vegetables and be low in fat. Antioxidants play an important role in preventing this disease, it is known that there is a relationship between red wine consumption and the low incidence of cardiovascular disease; this is due to the action of the antioxidants present in grapes. We recommend a daily intake of 375 mL of red wine to increase levels of high density lipoprotein HDL proteins, ie proteins responsible for transporting fat [70, 71]. Studies with another fruits can be determining its effectiveness in the prevention of arteriosclerosis.

Another fruit that has been investigated for its antioxidant and cardiovascular protective effects are blueberries. Studies realized in Arkansas State University, evaluated the effect on two groups of mice for twenty weeks. One group was used as a target, leading a normal diet; the other group was fed a blueberry base [72], found that mice with arterial lesions, a significant percentage decreased injuries, compared with the group of mice that did not eat blueberries. The researchers suggest incorporating blueberries to the diet to improve cardiovascular health and recommended as the ideal fruit for the treatment of hypercholesterolemia.

It is known that fruits such as cranberries have high antioxidant levels and tested their effectiveness in promoting cardiovascular health [73-75]. This study was supplemented to a group of men for two weeks with cranberry juice. Over time he found an increase in plasma antioxidant capacity and a decrease in LDL (low density lipoprotein) in addition to an increase in HDL in obese men. Work is to show whether supplementation based cranberry juice may have the same antioxidant capacity and the same protective benefit as red wine, if so would avoid alcohol.

In another study conducted at the University of Buffalo studied the effect of resveratrol as an antioxidant and its possible use in treating atherosclerosis. In this investigation were not used fruits or vegetables, but was used an extract of the plant. The extract containing resveratrol was administered at doses of 40 mg daily to a group of 10 people, another group of 10 people also served as a target. During the six weeks of the study, blood tests were performed on the results; researchers concluded that Polygonum cuspidatum extract has a therapeutic effect against oxidative stress. These results show that resveratrol, as already mentioned above, are effective to counteract the effect of free radicals, and in the case of arteriosclerosis, can also help prevent it [76].

5. Obesity and metabolic syndrome

The metabolic syndrome has been identified as a target for dietary therapies to reduce risk of cardiovascular disease; however, the role of diet in the etiology of the metabolic syndrome is poorly understood. The metabolic syndrome consists of a constellation of factors that increase the risk of cardiovascular disease and type 2 diabetes. The etiology of this syndrome is largely unknown but presumably represents a complex interaction between genetic, metabolic, and environmental factors including diet [77-79]. The studies endothelial function by assessing the vascular responses to L-arginine, the natural precursor of nitric oxide it´s characterized for the low-grade inflammatory state of patients with the metabolic syndrome by measuring circulating levels of high-sensitivity C-reactive protein (hs-CRP) as well as of interleukins 6 (IL-6), 7 (IL-7), and 18 (IL-18). These proinflammatory ILs have been prospectively associated with thrombotic cardiovascular events [80, 81] or have been suggested to be involved in plaque destabilization [82]. The diet designed to increase consumption of foods rich in phytochemicals, antioxidants, α-linolenic acid, and fiber prevent Metabolic Syndrome.

The diet rich in whole grains, fruits, vegetables, legumes, walnuts, and olive oil might be effective in reducing both the prevalence of the metabolic syndrome and its associated cardiovascular risk. One of the mechanisms responsible for the cardioprotective effect of such a diet may be through reduction of the low-grade inflammatory state associated with the metabolic syndrome. Although weight reduction remains a cornerstone of therapy for the metabolic syndrome, from a public health perspective adoption of a diet rich in phytochemicals, antioxidants, α-linolenic acid, and fiber may provide further benefit on cardiovascular risk, especially in patients who do not lose weight.

If antioxidants play a protective role in the pathophysiology of diabetes and cardiovascular disease, understanding the physiological status of antioxidant concentrations among people at high risk for developing these conditions, such as people with the metabolic syndrome, is of interest. However, little is known about this topic. Because the prevalence of obesity, which is associated with decreased concentrations of antioxidants [83], is high among people with the metabolic syndrome, they are probably more likely to have low antioxidant concentrations. Consequently, our purpose was to examine whether concentrations of several antioxidants are lower among those with than those without the metabolic syndrome.

For example a retinol from the liver, the main storage site for retinol is transported to peripheral tissues by retinol binding protein. Retinol may be released as a retinyl ester; however, when the ability of the liver to store retinol is exceeded or when liver function is impaired [84]. Thus, the higher retinyl ester concentrations among those who did not have the metabolic syndrome may indicate that they consumed larger amounts of vitamin A compared with people who have this syndrome. Our findings may have implications for people with the metabolic syndrome, health care professionals who care for them and researchers who study the metabolic syndrome. People with the metabolic syndrome are at increased risk for diabetes and cardiovascular disease, and a role for oxidative stress in the pathophysiology of these conditions has been postulated. Free radical species is one of the principal mechanisms of action of antioxidants, other mechanisms that affect the pathophysiology of diabetes and cardiovascular disease may be operating as well [83]. The effects of vitamins C and E have received a great deal of interest. Through effects on oxidation of LDL cholesterol concentration, leukocyte adhesion, and endothelial function, vitamins C and E may slow atherosclerosis [86, 87].

6. Liver cirrhosis

Currently the evidence supports the role of nutritional deficiency in Alcoholic Liver Disease (ALD) [88–95]. Lieber and colleagues show that progressive ALD proceeds despite adequate nutrition [96, 97]. The latter hypothesis was based primarily on the observation that baboons fed a nutritionally adequate liquid diet containing ethanol at 50% calories developed nearly the whole spectrum of ALD including cirrhosis. Studies demonstrated profound effects on ethanol-induced liver injury by intake of nutrients such as polyunsaturated fat and iron in quantities that were never thought to be important. The concept of 'sensitization' and 'priming' is currently considered fundamental to our pursuit for elucidation of pathogenetic mechanisms of ALD. The sensitization is a conditioning that makes the target cells, hepatocytes, more vulnerable to harmful effects triggered by ethanol and priming as the effect that promotes specific injurious mechanisms. The sensitizing and priming are rendered by the complex interactions of primary mechanistic factors and secondary risk factors. For example, intake of polyunsaturated fat in ethanol-fed rats, but not in pair-fed controls, results in a synergistic priming effect on induction of cytochrome P4502. E1 (CYP2E1) with consequent oxidative injury to the liver [98]. Conversely, saturated fat prevents this priming effect and abrogates depletion of a mitochondrial pool of glutathione (GSH) [99], one of the most crucial sensitization effects of ethanol on hepatocytes [100]. Iron is another example. Whereas a slight increase in hepatic iron content by dietary iron supplementation is harmless in control rats, it exacerbates alcoholic liver injury via accentuation of oxidative stress [101]. Further, increased iron storage in hepatic macrophages is a potential priming mechanism forenhanced expression of tumor necrosis factor a (TNF-a) in experimental ALD [102] Besides nutritional factors, female gender, age, concomitant intake of other drugs that can induce CYP2E1, hepatitis virus infection, and genetic predisposition are all considered risk factors. Even among the primary mechanistic factors that include acetaldehyde, oxidative stress, immune response, hypoxia, and membrane alterations, there are cross-interactive relationships to render sensitization or priming effects. For instance, acetaldehyde, a potent toxic metabolite of ethanol, induces liver injury via its covalent binding to structural or functional proteins of the cells [103] while promoting oxidative stress via consumption of GSH. In turn, deleterious effects of acetaldehyde-protein adduct formation may be accentuated by oxidative stress since malondialdehyde, a lipid peroxidation end product, can increase the binding affinity of acetaldehyde by 13-fold [104]. The resulting novel hybrid adducts are highly immunogenic and may incite immune response mediated liver injury [105, 106]. Although cellular immune response and inflammation are regarded as independent mechanisms of ALD, they can also lead to oxidative stress via the release of reactive oxygen species (ROS) by NADPH oxidase or action of TNF-a at the electron transport chain in target cells. The multifactorial nature and complex interaction among primary mechanistic factors and between primary and secondary factors appear to be the basis for the heterogeneous response that alcoholics exhibit for ALD. Elucidation of the sensitization and priming mechanisms involving cross-interactions of these factors should allow us to gain insight into the most fundamental question, which is why only a small fraction of alcoholics develop advanced ALD. The experimental models to use for control deletion and addition analyses in order to identify what primary and secondary factors are required for the expression of a particular aspect or whole spectrum of experimental ALD. It is need experts in various disciplines need to work together to provide cutting-edge science for elucidating the precise nature and mechanisms that underlie interactions.

Antioxidants represent a potential group of therapeutic agents for ALD. They likely provide beneficial effects on hepatocytes via desensitization against oxidant stress while inhibiting priming mechanisms for expression of proinflammatory and cytotoxic mediators via suppression of NF-kB [107, 108]. Potential approaches may include cell type-specific targeting of antioxidant therapy and development of modalities for more specific and selective regulation of NF-kB-mediated signaling.

The development of cirrhosis is usually associated with oxidative stress and lipid peroxidation (LPO). Studies in models of cirrhosis to use carbon tetrachloride (CCl_4) inhalation in the rat show several similarities with human cirrhosis. The metabolism of CCl_4 into trichloromethyl (CCl_3•) and peroxy trichloromethyl (•$OOCCl_3$) free radicals has been reported to cause hepatotoxic effects, like fibrosis, steatosis, necrosis, and hepatocarcinoma [109, 111].

Some compounds that have been studied as possible protectors against liver cirrhosis are known for their anti-inflammatory and antioxidant properties. Plants contain numerous polyphenols, which have been shown to reduce inflammation and thereby to increase resistance to disease [112]. Quercetin (Q), a polyphenolic flavonoid compound present in large amounts in vegetables, fruits, and tea, exhibits its therapeutic potential against many diseases, including hepatoprotection and the inhibition of liver fibrosis [113–114]. It contains a number of phenolic hydroxyl groups, which have strong antioxidant activity [116, 117]. The average intake varies between countries but is approximately 23 mg/day [118].

By increasing the endogenous antioxidant defenses, flavonoids can modulate the redox state of organisms. The major endogenous antioxidant systems include superoxide dismutase (SOD), catalase (CAT), glutathione reductase (GR), and glutathione peroxidase (GPx), which is essential for the detoxification of lipid peroxides [119-121].

http://dx.doi.org/10.5772/51994

7. Hypertension

Excessive reactive oxygen species (ROS) have emerged as a central common pathway by which disparate influences may induce and exacerbate hypertension. Potential sources of excessive ROS in hypertension include nicotinamide adenine dinucleotide phosphate (NADPH) oxidase, mitochondria, xanthine oxidase, endothelium-derived NO synthase, cyclooxygenase 1 and 2, cytochrome P450 epoxygenase, and transition metals. While a significant body of epidemiological and clinical data suggests that antioxidant-rich diets reduce blood pressure and cardiovascular risk, randomized trials and population studies using natural antioxidants have yielded disappointing results. The reasons behind this lack of efficacy are not completely clear, but likely include a combination of [122] ineffective dosing regimens, [123] the potential pro-oxidant capacity of some of these agents, [124] selection of subjects less likely to benefit from antioxidant therapy (too healthy or too sick), and inefficiency of nonspecific quenching of prevalent ROS versus prevention of excessive ROS production. Antioxidants as vitamins A, C and E, L-arginine, flavonoids, and mitochondria-targeted agents (Coenzyme Q10, acetyl-L-carnitine, and alpha-lipoic acid) can be use to treatment hypertension. Currently exist incomplete knowledge of the mechanisms of action of these agents, lack of target specificity, and potential interindividual differences in therapeutic efficacy preclude us from recommending any specific natural antioxidant for antihypertensive therapy at this time.

Reactive oxygen species (ROS) are generated by multiple cellular sources, including NADPH oxidase, mitochondria, xanthine oxidase, uncoupled endothelium-derived NO synthase, cycloxygenase, and lipoxygenase. The dominant initial ROS species produced by these sources is superoxide (O_2^-). Superoxide is short-lived molecule that can subsequently undergo enzymatic dismutation to hydrogen peroxide. Superoxide can oxidize proteins and lipids, or react with endothelium-derived nitric oxide (NO) to create the reactive nitrogen species peroxynitrite. Peroxynitrite and other reactive nitrogen species can subsequently oxidize proteins, lipids, and critical enzymatic cofactors that may further increase oxidative stress [125]. Hydrogen peroxide produced by enzymatic dismutation of O_2^- can be further convert to highly reactive hydroxyl radical (via Fenton chemistry) that can cause DNA damage. The balance between superoxide production and consumption likely keeps the concentration of O_2^- in the picomolar range and hydrogen peroxide in the nanomolar range [126]. These homeostatic levels of reactive oxygen species appear to be important in normal cellular signaling [127-132] and normal reactions to stressors [133, 134].

Randomized trials employing non-pharmacological dietary interventions emphasizing fruits, vegetables, whole grains, and nuts have shown impressive blood pressure lowering results in both hypertensive and normotensive subjects [135, 136]. Similar interventions demonstrated to reduce cardiovascular morbidity and mortality continue to maintain interest in the potential of isolating specific compounds enriched in these diets that may be responsible for the overall dietary benefits [137].

The dietary components in these studies are high in compounds known to have antioxidant properties leading many to ascribe the benefits of these diets to their increased content of natural antioxidants. However, prior randomized trials and population studies in healthy populations and patients at high risk for cardiovascular events that have employed combinations of some of these natural antioxidants as dietary supplements have, for the most part, shown disappointing results [138-145]. The reasons behind these disappointing results are not completely clear, but likely include a combination of 1) ineffective dosing and dosing regimens 2) the potential pro-oxidant capacity and other potentially deleterious effects of these some of these compounds under certain conditions [146-148], 3) selection of subjects less likely to benefit from antioxidant therapy (too healthy or too sick). Populations at intermediate cardiovascular risk may be better suitable to see effects of antioxidants in shorter term studies [149], 4) inefficiency of non-specific quenching of prevalent ROS versus prevention of excessive ROS production [150, 151].

When considering antioxidant therapy for hypertension, lessons from prior disappointing attempts to reduce blood pressure and cardiovascular risk with antioxidant therapy should be considered. The profile of an ideal agent is outlined in The importance of patient selection is being increasingly recognized in light of emerging data suggesting that antioxidant supplementation in healthy subjects may blunt the protective benefits of aerobic exercise training, suggesting ROS generation can be beneficial under certain circumstances.

Antioxidants neutralize the oxidative processes and modify levels in plasma	
↑ Lipid peroxidation	↑ MDA (TBAR), F2-isoprostane
↑ NO synthesis	↑ Nitrite, nitrate, nitrotyrosine
↓ Circulating antioxidants	↓ Uric acid, protein SH groups, Bilirubin (unconjugated)
	↓ Ascorbic acid, α-tocopherol, β-carotene, lycopene
	↓ Antioxidant enzymes (GSHPx)
	↓ Selenium, zinc
	↓ GSH
Xanthine oxidase activation	↑ Plasma xanthine oxidase

Table 1. Antioxidants neutralize the oxidative processes and modify levels in plasma. [150]

7.1. Antioxidant vitamins

7.1.1. Vitamin A precursors and derivatives

Vitamin A precursors and derivatives are retinoids that consist of a beta-ionone ring attached to an isoprenoid carbon chain. Foods high in vitamin A include liver, sweet potato, carrot, pumpkin, and broccoli leaf. Initial interest in vitamin A-related compounds focused primarily on beta-carotene, given initial promising epidemiological data with respect to its cardioprotective effects and some correlation with higher plasma levels to lower blood pressure in men. However, concerns about beta-carotene's pro-oxidative potential came to light with a report suggesting adverse mitochondrial effects of beta-carotene cleavage products. Further, adverse mortality data with respect to beta-carotene has limited interest in this compound as an effective antihypertensive agent.

Recently, interest in vitamin A derivatives has turned to lycopene, itself a potent antioxidant [152], found concentrated in tomatoes. One small study has shown a reduction in blood pressure with a tomato-extract based intervention (containing a combination of potential anti-oxidant compounds including lycopene) in patients with stage I hypertension, [153] although second study showed no effect in pre-hypertensive patients [154].

7.1.2. Ascorbic acid (Vitamin C)

L-ascorbic acid is a six-carbon lactone and, for humans, is an essential nutrient. In Western diets, commonly consumed foods that contain high levels of ascorbic acid include broccoli, lemons, limes, oranges, and strawberries. Toxicity potential of this compound is low, although an increased risk of oxalate renal calculi may exist at higher doses (exceeding 2 grams/day).

The initial purported mechanisms for the potential benefits of ascorbate supplementation were centered on quenching of single-electron free radicals. Subsequent research has demonstrated that the plasma concentrations of ascorbate required for this mechanism to be physiologically relevant are not attainable by oral supplementation [155]. However, vitamin C can concentrate in local tissues to levels an order of magnitude higher than that of plasma. At this ascorbate may to effectively compete for superoxide and reduce thiols [156]. Recent data also suggest potential suppressive effects of ascorbate on NADPH oxidase activity [157, 158]. Ascorbate appears to have limited pro-oxidant ability. [159].

Ascorbate's anti-hypertensive efficacy has been evaluated in multiple small studies [160-163] but not all, show modest reductions in blood pressure in both normotensive and hypertensive populations. These data also suggest that supplementation has limited effect on systemic antioxidant markers and little additional blood pressure benefits are seen beyond the 500 mg daily dose. Large scale randomized trial data specific to ascorbate supplementation and its effects on hypertension are currently lacking. Data from Heart Protection Study (HPS) suggest no significant mortality from supplementation with 250mg/day of ascorbate supplementation. However, the relatively low dose of ascorbate, use of combination therapy, and high-risk patient population studied in HPS leave unanswered the key questions of appropriate dosing and target. In the inflammatory processes follow next scheme in the therapy antioxidant [164].

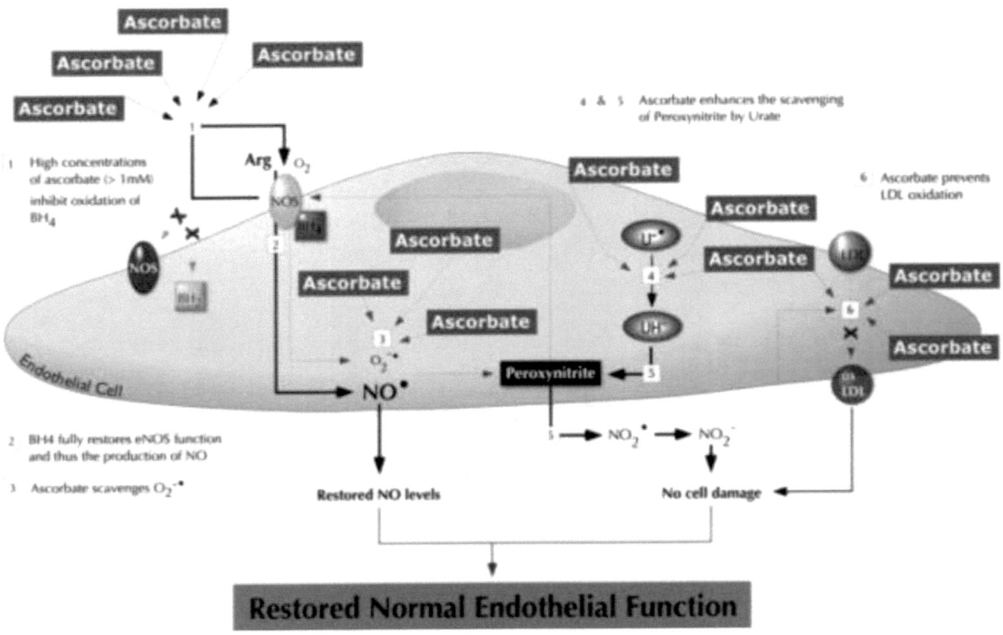

Figure 2. Restored normal endothelial function.

7.1.3. α-Tocopherol (Vitamin E)

Vitamin E is a generic term for a group of compounds classified as tocopherols and tocotrienols [165]. While there are four isomers in each class of Vitamin E compounds, the overwhelming majority of the active form is α-tocopherol. [166, 167]. Dietary sources high in vitamin E include avocados, asparagus, vegetable oils, nuts, and leafy green vegetables.

Vitamin E is a potent antioxidant that inhibits LDL and membrane phospholipid oxidation. Interestingly, inflammatory cells and neurons have binding proteins for α-tocopherol, the actions of which may include inhibition of NADPH oxidase, lipoxygenase, and cyclo-oxygenase, actions which may lower oxidative stress [168]. However, studies demonstrating vitamin E's pro-oxidant capacity under certain cellular conditions suggest that local condition may influence the vitamin E's redox activity [169]. Initial excitement for vitamin E supplementation was based on the reduction of cardiovascular events seen in the CHAOS study. However, follow-up studies have been largely disappointing [170-171]. While one small study that used vitamin E in combination with zinc, vitamin C, and beta-carotene showed a modest, significant reduction in blood pressure over 8 weeks of therapy, other small studies, show either no effect from vitamin E supplementation. Further, the more definitive HOPE trial, failed to show blood pressure or mortality benefit for patients at high risk for cardiovascular disease [172]. Vitamin E inhibits free radicals reactions.

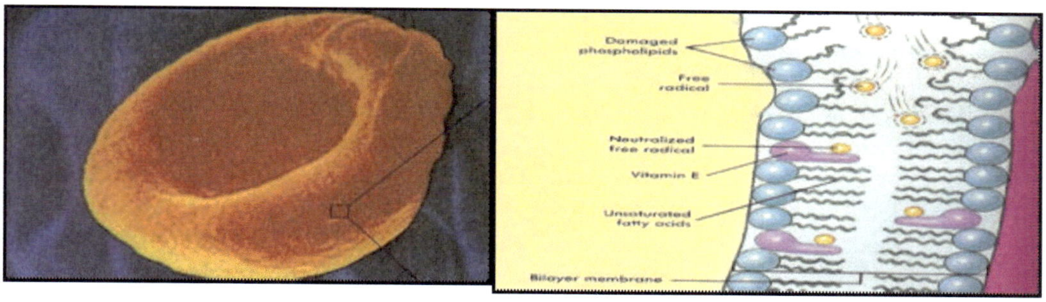

Figure 3. Antioxidants function in the organism.

7.1.4. L-Arginine

L-arginine is an amino acid and the main substrate for the production of NO from eNOS in a reaction that is dependent on tetrahydrobiopterin [173]. Potential dietary sources include milk products, beef, wheat germ, nuts, and soybeans. Reduced levels of tetrahydrobiopterin leads to uncoupling of reduced NADPH oxidation and NO synthesis, with oxygen as terminal electron acceptor instead of L-arginine, resulting in the generation of superoxide by eNOS [174-176]. Low cellular levels of L-arginine have been demonstrated in human hypertension. While L-arginine deficiency itself does not appear to lead to uncoupling of eNOS, [177] low levels of L-arginine may lead to reduced levels of bioavailable NO which could contribute to hypertension. Thus, L-arginine supplementation could theoretically reduce blood pressure by allowing for restoration of normal NO bioavailability, perhaps overcoming overall L-arginine deficiency as well as more successfully competing fo the eNOS active site with circulating asymmetric dimet hylarginine, a circulating competitor of L-arginine that may be increased in the setting of hypertension.

This concept is supported by studies demonstrating the anti-hypertensive effect of L-arginine supplementation in salt-sensitive rats, healthy human subjects, hypertensive diabetics, patients with chronic kidney disease, and diabetic patients in combination with N-acetylcysteine, a precursor of glutathione [178] L-arginine's anti-hypertensive response may be mediated in part by its suppressive effects on angiotensin II and endothelin-1, and its potentiating effects on insulin.

However, recent concerns about potential deleterious increases in homocysteine in the setting of L-arginine supplementation have been raised. The majority of L-arginine is processed into creatine, which leads increased homocysteine levels. Homocysteine can increase oxidative stress. A recent study confirms that this mechanism is relevant to L-arginine metabolism in humans [179] suggesting a potential mechanism for neutralizing the eNOS-related anti-oxidant effects of L-arginine.

7.1.5. Flavonoids

Flavonoids are polyphenolic compounds commonly found in concentrated amounts in multiple fruits, vegetables, and beverages, including apples, berries, grapes, onions, pomegranate, red wine, tea, cocoa, and dark chocolate. The exact structure and composition of the flavonoid compounds varies between food sources, and flavonoid content can be altered based on the manner of food preparation [180]. Interest in flavonoids as antioxidants therapy for cardiovascular disease originates from epidemiological data suggesting improved cardiovascular outcomes in individuals with high intake of food and beverages with high flavonoid content as well as cellular work suggesting a strong anti-oxidant effect of these compounds [181]. However, the limited oral bioavailability of flavonoids suggests cells signaling mechanism, rather than free radical quenching activity, is more likely to be root of sustained cardiovascular benefits from flavonoids [182, 183]. This concept is consistent with studies demondtrating that flavonoids can inhibit NADPH oxidase through ACE inhibition, increase eNOS-specific NO production through the estrogen receptor, and alter COX-2 expression [184]. Studies investigating the anti-hypertensive effects of flavonoids are inconclusive. While multiple small studies of short duration of dark chocolate therapy have demonstrated blood pressure lowering effects in hypertensives [185], studies in normotensive and pre-hypertensive individuals have demonstrated no benefit [186], further tea intake may, at least temporarily, increase blood pressure certain populations [187, 188]. The specific flavonoids and combination of flavonoids that exert the largest beneficial effects remain unknown. The follow table indicates a function of antioxidants in therapy.

Selenium	Septic ICU patients; major burns in combination with Cu and Zn; trauma patients	Ceiling "/>750 μg/day?
Zinc	Pneumonia in children: clinical course significantly shortened	Immune depression if doses"/>50 mg7day are provided
Cu-Se-Zn	Burns: trials showing reduction of infectious complication (pneumonia) and improved wound healing	Doses were calculated to compensate for the exudative losses
Vitamin E (α-tocoferol)	SIRS enteral supplementation	Convincing animal data
Vitamin C (ascorbic acid)	Burns, megadose during the first 24 h after injury; trauma, combined with vitamin E	Possible an endothelial mechanism (189)

Table 2. Antioxidants more indicated in treatments degeneratives chronics.

8. Conclusions

The antioxidants present in food playing an important role in preventing chronic diseases. A balanced diet can prevent diseases associated with oxidative stress and help keep the body in top condition.

Author details

Eva María Molina Trinidad, Sandra Luz de Ita Gutiérrez, Ana María Téllez López and Marisela López Orozco

Universidad Autónoma del Estado de Hidalgo UAEH, Instituto de Ciencias de la Salud IC-Sa, Área Académica de Farmacia. ExHacienda la Concepción, Tilcuautla, Hidalgo, México

References

[1] Huang, X. (2003). Iron overload and its association with cancer risk in humans: evidence for iron as a carcinogenic metal. Mutat.Res. , 533, 153-171.

[2] Markesbery WR, Lovell MA(2006). DNA oxidation in Alzheimer´s disease. Antioxid Redox Signal. , 8, 2039-2045.

[3] Halliwell, B. (2001). Role of free radicals in the neurodegenerative diseases: therapeutic implications for antioxidant treatment. Drugs Aging. , 18(9), 685-716.

[4] Vokurkova, M., Xu, S., & Touyz, R. M. (2007). Reactive oxygen species, cell growth, cell cycle progression and vascular remodeling in hypertension. Future Cardiol. Jan; , 3(1), 53-63.

[5] Herrera, E., Jimenez, R., Aruoma, O. I., Hercberg, S., Sanchez-Garcia, I., & Fraga, C. (2009). Aspects of antioxidant foods and supplements in health and disease. Nutr. Rev. 67 (Suppl. 1), SS144., 140.

[6] Dai, J., Jones, D. P., Goldberg, J., Ziegler, T. R., Bostick, R. M., Wilson, P. W., Manatunga, A. K., Shallenberger, L., Jones, L., & Vaccarino, V. (2008). Association between adherence to the Mediterranean diet and oxidative stress. Am. J. Clin. Nutr. , 88, 1364-1370.

[7] Organización Mundial de la Salud(2008). La lucha contra el cáncer tiene que ser una prioridad del desarrollo. Available: http://www.who.int/mediacentre/news/statements/2008/s09/es/index.html.Accessed 2010 December 23.

[8] American Cancer Society. American Cancer Society Cancer Facts and Figures ((2008). Available: http://www.cancer.org/downloads/STT/2008CAFFfinalsecured.pdf Accessed 2009 March 13.

[9] Valko, M., Leibfritz, D., Moncol, J., Cronin, M. T. D., Mazur, M., & Telser, J. (2007). Free radicals and antioxidants in normal physiological functions and human disease. Int. J. Biochem. Cell Biol. j. , 39, 44-84.

[10] Hercberg, S., Galan, P., Preziosi, P., Alfarez, M., & Vazquez, C. (1998). The potential role of antioxidant vitamins in preventing cardiovascular diseases and cancers. Nutrition j. , 14, 513-520.

[11] Borek, C. (2004). Dietary Antioxidants and Human Cancer. Integr. Cancer Ther. , 3, 333-341.

[12] Andreescu, S., et al. (2011). In Oxidative Stress: Diagnostics, Prevention, and Therapy. ACS Symposium Series. American Chemical Society: Washington, DC.

[13] Hakimuddin, F., Paliyath, G., & Meckling, K. (2006). Treatment of Mcf-7 Breast Cancer Cells with a Red Grape Wine Polyphenol Fraction Results in Disruption of Calcium Homeostasis and Cell Cycle Arrest Causing Selective cytotoxicity J. Agric. Food chem. j. 54: (20) 7912-7923.

[14] Noratto, G., Porter, W., Byrne, D., & Cisneros-Zevallos, L. (2009). Identifying peach and plum polyphenols with chemopreventive potential against estrogen-independent breast cancer cells J. Agric. Food chem. j. , 57, 5219-5226.

[15] Allouche, Y., Warleta, F., Campos, M., Sánchez-Quesada, C., Uceda, M., Beltrán, G., & Gaforio, J. J. (2011). Antioxidant, antiproliferative, and pro-apoptotic capacities of pentacyclic triterpenes found in the skin of olives on mcf-7 human breast cancer cells and their effects on DNA damage. J. Agric. Food chem. j. , 59, 121-130.

[16] [16]Available:http://support.dalton.missouri.edu/index.php/daltonnews/Breast_Cancer_Effectively_Treated_with_Chemical_Found_in_Celery_Parsley_by/. Accessed January 2012.

[17] Heidenreich, A., Aus, G., Bolla, M., Joniau, S., Matveev, V. B., Schmid, H. P., & Zattoni, F. (2008). EAU guidelines on prostate cancer. Eur. Urol. j. 53: (1) 68-80.

[18] Moorthy, H. K., & Venugopal, P. (2008). Strategies for prostate cancer prevention: Review of the literature. Indian J. Urol. j. 24: (3) 295-302.

[19] Singh, R. P., & Agarwal, R. (2006). Mechanisms of action of novel agents for prostate cancer chemoprevention. Endocr.-Related Cancer j. 13: (3) , 751 EOF-78 EOF.

[20] Gonzalez-Sarrias, A., Gimenez-Bastida, J. A., Garcia-Conesa, M. T., Gomez-Sanchez, M. B., Garcia-Talavera, N. V., Gil-Izquierdo, A., Sanchez-Alvarez, C., Fontana-Compiano, L. O., Morga-Egea, J. P., Pastor-Quirante, F. A., Martinez-Diaz, F., Tomas-Barberan, F. A., & Espin, J. C. (2010). Occurrence of urolithins, gut microbiota ellagic acid metabolites and proliferation markers expression response in the human prostate gland upon consumption of walnuts and pomegranate juice. Mol. Nutr. Food Res. j. 54: (3) 311-322.

[21] Gasmi, J., & Sanderson, Thomas. (2010). Growth Inhibitory, Antiandrogenic, and Pro-apoptotic Effects of Punicic Acid in LNCaP Human Prostate Cancer Cells. J. Agric. Food Chem. j. 58: (23) , 12149 EOF-12156 EOF.

[22] Seeram, N. P., Lee, R., Scheuller, H. S., & Heber, D. (2006). Identification of phenolics in strawberries by liquid chromatography electrospray ionization mass spectroscopy. Food Chem. j. , 97, 1-11.

[23] Zhang, Y., Seeram, N. P., Lee, R., Feng, L., & Heber, D. (2008). J. Agric. Food Chem. j. , 56, 670-675.

[24] Willett W C(1995). Diet, nutrition, and avoidable cancer. EnViron. Health Perspect. j. , 103, 165-170.

[25] Eberhardt M V, Lee C Y, Liu R H(2000). Antioxidant activity of fresh apples. Naturej. , 405, 903-904.

[26] Le -Marchand, L., Murphy, S. P., Hankin, J. H., Wilkens, L. R., & Kolonel, L. N. (2000). Intake of flavonoids and lung cancer. J. Natl. Cancer Inst. j. , 92, 154-160.

[27] Xing, N., Chen, Y., Mitchell, S. H., & Young, C. Y. F. (2001). Quercetin inhibits the expression and function of the androgen receptor in LNCaP prostate cancer cells. Carcinogenesisj. , 22, 409-414.

[28] Tsao, R., Yang, R., Xie, S., Sockovie, E., & Khanizadeh, S. (2005). J. Agric. Food Chem. j. , 53(12)

[29] Mc Rae, K. B., Lidster, P. D., de Marco, A. C., & Dick, A. (1990). J Comparison of the polyphenol profiles of the apple fruit cultivars by correspondence analysis. J. Sci. Food Agric. j. , 50, 329-342.

[30] Awad, M. A., de Jager, A., & van Westing, L. M. (2000). Flavonoid and chlorogenic acid levels in apple fruit: characterization of variation. Sci. Hortic. j. , 83, 249-263.

[31] Tsao, R., Yang, R., Young, J. C., & Zhu, H. (2003). Polyphenolic profiles in eight apple cultivars using high-performance liquid chromatography (HPLC). J. Agric. Food Chem. j. , 51, 6347-6353.

[32] Britton, G. (1995). Carotenoids 1: Structure and Properties of Carotenoids in Relation to Function. FASEB J. , 9, 1551-1558.

[33] Di Mascio, P., Kaiser, S., & Sies, H. (1989). Lycopene as the most efficient biological carotenoid singlet oxygen quencher. Arch. Biochem. Biophys. j. , 274, 532-538.

[34] Giovannucci, E., Ascherio, A., Rimm, E. B., Stampfer, M. J., Colditz, G. A., & Willett, Q. C. (1995). Intake of carotenoids and retinol in relation to risk of prostate cancer. J. Natl. Cancer Inst. j. , 87, 1767-1776.

[35] Giovannucci, E. ((1999).) Tomatoes, Tomato-based products, lycopene, and cancer: review of the epidemiological literature. J. Natl. Cancer Inst. j. ., 91, 317-331.

[36] Gann, P. H., Giovannucci, J., Willett, E., Sacks, W., Hennekens, F. M., Stampfer, C. H., & , M. J. (1999). Lower prostate cancer risk in men with elevated plasma lycopene levels: results of a prospective analysis. Cancer Res. j. , 59, 1225-1230.

[37] Reddivari, L., Vanamala, J., Chintharlapalli, S., Safe, S. H., Miller, J. C., & Jr , . (2007). Anthocyanin fraction from potato extracts is cytotoxic to prostate cancer cells through activation of caspase-dependent and caspase-independent pathways. Carcinogenesisj. , 28, 2227-2235.

[38] Available:http://www.cancer.gov/espanol/recursos/hojas-informativas/riesgo-causas/VPH-respuestas. AccessedFebruary (2012).

[39] Available, http://www.who.int/mediacentre/factsheets/fs297/es/index.html., & Accessed, . March (2012).

[40] Mcdougall, G. J., Ross, H. A., Ikeji, M., & Stewart, D. (2008). Berry Extracts Exert Different Antiproliferative Effects against Cervical and Colon Cancer Cells Grown in Vitro. J. Agric. Food Chem. j. , 56, 3016-3023.

[41] Forceville Xavier, Laviolle Bruno, Annane Djillali, Vitoux Dominique, Bleichner Gérard, Korach Jean Michel, Cantais Emmanuel, Georges Hug.(2007). Effects of high doses of selenium, as sodium selenite, in septic shock: a placebo-controlled, randomized, double-blind, phase II study. Critical Care. http://ccforum.com/content/11/4/R73,viewed on 27-07-2012., 1-10.

[42] Manzanares, W. ., & Hardy, . Selenium supplementation in the critically ill: posology and pharmacokinetics. G.Curr Opin Clin Nutr Metab Care (2009). , 12, 273-80.

[43] Friedman, M., Lee, K. R., Kim, H. J., Lee, I. S., & Kozukue, N. (2005). Anticarcinogenic effects of glycoalkaloids from potatoes against human cervical, liver, lymphoma, and stomach cancer cells. J. Agric. Food Chem. j. , 53, 6162-6169.

[44] World Health Organization.(1999). Definition, Diagnosis and Classification of Diabetes Mellitus and Its Complications. Part I: Diagnosis and Classification of Diabetes Mellitus; Geneva, Switzerland.

[45] Valko, M., Leibfritz, D., Moncol, J., Cronin, M. T. D., Mazur, M., & Telser, J. (2007). Free radicals and antioxidants in normal physiological functions and human disease. Int. J. Biochem. Cell Biol. j. , 39, 44-84.

[46] Rahimi, R., Nikfar, S., Larijani, B., & Abdollahi, M. (2005). A review on the role of antioxidants in the management of diabetes and its complications. Biomed. Pharmacother. j. , 59, 365-373.

[47] Chu, Y. F., Sun, J., Wu, X., & Liu, R. H. (2002). Antioxidant and antiproliferative activities of common vegetables. J. Agric. Food Chem. j. , 50, 6910-6916.

[48] Liu R H(2004). Potential synergy of phytochemicals in cancer prevention: Mechanism of action. J. Nutr. j. 134: , 3479S EOF-3485S EOF.

[49] Available, http://www.nlm.nih.gov/medlineplus/spanish/ency/article/003464.htm., & Accessed, . January (2012).

[50] Mc Dougall, G. J., Shpiro, F., Dobson, P., Smith, P., Blake, A., & Stewart, D. (2005). Different polyphenolic components of soft fruits inhibit R-amylase and R-glucosidase. J. Agric. Food Chem. j. , 53, 2760-2766.

[51] Tadera, K., Minami, Y., Takamatsu, K., & Matsuoka, T. (2006). Inhibition of R-glucosidase and R-amylase by flavonoids. J. Nutr. Sci. Vitaminol. j. , 52, 149-153.

[52] Mullen, W., Mcginn, J., Lean, M. E. J., Maclean, M. R., Gardner, P., Duthie, G. G., Yokota, T., & Crozier, A. (2002). Ellagitannins, flavonoids, and other phenolics in red raspberries and their contribution to antioxidant capacity and vasorelaxation properties. J. Agric. Food Chem. j. , 50, 5191-5196.

[53] Pinto, M. S., Kwon, Y. I., Apostolidis, E., Lajolo, F. M., Genovese, M. I., & Shetty, K. (2008). Functionality of bioactive compounds in Brazilian strawberry (Fragaria x ananassa Duch.) cultivars: evaluation of hyperglycemia and hypertension potential using in vitro models. J. Agric. Food Chem. j. , 56, 4386-4382.

[54] Matsui, T., Tanaka, T., Tamura, S., Toshima, A., Tamaya, K., Miyata, Y., Tanaka, K., & Matsumoto, K. (2007). R-Glucosidase inhibitory profile of catechins and theaflavins. J. Agric. Food Chem. j. , 55, 99-105.

[55] Kwon, Y. I., Vattem, D. A., & Shetty, K. (2006). Clonal herbs of Laminaceae species against diabetes and hypertension. Asia Pac. J. Clin. Nutr. j. , 15, 424-432.

[56] Genovese M I, Pinto M S, Gonc-alves A E S S, Lajolo F M(2008). Bioactive compounds and antioxidant capacity of exotic fruits and commercial frozen pulps from Brazil. Food Sci. Technol. Int. j. , 14, 207-214.

[57] de Souza, A. E., Gonc-alves, S., Lajolo, F. M., & Genovese, M. I. (2010). Chemical Composition and Antioxidant/Antidiabetic Potential of Brazilian Native Fruits and Commercial Frozen Pulps J. Agric. Food Chem. j. DOI:10.1021/jf903875u., 58, 4666-4674.

[58] Available, http://www.lenntech.es/vitaminas/biotina.htm., & Accessed, . February (2012).

[59] Available: http://www.nlm.nih.gov/medlineplus/spanish/druginfo/natural/313.html.

[60] Available: http://www.guia-diabetes.com/el-resveratrol-mejora-la-diabetes-con-su-accion-sobre-el cerebro.html. AccessedJanuary (2012).

[61] Available, http://www.cienciadirecta.com/espanol/web/noticias/ujalaurel9063.asp., & Accessed, . April (2012).

[62] Torissen, O., Hardy, R., & Shearer, K. (1989). Pigmentation of salmonoids carotenoid deposition and metabolism. CRC Crit. ReV. Aq. Sci. j. , 1, 209-225.

[63] Naito, Y., Uchiyama, K., Aoi, W., Hasegawa, G., Nakamura, N., Yoshida, N., Maoka, T., Takahashi, J., & Yoshikawa, T. (2004). Prevention of diabetic nephropathy by treatment with astaxanthin in diabetic db/db mice. Biofactors j. , 20, 49-59.

[64] Jacob, S., Hernrisken, E. J., Schiemann, A. L., et al. (1995). Enhancement of glucose disposal in patients with type 2 diabetes by alpha lipoic acid. Arzeneimittel- Forschung Drug Research j. 45: , 872 EOF-4 EOF.

[65] Lester Packer, Carol Colman(1999). The Antioxidant Miracle: Put Lipoic Acid, Pycogenol, and Vitamins E and C to Work for You J. ohn Wiley & sons: New York 0-47135-311-6

[66] Allan, E., Sosin, Beth. M., Ley-Jacobs, Julian. M., & Whitaker, . (1998). Alpha Lipoic Acid: Nature's Ultimate Antioxidant Kensington Books. New York. 157566366

[67] Burt Berkson ((1998).) Alpha Lipoic Acid Breakthrough: The Superb Antioxidant That May Slow Aging, Repair Liver Damage, and Reduce the Risk of Cancer, Heart Disease, and Diabetes. Three River Press. New York.

[68] lable:http://www.vitabasix.com/fileadmin/content/produktInfoPDFs/esPDF/Produktinfo_ALA_ES.pdf. AccessedMay, (2012).

[69] Available, http://www.nlm.nih.gov/medlineplus/spanish/ency/article/000171.htm., & Accesses, . June, (2012).

[70] Tsang, C., Higgins, S., Duthie, G. G., Duthie, S. J., Howie, M., Mullen, W., Lean, M. E., & Crozier, A. (2005). The influence of moderate red wine consumption on antioxidant status and indices of oxidative stress associated with CHD in healthy volunteers. Br. J. Nutr. j. , 93, 233-240.

[71] Zern T L, Fernandez T L(2005). Cardioprotective effects of polyphenols. J. Nutr. j. , 135, 2291-2294.

[72] Milner J A(2002). Foods and health promotion: The case for cranberry. Crit. ReV. Food Sci. Nutr. j. , 42, 265-266.

[73] Neto C C(2007). Cranberry and blueberry: Evidence for protective effects against cancer and vascular disease. Mol. Nutr. Food Res. j. , 51, 652-664.

[74] Ruel, G., Pomerleau, S., Couture, P., Lamarche, B., & Couillard, C. (2005). Changes in plasma antioxidant capacity and oxidized lowdensity lipoprotein levels in men after short-term CJ consumption. Metabolism j. , 54, 856-861.

[75] Ruel, G., Pomerleau, S., Couture, P., Lemieux, S., Lamarche, B., & Couillard, C. (2006). Favorable impact of low-calorie cranberry juice consumption on plasma HDL cholesterol concentrations in men. Br. J. Nutr. j. , 96, 357-364.

[76] Available, http://www.abajarcolesterol.com/resveratrol-previene-la-ateroesclerosis., & Accessed, . june (2012).

[77] Katherine Esposito, Raffaele Marfella, Miryam Ciotola, Carmen Di Palo, Francesco Giugliano, Giovanni Giugliano, Massimo D'Armiento, Francesco D'Andrea, Dario Giugliano(2004). Effect of a Mediterranean-Style Diet on Endothelial Dysfunction and Markers of Vascular Inflammation in the Metabolic Syndrome. JAMA. , 292(12), 1440-1446.

[78] Groop, L. (2000). Genetics of the metabolic syndrome. Br J Nutr. 83(suppl 1):SS48., 39.

[79] Lidfeldt, J., Nyberg, P., Nerbrand, C., et al. (2003). Socio-demographic and psychological factors are associated with features of the metabolic syndrome: the Women's Health in the Lund Area (WHILA) study. Diabetes Obes Metab. , 5, 106-112.

[80] Harris, T. B., Ferrucci, L., Tracy, R. P., et al. (1999). Association of elevated interleukin-6 and C-reactive protein levels with mortality in the elderly. Am J Med. , 106, 506-512.

[81] Blankenberg, S., Tiret, L., Bickel, C., et al. (2002). Interleukin-18 is a strong predictor of cardiovascular death in stable and unstable angina. *Circulation*, 106, 24-30.

[82] Damâs, J. K., Væhre, T., Yndestad, A., et al. (2003). Interleukin-7-mediated inflammation in unstable angina: possible role of chemokines and platelets. *Circulation*, 107, 2670-2676.

[83] Reitman, A., Friedrich, I., Ben-Amotz, A., & Levy, Y. (2002). Low plasma antioxidants and normal plasma B vitamins and homocysteine in patients with severe obesity. Isr Med Assoc J , 4, 590-593.

[84] Ballew, C., Bowman, Russell. R. M., Sowell, A. L., & Gillespie, C. (2001). Serum retinyl esters are not associated with biochemical markers of liver dysfunction in adult participants in the Third National Health and Nutrition Examination Survey (NHANES III), 1988-1994. Am J Clin Nutr , 73, 934-940.

[85] Visioli, F. (2001). Effects of vitamin E on the endothelium. Equivocal? Alphatocopherol and endothelial dysfunction. Cardiovasc Res , 51, 198-201.

[86] Carr, A. C., Zhu, B. Z., & Frei, B. (2000). Potential antiatherogenic mechanisms of ascorbate (vitamin C) and alpha-tocopherol (vitamin E). Circ Res , 87, 349-354.

[87] Cobo Abreu Carlos.(2001). Acido acetilsalicílico y vitamina E en la prevención de las enfermedades cardiovasculares. Rev Mex Cardiol; , 12(3), 128-133.

[88] Barak, A. J., Tuma, D. J., & Beckenhauer, J. L. (1971). Ethanol feeding and choline defiency as influences on hepatic choline uptake. J. Nutr. , 101, 533-538.

[89] French, S. W. (1966). Effect of chronic ethanol ingestion on liver enzyme changes induced by thiamine, riboflavin, pyridoxine or choline deficiency. J. Nutr. , 88, 291-302.

[90] Gomez-Dumm, C. L. A., Porta, E. A., Hartroft, W. S., & Koch, O. R. (1968). A new experimental approach in the study of chronic alcoholism. II. Effects of high alcohol intake in rats fed diets of various adequacies. Lab. Invest. , 18, 365-378.

[91] Hartfoft, W. S., & Porta, E. A. (1968). Alcohol, diet, and experimental hepatic injury. Can. J. Physiol. Pharmacol. Klatskin, G., Krehl, W. A., and Corn, H. (1954) Effect of alcohol on choline requirement. I changes in rat liver after prolonged ingestion of alcohol. J. Exp. Med. 100, 605-614., 46, 463-473.

[92] Mendenhall, C. L., Anderson, S., Weesner, R. E., Goldberg, S. J., & Crolic, K. A. (1984). Protein-calorie malnutrition associated with alcoholic hepatitis. Veterans Administration Cooperative Study Group on Alcoholic Hepatitis. Am. J. Med. , 76, 211-222.

[93] Porta, E. A., Hartroft, W. S., Gomez-Dumm, C. L. A., & Koch, O. R. (1967). Dietary factors in the progression and regression of hepatic alterations associated with experimental chronic alcoholism. Federation Proc. , 62, 1449-1457.

[94] Takeuchi, J., Takada, A., Ebata, K., Sawaw, G., & Okumura, Y. (1968). Effect of a single intoxication dose of alcohol on the livers of rats fed a choline-deficient diet or a commercial ration. Lab. Invest. , 19, 211-217.

[95] Lieber, C. S., Jones, D. P., Nendelson, J., & De Carli, L. M. (1963). Fatty liver, hyperlipemia and hyperuricemia produced by prolonged alcohol consumption, despite adequate dietary intake. Trans. Assoc. Am. Phys. , 76, 289-300.

[96] Lieber, C. S., De Carli, L. M., & Rubin, E. (1975). Sequential production of fatty liver, hepatitis, and cirrhosis in sub-human primates fed ethanol with adequate diets. Proc. Natl. Acad. Sci. USA , 72, 437-441.

[97] Nanji, A. A., Zhao, S., Lamb, R. G., Dannenberg, A. J., Sadrzadeh, S. M. H., & Wasman, D. J. (1994). Changes in cytochromes B1,4A, phospholipase A and C in intragastric feeding rat model for alcoholic liver disease: relationships to dietary fats and pathologic liver injury. Alcohol. Clin. Exp. Res. 18, 902-908, 4502E1.

[98] Colell, A., Kaplowitz, N., Tsukamoto, H., & Fernandez, Checa. J. C. (1997). Effects of dietary medium chain triglycerides (MCT) on ethanol induced mitochondrial GSH depletion in rat liver and pancreas. J. Hepatol. Suppl. 26, 127.

[99] Colell, A., Garcia-Ruiz, C., Miranda, M., Ardite, E., Mari, M., Morales, A., Corrales, F., Kaplowitz, N., & Fernandez-Checa, J. C. (1998). Selective glutathione depletion of mitochondria by ethanol sensitizes hepatocytes to tumor necrosis factor. Gastroenterology , 115, 1541-1551.

[100] Tsukamoto, H., Horne, W., Kamimura, S., Niemela, O., Parkkila, S., Yla-Herttuala, S., & Brittenham, G. M. (1995). Experimental liver cirrhosis induced by alcohol and iron. J. Clin. Invest. , 96, 620-630.

[101] Tsukamoto, H., Lin, M., Ohata, M., Giulivi, C., French, S., & Brittenham, G. (1999). Iron primes hepatic macrophages for NF-kB activation in alcoholic liver injury. Am. J. Physiol. 277, GG1250., 1240.

[102] Tuma, D. J., Newman, M. R., Donohue, T. M., & Sorrell, M. F. (1987). Covalent binding of acetaldehyde to proteins: participation of lysine residues. Alcohol. Clin. Exp. Res. , 579-584.

[103] Tuma, D. J., Thiele, G. M., Xu, D., Klassen, L. W., & Sorrell, M. F. (1996). Acetaldehyde and malondialdehyde administration. Hepatology , 23, 872-880.

[104] Thiele, G. M., Tuma, D. J., Willis, M. S., Miller, J. A., Mc Donald, T. L., Sorrell, M. F., & Klassen, L. W. (1998). Soluble proteins modified with acetaldehyde and malondialdehyde are immunogenic in the absence of adjuvant. Alcohol. Clin. Exp. Res. , 22, 1731-1739.

[105] Xu, D., Thiele, G. M., Beckenhauer, J. L., Klassen, L. W., Sorrell, M. F., & Tuma, D. J. (1998). Detection of circulation antibodies to malondialdehyde-acetaldehyde adducts in ethanol-fed rats. Gastroenterology , 115, 686-692.

[106] Ingelman-Sundberg, M., & Johansson, I. (1984). Mechanisms of hydroxyl radical formation and ethanol oxidation by ethanol-inducible and other forms of rabbit liver microsomal cytochrome J. Biol. Chem. 259, 6447-6458., 450.

[107] Castillo, T., Koop, D. R., Kamimura, S., Triafilopoulos, G., & Tsukamoto, H. (1992). Pole of cytochrome E1 in ethanol-carbon tetrachloride-and iron-dependent microsomal lipid peroxidation. Hepatology 16, 992-996., 450.

[108] Ekstrom, G., & Ingelman-Sundberg, M. (1989). Tat liver microsomal NADPH-supported oxidase activity and lipid peroxidation dependent on ethanol-inducible cytochrome P-4500IIE1). Biochem. Pharmacol. 38, 1313-1319., 450.

[109] Hill, D. B., Devalaraja, R., Joshi-Barve, S., & Mc Clain, C. J. (1999). Antioxidants attenuate nuclear factor-kappa B activation and tumor necrosis factor-alpha production in a alcoholic hepatitis patient monocytes and rat Kupffer cells, in vitr. o. Clin. Biochem , 32, 563-570.

[110] Mc Clain, C. J., Barve, S., Barve, S., Deaciuc, I., & Hill, D. B. (1998). Tumor necrosis factor and alcoholic liver disease.Alcohol. Clin. Exp. Res. 22, 248S-252S.

[111] Fang, H. L., & Lin, W. C. (2008). Lipid peroxidation products do not activate hepatic stellate cells. *Toxicology*, 253(1-3), 36-45.

[112] Perez, R., & Tamayo, . (1983). Is cirrhosis of the liver experimentally produced by CCl_4 an adequate model of human cirrhosis? Hepatology, , 3(1), 112-120.

[113] Bengmark, S., Mesa, M. D., Gil, A., & Hernández, . (2009). Plant-derived health-the effects of turmeric and curcuminoids. *Nutricion Hospitalaria*, 24(3), 273-281.

[114] Amália, P. M., Possa, M. N., Augusto, M. C., & Francisca, L. S. (2007). Quercetin prevents oxidative stress in cirrhotic rats,. *Digestive Diseases and Sciences*, 52(10), 2616-2621.

[115] González-Gallego, J., Sánchez-Campos, S., & Tuñón, M. J. (2007). Anti-inflammatory properties of dietary flavonoids,. Nutricion Hospitalaria, , 22(3), 287-293.

[116] Tieppo, J., Cuevas, M. J., Vercelino, R., Tuñón, M. J., Marroni, N. P., & González-Gallego, J. (2009). Quercetin administration ameliorates pulmonary complications of cirrhosis in rats. *Journal of Nutrition*, 139(7), 1339-1346.

[117] Martinez-Florez, S., González-Gallego, J., Culebras, J. M., & Tuñón, M. J. (2002). Flavonoids: properties and anti-oxidizing action. Nutrition Hospital, , 17(6), 271-278.

[118] Tokyol, C., Yilmaz, S., Kahraman, A., Çakar, H., & Polat, C. (2006). The effects of desferrioxamine and quercetin on liver injury induced by hepatic ischaemia-reperfusion in rats. *Acta Chirurgica Belgica*, 106(1), 68-72.

[119] Abilés, J., Moreno-Torres, R., Moratalla, G., et al. (2008). Effects of supply with glutamine on antioxidant system and lipid peroxidation in patients with parenteral nutrition," Nutricion Hospitalaria, 23(4), 332-339.

[120] Silvia Bona, Lidiane Isabel Filippin, F′abioCangeri Di Naso, Cintia de David,5 Bruna Valiatti,6 Maximiliano Isoppo Schaun, RicardoMachado Xavier, and Norma Possa-Marroni(2012). Effect of Antioxidant Treatment on Fibrogenesis in Rats with Carbon Tetrachloride-Induced Cirrhosis. International Scholarly Research Network ISRN Gastroenterology. , 2012

[121] Kizhakekuttu TJ, Widlansky ME(2010). Natural antioxidants and hypertension: promise and challenges. Cardiovasc Ther. 28(4):e, 20-32.

[122] Laursen, J. B., Somers, M., Kurz, S., et al. (2001). Endothelial regulation of vasomotion in apoE-deficient mice : Implications for interactions between peroxynitrite and tetrahydrobiopterin. *Circulation*, 103(9), 1282-1288.

[123] Munzel, T., Daiber, A., Ullrich, V., & Mulsch, A. (2005). Vascular consequences of endothelial nitric oxide synthase uncoupling for the activity and expression of the soluble guanylyl cyclase and the cGMP-dependent protein kinase. Arterioscler Thromb Vasc Biol, , 25(8), 1551-1557.

[124] Thomas, S. R., Chen, K., Keaney, J. F., & Jr , . (2002). Hydrogen peroxide activates endothelial nitric-oxide synthase through coordinated phosphorylation and dephosphorylation via a phosphoinositide 3-kinase-dependent signaling pathway. J Biol Chem. 22; , 277(8), 6017-6024.

[125] Drummond, G. R., Cai, H., Davis, Ramasamy. S., & Harrison, D. G. (2000). Transcriptional and posttranscriptional regulation of endothelial nitric oxide synthase expression by hydrogen peroxide. Circ Res. 18; , 86(3), 347-354.

[126] Stocker, R., Keaney, J. F., & Jr , . (2004). The role of oxidative modifications in atheroscle. rosis. Physiol Rev; , 84, 1381-1478.

[127] Chen, K., Thomas, S. R., Keaney, J. F., & Jr Beyond, . (2003). LDL oxidation: ROS in vascular signal transduction. Free Radic Biol Med, 15; , 35(2), 117-132.

[128] Moore TJ, Vollmer WM, Appel LJ, et al. (1999). Effect of dietary patterns on ambulatory blood pressure: results from the Dietary Approaches to Stop Hypertension (DASH) Trial. DASH Collaborative Research Group. *Hypertension*, 34(3), 472-477.

[129] Conlin, P. R., Chow, D., Miller, E. R. I. I. I., et al. (2000). The effect of dietary patterns on blood pressure control in hypertensive patients: results from the Dietary Approaches to Stop Hypertension (DASH) trial. Am J Hypertens.; , 13(9), 949-955.

[130] John, J. H., Ziebland, S., Yudkin, P., Roe, L. S., & Neil, H. A. (2002). Effects of fruit and vegetable consumption on plasma antioxidant concentrations and blood pressure: a randomised controlled trial. *Lancet*, 359(9322), 1969-1974.

[131] Parikh, A., Lipsitz, S. R., & Natarajan, S. (2009). Association between a DASH-like diet and mortality in adults with hypertension: findings from a population-based follow-up study. Am J Hypertens; , 22(4), 409-416.

[132] MRC/BHF(2002). Heart Protection Study of antioxidant vitamin supplementation in 20,536 high-risk individuals: a randomised placebo-controlled trial. Lancet; 6;, 360(9326), 23-33.

[133] Sesso HD, Buring JE, Christen WG, et al.(2008). Vitamins E and C in the prevention of cardiovascular disease in men: the Physicians' Health Study II randomized controlled trial. JAMA. 12;, 300(18), 2123-2133.

[134] Lee IM, Cook NR, Gaziano JM, et al.(2005). Vitamin E in the primary prevention of cardiovascular disease and cancer: the Women's Health Study: a randomized controlled trial. JAMA. 6; , 294(1), 56-65.

[135] Bjelakovic, G., Nikolova, D., Gluud, L. L., Simonetti, R. G., & Gluud, C. (2008). Antioxidant supplements for prevention of mortality in healthy participants and patients with various diseases. Cochrane Database Syst Rev. (2):, CD007176 EOF.

[136] Weinberg RB, VanderWerken BS, Anderson RA, Stegner JE, Thomas MJ.(2001). Prooxidant effect of vitamin E in cigarette smokers consuming a high polyunsaturated fat diet. Arterioscler Thromb Vasc Biol. , 21(6), 1029-1033.

[137] Salonen, J. T., Nyyssonen, K., Salonen, R., et al. (2000). Antioxidant Supplementation in Atherosclerosis Prevention (ASAP) study: a randomized trial of the effect of vitamins E and C on 3-year progression of carotid atherosclerosis. J Intern Med; , 248(5), 377-386.

[138] Münzel, T., Keaney, J. F., & Jr , . (2001). Are ACE-inhibitors a "magic bullet" against oxidative stress? *Circulation*, 104(13), 1571-1574.

[139] Ristow, M., Zarse, K., Oberbach, A., et al. (2009). Antioxidants prevent health-promoting effects of physical exercise in humans. Proc Natl Acad Sci U S A 26;, 106(21), 8665-8670.

[140] Stamler, J., Liu, K., Ruth, K. J., Pryer, J., & Greenland, P. (2002). Eight-year blood pressure change in middle-aged men: relationship to multiple nutrients. *Hypertension*, 39(5), 1000-1006.

[141] Siems, W., Sommerburg, O., Schild, L., Augustin, W., Langhans, C. D., & Wiswedel, I. (2002). Beta-carotene cleavage products induce oxidative stress in vitro by impairing mitochondrial respiration. FASEB J. , 16(10), 1289-1291.

[142] Upritchard JE, Sutherland WH, Mann JI. (2000). Effect of supplementation with tomato.juice, vitamin E, and vitamin C on LDL oxidation and products of inflammatory activity in type 2 diabetes. *Diabetes Care*, 23(6), 733-738.

[143] Engelhard, Y. N., Gazer, B., & Paran, E. (2006). Natural antioxidants from tomato extract reduce blood pressure in patients with grade-1 hypertension: a double-blind, placebo-controlled pilot study. Am Heart J. 151(1):100.

[144] Ried, K., Frank, O. R., & Stocks, N. P. (2009). Dark chocolate or tomato extract for prehypertension: a randomised controlled trial. BMC Complement Altern Med. 9:22.

[145] Chen, X., Touyz, R. M., Park, J. B., & Schiffrin, E. L. (2001). Antioxidant effects of vitamins C and E are associated with altered activation of vascular NADPH oxidase and superoxide dismutase in stroke-prone SHR. HypertensionPt 2):, 606 EOF-11 EOF.

[146] Ulker, S., Mc Keown, P. P., & Bayraktutan, U. (2003). Vitamins reverse endothelial dysfunction through regulation of eNOS and NAD(P)H oxidase activities. Hypertension. , 41(3), 534-539.

[147] Muhlhofer, A., Mrosek, S., Schlegel, B., et al. (2004). High-dose intravenous vitamin C is not associated with an increase of pro-oxidative biomarkers. Eur J Clin Nutr. , 58(8), 1151-1158.

[148] Fotherby, Williams. J. C., Forster, L. A., Craner, P., & Ferns, G. A. (2000). Effect of vitamin C on ambulatory blood pressure and plasma lipids in older persons. *Journal of Hypertension*, 18, 411-415.

[149] Mullan, Young. I. S., Fee, H., & Mc Cance, D. R. (2002). Ascorbic Acid reduces blood pressure and arterial stiffness in type 2 diabetes. *Hypertension*, 40(6), 804-809.

[150] Darko, D., Dornhorst, A., Kelly, F. J., Ritter, J. M., & Chowienczyk, P. J. (2002). Lack of effect of oral vitamin C on blood pressure, oxidative stress and endothelial function in Type II diabetes. Clin Sci (Lond) , 103(4), 339-344.

[151] McDermott JH.(2000). Antioxidant nutrients: current dietary recommendations and research update. J Am Pharm Assoc (Wash) , 40(6), 785-799.

[152] Upston, J. M., Witting, P. K., Brown, A. J., Stocker, R., Keaney, J. F., & Jr , . (2001). Effect of vitamin E on aortic lipid oxidation and intimal proliferation after arterial injury in cholesterol-fed rabbits. Free Radic Biol Med. 15;, 31(10), 1245-1253.

[153] Azzi, A., Ricciarelli, R., & Zingg, J. M. (2002). Non-antioxidant molecular functions of alpha-tocopherol (vitamin E) FEBS Lett. 2002 May 22;519(1-3):8-10.

[154] Yusuf, S., Dagenais, G., Pogue, J., Bosch, J., & Sleight, P. (2000). Vitamin E supplementation and cardiovascular events in high-risk patients. The Heart Outcomes Prevention Evaluation Study Investigators. N Engl J Med. , 342, 154-160.

[155] Miller, E. R. I. I. I., Pastor-Barriuso, R., Dalal, D., Riemersma, R. A., Appel, L. J., & Guallar, E. (2005). Meta-analysis: high-dosage vitamin E supplementation may increase all-cause mortality. Ann Intern Med. 2005 January 4;, 142(1), 37-46.

[156] Lonn, E., Bosch, J., Yusuf, S., et al. (2005). Effects of long-term vitamin E supplementation on cardiovascular events and cancer: a randomized controlled trial. JAMA. , 293(11), 1338-1347.

[157] Palumbo, G., Avanzini, F., Alli, C., et al. (2000). Effects of vitamin E on clinic and ambulatory blood pressure in treated hypertensive patients. Collaborative Group of the Primary Prevention Project (PPP)--Hypertension study. Am J Hypertens. 13(5 Pt 1):, 564 EOF-7 EOF.

[158] Ward NC, Wu JH, Clarke MW, et al.(2007). The effect of vitamin E on blood pressure in individuals with type 2 diabetes: a randomized, double-blind, placebo-controlled trial. J Hypertens. , 25(1), 227-234.

[159] Tiefenbacher CP.(2001). Tetrahydrobiopterin: a critical cofactor for eNOS and a strategy in the treatment of endothelial dysfunction? Am J Physiol Heart Circ Physiol. 280 (6):HH2488., 2484.

[160] Govers, R., & Rabelink, T. J. (2001). Cellular regulation of endothelial nitric oxide synthase. Am J Physiol Renal Physiol. 280(2):FF206., 193.

[161] Katusic ZS(2001). Vascular endothelial dysfunction: does tetrahydrobiopterin play a role? Am J Physiol Heart Circ Physiol. 281(3):HH986., 981.

[162] Schlaich MP, Parnell MM, Ahlers BA, et al. (2004). Impaired L-arginine transport and endothelial function in hypertensive and genetically predisposed normotensive subjects. *Circulation*, 110(24), 3680-3686.

[163] Wang, D., Strandgaard, S., Iversen, J., & Wilcox, C. S. (2009). Asymmetric dimethylarginine, oxidative stress, and vascular nitric oxide synthase in essential hypertension. Am J Physiol Regul Integr Comp Physiol. 296(2):RR200., 195.

[164] Hidalgo Ponce Alejandro.(2007). Terapia Antioxidante. Foco en la microcirculación. Critical Care Medicine-Suppl., 35(9)

[165] Bevers, L. M., Braam, B., Post, J. A., et al. ((2006).) Tetrahydrobiopterin, but not L-arginine, decreases NO synthase uncoupling in cells expressing high levels of endothelial NO synthase. Hypertension. ., 47(1), 87-94.

[166] Matsuoka, H., Itoh, S., Kimoto, M., et al., Asymmetrical, dimethylarginine., an, endogenous., nitric, oxide., synthase, inhibitor., in, experimental., & hypertension, . Hypertension. (1997). January;29(1 Pt 2):242-247.

[167] Siani, A., Pagano, E., Iacone, R., Iacoviello, L., Scopacasa, F., & Strazzullo, . (2000). May;P. Blood pressure and metabolic changes during dietary L-arginine supplementation in humans. Am J Hypertens. 13(5 Pt 1):547-551.

[168] Martina, V., Masha, A., Gigliardi, V. R., et al. (2008). Long-term N-acetylcysteine and L-arginine administration reduces endothelial activation and systolic blood pressure in hypertensive patients with type 2 diabetes. *Diabetes Care*, 31(5), 940-944.

[169] Kelly, Alexander. J. W., Dreyer, D., et al. (2001). Oral arginine improves blood pressure in renal transplant and hemodialysis patients. JPENJ Parenter Enteral Nutr. , 25(4), 194-202.

[170] Gokce, N., & et, al. . (2004). L-arginine amd hypertension. J Nutr 134(10 Suppl) 2807S-28011S.

[171] Loscalzo, J. (2003). Adverse effects of supplemental L-arginine in atherosclerosis: consequences of methylation stress in a complex catabolism? Arterioscler Thromb Vasc Biol. 1; , 23(1), 3-5.

[172] Persky AM, Brazeau GA.(2001). Clinical pharmacology of the dietary supplement creatine monohydrate. Pharmacol Rev. , 53(2), 161-176.

[173] Tyagi, N., Sedoris, K. C., Steed, M., Ovechkin, A. V., Moshal, K. S., & Tyagi, S. C. (2005). Mechanisms of homocysteine-induced oxidative stress. Am J Physiol Heart Circ Physiol. 289(6):HH2656., 2649.

[174] Jahangir, E., Vita, J. A., Handy, D., et al., & (200, . (2009). The effect of l-arginine and creatine on vascular function and homocysteine metabolism. Vasc Med. , 14(3), 239-248.

[175] Peters, U., Poole, C., & Arab, L. (2001). Does tea affect cardiovascular disease? a meta-analysis. Am J Epidemiol. 15; , 154(6), 495-503.

[176] Bazzano, L. A., He, J., Ogden, L. G., et al. (2002). Fruit and vegetable intake and risk of cardiovascular disease in US adults: the first National Health and Nutrition Examination Survey Epidemiologic Follow-up Study. Am J Clin Nutr. , 76(1), 93-99.

[177] Aviram, M., & Fuhrman, B. (2002). Wine flavonoids protect against LDL oxidation and atherosclerosis. Ann N Y Acad Sci. , 957, 146-161.

[178] Lotito, S. B., & Frei, B. (2006). Consumption of flavonoid-rich foods and increased plasma antioxidant capacity in humans: cause, consequence, or epiphenomenon? Free Radic Biol Med. , 41(12), 1727-1746.

[179] Aviram, M., & Dornfeld, L. (2001). Pomegranate juice consumption inhibits serum angiotensin converting enzyme activity and reduces systolic blood pressure. *Atherosclerosis*, 158(1), 195-198.

[180] Aviram, M., Rosenblat, M., Gaitini, D., et al. (2004). Pomegranate juice consumption for 3 years by patients with carotid artery stenosis reduces common carotid intima-media thickness, blood pressure and LDL oxidation. Clin Nutr. , 23(3), 423-433.

[181] Anter, E., Thomas, S. R., Schulz, E., Shapira, O. M., Vita, J. A., Keaney, J. F., & Jr , . (2004). Activation of eNOS by the MAP kinase in response to black tea polyphenols. J Biol Chem. 45:46637-46643., 38.

[182] Diebolt, M., Bucher, B., & Andriantsitohaina, R. Wine polyphenols decrease blood pressure, improve NO vasodilatation, and induce gene expression. Hypertension(2001). August;, 38(2), 159-165.

[183] Taubert, D., Berkels, R., Roesen, R., & Klaus, W. (2003). Chocolate and blood pressure in elderly individuals with isolated systolic hypertension. JAMA. , 290(8), 1029-1030.

http://dx.doi.org/10.5772/51994

[184] Grassi, D., Lippi, C., Necozione, S., Desideri, G., & Ferri, C. (2005). Short-term administration of dark chocolate is followed by a significant increase in insulin sensitivity and a decrease in blood pressure in healthy persons. Am J Clin Nutr. , 81(3), 611-614.

[185] Taubert, D., Roesen, R., Lehmann, C., Jung, N., & Schomig, E. (2007). Effects of low habitual cocoa intake on blood pressure and bioactive nitric oxide: a randomized controlled trial. JAMA. , 298(1), 49-60.

[186] Grassi, D., Desideri, G., Necozione, S., et al. (2008). Blood pressure is reduced and insulin sensitivity increased in glucose-intolerant, hypertensive subjects after 15 days of consuming high-polyphenol dark chocolate. J Nutr. , 138(9), 1671-1676.

[187] Zilkens, R. R., Burke, V., Hodgson, J. M., Barden, A., Beilin, L. J., & Puddey, I. B. (2005). Red wine and beer elevate blood pressure in normotensive men. *Hypertension*, 45(5), 874-879.

[188] Taubert, D., Roesen, R., & Schomig, E. (2007). Effect of cocoa and tea intake on blood pressure: a meta-analysis. Arch Intern Med. , 167(7), 626-634.

Chapter 16

The Role of Natural Antioxidants in Cancer Disease

Carmen Valadez-Vega, Luis Delgado-Olivares,
José A. Morales González, Ernesto Alanís García,
José Roberto Villagomez Ibarra,
Esther Ramírez Moreno , Manuel Sánchez Gutiérrez,
María Teresa Sumaya Martínez,
Zuñiga Pérez Clara and Zuli Calderón Ramos

Additional information is available at the end of the chapter

http://dx.doi.org/10.5772/51503

1. Introduction

Cell oxidation can lead to the onset and development of a wide range of diseases including Alzheimer and Parkinson, the pathologies caused by diabetes, rheumatoid arthritis, neurodegeneration in motor neuron diseases, and cancer. Reactive species (RS) of various types are powerful oxidizing agents, capable of damaging DNA and other biomolecules. Increased formation of RS can promote the development of malignancy, 'normal' rates of RS generation may account for the increased risk of cancer development.

Oxidants and free radicals are inevitably produced during the majority of physiological and metabolic processes and the human body has defensive antioxidant mechanisms; these mechanisms vary according to cell and tissue type and may act antagonistically or synergistically. They include natural enzymes like Superoxide dismutase (SOD), Catalase (CAT), and Glutathione peroxidase (GPx), as well as antioxidants such as vitamins, carotenoids, polyphenols, and other natural antioxidants, which have attracted great interest in recent years.

There has been a great deal of interest of late in the role of complementary and alternative drugs for the treatment of various acute and chronic diseases. Among the several classes of phytochemicals, interest has focused on the anti-inflammatory and antioxidant properties of the polyphenols that are found in various botanical agents. Plant vegetables and spices used in folk and traditional medicine have gained wide acceptance as one of the main sources of prophylactic and chemopreventive drug discoveries and development.

Recently, researches on medicinal plants has drawn global attention; large bodies of evidence have accumulated to demonstrate the promising potential of medicinal plants used in various traditional, complementary, and alternate treatment systems of human diseases. The plants are rich in a wide variety of secondary metabolites, such as tannins, terpenoids, alkaloids, flavonoids, etc., which have been screened *in vivo* and *in vitro* and have indicated antioxidant and anticarcinogenic properties and which are used to developed drugs or dietary supplements.

Evidence suggests that the plant kingdom is considered a good candidate for chemoprevention and cancer therapy due to the high concentration and wide variety of antioxidants such as resveratrol, genestein, beicalein, vitamin A, vitamin C, polyphenols, (–)–Epigallocatechin 3-gallate, flavonoids, polyphenols, gallic acid, glycosides, verbascoside, calceorioside, epicatechin, quercetin, curcumin, lovastatin, and many other types of compounds with the capability to inhibit the cell proliferation of different cancer cells *in vitro* and *in vitro*, such as colon cancer (HT-29, SW48, HCT116), breast (MCF7, MDA), cervix (HeLa, SiHa, Ca-Ski, C33-A), liver (Hep G2), skin (A 431), fibroblasts (3T3 SV40), and many other malignant cells; studies have indicated that antioxidants can be employed efficiently as chemopreventives and as effective inhibitors of cell proliferation, promoting cell apoptosis, and increasing detoxification enzymes, and inhibiting gene expression and scavenger Reactive oxygen species (ROS). Thus, many researchers are working with different types of natural antioxidants with the aim of finding those with the greatest capacity to inhibit the development of cancer both *in vitro* as well as *in vivo*, because these compounds have exhibited high potential for use not only in the treatment of this disease, but they also act as good chemoprotective agents.

2. Antioxidants

The production of ROS during metabolism is an inevitable phenomenon associated with the process of aerobic metabolism; on the other hand, we are exposed at all times to several exogenous sources of oxidant molecules, for example, environmental and pollutant factors and many dietary compounds, which increase their levels. ROS participate in different cellular processes; their intracellular levels are relatively low. However, because ROS are highly toxic when their concentration increases, the phenomenon denominated Oxidative stress (OS) is produced [123], which can injure various cellular biomolecules, causing serious damage to tissues and organs and resulting in chronic diseases [24]. Oxidative damage can be prevented by antioxidants, which are present within the cell at low concentrations compared with oxidant molecules [141, 50].

Antioxidants are capable of donating electrons to stabilize ROS and to inhibit their detrimental effects, including both endogenous (synthesized by the body itself) and exogenous molecules (those from external sources to the body) [141]. Endogenous antioxidants include Superoxide dismutase (SOD), which catalyzes the dismutation reaction of superoxide ($O2\bullet^-$) into hydrogen peroxide (H_2O_2), which is in turn transformed into oxygen and water for the Catalase (CT), and in addition Glutathione peroxidase (GPx) can catalyze its reduction; however, if in the presence of transition metals such as iron, H_2O_2, by means of the Fenton reaction, can produce the hydroxyl radical (OH•⁻); wich is of more reactive the ROS, capable to produce the majority of oxidative damage [24]. On the other hand, exogenous antioxidants can be from animal and plant sources; however, those of plant origin are of great interest because they can contain major antioxidant activity [19]. Different reports show that persons with a high intake of a diet rich in fruit and vegetables have an important risk reduction of developing cancer, mainly due to their antioxidant content [70]. Among the vegetable antioxidants are vitamins E and C, and ß-carotene, which are associated with diminished cardiovascular disease and a decreased risk of any cancer [48]. In particular, ß-carotene and vitamin E can reduce the risk of breast cancer, vitamin C, ß-carotene, and lutein/zeaxanthin possess a protector effect against ovarian cancer, and vitamin C, ß-carotene, and rivoflavin prevent colorectal cancer [70], while flavonoids such as plant phenolics and wine phenolics can inhibit lipid peroxidation and lipoxygenase enzymes. In addition, any microelement, such as Se, Zn, Mn, and Cu, can exhibit antioxidant activity [48, 24].

In recent years, interest has grown in the use of natural antioxidants for the prevention or treatment of different diseases related with OS; however despite the widespread information of the beneficial effects of antioxidants in the prevention of cancer, their use remains questionable, because different reports have shown that reducing the levels of ROS may have counterproductive effects because due to raising the risk of cancer; the latter may be due to that ROS can produce apoptosis in malignant cells [38, 101].

3. Molecular Studies of Natural Antioxidants

Different types of natural antioxidants are present in fruit and vegetables; they have synergistic interactions that are important due to their activity and regenerative potential. For example, ascorbate can regenerate into α-tocopherol [53], and the ascorbate radical is regenerated into other antioxidants via the thiol redox cycle. Taken together, all of these interactions are known as the "antioxidant network".

Vitamin E is an antioxidant that penetrates rapidly through the skin and is incorporated into the cellular membranes, inhibiting lipid peroxidation; specifically, α-tocotrienol, the vitamin E isoform, demonstrates greatest protection. Additionally, vitamin E possesses antiproliferative properties that interfere in signal transduction and in inducing cell cycle arrest.

Tumor necrosis factor-alpha (TNF-α) is a cytokine that, under normal conditions, induces inflammation, tumor inhibition, and apoptotic cell death. However, when the former undergoes deregulation, it acts as a breast tumor promoter, enhancing the proliferation of chemically induced mammary tumors [113]. Phenolic antioxidants can block the increase of TNF-α at the transcriptional level in the nucleus, which suggests the molecular mechanism of phenolic antioxidants through control of cytokine induction [81].

4. Oxidative Stress and Diseases

The ROS, as the superoxide anion ($O_2•^-$), hydrogen peroxide (H_2O_2), and the hydroxyl radical (OH•), are produced during cell metabolism in the lysosomes, peroxisomes, endoplasmic reticulum in the process carried out to obtain energy such as Adenosine triphosphate (ATP) [108]. There are other sources of oxidant molecules, such as pollution, the environment, and certain foods. During recent years, it has been discovered that during aging, the mitochondria increase the levels of ROS production and antioxidant endogens are diminished [98, 13]. ROS play an important role in the physiological process; however, due to their toxicity, their levels must be controlled by the endogenous antioxidant system. But when ROS formation is increased, an imbalance is promoted between these and the antioxidant molecules; phenomenon known as Oxidative stress (OS) [123]. OS can cause oxidative damage of proteins, lipids, and nucleic acids, macromolecules involved in the cell function, membrane integrity, or in maintaining genetic information (nucleic acids) [44, 45, 65].

Proteins are responsible for different cell processes (enzymatic, hormonal, structural support). The oxidation of proteins produces disulfide crosslinks, nitration, or tyrosine residues, and carbonylation, resulting in the loss of the structure and function of proteins and fragmentation [11, 97]. But because the chaperones are susceptible to oxidative damage, allowing the accumulation of misfolding proteins and increasing their susceptibility to protease degradation [115], however, the proteasome also undergoes oxidation and its activity is diminished, which makes the aggregates accumulate in the cell wich have been associated with aging and various pathologies, such as cancer and neurodegenerative disorders, such as Parkinson, Huntington, and Alzheimer disease [98].

The brain is the organ with the highest oxygen consumption; it has high levels of fatty acids, iron, and low antioxidant defenses. This is an organ with major susceptibility to oxidative damage [141], producing neurodegeneration that results in different diseases such as Parkinson disease, Alzheimer disease, Down syndrome, autism, bipolar disorder, and epilepsy [23, 24], and the cognitive alteration known as Mild cognitive impairment (MCI), which is produced preferentially in regions of the brain involved in regulating cognition, contributing to the development of dementia [65]. Similar processes occur during aging, resulting in the genetic response of increasing levels of antioxidant enzymes and chaperone proteins [73]. Reduction of OS causes improvement of the long-term memory [102].

Polyunsaturated fatty acids (mainly compounds of the membranes) are susceptible to peroxidation, which affects the integrity of the membranes of organelles of the cell membrane and the respiratory chain, in turn affecting cell viability. Lipid peroxidation produces aldehydes such as 4-hydroxy-2 *E*-nonenal, which is toxic and is involved in alterations in Alzheimer disease and DNA damage, causing mutations associated with the development of cancer [38, 20].

Ribosomal RNA and transfer RNA constitute the majority of stable species of cellular RNA, which possess a greater oxidation rate than DNA. The major modification for oxidation into RNA comprises 8-hydroxyGuanine (8-oxoG), which under normal conditions is present three times more in non-ribosomal that in ribosomal RNA; however, when the cell is exposed to H_2O_2, the concentration of 8-oxoG in ribosomal RNA increases at the same levels in both RNA [97]. RNA oxidation can diminish the capacity of replacement oxidation of proteins [65, 44] and the inhibition of protein synthesis, cell cycle arrest, and cell death. Oxidation of RNA is involved in the development of cancer, viral infections, AIDS, hepatitis (VIH-1; HCV; 107, 148], and neurological diseases. It has been reported that each neurological disease, present a damage oxidative of RNA in a specific region on the brain, for example in Alzheimer disease, there are increased RNA oxidation in the hippocampus and cerebral neocortex, while in Parkinson disease, RNA oxidation is localized in the *sustancia nigra* [97].

On the other hand, high-fat diets induce obesity and insulin resistance, resulting in increased ROS production, which modifies sympathetic brain activity, which in turn contributes to the rise in blood pressure, increase in insulin resistance, and obesity [6]. Obesity is the principal factor in the development of the metabolic syndrome, due to that persons with obesity have deficient antioxidant defense and increased production of ROS [126, 30, 75], which leads to spoilage and subsequently cell death, resulting in tissue and organ damage, to tissues causing serious health problems such as insulin resistance [7], diabetes mellitus, and hypertension [82]. Moreover, in the metabolic syndrome, NAD(P)H oxidase, the major source of ROS in several tissues, is up-regulated, resulting in an increase of ROS production and the down-regulation of several antioxidant enzymes (SOD isoforms, GPx, and heme oxygenase) [114]. This enzyme, specifically in the type 4 isoform (NOX4), is implicated in the damage due to OS during cerebral ischemia [67].

The scientific literature has shown that oxidative stress is involved in the development of a wide range of disease, such as heart diseases, Hutchinson-Gilford syndrome or progeria, hypertensive brain injury, muscular dystrophy, multiple sclerosis, congenital cataract, retinal degeneration, retinopathy of the premature, autoimmune diseases, cardiovascular abnormalities, nephrological disorders, emphysema, stroke, rheumatoid arthritis, anemia, hepatitis, pancreatitis, aging, premature wrinkles and dry skin, endothelial dysfunction, and dermatitis, among others [83, 7, 137, 91, 23, ¸102].

However the most important damage caused by OS are the DNA modifications, which can result in permanent mutations, due to that oxidative damage also affects the proteins involved in repairing the harm or reducing the OS (the endogenous antioxidant); thus, oxidative damage to DNA can be the cause of the development of various diseases, such as cancer [13, 51].

5. Cancer

Cancer is unnatural cell growth, in which cells can lose their natural function and spread throughout the blood in the entire body. Breast cancer is the most commonly diagnosed cancer in industrialized countries and has the highest death toll [88]. OS is involved in the process of the development of cancer and tumors, due to that ROS can damage the macromolecules as lipids, which react with metals (such as free iron and copper) and produce aldehydes and synthesize malondialdehyde-inducing mutations [96] or cause breaks in the double chain, produce modifications in guanine and thymine bases, and sister chromatid exchanges [16], which can affect the activities of signal transduction, transcription factors, and gene tumor suppressors such as *p53*, which is a gene important in apoptosis and in cell cycle control. This inactivation can increase the expression of proto-oncogenes [96] which can produce major damage. Oxidative damage or genetic defects that result in some defective enzymes are incapable of repairing the mutations increase the incidence of age-dependent cancer [51].

On the other hand, treatments with anticancer drugs and radiation increase ROS and decrease antioxidants content, producing a state of severe oxidative stress and causing apoptosis, resulting in side effects [96], while persistent oxidative stress at sublethal levels can result in resistance to apoptosis [16].

Some microorganisms, as bacteria and viruses, are involved, via OS, in the process of the production of certain cancers such as, for example *Helicobacter pylori*, inducing gastric cancer and colon cancer through the production of $SO\bullet^-$ [96]. It has been proposed that lower antioxidant activity increases the risk of developing cancer; thus, ingestion of antioxidants can prevent cancerogenesis. However is not clear the decrease of antioxidants levels is not clear, in as much as in freshly cancerous tissue, MnSOD levels are elevated; therefore, some investigators have proposed that this antioxidant enzyme is involved in tumor invasion; thus, it is possible that antioxidants have a role as pro-oxidants. Another point to consider is that when the 8-oxodG level in DNA increases, cancer rates do not increase [96, 51]. However, OS is a factor for cancer and other diseases, but not the sole factor for diseases, because others, such as genetic factors (genetic predisposition) are involved.

6. Antioxidants and Cancer

Humans are constantly bombarded by exogenous factors such as Ultraviolet (UV) rays, tobacco smoke, and many others agents that cause OS. Such stress can also arise from the drugs that are employed in medical practice. On the other hand, under physiological conditions, normal aerobic metabolism gives rise to active and potentially dangerous oxidants in cells and tissues; these endogenous sources of OS include those derived from the activities of mitochondria or microsomes and peroxisomes in the electron transfer system and from the activities of the NADPH enzyme present in macrophages and neutrophils as a mechanism of protection against infection. Various reducing substances in the human body control the status of oxidation-reduction (redox), and a continuing imbalance in favor of oxidation causes several problems when it exceeds the capacity of such a control [96].

Otto Warburg was the first scientist to implicate oxygen in cancer [147] as far back as the 1920s. However, the underlying mechanism by which oxygen might contribute to the carcinogenic process was undetermined for many years. The discovery of superoxide dismutase in 1968 by [90] led to an explosion of research on the role of reactive oxygen in the pathologies of biological organisms. Reactive oxygen has been specifically connected with not only cancer, but also many other human diseases [5, 57]. For many years, research on OS focused primarily on determining how ROS damage cells through indiscriminate reactions with the macromolecular machinery of a cell, particularly lipids, proteins, and DNA. It is well known and in great detail the manner in which ROS react with lipids, leading to the peroxidation of biological membranes and resulting in necrotic lesions [43] and the way ROS react with the nucleotides of DNA, leading to potential mutations [17, 43, 139].

When produced in excess, ROS (some of which are free radicals) can seriously alter the structure of biological substrates such as proteins, lipids, lipoproteins, and Deoxyribonucleic acid (DNA). They possess a huge range of potential actions on cells, and one could easily envisage them as anti-cancer (e.g., by promoting cell-cycle stasis, senescence, apoptosis, necrosis or other types of cell death, and inhibiting angiogenesis), or as pro-cancer (promoting proliferation, invasiveness, angiogenesis, metastasis, and suppressing apoptosis).

Active oxygen may be involved in carcinogenesis through two possible mechanisms: induction of gene mutations that result from cell injury [34], and the effects on signal transduction and transcription factors. Which mechanism it follows depends on factors such as the type of active oxygen species involved and the intensity of stress [86]. Cellular targets affected by oxidative stress include DNA, phospholipids, proteins, and carbohydrates on the cell membrane. Oxidized and injured DNA has the potential to induce genetic mutation. That some telomere genes are highly susceptible to mutation in the presence of free radicals is now apparent, and it is known that tumor suppressor genes such as *p53* and cell cycle-related genes may undergo DNA damage. In addition, oxidized lipids react with metals to produce active substances (e.g., epoxides and aldehydes) or synthesize malondialdehyde, which has the potential to induce mutation. Active oxygen species act directly or indirectly via DNA damage on gene expression (DNA binding of transcription factors) and signaling at the cellular level.

Markers for OS can be divided into three categories:

1. formation of modified molecules by free radical reactions;
2. consumption or induction of antioxidant molecules or enzymes, and
3. activation or inhibition of transcription factors.

Targets of free radicals include all types of molecules in the body. Among these, lipids, nucleic acids, and proteins are the major targets. Because free radicals are usually generated near membranes (cytoplasmic membrane, mitochondria, or endoplasmic reticulum), lipid peroxidation is the first reaction to occur. Lipid peroxidation products can be detected as classical Thiobarbituric acid (TBA)-reactive substances. Recently, the detection of 4-Hydroxy-2-nonenal (HNE) or Malondialdehyde (MDA) is favored due to their high specificity [32], aldehydes are end-products of lipid peroxidation but continue to be reactive with cell proteins [136].

Exposure to free radicals from a variety of sources has led organisms to develop a series of defense mechanisms that involve the following:

1. preventative mechanisms;
2. repair mechanisms;
3. physical defenses, and

4. antioxidant defenses.

Enzymatic antioxidant defenses include Superoxide dismutase (SOD), Glutathione peroxidase (GPx), and Catalase (CAT). Non-enzymatic antioxidants are represented by ascorbic acid (vitamin C), α-tocopherol (vitamin E), Glutathione (GSH), carotenoids, flavonoids, tannins, triterpepenoids, saponins, glycosides, steroids, and other antioxidants [46]. Under normal conditions, there is a balance between both the activities and the intracellular levels of these antioxidants: this equilibrium is essential for the survival of organisms and their health

7. Antioxidants in Cancer Assays

Humans have evolved with antioxidant systems for protection against free radicals and ROS. These systems include some antioxidants produced in the body (endogenous) and others obtained from the diet (exogenous) [21]. The former include

1. enzymatic defenses, such as *Se*-glutathione peroxidase, catalase, and superoxide dismutase, which metabolize superoxide, hydrogen peroxide, and lipid peroxides, thus preventing the majority of the formation of toxic HO•, and

2. non-enzymatic defenses, such as glutathione, histidine peptides, the iron-binding transfer proteins and ferritin, and dihydrolipoic acid, reduced Coenzyme Q10, melatonin, urate, and plasma protein thiols, with the latter two accounting for the major contribution to the radical-trapping capacity of plasma.

The various defenses are complementary to each other because they act against different species in different cellular compartments. However, despite these defense antioxidants (able either to suppress free radical formation and chain initiation or to scavenge free radicals and chain propagation), some ROS escape to cause damage. Thus, the body's antioxidant system is also provided with repair antioxidants (able to repair damage) and based on proteases, lipases, transferases, and DNA repair enzymes [145, 103].

Owing to the incomplete efficiency of our endogenous defense systems and the existence of some physiopathological situations (cigarette smoke, air pollutants, UV radiation, a high, polyunsaturated fatty acid diet, inflammation, ischemia/reperfusion, etc.) in which ROS are produced in excess and at the wrong time and place, dietary antioxidants are required to diminish the cumulative effects of oxidative damage throughout the human lifespan [149, 47). Well known natural antioxidants derived from the diet, such as vitamins C, E, and A and the carotenoids, have been studied intensively [124]. In addition to these, antioxidants in plants might account for at least part of the health benefits associated with vegetable and fruit consumption [103].

The plants, vegetables, and spices used in folk and traditional medicine have gained wide acceptance as one of the main sources of prophylactic and chemopreventive drug discovery and development [85, 29].

Some reports indicate that the prevalence of use of complementary and alternative medicine by patients with cancer has been estimated at a range of 7–64% [3, 4, 58]. At present, many patients with cancer combine some forms of complementary and alternative therapy with their conventional therapies [4, 58]. A recent survey of patients at a comprehensive cancer center placed the use of vitamin and minerals at 62.6%; of these patients, 76.6% combined the use of vitamins and minerals with conventional chemotherapy [58, 27].

These types of patients employ complementary and alternative therapies for a variety of reasons [31, 14]: to improve quality of life (77%); to improve immune function (71%); to prolong life (62%), or to relieve symptoms (44%) related with their disease [31]. Only 37.5% of the patients surveyed expected complementary and alternative therapies to cure their disease. Whatever the reasons, alternative therapy use is on the rise and this includes the use of megavitamins, minerals, and cocktails of natural substances during chemotherapy administration; these cocktails include antioxidants such as the commonly consumed antioxidants vitamin E (mixed tocopherols and tocotrienols), vitamin C, β-carotene (natural mixed carotenoids), polyphenols, tannins, terpenoids, alkaloids, flavonoids, vitamin A, and many others. Controversy exists concerning the use of antioxidants with chemotherapy, but increasing evidence suggests a benefit when antioxidants are added to chemotherapy [111, 112, 106, 151, 117, 105, 22, 27].

It is widely accepted that diets rich in fruits and plants are rich sources of different types of antioxidants; phenolic compounds are the most studied of these and have been recognized to possess a wide range of properties including antioxidant, antibacterial, anti-inflammatory, hepatoprotective, and anticarcinogenic actions [3, 4, 63]. Many of the biological functions of flavonoid, phenolic, catechin, curcumin, resveratrol, and genistein compounds have been attributed to their free-radical scavenging, metal-ion chelating, and antioxidant activities [118, 152]. Antioxidant phenolic agents have been implicated in the mechanisms of chemoprevention, which refers to the use of chemical substances of natural or of synthetic origin to reverse, retard, or delay the multistage carcinogenic process [29].

It has been shown that dietary phytochemicals can interfere with each stage of the development of carcinogenesis [130, 93]. As in the case of direct antioxidant effects, dietary polyphenols are most likely to exert their chemopreventive effects on the gastrointestinal tract, where they are present at highest concentrations [52, 49, 84, 75]. Indeed, studies have shown that various polyphenol-rich fruits and vegetables are particularly effective in protecting against several types of cancer development [84, 75, 59]. Dietary polyphenols may exert their anticancer effects through several possible mechanisms, such as removal of carcinogenic agents, modulation of cancer cell signaling and antioxidant enzymatic activities, and induction of apoptosis as well as of cell cycle arrest. Some of these effects may be related, at least partly, with their antioxidant activities [59]. They may exert protective effects against cancer development, particularly in the gastrointestinal tract, where they will be at their highest concentration. In fact, many studies have shown that various polyphenol-rich fruits and vegetables are particularly effective in protecting against colon cancer development [84, 75].

At the cellular level, there is good evidence that polyphenols present in tea, red wine, cocoa, fruit juices, and olive oil; at some level, they are able to stimulate carcinogenesis and tumor development [93]. For example, they may interact with reactive intermediates [28] and activated carcinogens and mutagens [18], they may modulate the activity of the key proteins involved in controlling cell cycle progression [104], and they may influence the expression of many cancer-associated genes [142]. Perhaps most notably, the anticancer properties of green tea flavanols have been reported in animal models and in human cell lines (Takada *et al.*, 2002], as well as in human intervention studies [60]. On the other hand, green tea consumption has been proposed as significantly reducing the risk of cancer of the biliary tract [133], bladder [110], breast [74], and colon [72]. Many of the anti-cancer properties associated with green tea are thought to be mediated by the flavanol Epigallocatechin gallate (EGCG), which has been shown to induce apoptosis and inhibit cancer cell growth by altering the expression of cell cycle regulatory proteins and the activity of signaling proteins involved in cell proliferation, transformation, and metastasis [66]. In addition to flavonoids, phenolic alcohols, lignans, and secoiridoids (all found at high concentrations in olive oil) are also thought to induce anti-carcinogenic effects [99] and have been reported in large intestinal cancer cell models [79], in animals [10, 128], and in humans [99]. These effects may be mediated by the ability of olive oil phenolics to inhibit initiation, promotion, and metastasis in human colon adenocarcinoma cells [42, 55] and to down-regulate the expression of COX-2 and Bcl-2 proteins, which play a crucial role in colorectal carcinogenesis [79, 146].

In vivo studies have demonstrated that many natural compounds found in plants and fruits have the capability to inhibit many types of human and animal cancer. Vitamins such as C, E, and A have shown that they can diminish cervical, bladder, prostate, intestinal, skin, and other gastrointestinal cancer types and that they have the capability to inhibit ROS production in patients [36, 37, 89, 134, 131, 62, 127]. In addition, it was demonstrated that these vitamins can inhibit progression and pathogenesis in colorectal cancer [12]. In animal models, vitamins showed promise for chemopreventive agents against several types of gastrointestinal cancer [62].

With the use of a combination of vitamins, selenium, β-carotene, essential fatty acids, and coenzyme Q10 in patients with breast cancer, it was observed that during the study no patient died, no patient showed signs of further distant metastasis, quality of life improved, and six patients showed apparent partial remission [80]. Human studies demonstrated that consumption of total antioxidants in the diet (fruits and vegetables) is inversely associated with the risk of distal gastric cancer [87]. Antioxidants, especially polyphenols, have been found to be promising agents against cervical cancer, including induction of apoptosis, growth arrest, inhibition of DNA synthesis, and modulation of signal transduction pathway; additionally, polyphenols can interfere with each stage of carcinogenesis initiation, promotion, and progression for the prevention of cancer development [26].

Camelia sinensis tea, which contains a great quantity of polyphenols (epichatechin, (–)–epigallocatechin-3-gallate) is the most widely consumed beverage worldwide, and it was demonstrated that consumption of this beverage has shown to afford protection against chemical carcinogen-induced stomach, lung, esophagus, duodenum, pancreas, liver, breast, and colon carcinogenesis in specific bioassay models. The properties of the tea's polyphenols make them effective chemopreventive agents against the initiation, promotion, and progression stages of multistage carcinogenesis [64]. Rosmanic acid had demonstrated to possess potent anticancer and apoptotic effect in mouse-induced skin cancer [121], curcumin, (–)–epigallocatechin-3-gallate, and lovastatin in combination were able to suppress esophageal cancer in mouse [154], and melatonin demonstrated diminishing the development and mortality of mouse implanted with murine hepatoma cells MN22a [39]. It was demonstrated that beta-ionone, a precursor of carotenoids, ameliorated lung carcinogenesis; the latter is attributed to the antiproliferative and antioxidant potential of beta-ionone through free radical scavenging properties [9]. A-tocopherol showed down-regulation of the expression of the stress-activated genes *PKC-α*, *c-Myc*, and *Lactate dehydrogenase A (LDHA)* in cancerous mice, decreasing cancer cell proliferation [120]. It has been suggested that rosmanic acid suppresses oral carcinogenesis by stimulating the activities of detoxification enzymes, improving the status of lipid peroxidation and antioxidants, and down-regulating the expression of *p53* and *bcl-2* during 7,12 dimethylbenz(a)anthracene-induced oral carcinogenesis in hamster [8]. In the same manner, the methanolic extract of fennel seed exhibited an antitumoral affect by modulating lipid peroxidation and augmenting the antioxidant defense system in Ehrlich ascites carcinoma- bearing mice with or without exposure to radiation [94]. Silymarin, a natural flavonoid from the milk thistle seed, displayed chemopreventive action against 1,2-dimethylhydrazine plus dextran sodium sulfate-induced inflammation associated with colon carcinogenesis [135]. Quercetin, a flavonoid found in many natural foods, demonstrated to exert a direct oro-apoptotic affect on tumor cells and can indeed block the growth of several human cancer-cell lines in different cell-cycle phases, which have been demonstrated in several animal models [41]. The methanolic extract of *Indigofera cassioides* was evaluated in terms of their antitumor activity on Ehrlich ascites carcinoma- bearing mice; the extract showed a potent antitumoral effect against tumor cells due its preventing lipid peroxidation and promoting the enzymatic antioxidant defense system in animals [69]. Brucine, a natural plant alkaloid, was reported to possess cytotoxic and antiproliferative activities and also had showed to be a potential anti-metastatic and -angiogenic agent [2].

An *in vitro* assay demonstrated that the mechanism's antioxidant action, according to Halliwell [52], can include the following:

1. suppressing ROS formation either by inhibiting the enzymes or chelating the trace elements involved in free radical production;

2. scavenging ROS, and

3. up-regulating or protecting antioxidant defenses.

Flavonoids have been identified as fulfilling the majority of the criteria previously described. Thus, their effects are two-fold as follows:

1. Flavonoids inhibit the enzymes responsible for superoxide anion production, such as xanthine oxidase [54] and Protein kinase C (PKC) [140], and

2. Flavonoids have also shown to inhibit cyclo-oxygenase, lipoxygenase, microsomal mono-oxygenase, glutathione *S*-transferase, mitochondrial succinoxidase, and (Nicotinamide adenine denucleotide (NADH) oxidase, all of which are involved in ROS generation [68, 15].

A number of flavonoids efficiently chelate trace metals, which play an important role in oxygen metabolism. Free iron and copper are potential enhancers of ROS formation, as exemplified by the reduction of hydrogen peroxide with the generation of the highly aggressive hydroxyl radical [103].

On the other hand, *in* vitro studies showed that the compounds present in fruits and vegetables, such as resveratrol, genestein, baicalein, and many others are attractive candidates for improved chemotherapeutic agents [35]. Resveratrol in combination with platinum drugs and oxaliplatin demonstrated that resveratrol administered 2 h prior to platinum drugs may sensitize ovarian cancer cells to platinum, inducing apoptosis and providing a means of overcoming resistance [95].

Ren [109] demonstrated that (–)–epigallocatechin-3-gallate induces reduction in IM9 myeloma cells and that its activity was dose- and time-dependent on the induction of apoptotic cell death; additionally, this natural metabolite combined with curcumin and lovastatin possessed the ability to suppress esophageal cancer-cell growth [154]. In multilla berries, it was found that their high levels of polyphenols, flavonoids, and flavonols and their antioxidants have a strong ability to reduce the viability of colon-cancer HT-29 and SW480 cell lines [33]. The anticancer activity of baicalein, a flavonoid found in several plants, was evaluated in a cutaneous squamous carcinoma-cell line, A431; it was found that this compound reduced the migration and invasiveness of the cells through inhibition of ezrin expression, which leads to the suppression of tumor metastasis [153].

In beans, it was found that these contain several compounds with cytotoxic activity on animals and human cell lines (C33-A, SW480, and 3T3), which can be attributed to the antioxidants and damage to DNA caused by tannins, saponins, lectins, and others compounds found in the seed [143, 144].

Melastoma malabathricum showed to have the ability to inhibit the proliferation of Caov-3, HL-60, CEM-SS, MCF-7, HeLa, and MDA-MB-231 cell lines, indicating that the leaves of this plant possess potential antiproliferative and antioxidant activities that could be attributed to its high content of phenolic compounds [122]. Melatonin, a naturally occurring compound, showed cytotoxic activity toward transformed 3T3-SV40 fibroblasts [143] and murine hepatoma cells MN22a, and it was shown that the sensitivities of both cell types to lysis by killer cells fell sharply [139].T he potent antioxidant activity of *Kalanchoe gracilis* (L.) DC stems due to that the polyphenolic compound found in this medicinal plant showed to have the ability to inhibit HepG2 cell proliferation [171], and the flavonoids found in *Rosa canina* L. are responsible for the antiproliferative activity in HeLa, MCF7, and HT-29 cancer-cell lines [138]. Analysis of the fruit of *Phelaria macrocarpa* (Boerl.) Scheff and of *Olea europaea* L. indicated that all parts of the fruit possess cytotoxic activity against HT-29, MCF-7, HeLa, BPH-1, and Chang cells, indicating that these fruits are a sources of bioactive compounds that are as potent as antioxidants and antioxidant agents, suggesting its possible use as an adjuvant agent in the treatment of cancer [56, 1].

The extract of *Calluna vulgaris* exhibited a photoprotective effect on human keratinocytes (HaCaT) exposed to Ultraviolet B (UVB) radiation [100]. *Cachrys pungens Jan* was analyzed in a human tumor- cell line, amelanotic melanoma, and it was found that its extract contains antioxidants, such as coumarins, which are responsible for their cytotoxicity in A375 cells [92]. *Inonotus obliquus* and *Peperomia pellucida*, plants employed as folk remedies for cancer treatment, were evaluated in several tumor cell-line types and it was found that these plants contains several antioxidants, such as lanosterol, inotodiols, ergosterol, phytol, 2-naphthalenol, decahydro hexadecanoic acid, methyl ester, and 9,12 octadecadienoic acid, indicating that these antioxidant compounds are responsible for the anticarcinogenic activity of the plant extract [129, 150]. The extract of *Indigofera cassioides* indicated the presence antioxidant activity, preventing lipid peroxidation and promoting the enzymatic antioxidant defense system, and also showed potent antitumoral and cytotoxic affect against EAC, DLA, HeLa, Hep-2, HepG-2, MCF-7, Ht-29, and NIH 3T3 cells [69].

Hesperetin, hesperetin analog, carnocine, and resveratrol were evaluated for their antioxidant and anticarcinogenic activity on HT-29, HCT116, and mouse skin carcinogenesis; their studies demonstrated that these compounds can inhibit cell proliferation, induce apoptosis, affect glycolysis, and decrease tumoration [125, 161, 40]. Honey, a natural product commonly used throughout the world, contains antioxidant properties and exerts a preventive effect against disease. Chrysin is a natural flavone commonly found in honey, and it was demonstrated that this compound induced apoptosis in PC-3 cells [116], fennel seeds (*Foeniculum vulgare*) are present in antioxidants that have an anticancer potential against HepG2 and MCF-7 cell lines [94). It was indicated that compounds such as quercetin, flavonoids, and brucine have chemopreventive action against the osteosarcoma cell line (MG63), C6 glioma cells, and Ehrlich ascites cells, and that they can be used as anticancer, antigenotoxic agents and can induce apoptosis [135, 119, 2].

8. Conclusion

Oxidative stress causes injury to cells, induces gene mutation, and is involved in carcinogenesis and other degenerative diseases by directly or indirectly influencing intracellular signal transduction and transcription factors. The state of OS under carcinogenesis and tumor-bearing conditions is an intricate one in which various substances are involved in complex interactions.

The data discussed in this paper show that the biological effects of antioxidants on humans and animals can be controversial. Due to that the action of antioxidants depends on the oxidative status of cells, antioxidants can be protective against cancer; because ROS induce oxidative carcinogenic damage in DNA, antioxidants can prevent cancer in healthy persons harboring increased ROS levels.

Oxidative stress as cause and effect is not the sole factor in the development of cancer. It is important to take into account that there are other factors involved in its development, such as genetic predisposition, eating habits, environment, etc. Because ROS at moderate concentrations act as indispensable mediators of cancer-protective apoptosis and phagocytosis, an excess of antioxidants in persons with low ROS levels can block these cancer-preventive mechanisms. High doses of antioxidants can reduce the ROS level in persons who overproduce ROS and protect them against cancer and other ROS-dependent morbid conditions.

For individuals with low ROS levels, high doses of antioxidants can be deleterious, suppressing the already low rate of ROS generation and ROS-dependent cancer-preventive apoptosis. Screening and monitoring the human population regarding their ROS level can transform antioxidants into safe and powerful disease-preventive tools that could significantly contribute to the nation's health.

Many *in vivo* and *in vitro* studies performed to evaluate the capability of antioxidants against cancer, such as chemopreventive or therapeutic agents, were conduced employing natural antioxidants from fruits and vegetables; these are mainly supplied through food, which often do not provide sufficient input for these to function as chemoprotectors. Thus, humans are forced to consume antioxidants in a more direct manner, either in the form of a tablet, a pill, or any other form in order to supply the levels that the body requires of these compounds to protect it against cell damage caused by oxidation reactions, thus reducing the risk of certain cancer types, especially those of the epithelial surface and in the upper part of the body, such as breast, lung, kidney, liver, intestine, and many others that have been well documented. However, further investigations are expected before our better understanding of the function of many antioxidants and their utilization in the prevention and treatment of cancer and other degenerative diseases.

Author details

Carmen Valadez-Vega[1], Luis Delgado-Olivares[1], José A. Morales González[1], Ernesto Alanís García[1], José Roberto Villagomez Ibarra[2], Esther Ramírez Moreno [1], Manuel Sánchez Gutiérrez[1], María Teresa Sumaya Martínez[3], Zuñiga Pérez Clara[1] and Zuli Calderón Ramos[1]

1 Institute of Health Sciences, Autonomous University of Hidalgo State, Ex-Hacienda de la Concepción, Tilcuautla, Hgo, Mexico. C.P.42080., Mexico

2 Institute of Basic Sciences, Autonomous University of Hidalgo State, Km 4.5 Carretera Pachuca-Tulancingo, Ciudad del Conocimiento, Mineral de la Reforma Hidalgo, C.P. 42076, Mexico

3 Secretary of Research and Graduate Studies, Autonomous University of Nayarit, Ciudad de la Cultura "Amado Nervo", Boulevard Tepic-Xalisco S/N. Tepic, Nayarit, Mexico

References

[1] Acquaviva, R, Di Giacomo, C, Sorrenti, V, Galvano, F, Santangelo, R, Cardile, V, Gangia, S, D'Orazio, N, Abraham, NG, & Vanella, L. (2012). Antiproliferative effect of oleuropein in prostate cell lines. *International Journal of Oncology*, Print, 1791-2423, Online, 1019-6439, 41, 31-38.

[2] Agrawal, S. S., Saraswati, S., Mathur, R., & Pandey, M. (2011). Cytotoxic and antitumor effects of brucine on Ehrlich ascites tumor and human cancer cell line. *Life Science*, 89, 147-158, 0024-3205.

[3] Akah, P. A., & Ekekwe, R. K. (1995). Ethnopharmacology of some of the asteraceae family used in the Nigerian tradition al medicine. *Fitoterapia*, 66, 352-355, 0036-7326 X.

[4] Akinpelu, D. A. (1999). Antimicrobial activity of Vernonia amygdalina leaves. *Fitoterapia*, 70, 232-234, 0036-7326 X.

[5] Allen, R. G., & Tresini, M. (2000). Oxidative stress and gene regulation. *Free Radical Biology & Medicine*, 28, 463-499, 0891-5849.

[6] Ando, K., & Fujita, T. (2009). Metabolic Syndrome and Oxidative Stress. *Free Radical Biology & Medicine*, 47, 213-218, 0891-5849.

[7] Andreazza, AC, Kapczinski, F, Kauer-Sant'Anna, M, Walz, JC, Bond, DJ, Gonçalves, CA, Young, LT, & Yatham, LN. (2009). 3-Nitrotyrosine and glutathione antioxidant system in patients in the early and late stages of bipolar disorder. *Journal of Psychiatry and Neuroscience*, 1488-2434, 4, 263-271.

[8] Anusuya, C, & Manoharan, S. (2011). Antitumor initiating potential of rosmarinic acid in 7,12-dimethylbenz(a)anthracene-induced hamster buccal pouch carcinogenesis. *Journal of Environmental Pathology, Toxicology and Oncology*, Print, 0731-8898, Online, 2162-6537, 30, 199-211.

[9] Asokkumar, S., Naveenkumar, C., Raghunandhakumar, S., Kamaraj, S., Anandakumar, P., Jagan, S., & Devaki, T. (2012). Antiproliferative and antioxidant potential of beta-ionone against benzo(a)pyrene-induced lung carcinogenesis in Swiss albino mice. *Molecular and Cellular Biochemistry*, 363, 335-345, 0300-8177, Print, 1573-4919, (Online).

[10] Bartoli, R., Fernandez-Banares, F., Navarro, E., Castella, E., Mane, J., Alvarez, M., Pastor, C., Cabre, E., & Gassull, M.A. (2000). Effect of olive oil on early and late events of colon carcinogenesis in rats: Modulation of arachidonic acid metabolism and local prostaglandin E(2) synthesis. *Gut*, 46, 191-199, 0017-5749, Print, 1468-3288, (Online).

[11] Berlett, BS, & Stadtman, E. R. (1997). Protein Oxidation in Aging, Disease, and Oxidative Stress. *The Journal Of Biological Chemistry*, 272(33), 20313-20316, 0021-9258, Print, 1083-351X, (Online).

[12] Bhagat, S. S., Ghone, R. A., Suryakar, A. N., & Hundekar, P. S. (2011). Lipid peroxidation and antioxidant vitamin status in colorectal cancer patients. *Indian Journal Physiology and Pharmacology*, 55, 72-76, 0019-5499.

[13] Bohr, V., Anson, S., Mazur, R. M., & Dianov, G. (1998). Oxidative DNA damage processing and changes with aging. *Toxicology Letters Vols*, 102-103, 47-52, 0378-4274.

[14] Boon, H., Stewart, M., Kennard, MA, Gray, R., Sawka, C., Brown, J. B., Mc William, C., Garvin, A., Baron, R. A., Aaron, D., & Haines-Kamka, T. (2000). Use of complementary/alternative medicine by breast cancer survivors in Ontario: prevalence and perceptions. *Journal of Clinical Oncology*, 8, 2515-2521, 0073-2183 X.

[15] Brown, J. E., Khodr, H., Hider, R. C., & Rice-Evans, C. (1998). Structural dependence of flavonoid interactions with Cu2+ ions: implications for their antioxidant properties. *Biochemical Journal*, 330, 1173-1178, 0264-6021, Print, 1470-8728, (Online).

[16] Brown, N. S., & Bicknell, R. (2001). Hypoxia and oxidative stress in breast cancer: Oxidative stress: its effects on the growth, metastatic potential and response to therapy of breast cancer. *Breast Cancer Research*, 3, 323-327, 0167-6806, Print, 1573-7217, (Online).

[17] Cadet, J., Douki, T., & Ravanat, J. L. (1997). Artifacts associated with the measurement of oxidized DNA bases. *Environmental Health Perspectives*, 105, 1034-1039, 0091-6765.

[18] Calomme, M., Pieters, L., Vlietinck, A., & Vanden, Berghe. D. (1996). Inhibition of bacterial mutagenesis by Citrus flavonoids. *Planta Medica*, 62, 222-226, 0032-0943.

[19] Carlsen, M. H., Halvorsen, B. L., Holte, K., Bøhn, S. K., Dragland, S., Sampson, L., Willey, C., Senoo, H., Umezono, Y., Sanada, C., Barikmo, I., Berhe, N., Willett, W. C., Phillips, K., Jacobs, D. R. Jr, & Blomhoff, R. (2010). The total antioxidant content of more than 3100 foods, beverages, spices, herbs and supplements used worldwide. *Nutrition Journal*, 9(3), http://www.nutritionj.com/content/9/1/3, 1475-2891.

[20] Cejas, P., Casado, E., Belda-Iniesta, C., De Castro, J., Espinosa, E., Redondo, A., Sereno, M., García-Cabezas, M. A., Vara, J. A., Domínguez-Cáceres, A., Perona, R., & González-Barón, M. (2004). Implications of oxidative stress and cell membrane lipid peroxidation in human cancer (Spain). *Cancer Causes and Control*, 15, 707-719, 0957-5243, Print, 1573-7225, (Online).

[21] Chen, L, Hu, JY, & Wang, SQ. (2012). The role of antioxidants in photoprotection: A critical review. *Journal of the American Academy of Dermatology*, 10.1016/j.jaad.2012.02.009, [Epub ahead of print], 0190-9622, 0190-9622.

[22] Chinery, R., Brockman, J. A., Peeler, M. O., Shyr, Y., Beauchamp, R. D., & Coffey, R. J. (1997). Antioxidants enhance the cytotoxicity of chemotherapeutic agents in colorectal cancer: a p53-independent induction of p21WAF1/CIP1 via C/EBP. *Nature Medicine*, 3, 1233-1241, 1078-8956.

[23] Dal-Pizzol, F., Ritter, C., Cassol Jr, Oj., Rezin, G. T., Petronilho, F., Zugno, A. I., Quevedo, J., & Streck, E. L. (2009). Oxidative Mechanisms of Brain Dysfunction During Sepsis. *Neurochemical Research*, 35, 1-12, DOI: s11064-009-0043-4, 0364-3190, Print, 1573-6903, (Online).

[24] Delgado, O. L., Betanzos, C. G., & Sumaya, M. M. T. (2010). Importancia de los antioxidantes dietarios en la disminución del estrés oxidativo. *Investigación y Ciencia*, 50, 10-15, 1665-4412.

[25] De Mejia, E. G., Valadez-Vega, M. C., Reynoso-Camacho, R., & Loarca-Pina, G. (2005). Tannins, trypsin inhibitors and lectin cytotoxicity in tepary (Phaseolus acutifolius) and common (Phaseolus vulgaris) beans. *Plant Foods Hum Nutr*, 60, 137-145, 0921-9668.

[26] Di Domenico, F, Foppoli, Coccia, C, R, & Perluigi, M. (2012). Antioxidants in cervical cancer: Chemopreventive and chemotherapeutic effects of polyphenols. *Biochimica et Biophysica Acta*, 0005-2736, 1822, 737-747.

[27] Drisko, J. A., Chapman, J., & Hunter, V. J. (2003). The use of antioxidants with first-line chemotherapy in two cases of ovarian cancer. *Journal of the American College of Nutrition*, 22, 118-123, 1665-4412.

[28] Duthie, S. J., & Dobson, V. L. (1999). Dietary flavonoids protect human colonocyte DNA from oxidative attack in vitro. *European Journal of Nutrition*, 38, 28-34, 0022-3166, Print, 1541-6100, (Online).

[29] Ebenezer, O., Farombi, A., & Olatunde. (2011). Antioxidative and chemopreventive properties of Vernonia amygdalina and Garcinia biflavonoid. *International Juornal of Environment Researc and Public Health*, 8, 2533-2555, 1661-7827, Print, 1660-4601, (Online).

[30] Echart, M. A. M., Barrio, L. J. P., Maria, Gabriela., Valle, G. M. G., Augustin, S. C. H., Ugalde Marques da Rocha, MI, Manica-Cattani, MF, Feyl dos Santos, G, & Manica da Cruz, IB. (2009). Association between manganese superoxide dismutase (MnSOD). gene polymorphism and elderly obesity. *Molecular and Cellular Biochemistry*, 328, 33-40, 0300-8177, Print, 1573-4919, (Online).

[31] Ernst, E., & Cassileth, B. R. (1998). The prevalence of complementary/alternative medicine in cancer: a systematic review. *Cancer*, 83, 777-782, 0000-8543 X, (Print), 1097-0142, (Online).

[32] Esterbauer, H., Schauur, J. S., & Zollner, H. (1991). Chemistry and biochemistry of 4-hydroxynonenal, malonaldehyde and related aldehydes. *Free Radical Biology & Medicine*, 11, 81-128, 0891-5849.

[33] Flis, S., Jastrzebski, Z., Namiesnik, J., Arancibia-Avila, P., Toledo, F., Leontowicz, H., Leontowicz, M., Suhaj, M., Trakhtenberg, S., & Gorinstein, S. (2012). Evaluation of inhibition of cancer cell proliferation in vitro with different berries and correlation with

their antioxidant levels by advanced analytical methods. *Journal of Pharmaceutical Biomedical Analysis*, 62, 68-78, 0731-7085.

[34] Floyd, R. A., Watson, J. J., & Wong, P. K. (1986). Hydroxyl free radical adduct of deoxyguanosine: sensitive detection and mechanisms of formation. *Free Radical Research Communications*, 1, 163-172, 8755-0199.

[35] Fox, J. T., Sakamuru, S., Huang, R., Teneva, N., Simmons, S. O., Xia, M., Tice, R. R., Austin, , & Myung, K. (2012). High-throughput genotoxicity assay identifies antioxidants as inducers of DNA damage response and cell death. *Proceedings of the National Academy of Sciences of United States of America*, 109, 5423-5428.

[36] Fuchs-Tarlovsky, V., Bejarano-Rosales, M., Gutierrez-Salmeán, G., Casillas, MA, López-Alvarenga, J. C., & Ceballos-Reyes, G. M. (2011). Effect of antioxidant supplementation over oxidative stress and quality of life in cervical cancer. *Nutrición Hospitalaria*, 26, 819-826, 0212-1611.

[37] Fukumura, H, Sato, M, Kezuka, K, Sato, I, Feng, X, Okumura, S, Fujita, T, Yokoyama, U, Eguchi, H, Ishikawa, Y, & Saito, T. (2012). Effect of ascorbic acid on reactive oxygen species production in chemotherapy and hyperthermia in prostate cancer cells. *The Jornal of Physiological Sciences*, 1880-6546, (Print), 1880-6562, Online, 62, 251-257.

[38] Gago-Dominguez, M., Jiang, X., & Castelao, J. E. (2007). Lipid peroxidation, oxidative stress genes and dietary factors in breast cancer protection: a hypothesis. *Breast Cancer*, 9, 1-11, 10.1186/bcr1628, http://breast-cancer-research.com/content/9/1/201, 0146-5542 X.

[39] Gamaleĭ, I. A., Kirpichnikova, K. M., & Filatova, N. A. (2011). Effect of melatonin on the functional properties of transformed cells. *Vopr Onkol*, 57, 481-485, 0507-3758.

[40] George, J, Singh, M, Srivastava, AK, Bhui, K, Roy, P, Chaturvedi, PK, & Shukla, Y. (2011). Resveratrol and black tea polyphenol combination synergistically suppress mouse skin tumors growth by inhibition of activated MAPKs and p53. *PLoS One*, 1932-6203, 6, 23395-23408.

[41] Gibellini, L., Pinti, M., Nasi, M., Montagna, J. P., De Biasi, S., Roat, E., Bertoncelli, L., Cooper, E. L., & Cossarizza, A. (2011). Quercetin and cancer chemoprevention. *Evidence-Based Complementary and Alternative Medicine*, 59, 1356-1365, 0174-1427, Print, 1741-4588, (Online).

[42] Gill, C. I., Boyd, A., Mc Dermott, E., Mc Cann, M., Servili, M., Selvaggini, R., Taticchi, A., Esposto, S., Montedoro, G., Mc Glynn, H., & Rowland, I. (2005). Potential anticancer effects of virgin olive oil phenols on colorectal carcinogenesis models in vitro. *International Journal of Cancer*, 117, 1-7, 0020-7136, 1097-0215, (Online).

[43] Gille, G, & Sigler, K. (1995). Oxidative stress and living cells. *Folia Microbiological*, 0015-5632, (Print), 1874-9356, (Online), 40, 131-152.

[44] Gong, G., Waris, G., Tanveer, R., & Siddiqui, A. (2001). Human hepatitis C virus NS5A protein alters intracellular calcium levels, induces oxidative stress, and acti-

vates STAT-3 and NF-B. *Proceedings of the National Academy of Sciences of United States of America*, 98(17), 9599-9604, 0027-8424.

[45] Grimsrud, P. A., Xie, H., Griffin, T. J., & Bernlohr, D. A. (2008). Oxidative Stress and Covalent Modification of Protein with Bioactive Aldehydes. *Journal of Biological Chemistry*, 283(32), 21837-21841, 0021-9258, (Print), 1083-351X, (Online).

[46] Gupta, V, & Sharma, M. (2012). Phytochemical Analysis and Evaluation of Antioxidant Activities of Methanolic Extracts of Maytenus emarginata. 1536-2310, (Print), 1557-8100, Online, 16(5), 257-262.

[47] Halliwell, B. (1994). Free radicals, antioxidants, and human disease: curiosity, cause, or consequence? *The Lancet*, 344, 721-724, 1040-6736.

[48] Halliwell, B. (1996). Antioxidants in Human Health and Disease. Annual Reviews, 1550-8382 Online , 16, 33-50.

[49] Halliwell, B. (2000). The antioxidant paradox. *The Lancet*, 1, 1179-1180, 1040-6736.

[50] Halliwell, B., & Gutteridge, J. M. C. (2006). Free Radicals in Biology and Medicine. Ed 4. Clarendon Press, Oxford.

[51] Halliwell, B. (2007). Oxidative stress and cancer: have we moved forward? *Biochemical Journal*, 401, 1-11, 0264-6021, Print, 1470-8728, (Online).

[52] Halliwell, B. (2008). Are polyphenols antioxidants or pro-oxidants? What do we learn from cell culture and in vivo studies? *Archives of Biochemistry and Biophysics*, 476, 107-112, 0003-9861, Print, 1096-0384, (Online).

[53] Han, R.M., Tian, Y.X., Becker, E.M., Andersen, M.L., Zhang, J.P., & Skibsted, L.H. (2007). Puerarin and conjugate bases as radical scavengers and antioxidants: molecular mechanism and synergism with beta-carotene. *Journal of Agricultural and Food Chemistry*, 0021-8561, Print, 1520-5118, Online, 55, 2384-2389.

[54] Hanasaki, Y., Ogawa, S., & Fukui, S. (1994). The correlation between active oxygens scavenging and antioxidative effects of flavonoids. *Free Radical Biology & Medicine*, 16, 845-850, 0891-5849.

[55] Hashim, Y. Z., Rowland, I. R., Mc Glynn, H., Servili, M., Selvaggini, R., Taticchi, A., Esposto, S., Montedoro, G., Kaisalo, L., Wahala, K., & Gill, C. I. (2008). Inhibitory effects of olive oil phenolics on invasion in human colon adenocarcinoma cells in vitro. *International Journal of Cancer*, 122, 495-500, 0020-7136, Print, 10970215, Online.

[56] Hendra, R., Ahmad, S., Oskoueian, E., Sukari, A., & Shukor, M. Y. (2011). Antioxidant, anti-inflammatory and cytotoxicity of Phaleria macrocarpa (Boerl.) Scheff Fruit. *BMC Complemententary & Alternative Medicine*, 11, 110-121, 1472-6882.

[57] Hippeli, S., Heiser, I., & Elstner, E. F. (1999). Activated oxygen and free oxygen radicals in pathology: New insights and analogies between animals and plants. *Plant Physiology Biochemistry*, 37, 167-178, 0981-9428.

[58] Hladik, C., Krief, S., & Haxaire, C. (2005). Ethnomedicinal and bioactive properties of plants ingested by wild chimpanzees in Uganda. *Journal Ethnopharmacology*, 101, 1-5, 0378-8741.

[59] Hu, M.L. (2011). Dietary Polyphenols as Antioxidants and Anticancer Agents: More Questions than Answers. *Chang Gung Medical Journa*, 2072-0939, 34, 449-459.

[60] Inoue, M., Tajima, K., Mizutani, M., Iwata, H., Iwase, T., Miura, S., Hirose, K., Hamajima, N., & Tominaga, S. (2001). Regular consumption of green tea and the risk of breast cancer recurrence: Follow-up study from the Hospital-based Epidemiologic Research Program at Aichi Cancer Center (HERPACC), Japan. *Cancer Letters*, 167, 175-182, 0304-3835.

[61] Iovine, B., Iannella, M. L., Nocella, F., Pricolo, M. R., & Bevilacqua, MA. (2012). Carnosine inhibits KRAS-mediated HCT116 proliferation by affecting ATP and ROS production. *Cancer Letters*, 28, 122-128, 0304-3835.

[62] Jayaprakash, V., & Marshall, J. R. (2011). Selenium and other antioxidants for chemoprevention of gastrointestinal cancers. *Best Practtice & Research Clinical Gastroenterology*, 25, 507-518, 1521-6918.

[63] Jisaka, M., Ohigashi, H., Takegawa, K., Hirota, M., Irie, R., Huffman, MA, & Koshimizu, K. (1993). Steroid glucosides from Vernonia amygdalina, a possible chimpanzee plant. *Phytochemistry*, 34, 409-413, 0031-9422.

[64] Katiyar, S. K., & Mukhtar, H. (1997). Tea antioxidants in cancer chemoprevention. *Journal of Cellular Biochemistry*, 27, 59-67, 1097-4644.

[65] Keller, J.N. (2006). Interplay Between Oxidative Damage, Protein Synthesis, and Protein Degradation in Alzheimer's Disease. *Journal of Biomedicine and Biotechnology*, http://www.ncbi.nlm.nih.gov/pmc/articles/PMC1510934/pdf/JBB2006-12129.pdf, 1110-7243, Print, 1110-7251, Online, 2006, 1-3.

[66] Khan, N., Afaq, F., Saleem, M., Ahmad, N., & Mukhtar, H. (2006). Targeting multiple signaling pathways by green tea polyphenol (−)-epigallocatechin-3-gallate. *Cancer Research*, 66, 2500-2505, 0008-5472, Print, 1538-7445, Online.

[67] Kleinschnitz, C., Grund, H., Wingler, K., Armitage, ME, Jones, J., Mittal, M., Barit, D., Schwarz, T., Geis, C., Kraft, P., Barthel, K., Schuhmann, M. K., Herrmann, A. M., Meuth, S. G., Stoll, G., Meurer, S., Schrewe, A., Becker, L., Gailus-Durner, V., Fuchs, H., Klopstock, T., Hrabe′ de Angelis, M., Jandeleit-Dahm, K., Shah, A. M., Weissmann, N., & Schmidt, H. H. H. W. (2010). Post-Stroke Inhibition of Induced NADPH Oxidase Type 4Prevents Oxidative Stress and Neurodegeneration. *PloS Biology*, 8(9), http://www.plosbiology.org/article/info%3Adoi%2F10.1371%2Fjournal.pbio.1000479, 1545-7885, 1544-9173.

[68] Korkina, L. G., & Afanas'ev, I. B. (1997). *Antioxidants in Disease Mechanisms and Therapy*, Sies, H., Ed.; Academic Press: San Diego, 151-163.

[69] Kumar, RS, Rajkapoor, B, & Perumal, P. (2011). In vitro and in vivo anticancer activity of Indigofera cassioides Rottl. *Ex. DC. Asian Pacific Journal of Tropical Medicine*, 1995-7645, 4, 379-385.

[70] La Vecchia, C., Altieri, A., & Tavani, A. (2001). Vegetables, fruit, antioxidants and cancer: a review of Italian studies. *European Journal of Nutrition*, 40, 261-267, 1436-6207, Print, 1436-6215, Online.

[71] Lai, Z. R., Ho, Y. L., Huang, S. C., Huang, T. H., Lai, S. C., Tsai, J. C., Wang, C. Y., Huang, G. J., & Chang, Y. S. (2011). Antioxidant, anti-inflammatory and antiproliferative activities of Kalanchoe gracilis (L.) DC stem. *The American Journal of Chinese Medicine*, 39, 1275-1290, 0019-2415 X, Print, 1793-6853, Online.

[72] Larsen, CA, & Dashwood, R. H. (2009). Suppression of Met activation in human colon cancer cells treated with (−)-epigallocatechin-3-gallate: Minor role of hydrogen peroxide. *Biochemical and Biophysical Research Communnications*, 389, 527-530, 0000-6291 X.

[73] Lee, C. K., Weindruch, R., & Prolla, T. A. (2000). Gene-expression profile of the ageing brain in mice. *Nature Genetics*, 25, 294-297, 1061-4036.

[74] Leong, H., Mathur, P. S., & Greene, G. L. (2008). Inhibition of mammary tumorigenesis in the C3(1)/SV40 mouse model by green tea. *Breast Cancer Research and Treatment*, 107, 359-369, 0167-6806, Print, 0167-6806, Print, 1573-7217, Online.

[75] Li, Q., Zhao, H. F., Zhang, Z. F., Liu, Z. G., Pei, X. R., Wang, J. B., Cai, M. Y., & Li, Y. (2009). Long-term administration of green tea catechins prevents age-related spatial learning and memory decline in C57BL/6 J mice by regulating hippocampal cyclic AMP-response element binding protein signaling cascade. *Neuroscience*, 159, 1208-1215, 0306-4522.

[76] Li, W., Shi, Y. H., Yang, R. L., Cui, J., Xiao, Y., Wang, B., & Le , G. W. (2010). Effect of somatostatin analog on high-fat diet-induced metabolic syndrome: Involvement of reactive oxygen species. *Peptides*, 31(4), 625-629, 0196-9781.

[77] Liang, W., Li, X., Li, C., Liao, L., Gao, B., Gan, H., Yang, Z., Liao, L., & Chen, X. (2011). Quercetin-mediated apoptosis via activation of the mitochondrial-dependent pathway in MG-63 osteosarcoma cells. *Molecular Medicine Reports*, 4, 1017-1023, 1791-2997, Print, 1791-3004, Online.

[78] Liu, M., Gong, X., Alluri, R. K., Wu, J., Sablo, T., & Li, Z. (2012). Characterization of RNA damage under oxidative stress in Escherichia coli. *Biol Chem*, 393(3), 123-132, 1437-4315.

[79] Llor, X., Pons, E., Roca, A., Alvarez, M., Mane, J., Fernandez-Banares, F., & Gassull, M. A. (2003). The effects of fish oil, olive oil, oleic acid and linoleic acid on colorectal neoplastic processes. *Clinical Nutrition*, 22, 71-79, 0261-5614.

[80] Lockwood, K., Moesgaard, S., Hanioka, T., & Folkers, K. (1994). Apparent partial remission of breast cancer in 'High Risk' patients supplemented with nutritional anti-

oxidants, essential fatty acids and Coenzyme Q_{10} . *Biochemical and Biophysical Research Communications*, 15, 231-s240, 0000-6291 X.

[81] Ma, Q, & Kinneer, K. (2002). Chemoprotection by phenolic antioxidants. Inhibition of tumor necrosis factor alpha induction in macrophages. *Journal of Biological Chemistry*, 0021-9258, Print, 1083-351X, Online, 277, 2477-2484.

[82] Maritim, A. C., Sanders, R. A., & Watkins, I. I. I. J. B. (2003). Diabetes, Oxidative Stress, and Antioxidants: A Review. *Journal of Biochememical Molecular and Toxicology*, 17, 24-38, 1095-6670, Print, 1099-0461, Online.

[83] Markesbery, W. R. (1997). Oxidative Stress Hypothesis In Alzheimer's Disease. *Free Radical Biology & Medicine*, 23(1), 134-147, 0891-5849.

[84] Martinez, M. E. (2005). Primary prevention of colorectal cancer: Lifestyle, nutrition, exercise. *Recent Results in Cancer Research*, 166, 177-211, 0080-0015.

[85] Matés, JM, Segura, JA, Alonso, FJ, & Márquez, J. (2011). Anticancer antioxidant regulatory functions of phytochemicals. *Current Medicinal Chemistry*, 0929-8673, Print, 1875-533X, Online, 18, 2315-2338.

[86] Mates, J. M., Perez-Gomez, C., & Nunez de Castro, I. (1999). Antioxidant enzymes and human diseases. *Clinical Biochemistry*, 32, 595-603, 0009-9120.

[87] Mauro, S., Rino, B., Alicja, W., & Anna, (2002. (2002). Total antioxidant potential of fruit and vegetables and risk of gastric cancer. *Gastroenterology*, 123, 985-991, 0016-5085.

[88] Maxmen, A. (2012). The Hard Facts. *Nature*, 485, S50-S51, 0028-0836.

[89] Mazdak, H., & Zia, H. (2012). Vitamin e reduces superficial bladder cancer recurrence: a randomized controlled trial. *International Journal of Preventive Medicine*, 3, 110-115.

[90] McCord, J. M., & Fridovich, I. (1968). The reduction of cytochrome c by milk xanthine oxidase. *The Journal of Bioogical Chemistry*, 2008-7802, Print, 2008-8213, Online, 243, 5753-5760.

[91] Medina-Ceja, L., Guerrero-Cazares, H., Canales-Aguirre, A., Morales-Villagrán, A., & Feria-Velasco, A. (2007). Características estructurales y funcionales de los transportadores de glutamato: su relación con la epilepsia y el estrés oxidativo. *Revista de Neurología*, 45(6), 341-352.

[92] Menichini, G, Alfano, C, Provenzano, E, Marrelli, M, Statti, GA, Menichini, F, & Conforti, F. (2012). Cachrys pungens Jan inhibits human melanoma cell proliferation through photo-induced cytotoxic activity. *Cell Proliferation*, 0960-7722, Print, 1365-2184, Online, 45, 39-47.

[93] Middleton, E. Jr, Kandaswami, C., & Theoharides, T. C. (2000). The effects of plant flavonoids on mammalian cells: Implications for inflammation, heart disease, and cancer. *Pharmacological Reviews*, 52, 673-751, 0031-6997.

[94] Mohamad, R. H., El -Bastawesy, A. M., Abdel-Monem, M. G., Noor, A. M., Al-Mehdar, H. A., Sharawy, S. M., & El -Merzabani, MM. (2011). Antioxidant and anticarcinogenic effects of methanolic extract and volatile oil of fennel seeds (Foeniculum vulgare). *Journal of Medicine Food,*, 14, 986-1001, 0109-6620, Print, 1557-7600, Online.

[95] Nessa, M. U., Beale, P., Chan, C., Yum, J. Q., & Huq, F. (2012). Combinations of resveratrol, cisplatin and oxaliplatin applied to human ovarian cancer cells. *Anticancer Res*, 32, 53-59, 0250-7005, Print, 1791-7530, Online.

[96] Noda, N., & Wakasugi, H. (2000). Cancer and oxidative stress. *Journal of the Japan Medical Association*, 124(11), 1571-1574, 1356-8650.

[97] Nunomura, A., Honda, K., Takeda, A., Hirai, K., Zhu, X., Smith, M. A., & Perry, G. (2006). Oxidative Damage to RNA in Neurodegenerative Diseases. *Journal of Biomedicine and Biotechnology* [82323], 1-6, 1110-7243, Print, 1110-7251, Online.

[98] Nyström, N. (2005). Role of oxidative carbonylation in protein quality control and senescence. *EMBO Journal*, 0261-4189, Print, 1460-2075, Online, 24, 1311-1317.

[99] Owen, R. W., Giacosa, A., Hull, W. E., Haubner, R., Spiegelhalder, B., & Bartsch, H. (2000). The antioxidant/anticancer potential of phenolic compounds isolated from olive oil. *European Journal Cancer*, 36, 1235-1247, 0959-8049.

[100] Perde-Schrepler, M, Chereches, G, Brie, L, Virag, P, Barbo,s, O, Soritau, O, Tatomir, C, Fischer-Fodor, E, Filip, A, Vlase, L, & Postescu, ID. (2011). Photoprotective effect of Calluna vulgaris extract against UVB-induced phototoxicity in human immortalized keratinocytes. *Journal of Environment Pathology Toxicology and Oncology*, 0371-8898, Print, 2162-6537, Online, 30, 323-331.

[101] Perera, R. M., & Bardeesy, N. (2011). When antioxidants are bad. *Nature*, 4, 4, 0028-0836.

[102] Pietá, D., Martins De, Lima. M. N., Presti-Torres, J., Dornelles, A., Garcia, V. A., Siciliani, S. F., Rewsaat, M. G., Constantino, L., Budni, P., Dal-Pizzol, F., & Schrödera, N. (2007). Memantine Reduces Oxidative Damage And Enhances Long-Term Recognition Memory In Aged Rats. *Neuroscience*, 146, 1719-1725, 0306-4522.

[103] Pietta, P. G. (2000). Flavonoids as Antioxidants. *Journal of Natural Produts*, 1035-1042, 0163-3864, Print, 1520-6025, Online.

[104] Plaumann, B., Fritsche, M., Rimpler, H., Brandner, G., & Hess, R. D. (1996). Flavonoids activate wild-type p53. *Oncogene*, 13, 1605-1614, 0950-9232.

[105] Prasad, K. N., Kumarm, A., Kochupillaim, V., & Colem, W. C. (1999). High doses of multiple antioxidant vitamins: essential ingredients in improving the efficacy of standard cancer therapy. *Journal of the American College of Nutrition*, 18, 13-25, 0731-5724, Print, 1541-1087, Online.

[106] Prasad, K. N., Cole, W. C., Kumar, B., & Prasad, K. C. (2001). Scientific rationale for using high-dose multiple micronutrients as an adjunct to standard and experimental

cancer therapies. *Journal of the American College of Nutrition*, 20, 450S-463S, 0731-5724, Print, 1541-1087, Online.

[107] Price, T. O., Ercal, N., Nakaoke, R., & Banks, W. A. (2005). HIV-1viralproteins gp120 and Tatinduceoxidativestress in brain endothelial cells. *Brain Research*, 1045, 57-63, 0006-8993.

[108] Rabek, J. P., Boylston, I. I. I. W. H., & Papaconstantinou, J. (2003). Carbonylation of ER chaperone proteins in aged mouse liver. *Biochemical and Biophysical Research Communications*, 305, 566-572, 0000-6291 X.

[109] Ren, L., Yang, H. Y., Choi, H. I., Chung, K. J., Yang, U., Lee, I. K., Kim, H. J., Lee, , Park, B. J., & Lee, T. H. (2011). The role of peroxiredoxin V in (-)-epigallocatechin 3-gallate-induced multiple myeloma cell death. *Oncology Research*, 19, 391-398, 0965-0407.

[110] Rieger-Christ, KM, Hanley, R, Lodowsky, C, Bernier, T, Vemulapalli, P, Roth, M, Kim, J, Yee, AS, Le, SM, Marie, PJ, Libertino, JA, & Summerhayes, IC. (2007). The green tea compound, (−)-epigallocatechin-3-gallate downregulates N-cadherin and suppresses migration of bladder carcinoma cells. *Journal of Cellular Biochemistry*, 0730-2312, Print, 1097-4644, Online, 102, 377-388.

[111] Riordan, N. H., Riordan, H. D., Meng, Y. L., & Jackson, J. A. (1995). Intravenous ascorbate as a tumor cytotoxic. chemotherapeutic agent. *Medical Hypotheses*, 44, 207-213, 0306-9877.

[112] Riordan, N. H., Riordan, H. D., & Casciari, J. P. (2000). Clinical and experimental experiences with intravenous vitamin C. *Journal of Orthomolecular Medicine*, 5, 201-213, 0317-0219.

[113] Rivas, MA, Carnevale, R. P., Proietti, C. J., Rosemblit, C., Beguelin, W., Salatino, M., Charreau, E. H., Frahm, I., Sapia, S., Brouckaert, P., Elizalde, P. V., & Schillaci, R. (2008). TNF alpha acting on TNFR1 promotes breast cancer growth via P42 P44 MAPK, JNK, Akt and NF-kappa B-dependent pathways. *Experimental Cell Research*, 314(3), 509-29, 0014-4827.

[114] Roberts, C. K., Barnarda, R. J., Sindhub, R. K., Jurczak, M., Ehdaieb, A., & Vaziri, N. D. (2006). Oxidative stress and dysregulation of NAD(P)H oxidase and antioxidant enzymes in diet-induced metabolic syndrome. *Metabolism Clinical and Experimental*, 55, 928-934, 1532-8600.

[115] Roche, CE, & Romero, A. D. (1994). Estrés oxidativo y degradación. de proteínas. *Medicina clínica*, 103(5), 189-196, 0025-7753.

[116] Samarghandian, S, Afshari, JT, & Davoodi, S. (2011). Chrysin reduces proliferation and induces apoptosis in the human prostate cancer cell line pc-3. *Clinics (Sao Paulo)*, 1807-5932, Print, 1980-5322, Online, 66, 1073-1079.

[117] Schmitt, CA, & Lowe, S. W. (1999). Apoptosis and therapy. *The Journal of Pathololy*, 187, 127-137.

[118] Seef, L. B., Lindsay, K. L., Bacon, B. R., Kresina, F., & Hoofnagle, H. (2001). Complementary and alternative medicine in chronic liver disease. *Hepatology*, 34, 595-603, 1096-9896, Online.

[119] Seibert, H, Maser, E, Schweda, K, Seibert, S, & Gülden, M. (2011). Cytoprotective activity against peroxide-induced oxidative damage and cytotoxicity of flavonoids in C6 rat glioma cells. *Food and Chemical Toxicology*, 0278-6915, 49, 2398-2407.

[120] Sharma, R., & Vinayak, M. (2012). Antioxidant α-tocopherol checks lymphoma promotion via regulation of expression of protein kinase C-α and c-Myc genes and glycolytic metabolism. *Leukemia & Lymphoma*, 1042-8194, Print, 1029-2403, Online, 53(6), 1203-1210.

[121] Sharmila, R., & Manoharan, S. (2012). Anti-tumor activity of rosmarinic acid in 7,12-dimethylbenz(a) anthracene (DMBA) induced skin carcinogenesis in Swiss albino mice. *Indian Journal of Experimental Biology*, 50, 187-194, 0975-1009, Print, 0019-5189, Online.

[122] Zakaria, Z. A., Rofiee, MS, Mohamed, A. M., the, L. K., & Salleh, M. Z. (2011). In vitro antiproliferative and antioxidant activities and total phenolic contents of the extracts of Melastoma malabathricum leaves. *Journal of Acupuncture and Meridian Studies*, 4(4), 248-256, 0000-0020.

[123] Sies, H. (1997). Antioxidants in Disease Mechanisms and Therapy. *Advances in Pharmacology*, 38, Academic Press: San Diego.

[124] Sies, H. (1997). Oxidative Stress: Oxidants And Antioxidants. *Experimental Physiology*, 82, 291-295, 0958-0670, Print, 1469-445X, Online.

[125] Sivagami, G., Vinothkumar, R., Preethy, CP, Riyasdeen, A., Akbarsha, MA, Menon, V. P., & Nalini, N. (2012). Role of hesperetin (a natural flavonoid) and its analogue on apoptosis in HT-29 human colon adenocarcinoma cell line- A comparative study. *Food and Chemical Toxicology*, 50, 660-671, 0278-6915.

[126] Skalicky, J., Muzakova, V.,., Roman, Kandar. R., Meloun, M., Rousar, T., & Palicka, V. (2008). Evaluation of oxidative stress and inflammation in obese adults with metabolic syndrome. *Clinical Chemistry and Laboratory Medicine*, 46(4), 499-505, 1434-6621, Print, 1437-4331, Online.

[127] Slaga, T. J. (1995). Inhibition of the induction of cancer by antioxidants. *Advances in Experimental Medicine and Biology*, 369, 167-174, 0065-2598.

[128] Solanas, M, Hurtado, A, Costa, I, Moral, R, Menendez, J.A., Colomer, R., & Escrich, E. (2002). Effects of a high olive oil diet on the clinical behavior and histopathological features of rat DMBA-induced mammary tumors compared with a high corn oil diet. *International Journal of Oncology*, 1791-2423, 21, 745-753.

[129] Sun, Y, Yin, T, Chen, XH, Zhang, G, Curtis, RB, Lu, ZH, & Jiang, JH. (2011). In vitro antitumor activity and structure characterization of ethanol extracts from wild and

cultivated Chaga medicinal mushroom, Inonotus obliquus (Pers.:Fr.) Pilát (Aphyllophoromycetideae). *International Journal of Medical Mushrooms,* 1521-9437, Print, 1940-4344, Online, 13, 121-130.

[130] Surh, YJ. (2003). Cancer chemoprevention with dietary phytochemicals. *Nature Review Cancer,* 3, 768-780, 1097-0142, Online.

[131] Szpetnar, M., Matras, P., Kiełczykowsk,a, M., Horecka, A., Bartoszewska, L., Pasternak, K., & Rudzki, S. (2012). Antioxidants in patients receiving total parenteral nutrition after gastrointestinal cancer surgery. *Cell Biochemistry and Funciont,* 30, 211-216, 1099-0844, Online.

[132] Takada, M, Ku, Y., Habara, K, Ajiki, T., Suzuki, Y., & Kuroda, Y. (2002). Inhibitory effect of epigallocatechin-3-gallate on growth and invasion in human biliary tract carcinoma cells. *World Journal of Surgery,* 0364-2313, Print, 1432-2323, Online, 26, 683-686.

[133] Takada, M., Nakamura, Y., Koizumi, T., Toyama, H., Kamigaki, T., Suzuki, Y., Takeyama, Y., & Kuroda, Y. (2002). Suppression of human pancreatic carcinoma cell growth and invasion by epigallocatechin-3-gallate. *Pancreas,* 25, 45-48, 0885-3177, Print, 1536-4828, Online.

[134] Thapa, D., & Ghosh, R. (2012). Antioxidants for prostate cancer chemoprevention: Challenges and opportunities. *Biochemical Pharmacology,* 83, 1319-1330, 0006-2952.

[135] Toyoda-Hokaiwado, N, Yasui, Y, Muramatsu, M, Masumura, K, Takamune, M, Yamada, M, Ohta, T, Tanaka, T, & Nohmi, T. (2011). Chemopreventive effects of silymarin against 1,2-dimethylhydrazine plus dextran sodium sulfate-induced inflammation-associated carcinogenicity and genotoxicity in the colon of gpt delta rats. *Carcinogenesis,* 0143-3334, Print, 1460-2180, Online, 32, 1512-1517.

[136] Toyokuni, MD. (1998). Oxidative Stress and Cancer: The Role of Redox Regulation Shinya. *Biotherapy,* 11, 147-154, 0092-1299 X, Print, 1573-8280, Online.

[137] Tsaluchidu, S., Cocchi, M., Tonello, L., & Puri, B. K. (2008). Fatty acids and oxidative stress in psychiatric disorders. *BMC Psychiatry,* 8(1), S1-S5, 0147-1244 X, Online.

[138] Tumbas, V. T., Canadanović-Brunet, J. M., Cetojević-Simin, D. D., Cetković, G. S., Ethilas, S. M., & Gille, L. (2012). Effect of rosehip (Rosa canina L.) phytochemicals on stable free radicals and human cancer cells. *Journal of the Science of Food and Agriculture,* 92, 1273-1281, 0022-5142, Print, 1097-0010, Online.

[139] Upham, B. L., & Wagner, J. G. (2001). Toxicological Highlight Toxicant-Induced Oxidative Stress in Cancer. *Toxicological sciences,* 64, 1-3, 1096-6080, Print, 1096-0929, Online.

[140] Ursini, F., Maiorino, M., Morazzoni, P., Roveri, A., & Pifferi, G. (1994). A novel antioxidant flavonoid (IdB 1031) affecting molecular mechanisms of cellular activation. *Free Radical Biology & Medicine,* 16, 547-553, 0891-5849.

[141] Uttara, B., Singh, A. V., Zamboni, P., & Mahajan, R. T. (2009). Oxidative Stress and Neurodegenerative Diseases: A Review of Upstream and Downstream Antioxidant Therapeutic Options. *Current Neuropharmacology*, 7, 65-74, 0157-0159 X.

[142] Van Erk, M. J., Roepman, P., van der Lende, T. R., Stierum, R. H., Aarts, J. M., van Bladeren, P. J., & van Ommen, B. (2005). Integrated assessment by multiple gene expression analysis of quercetin bioactivity on anticancer-related mechanisms in colon cancer cells in vitro. *European Journal of Nutrition*, 44, 143-156, 1436-6207, Print, 1436-6215, Online.

[143] Valadez-Vega, C., Guzmán-Partida, A. M., Soto-Cordova, F. J., Alvarez-Manilla, G., Morales-González, J. A., Madrigal-Santillán, E., Villagómez-Ibarra, J. R., Zúñiga-Pérez, C., Gutiérrez-Salinas, J., & Becerril-Flores, MA. (2011). Purification, biochemical characterization, and bioactive properties of a lectin purified from the seeds of white tepary bean (phaseolus acutifolius variety latifolius). *Molecules*, 21, 2561-2582, 1420-3049.

[144] Valadez-Vega, C., Alvarez-Manilla, G, Riverón-Negrete, L, García-Carrancá, A, Morales-González, JA, Zuñiga-Pérez, C, Madrigal-Santillán, E, Esquivel-Soto, J, Esquivel-Chirino, C, Villagómez-Ibarra, R, Bautista, M, & Morales-González, A. (2011). Detection of cytotoxic activity of lectin on human colon adenocarcinoma (Sw480) and epithelial cervical carcinoma (C33-A). *Molecules*, 1420-3049, 2, 2107-2118.

[145] Varma, SD, Devamanoharan, S., & Morris, SM. (1995). Prevention of cataracts by nutritional and metabolic antioxidants. *Crititical Reviews in Food Science and Nutrition*, 35, 111-129, 1040-8398, Print, 1549-7852, Online.

[146] Vauzour, D, Rodriguez-Mateos, A, Corona, G, Oruna-Concha, MJ, & Spence, JPE. (2010). Polyphen ols and Human Health: Prevention of Disease and Mechanisms of Action. *Nutrients*, 2072-6643, 2, 1106-1131.

[147] Warburg, O. (1956). On the origin of cancer cells. *Science*, 123, 309-314, 0036-8075, Print, 1095-9203, Online.

[148] Waris, G., & Siddiqui, A. (2005). Hepatitis C virus stimulates the expression of cyclooxygenase-2 via oxidative stress: role of prostaglandin E2 in RNA replication. *Journal of Virology*, 79, 9725-34, 0002-2538 X, Print, 1098-5514, Online.

[149] Wayner, D. D. M., Burton, G. W., Ingold, K. U., Barclay, L. R. C., & Locke, S. J. (1987). The relative contributions of vitamin E, urate, ascorbate and proteins to the total peroxyl radical-trapping antioxidant activity of human blood plasma. *Biochemica et Biophysica Acta*, 924, 408-419, 0006-3002.

[150] Wei, L. S., Wee, W., Siong, J. Y., & Syamsumir, D. F. (2011). Charactetization of anticancer, antimicrobial, antioxidant properties and chemical composition of Peperomia pellucid. *Acta Medica Iranica*, 49, 670-674, 0044-6025, Print, 1735-9694, Online.

[151] Weijl, N. I., Cleton, F. J., & Osanto, S. (1997). Free radicals and antioxidants in chemotherapy induced toxicity. *Cancer Treatment Reviews*, 23, 209-240, 0305-7372, Print.

[152] Winslow, LC, & Krol, DJ. (1998). Herbs as medicines. *Archives Internal Medicine*, 0003-9926, Print, 1538-3679, Online, 1258, 2192-219.

[153] Wu, B, Li, J, Huang, D, Wang, W, Chen, Y, Liao, Y, Tang, X, Xie, H, & Tang, F. (2011). Baicalein mediates inhibition of migration and invasiveness of skin carcinoma through Ezrin in A431 cells. 1471-2407, 11, 527-536.

[154] Ye, F, Zhang, GH, Guan, BX, & Xu, XC. (2012). Suppression of esophageal cancer cell growth using curcumin, (-)-epigallocatechin-3-gallate and lovastatin. *World Journal of Gastroenterology*, 1007-9327, Print, 2219-2840, Online, 18, 126-135.

Chapter 17

Emerging Role of Natural Antioxidants in Chronic Disease Prevention with an Emphasis on Vitamin E and Selenium

Manuel Soriano García

Additional information is available at the end of the chapter

http://dx.doi.org/10.5772/51610

1. Introduction

The possibility has arisen within the last three decades that major diseases that directly affect humankind worldwide may be preventable by the simple improving the dietary intake of those nutrient substances that have become called "antioxidant nutrients".

There is no doubt that successful prevention is the key to controlling morbidity and mortality from chronic diseases affecting humankind. Prevention provides: the methods to avoid occurrence of disease and most population-based health promotion efforts are of this type; methods to diagnose and treat extant disease in early stages before it causes significant morbidity; methods to reduce negative impact of extant disease by restoring function and reducing disease-related complications; and finally, the methods to mitigate or avoid results of unnecessary or excessive interventions in the health system.

The quality and quantity of diet with respect to the intake of fresh food (fruits, seeds and vegetables) may improve our health and consequently decrease the risk of any disease. Currently, the antioxidant nutrients are the vitamins C and E and β-carotene. However, it is worthy to mention that these compounds are involved in other functions a part from being antioxidant nutrients.

Selenium (Se), a trace mineral. Is the 34th element and is located between sulfur and tellurium in Group 16 in the periodic table. It is a nonmetallic element and its properties are intermediate between adjacent sulfur and tellurium. It was originally discovered by a German chemist Martin Heinric Klaproth, but misidentified as tellurium. Later, in 1818 a Swedish chemist Jons Jacob Berzelius discovered selenium and was named after the Greek goddess of the moon, Selene [1] and its name was associated with tellurium, a name for earth. He observed the element as a deposit following the oxidation of sulfur dioxide from cooper pyrites. It ranks seventieth in abundance among the elements and is distributed in the Earth's crust at concentrations averaging 0.09 mg/kg [2]. Selenium has six major stable isotopes have been reported and the most abundant in nature are 80Se (49.6%) and 78Se (23.8%) [3]. In general, selenium is present in the environment in elemental form or in the form of selenide (Se^{2-}), selenate (SeO_4^{2-}), or selenite (SeO_3^{2-}). The identity and amounts of the various oxidation-state species in soils depends enormously on the redox-potential conditions. The lower oxidation states predominate in anaerobic conditions, acidic soils, and the higher oxidation states are favored in alkaline and aerobic conditions. Both selenites and selenates are taken up by plants and converted to protein-bound selenocysteine and selenomethionine, soluble inorganic forms, several free amino acids, and volatile organoselenium compounds. The elemental form of selenium, selenium dioxide, and volatile organoselenium compounds produced by industries and plants are incorporated in the environment. Selenium occurs naturally in water in trace amounts as a result of geochemical processes, such as weathering of rocks and erosion of soils, and is usually present in water as selenate or selenite; however the elemental form may be carried in suspension [4].

Interest in selenium and health was focused primarily on the potentially toxic effects of high intakes in humans, stimulated by reports of alkali disease in livestock raised in seleniferous areas, in the last century [5]. Selenium is a trace mineral that is essential to good health but required only small amounts [6,7]. Selenium is considered as essential human micronutrient and is incorporated into proteins to make selenoproteins. Selenium is present in the selenoproteins, as the aminoacid selenocysteine (Se-Cys) [8-12].

Dietary levels of the desired amount of Se are in a very narrow range: consumption of foods containing less than 0.1 mg kg-1 of this element will result in Se deficiency, whereas dietary levels above 1 mg kg-1 will lead to toxic manifestations [13]. Se status varies significantly across different populations and different ethnic groups [14-15].

Selenium enters the food chain through plants, and the amount and bioavailability of selenium in the soil typically reflects the plant level. Selenium is provided by the diet in humans, but may also be provided from drinking water, environmental pollution, and in recent years through supplementation [16,17]. Plants convert Se mainly into selenomethionine (Se-Met) and incorporated it into protein place of methionine. More than 50% of the total Se content of the plant exist as Se-Met, the rest exist as selenocysteine (Se-Cys), methyl-Se-Cys and c-glutamyl-Se-methyl-Cys. The later compounds are not significantly incorporated into plant protein. Higher animals are unable to synthesize Se-Met and only Se-Cys was detected in rats supplemented with Se as selenite [18]. Animals that eat grains (Brazil nuts, sunflower seeds, walnuts and grains) that were grown in selenium rich soil have higher levels of selenium in their muscles, liver, kidney, heart, spleen and fingernails. Other natural selenium sources are butter, eggs, brewer's yeast, wheat germ, garlic, raspberry leaf, radish, horseradish, onions, shellfish, broccoli, fennel seed and ginseng, among other sources.

Most ingested forms of selenium ultimately are metabolized to low molecular weight inorganic and organic compounds that play a central role in human health either via incorporation into selenoproteins or binding to selenium binding proteins [19]. Therefore, a tremendous effort has been directed toward the synthesis of stable organoselenium compounds that could be used as antioxidants, enzyme modulators, antitumor, antimicrobials, antihypertensive agents, antivirals and cytokine inducers. Several excellent books and reviews appeared in literature describing the biological function of organoselenium compounds [20-22].

The role of organoselenium compounds as antioxidants, as enzyme modulators, photo-chymotherapeutic agents, cytokine inducers and immunomodulators, and antihypertensive and cardiotonic agents have been recently described in literature [23].

The essentiality of selenium results as a necessary component of the active center of a number of selenoenzymes. Selenium functions as a redox center. The term selenoprotein is any protein that includes in its primary sequence of amino acids, the selenocysteine (Se-Cys) residue [24]. There are at least 30 selenoproteins that have been identified in mammals, and it has been estimated that humans have about 25 selenoproteins, including glutathione peroxidase, thioredoxin reductase, iodothyronine, deiodinase, and selenoproteins P, W, and R [25-27]. GPx accounts for 10–30% of plasma selenium, and selenoprotein P accounts for another 50% [28]. These enzymes protect cells from free radical damage and regulate DNA transcription and cell proliferation. The glutathione and thioredoxin systems in particular have long been considered the major pathways through which selenium exerts its potential chemopreventive effect [24], while some investigations have also suggested growth inhibitory, proapoptotic activity for selenometabolites in premalignant cells [29]. Selenium is also involved in thyroid function, T cell immunity, and spermatogenesis [28], and is a competitive antagonist of potentially carcinogenic heavy metals such as arsenic and cadmium [30].

The organism has several biological defense mechanisms against intracellular oxidative stress such as superoxide dismutase, catalase, glutathione peroxidase and nonenzymatic antioxidants such as glutathione, vitamins A, C and E, riboflavin, a B vitamin and selenium can also contribute to overcome oxidative stress [31].

Vitamin E is a fat-soluble vitamin known for its antioxidant capacity that is why it is well known as a lipophilic antioxidant that protects membranes from being oxidatively damaged as an electron donor to free radicals [32]. Vitamin E belongs to a group of the eight naturally occurring vitamer forms, four tocopherols (α, β, γ, δ) and four tocotrienols (α, β, γ, δ) based on the hydroxyl and methyl substitution in their phenolic rings, all of which have saturated and three double bonds in their phytyl tails. α-tocopherol (from the Greek *tokos* = child, *phero* = to bear and *ol* indicating that the substance is an alcohol, is the most abundant form in nature; it's the most active and corrects human E deficiency symptoms [33]. The most abundant sources of vitamin E are vegetable oils, which typically contain all four tocopherol (α, β, γ, δ) in varying proportions, Other important source are nuts and seeds such as sunflower and amaranth seeds.

It is well known that all forms of vitamin E are lipid soluble they easily absorbed from the intestinal lumen after dietary intake via micelles created by biliary and pancreatic secretions [34-35]. Vitamin E is then incorporated into chylomicrons and secreted into the circulation where, transported by various lipoproteins, it travels to the liver [36]. Plasma α-tocopherol concentrations in humans range from 11 to 37 μmol/L, whereas γ-tocopherol are between 2 and 5 μmol/L. The liver plays a central role in regulating α-tocopherol levels by directly acting on the distribution, metabolism, and excretion of this vitamin [37]. The major hepatic regulatory mechanism is the α-tocopherol transfer protein, α-TTP, which has been identified in a variety of mammals, including humans [38]. This protein facilitates secretion of α-tocopherol from the liver into the bloodstream, by acquiring it from endosomes and then delivering it to the plasma membrane where it is released and promptly associates with the different nascent lipoproteins [39]. Plasma concentration of vitamin E depends completely on the absorption, tissue delivery, and excretion rate. The estimated α-tocopherol half-life in plasma of healthy individuals is ~ 48 to 60 H, which is much longer than the half-life of γ-tocopherol approximately 15 H. These kinetic data underscore an interesting concept that while α-tocopherol levels are maintained, the other forms of vitamin E are removed much more rapidly [40].

2. Selenium and Health

Selenium deficiency is associated with the pathogenesis of wide variety of processes that affects our health and disease including the antioxidant activity, depression, allergies, preventing oxidative stress, HIV infection, in the brain, thyroid metabolism, cancer, diabetes mellitus, male fertility, asthma, cardiovascular disorders, rheumatoid arthritis, pre-eclampsia, in immune function, in alleviate bone impairments, aging, gastrointestinal problems, selenium interactions and toxicity, anti-inflammatory effects, and hypertension. The list of clinical disorders expected to be influenced by Se deficiency is rapidly growing with time. Some selected issues regarding the role of Se in health and disease have been briefly outlined as follows:

2.1. Se and antioxidant activity

Selenocysteine is recognized as the 21st amino acid, and it forms a predominant residue of selenoproteins and selenoenzymes in biological tissues. The molecular structure of selenocystiene is an analogue of cysteine where a sulphur atom is replaced by Se. Even though Se and sulphur share some similar chemical properties, there are also some differences. The R-SeH with a pKa 5.2 is more is more acidic than R-SH with a pKa 8.5, and readily dissociated at physiological pH, which may contribute to its biological reactivity. In the body, both organic [selenocysteine(SeCys) and selenomethionine (SeMet)] and inorganic (selenite, selenate) Se compounds are readily metabolized to various forms of Se metabolites [41]. Of particular importance during this metabolic process is the formation of hydrogen selenide (H_2Se) from selenite after the action of glutathione-coupled reactions *via* selenodiglutathione (GS-Se-SG) and glutathione selenopersulfide (GS-SeH). H_2Se is further metabolized and involved in the formation of methylselenol and dimethylselenide, which are exhaled or secreted *via* the skin. Selenium is also excreted in urine as trimethylselenonium ion and selenosugar compounds [42]. The selenoproteins are classified on the basis of their biological function [25]. The first identified selenoprotein was glutathione peroxidase 1 (GPx1). The selenoenzymes with strong antioxidant activity are GPx, GPx1, GPx3, GPx4, GPx5 and GPx6. In Humans GPx1 through GPx4 and GPx6 are selenocysteine containing enzymes. These GPx play a significant role in protecting cells against oxidative damage from reactive oxygen species (ROS) and reactive nitrogen species (RNS), which include superoxide, hydrogen peroxide, hydroxyl radicals, nitric oxide and peroxynitrite [43-44]. The other essential antioxidant selenoenzymes are the thioredoxin reductase (TrxR) where they use thioredoxin (Trx) as a substrate to maintain a Trx/TrxR system in a reduced state for removal of harmful hydrogen peroxide and there are three types of TrxR. Iodothyronine deiodinase (DIO) have three subtypes, DIO 1, 2, and 3 [45].

2.2. Se and depression

In [46] selenium's function as an antioxidant, and as a constituent of selenoproteins that are important in redox homeostasis, warrants further investigation as a risk factor for depression, and suggest a potentially novel modifiable factor in the primary prevention and management of depression. Depression is becoming recognized as an inflammatory disorder, accompanied by an accumulation of highly reactive oxygen species that overwhelm usual defensive physiological processes [47-51]. Several indicators support a role for selenium in normal brain function. During times of selenium deficiency, there is preferential storage of selenium in the brain [52]. Selenium has significant modulatory effects on dopamine [53] and dopamine plays a role in the pathophysiology of depression and other psychiatric illnesses [54]. Diminished levels of selenium in the brain are associated with cognitive decline [55] and Alzheimer's disease [56]. Selenium supplementation has been linked with improvements in mood [57] and protection against postpartum depression [58]. What is unclear is if low dietary selenium is a risk factor for the development of depression. In recognition of selenium's biological activity, it has been hypothesized that low levels of dietary selenium would be associated with an increased risk of major depressive disorder (MDD) in a representative population-based sample of women.

Alterations in redox biology are established in depression; however, there are no prospective epidemiological data on redox-active selenium in depression. It is known that selenium's function as an antioxidant, and as a constituent of selenoproteins that are important in redox homeostasis, warrants further investigation as a risk factor for depression, and suggest a potentially novel modifiable factor in the primary prevention and management of depression.

2.3. Selenium and allergies

The International Study of Asthma and Allergies in Childhood (ISAAC) found that one in four New Zealand children aged 6–7 years had experienced asthma symptoms, which placed New Zealand in the top four countries for asthma prevalence [59]. The reasons for the high prevalence and severity of this condition or the increased prevalence of asthma over the last 20 years are not well understood. One of a number of environmental factors that have been proposed as a reason for the escalation in asthma prevalence is a decreasing intake of dietary antioxidants [60]. It is well known that selenium is essential for the optimal functioning of the selenoenzymes glutathione peroxidases (GPx) and thioredoxin reductases, powerful antioxidants, and is found abundantly in lung tissue and the extracellular fluid of the respiratory system [61]. Selenium has been implicated in inflammation by reducing the severity of the inflammatory response through modulation of the pro-inflammatory leukotrienes, important mediators of acute asthmatic reactions as well as sustaining the inflammatory process causing a late allergic reaction metabolism [62]. Evidence from randomized controlled trials [63] and basic mechanistic work investigating the effect of selenium on markers of inflammation and oxidative stress [62]. Evidences have supported a protective role for selenium in asthma, although other studies have not [64-66]. The ISAAC study does not support a strong association between selenium status and the high incidence of asthma in New Zealand. However, there was a modest association between lower plasma selenium and whole blood glutathione peroxidase activity and higher incidence of persistent wheeze [67].

2.4. Selenium in preventing oxidative stress

The reactivity of organoselenium compounds [22,68] characterized by high nucleophilicity and antioxidant potential, and provides the basis for their pharmacological activities in mammalian models. Organochalcogens have been widely studied given their antioxidant activity, which confers neuroprotection, antiulcer, and antidiabetic properties. Given the complexity of mammalian models, understanding the cellular and molecular effects of organochalcogens has been hampered. In reference [69] the nematode worm *Caenorhabditis elegans* is an alternative experimental model that affords easy genetic manipulations, green fluorescent protein tagging, and in vivo live analysis of toxicity. Manganese (Mn)-exposed worms exhibit oxidative-stress-induced neurodegeneration and life-span reduction. Diethyl-2-phenyl-2-tellurophenyl vinyl phosphonate (DPTVP) and 2-Phenyl-1,2-benzoisoselenazol-3-(2H)-one (Ebselen) were tested for reversing the Mn-induced reduction in survival and lifespan in this nematode. DPTVP was the most efficacious compound as compared to Ebselen in reversing the Mn-induced toxicity and increasing in survival and life span. DPTVP and ebselen act as antiaging agents in a model of Mn-induced toxicity and aging by regulating DAF-16/FOXO signaling and attenuating oxidative stress.

Bone is a specialized connective tissue, which forms the framework of the body. Various physiological conditions can adversely affect femoral bone metabolism. These physiological conditions could be food deprivation [70], and iodine and/or selenium (Se) deficiency [71,72] and antithyroid drugs [73] affects bone maturation. Selenium is an important protective element that may be used as a dietary supplement protecting against oxidative stress, cellular damage and bone impairments [74].

2.5. Selenium in HIV infection

The HIV pandemic has placed a great demand upon the scientific community to develop effective prevention and treatment methods. Since the beginning of the pandemic in 1981, over 25 million people are estimated to have died from the disease [75]. It is currently a leading cause of death in many parts of the world, and a disease that disproportionately affects

http://dx.doi.org/10.5772/51610

the marginalized and socially disadvantaged. It is currently a leading cause of death in many parts of the world, and a disease that disproportionately affects the marginalized and socially disadvantaged. Many of those affected also suffer from chronic food insecurity and malnutrition, so therapies that could potentially target both HIV disease and malnutrition, such as multivitamins, have been extensively researched for potential benefits [76]. Among such therapies, the antioxidant micronutrients theorized to have potential benefits in HIV disease, apart from correcting deficiencies, have been examined frequently [77,78].

Selenium has an inhibitory effect on HIV in vitro through antioxidant effects of glutathione peroxidase and other selenoproteins. Numerous studies have reported low selenium status in HIV-infected individuals, and serum selenium concentration declines with disease progression. Some cohort studies have shown an association between selenium deficiency and progression to AIDS or mortality. In several randomized controlled trials, selenium supplementation has reduced hospitalizations and diarrheal morbidity, and improved CD4+ cell counts, but the evidence remains mixed. Additional trials are recommended to study the effect of selenium supplementation on opportunistic infections, and other HIV disease-related comorbidities in the context of highly active antiretroviral therapy in both developing and developed countries [79].

There is a historical record showing that organoselenium compounds can be used as antiviral and antibacterial agents. This topic has been reviewed by [22,23].

2.6. Selenium in the brain

In addition to the well-documented functions of Se as an antioxidant and in the regulation of the thyroid and immune function [80]. Recent advances have indicated a role of Se in the maintenance of brain function [81]. Selenium is widely distributed throughout the body, but is particularly well maintained in the brain, even upon prolonged dietary Se deficiency [82]. In the brain, the highest concentration of Se is found in the gray matter, an area responsible for chemical synaptic communication [83]. It has been shown that rats on a Se-deficient diet for thirteen weeks retained Se in their brain, while their plasma Se concentrations were depleted [84]. After intraperitoneal injection of $^{75}SeO3^{2-}$ into Se-deficient rats, the brain rapidly sequesters a large portion of the available Se [85]. In the brain, it was found that the cerebellum accumulated the highest concentration of Se, followed by the cortex, medulla oblongata, cerebral hemisphere, and the spinal cord. Interestingly, Se retention in the brain depends on Selenoprotein P expression [86]. Because the body preferentially allocates available Se to the brain during Se deficiency, Se may play an essential role in the brain. More evidence for the brain being at the apex of Se retention is provided by a study showing that a six generation Se deficiency in rats caused a more than 99% reduction of Se concentration in the liver, blood, skeletal tissue, and muscle, while the brain retained a 60% of the Se [87]. Se concentration in Alzheimer's brains was found to be 60% of the age-matched control individuals [88]. Accumulated lines of evidence indicate important roles of selenoproteins in the maintenance of optimal brain functions via redox regulation. Decreased expression of several selenoproteins is associated with the pathologies of a few age-associated neurodisorders, including Parkinson's disease, Alzheimer's disease and epilepsy [81].

Oxidative stress and generation of reactive oxygen species are strongly implicated in a number of neuronal and neuromuscular disorders, including epilepsy. The functions of selenium as an antioxidant trace element are believed to be carried out by selenoproteins that possess antioxidant activities and the ability to promote neuronal cell survival [89]. It is known the role of selenium in a detoxifying enzyme, glutathione peroxidase, this element has been demonstrated to have a positive biological function in various aspects of human health [90]. Oxidative stress and generation of reactive oxygen species are strongly implicated in a number of neurologic disorders including seizure disorders. Oxidative phosphorylation occurring in the mitochondria produces oxygen radicals routinely in all tissues as well as the nervous system. One important defense may be to remove the oxygen radicals. Selenium-requiring processes are involved in normal maintenance of cell function. However, when the system is overused or chronically activated beyond its normal state, such as recurrent or intractable seizures, abnormal increases in by-products can produce neuronal cell damage. Selenium provides protection from reactive oxygen species–induced cell damage. The proposed mechanisms are mainly through the functions of seleno-dependent enzymes and selenoproteins [82,91]. It seems that selenium plays an important role in stopping the vicious cycle of oxidative stress and neuronal damage in patients with intractable seizures by restoring the defense mechanism.

2.7. Selenium and the thyroid

Some selenoproteins of the human selenoproteome display multiple genes performing similar functions. The main selenoprotein families are the glutathione peroxidases (GPxs; seven genes), the thioredoxin reductases (TRxs; three genes) and the iodothyronine deiodinases (DIs; three genes) [92,93]. The GPxs, which possess oxidoreductase functions, protect the cell from oxidative stress. The TRxs form a cellular redox system, existing in many organisms, which is essential for cell development and proliferation. The DIs that catalyzes the conversion of T4 to T3 provides the sources of T3 production. It may thus be hypothesized that the essential micronutrient selenium, in the form of Se-Cys, modulates redox-sensitive signaling pathways and thereby potentially modifies selenoprotein gene expression. These findings have aroused growing interest of the scientific community in this multifaceted element. In this context, whereas selenium administration for cancer chemoprevention produced questionable results, those of selenium supplementation in patients with autoimmune thyroid disease have been more encouraging. In [94] comprises an in-depth discussion of the link between selenium and thyroid function; it provides a critical analysis of the data contained in recent studies, an update and evaluation of current knowledge with regard to the mechanisms of action of selenium, and reflections on the prospects for selenium supplementation in thyroid pathology.

Evidence in support of selenium supplementation in thyroid autoimmune disease is evaluated; the results herein presented demonstrating the potential effectiveness of selenium in reducing the antithyroid peroxidase titer and improving the echostructure in the ultrasound examination. However, considerable discord remains as to who should comprise target groups for selenium treatment, who will most benefit from such treatment, the precise impact of the basal antithyroid peroxidase level, and the effect of disease duration on the treatment outcome. Clearly, further in-depth studies and evaluation are required concerning the mechanism of action of selenium as well as the choice of supplements or dietary intake.

2.8. Selenium in cancer

The reactive oxygen species (ROS) are derived from cellular oxygen metabolism and from exogenous sources. An excess of ROS results in oxidative stress and may eventually cause cell death. ROS levels within cells and in extracellular body fluids are controlled by concerted action of enzymatic and non-enzymatic antioxidants. The essential trace element selenium exerts its antioxidant function mainly in the form of selenocysteine residues as an integral constituent of ROS-detoxifying selenoenzymes such as glutathione peroxidases (GPx), thioredoxin reductases (TrxR) and possibly selenoprotein P (SeP). In particular, the dual role of selenoprotein P as selenium transporter and antioxidant enzyme is highlighted herein. A cytoprotective effect of selenium supplementation has been demonstrated for various cell types including neurons and astrocytes as well as endothelial cells. Maintenance of full GPx and TrxR activity by adequate dietary selenium supply has been proposed to be useful for the prevention of several cardiovascular and neurological disorders. On the other hand, selenium supplementation at supranutritional levels has been utilized for cancer prevention: antioxidant selenoenzymes as well as prooxidant effects of selenocompounds on tumor cells are thought to be involved in the anti-carcinogenic action of selenium [95,96].

Among various antioxidant minerals, selenium it may prove to be of major significance as a prophylactic agent against cancer. Low blood selenium concentration and incidence of carcinogenesis have been well observed in both animals [97] as well as in human studies [98]. In addition, it has been demonstrated in a double blind randomized cancer prevention trial in humans that increased selenium intake has a significant role in the treatment of cancer [99]. A similar prospective study could also be designed for other cancers to determine the chemopreventive effect of Se. Selenium has also been reported to have a beneficial effect on the incidence of gastrointestinal and bladder cancers [100,101].

Although selenium is reported to play a significant role in cancer development, its exact anticancer mechanism of action at molecular levels is not fully understood. However, it has been hypothesized that the most possible mechanistic action of Se as chemoprevention is its role in the antioxidant defense systems to reduce oxidative stress and limit DNA damage [24,102]. Experiments carried out within the framework of a canine model using male beagle dogs to mimic prostate cancer in humans showed that the damage to DNA was significantly reduced when the animals were exposed to increased Se dietary supplements [103]. The effectiveness of Se in the prevention of DNA damage, however, depends on its chemical forms. In an *in vitro* study [104] found that selenocysteine inhibited DNA damage more strongly than the selenomethionine. Other possible anticancer mechanisms of Se include the induction of apoptosis, cell-cycle arrest and DNA-repair genes, inhibition of protein kinase C activity and cell growth and effect on estrogen- and androgen-receptor expression [102].

In [105] knowledge of the plasma selenium levels are associated with optimized concentration or activity of specific selenoproteins can provide considerable insights from epidemiological data on the possible involvement of those selenoproteins in health, most notably with respect to cancer. For cohort studies, if selenoproteins such as glutathione peroxidase and selenoprotein P are relevant to cancer, one might only expect to see an effect on risk when the concentrations in the cohort range from below, to above, the level needed to optimize the activity or concentration of these enzymes. Similarly, trials would only show a beneficial effect of supplementation if selenium status were raised from below, to above, the optimal concentration for the selenoproteins likely to be implicated in cancer risk, as occurred in the Nutritional Prevention of Cancer (NPC) trial but not in Selenium and Vitamin E Cancer Prevention Trial (SELECT). The most powerful evidence for the involvement of selenoproteins in human health comes from epidemiological studies that have related single nucleotide polymorphisms in selenoproteins to disease risk. The totality of the evidence currently implicates GPx1, GPx4, SEPS1, Sep15, SEPP1 and TXNRD1 in conditions such as cardiovascular disease, pre-eclampsia and cancer. Future studies therefore need to determine not only selenium status, but genotype, both in selenoproteins and related pathways, when investigating the relationship of selenium with disease risk.

2.9. Selenium in diabetes

The evidence supporting an effect of selenium on the risk of diabetes is variable, occasionally conflicting, and limited to very few human studies. Following a trial investigating the effect of selenium supplementation (200 μg/day) on skin cancer, subsequent analysis showed that there was an increased risk of developing type 2 diabetes in the supplemented group. Evidence from analysis of NHANES III [106] supports these findings; the adjusted mean serum selenium concentrations were slightly, but significantly, higher in diabetics compared with those without the disease. This study, conducted in an elderly French population, found a sex-specific protective effect of higher selenium status at baseline on later occurrence of dysglycemia; that is, risk of dysglycemia was significantly lower in men with plasma selenium, but no significant relationship was observed in women [107].

The role of selenium as an antioxidant, particularly within the GPxs, selenium is likely to be important in reducing oxidative stress, an important risk factor for developing diabetes. There are also plausible suggestions that selenium can influence glucose metabolism. However, at high intakes it is also conceivable that reactive oxygen species could be generated or selenium may accumulate in the organs associated with glucose metabolism [108]. In patients with diabetes, selenium supplementation (960 μg/day) reduced NF-κB levels to those comparable with nondiabetic controls [109]. In addition, further analysis of the Nutritional Prevention of Cancer trial data has shown an increased risk of self-reported Type-2 diabetes in those supplemented with Se, though the effect was significant only in those in the top tertile of plasma Se at baseline [110].

2.10. Selenium and male fertility

Selenoprotein P transports selenium particularly to testis and brain [111]. Among the five enzymes of GPx, GPx1 prevents apoptosis induced by oxidative stress and GPx4 acts directly on membrane phospholipid hydroperoxides and detoxifies them. Selenium as GPx, is present in spermatids and forms the structural part in the mid piece of mature spermatozoa. Some well known effects of selenium deficiency include instability of the middle piece leading to defective sperm motility [112], low reproductive ability and abnormal development of spermatozoa [113]. Selenium is also required for testosterone synthesis and sequential development of flagella [114]. It can restore the physiological constitution of polyunsaturated fatty acid in the cell membrane [115]. Testes are extremely resistant to Se depletion and have high Se content. Recent studies have shown that sperm and testicular Se was unaffected by the supplementation, suggesting that testes are protected from Se excess as well as from Se deficiency [116].

2.11. Selenium in asthma

Se status is decreased in patients with asthma, as is activity of glutathione peroxidase in platelets and erythrocytes. There is an associated marked oxidant/antioxidant imbalance in the blood of asthmatics, which reflects poor antioxidant status and enhanced inflammatory mediated oxidative stress [117]. According to the University of Maryland Medical Center, a 2004 study of 24 asthmatics that were given selenium supplements for 14 weeks had significant improvement in their symptoms when compared to a control group given a placebo. Although this is a small study done over a short amount of time, it's encouraging [118].

2.12. Selenium in cardiovascular disorders

Free radicals are toxic to the myocardium and can cause tissue damage that leads to extensive necrosis, myocytolysis and cellular edema [119]. Atherosclerotic plaque formation may be a reflection of sub-optimal GPx4 activity in the prevention of LDL oxidation, with subsequent uptake by endothelial cells and macrophages in arterial blood vessels [120]. Selenium via GPx reduces phospholipids, hydro peroxides and cholestryl esters associated with lipoproteins and may therefore, not only reduce the accumulation of oxidized LDL in arterial wall but also reduce platelet aggregation and activation of monocyte and macrophages [121]. Selenium owing to its antithrombotic effect on the interaction between platelets and endothelial cells via GPx, also provides concrete evidence in the prevention of atherosclerosis [122].

The study on acute myocardial infarction (AMI) patients, it was observed that selenium dependent GPx level decreases significantly in AMI patients and explained it as an imperative consequent of GPx activity in annihilating oxygen toxicity by metabolizing H_2O_2 and inhibiting further free oxygen radical production in early phase of myocardial infarction [123].

2.13. Selenium in rheumatoid arthritis

Scientific research shows that people with rheumatoid arthritis have low levels of selenium. A study suggests, it is part of the body's defense mechanism [124]. In reference [125] the authors found lower selenium levels in patients with rheumatoid arthritis who were treated with arthritis medication compared with people without the condition. In people without rheumatoid arthritis or a family history of the condition, low levels of the mineral may increase the risk of developing rheumatoid arthritis [126].

2.14. Selenium in pre-eclampsia

In reference [124], pre-eclampsia (pregnancy induced hypertension; PIH), is an important cause of maternal morbidity and mortality with essentially unknown etiology. However, the precise factors involved in the pathogenesis of PIH are still unknown [127]. It has been conceived that free radical mediated oxidative stress may contribute to the development of pre-eclampsia. Selenium and its related enzymes especially GPx play a crucial role in annihilating oxygen toxicity and there by controlling the progression of disease [128]. In addition, selenium deficiency in women may result in infertility, miscarriages and retention of the placenta [129].

2.15. Selenium in immunity

In reference [130], the generation of ROS in a limited dose is one of the processes induced by the immune system to destroy microbial pathogens and viruses. However, the over-production of ROS can also cause damage to the host cells that need to be protected by Se at various stages in the immune system. Keshan disease, an endemic cardiomyopathy in China that develops as a result of Se deficiency, may also be complicated with viral infection, and this has led to the investigation of the effects of viruses, such as coxsackievirus, on Se-deficient animals [131,132]. Results from animal studies have demonstrated that Se deficiency can lead to an impairment of immune functions that result in the inability of phagocytic neutrophils and macrophages to destroy antigens. A low Se status in humans has been reported to cause a decreased immune response to poliovirus vaccination [133]. This study also demonstrated that the subjects supplemented with Se showed fewer mutations in poliovirus than those who received a placebo. The involvement of Se in the immune system may be associated with a number of mechanisms, including the increased activity of natural killer (NK) cells, the proliferation of T-lymphocytes, increased production of interferon c, increased high-affinity interleukin-2 receptors, stimulation of vaccine-induced immunity and increased antibody-producing B-cell numbers [134,135].

2.16. Selenium in bone impairments

Osteoblasts (bone-forming cells) and osteoclasts (bone-resorption cells) are involved in bone remodeling. Therefore, any loss of osteoblastic activity or an increase in osteoclastic activity could lead to a decrease in bone-mineral densities (BMD), bone mass, and make the bones more likely to osteoporosis, and ultimately to fractures [136]. In addition, high levels of reactive oxygen species (ROS) and many other factors such as genetic race, hormonal, mechanical, and nutritional statues are involved in bone weakness and fractures. ROS shift cells into a state of oxidative stress [137] which contributes to the etiology of various degenerative diseases that cause tissue injury [136,137]. Studies have demonstrated that the ischemia-reperfusion processes that occur after a fracture are associated with oxidative stress development [136,138]. It is believed that bone markers such as osteocalcin and alkaline phosphatase as

http://dx.doi.org/10.5772/51610

well as antioxidant enzymes play a significant role in fracture healing. However, to the best of our knowledge, there are no reports about the use of vitamins A, C, E, and selenium as antioxidant therapy to explore their effects in the levels of bone-healing markers and oxidative stress parameters of osteoporotic patients. In [139] suggests that selenium is an important protective element that may be used as a dietary supplement protecting against bone impairments.

3. Vitamin E and Health

3.1. Vitamin E

All forms of vitamin E meet the chemical definition of an antioxidant moiety: "chain-breaking free radical scavenger." Indeed consistent data have shown that all isoforms act as potent antioxidants in conventional in vitro paradigms. The free hydroxyl group on the aromatic ring is thought to be responsible for this property, and a relatively stable form of the original vitamin E is formed when hydrogen from this group is donated to a free radical. Yet, definitive proof that vitamin E possesses antioxidant properties has been hampered for a long time because of a lack of sensitive and specific analytical techniques to measure this biologic event *in vivo* [36]. Apart from antioxidant properties, more recent studies have clearly demonstrated that vitamin E also possesses important non-antioxidant cellular and molecular functions. One of the first roles of α-tocopherol in cell signaling was the report that it inhibits smooth muscle cell proliferation, decrease protein kinase C activity, and controls expression of the α-tropomyosin gene [140].

One of the major vitamin E-deficiency symptoms are neurological disorders. Furthermore, vitamin E deficiency is related to female infertility. The frame to pinpoint the physiological action of vitamin E is set by its chemical nature: (i) It is a redox-active compound prone to undergo 1- and 2-electron transitions and (ii) it is highly lipophilic, although this property may be modulated by phosphorylation [141]. In [142] oxidative stress is a developing research field and is being examined in female infertility. Prooxidants, also called free radicals or reactive oxygen species (ROS), and their neutralizing agents the antioxidants are the main chemicals of the oxidation mechanism. The term oxidative stress refers to the dysequilibrium between the free radicals and the antioxidants in favor of the free radicals. In actuality, free radicals are not so frightening, since they are necessary for the adequate reproductive functions within the ovary and the endometrium. Vit E administration may improve the endometrial response in unexplained infertile women *via* the likely antioxidant and the anticoagulant effects. It may also modulate the antiestrogenic effect of clomiphene citrate and the problem of a thin endometrium in these cycles may be adjusted.

Some non-antioxidant properties of vitamin E could play a key role in neuroprotection. It has been recently shown that α-tocotrienol, at nanomolar concentrations, protects mouse hippocampal and cortical neurons from cell death by modulating neurodegenerative signaling cascades. Furthermore, it has been shown that α-tocotrienol modulates 12-lipoxygenase and phospholipase A2 activities, which are implicated in glutamate-induced neuronal cell death [143]. Some vitamin E forms (α- and γ-tocopherol, tocotrienols) also exhibit potent anti-inflammatory properties [144,145]. The introduction of the free radical theory of brain aging has propelled a renewed interest in this vitamin. As result, by preventing and/or minimizing the oxidative stress dependent brain damage, this vitamin plays important role in brain aging, cognition, and Alzheimer's dementia.

Vitamin E is a potent peroxyl radical scavenger that prevents lipid peroxidation [146] and is found in high concentrations in immune cells [147]. Deficiency in vitamin E is associated with increased oxidative stress [148] and impaired immune function, including both humoral and cell-mediated immunity, phagocyte function, and lymphocyte proliferation [149]. Age-related declines in immune function can be restored by vitamin E supplementation [150]. This vitamin is an exogenous, lipidsoluble antioxidant molecule. It is thought to be a direct free radical scavenger by activating the intracellular antioxidant enzymes and saving the cell membranes from lipid peroxidation, which was demonstrated on sperm membrane components [151]. Its antioxidant effect was concluded in cancer therapy, high-risk pregnancy and male infertility [152-154].

Vitamin E (α–tocopherol acetate) is found within the phospholipid bilayer of cell membranes where it functions as an electron donor to free radicals. It has been recognized as one of the body's major natural antioxidants. Another antioxidant, Se appears to function as an antimutagenic agent, preventing the malignant transformation of normal cells. Its protective effects seem to be primarily associated with its presence in the seleno-enzymes which are known to protect DNA and other cellular components from oxidative damage [44].

Selenium, vitamin A (retinol) and vitamin E (α-tocopherol) are essential micronutrients for human health. Both selenium and vitamin E are important in host antioxidant defense and immune function. It has been reported that deficiency of selenium and vitamins may promote peroxidation events leading to the release of free radicals. All have free-radical-scavenging properties that allow them to function as physiologic antioxidants in protecting a number of chronic diseases, such as cancer and cardiovascular disease. In addition to its antioxidant capacity, α-tocopherol regulates expression of genes involved in a wide range of cell functions, including cell cycle regulation, inflammation and cell adhesion, cell signaling, and lipid uptake [155].

Selenium also has an important role in antioxidant defense and immune function. Due to its incorporation as selenocysteine into glutathione peroxidase (GPX) [156] and thioredoxin reductase [157], selenium is important for the control of oxidative stress and, therefore, the redox tone of the cell. In total, there are 25 identified selenoproteins (24 in rodents), many with unknown function [25]. Selenium is important for cytotoxic T-lymphocyte and natural killer cell activity [158], respiratory burst [159], and protection against endotoxin-induced oxidative stress [160]. Multiple studies have shown that NF-kB activation can be affected by selenium status [161,162], and selenium deficiency can alter chemokine and cytokine expression during viral infections [163]. Various investigators have reported the role of selenium as an inhibitor of carcinogenesis in various organs including liver, skin, stomach, mammary gland, gastrointestinal and oral cavity [164,165].

http://dx.doi.org/10.5772/51610

In [166] blader cancer represents an important cause of morbidity and mortality. In 2010 it was again the second most common genitourinary cancer in the United States with an established 70,530 new cases and 14,680 deaths [167]. Currently it is estimated that more than 500,000 men and women in the United States have a history of bladder cancer. The etiology of most bladder urothelial carcinoma is associated with tobacco exposure, occupational exposure to aromatic amines, and exposure to the chemical and rubber industries [168]. Bladder cancer is the most expensive cancer in the United States, accounting for almost $3.7 billion (2001 value) in direct costs [169]. There is substantial epidemiological and biological evidence that selenium and vitamin E may prevent bladder cancer. A recent meta-analysis of 7 published epidemiological studies, including 3 case-control, 3 nested case-control and 1 case cohort series, examined the association between selenium levels and bladder cancer [170]. In the analysis stratified by gender only women showed a significantly decreased risk associated with selenium. An opposite gender pattern, with protective effects in men but not in women, was reported in a meta-analysis of selenium supplementation, primary cancer incidence and mortality [171]. Epidemiological and biological evidence suggests a preventive effect of selenium and vitamin E on bladder cancer. These researches assessed the effect of selenium and/or vitamin E on bladder cancer development.

3.2. Selenium and vitamin E

Selenium and vitamin E are essential components of the human diet and have been studied as antioxidants and/or potential agents for a variety of human diseases. Various formulations of both selenium and vitamin E have been shown to possess a therapeutic and preventive effect against prostate cancer. The Selenium and Vitamin E Cancer prevention Trial (SELECT) started in 2001 and was a phase III, randomized placebo/controlled human trial to investigate the prostate cancer chemopreventive effects of selenium and vitamin E or their combination [172,173].

Sselenium an essential trace element, and vitamin E, a lipid soluble antioxidant, are important mediators for protection against oxidative stress. Deficiencies in either Se or vitamin E result in increased viral pathogenicity and altered immune responses. Furthermore, deficiencies in either Se or vitamin E results in specific viral mutations, changing relatively benign viruses into virulent ones. Thus, host nutritional status should be considered a driving force for the emergence of new viral strains or newly pathogenic strains of known viruses [174].

Several studies have evaluated the possible association between antioxidants vitamins or selenium supplement and the risk of prostate cancer, but the evidence is still inconsistent. We systematically searched PubMed, EMBASE, the Cochrane Library, Science Citation Index Expanded, Chinese biomedicine literature database, and bibliographies of retrieved articles up to January 2009. We included 9 randomized controlled trials with 165,056 participants; methodological quality of included trials was generally high. Meta-analysis showed that no significant effects of supplementation with β-carotene (3 trials), vitamin C (2 trials), vitamin E (5 trials), and selenium (2 trials)versus placebo on prostate cancer incidence. The mortality of prostate cancer did not differ significantly by supplement of β-carotene (1 trial), vitamin

C (1 trial), vitamin E (2 trials), and selenium (1 trial). This study indicates that antioxidant vitamins and selenium supplement did not reduce the incidence and mortality of prostate cancer; these data provide no support for the use of these supplements for the prevention of prostate cancer [175].

Epidemiological studies demonstrated that human exposure to methylmercury (MeHg) may contribute to the development and progression of metabolic and cardiovascular disorders. However, the mechanisms involved and the role of selenium (Se) and vitamin E (VE) supplementation in modulating MeHg cardiovascular toxicities remain unclear. The effects of Se and VE supplementation on MeHg-mediated systemic oxidative stress, antioxidant defense, inflammation, and endothelial dysfunction are carried out in an animal model. Male Sprague–Dawley rats were fed a starch-based casein diet or the same diet supplemented with 1 or 3 mg Se/kg diet and with or without 250 or 750 mg VE/kg diet. After 28 days of dietary treatment, rats were gavaged with 0 or 3 mg MeHg/kg BW for 14 consecutive days. Results suggested that exposure to MeHg may increase the risk of cardiovascular disease by decreasing circulating paraoxonase-1 activities, increasing serum oxidized low density lipoprotein levels, and associated systemic inflammation and endothelial dysfunction as reflected by increased leukocyte counts and serum levels of intercellular adhesion molecule-1 and monocyte chemotactic protein-1. Se and VE supplementation may either alleviate or augment the effects of MeHg, depending on their doses and combinations [176].

The analysis of the hepatotoxic effect of malathion in adult male rats and evaluate the possible hepatoprotective effect of vitamin E and/or selenium. Oral administration of malathion for 45 days significantly induced marked hepatic injury as revealed by increased activity of the plasma enzymes (alanine aminotransferase (ALT), aspartate aminotransferase (AST), lactate dehydrogenase (LDH) and gamma-glutamyl transferase GGT). Oral administration of vitamin E and selenium in combination with malathion exhibited a significant protective effect by lowering the elevated plasma levels of the previous enzymes. Light microscopic investigation revealed that malathion exposure was associated with necrosis of hepatocytes, marked changes of liver tissues in the form of dilated veins, hemorrhagic spots and some degenerative signs of hepatocytes [177].

4. Conclusion

Research on Se during the last few years has produced a great deal of evidence demonstrating the important role that Se and its metabolites play in human diseases. In particular, our knowledge of the functional roles of the GPx and TrxR groups as essential antioxidant selenoenzymes in protecting cells from oxidative stress has greatly increased, as has the link between these enzymes and various diseases. However, there are still areas of research that require in-depth study, including the mechanistic modes of action of Se in cancer etiology, how Se delivers its anticancer activity at the molecular and genetic levels, and what biomarkers can be used to accurately measure the efficacy of Se for use in chemoprevention. It is not well understood the specific mechanism by which Se protects cells and tissue at the cellular level from damage due to oxidative stress; this is particularly relevant in heart diseases, which are still a major cause of death worldwide. Given the number of Se cancer preventive trials that are currently being undertaken in many countries, the significant outcomes of these trials will not only provide us with more information on optimal Se intake for the treatment and prevention of cancer, but they will also provide us with strategies in the management of other potential human diseases associated with low Se status. Until the specific biomarkers are identified that will directly link Se with disease prevention and treatment, its use as supplements in health therapy should be taken with caution.

The Selenium and Vitamin E Cancer prevention Trial (SELECT) failed to show an effect in human population. However, pre- and pro-SELECT studies are still supporting the potential usefulness of selenium and/or vitamin E for prevention of prostate cancer and possibly other conditions. Much remains to be understood about the absorption, metabolism and physiologic chemistry of these agents. Nonetheless, the existing evidence supporting selenium and vitamin E as potential prostate cancer chemopreventive agents is possibly enough to justify further efforts in this direction.

My goal in putting this review together was to provide a wide range of subjects dealing with selenium and vitamin E supplementation, that are used in chronic disease prevention, due to their antiradical activities indicating that the combine effects of Se and vitamin E could provide an important dietary source of antioxidants and/or potential agents for a variety of human diseases. It is my hope that readers will find this chapter to be useful in further studies dealing with this subject.

Author details

Manuel Soriano García*

Chemistry of Biomacromolecules Department, Chemistry Institute, National Autonomus University of México, University City, Mexico

References

[1] Berzelius, J. J. (1818). Afhandl. Fys. Kemi Minerag. 6, 42

[2] Shamberger, R. J. (1984). Selenium. *In Biochemistry of the Essential Ultratrace Elements*, Earl Frieden, Ed.; Plenum Press; New York, NY, 3, 201-237.

[3] Brasted, R. C. (1961). *In Comprehensive Inorganic Chemistry: Sulfur. Selenium, Tellurium, Polonium and Oxygen*, Robert C. Brasted, Ed.; D. Van Nostrand Co.; Princeton, NJ, 2.

[4] Merian, E. (1984). Introduction on environmental chemistry and global cycles of chromium, nickel, cobalt beryllium, arsenic, cadmium and selenium, and their derivatives. *Toxicological and Environmental Chemistry*, 8, 9-38.

[5] Smith, M., Franke, K. W., & Westfall, B. B. (1936). The selenium problem in relation to public health. A preliminary survey to determine the possibility of selenium intoxication in the rural population living in seleniferous soil. *US Public Health Reports*, 51, 1496-1505.

[6] Thomson, C. D. (2004). Assessment of requirements for selenium and adequacy of selenium status: a review. *European Journal of Clinical Nutrition*, 58, 391-402.

[7] Goldhaber, S. B. (2003). Trace element risk assessment vs toxicity. *Regulatory Toxicology and Pharmacology*, 38, 232-242.

[8] Castellano, S., Gladyshev, V. N., Guigo, R., & Berry, M. J. (2008). SelenoDB 1.0: a database of selenoprotein genes, proteins and SECIS elements. *Nucleic Acids Research*, 36, 332D-338D.

[9] Allmang, C., Wurth, L., & Krol, A. (2009). The selenium to selenoprotein pathway in eukaryotes: more molecular partners than anticipated. *Biochemica et Biophysica Acta*, 1790, 1415-1423.

[10] Arbogast, S., & Ferreiro, A. (2009). Selenoproteins and protection against oxidative stress selenoprotein N as a novel player at the crossroads of redox signaling and calcium homeostasis. *Antioxidants & Redox Signaling*, 12, 893-904.

[11] Kipp, A., Banning, A., Van Schothorst, E. M., Meplan, C., Schomburg, L., Evelo, C., Coort, S., Gaj, S., Keijer, J., Hesketh, J., & Brigelius-Flohe, R. (2009). Four selenoproteins, protein biosynthesis, and Wnt signaling are particularly sensitive to limited selenium intake in mouse colon. *Molecular Nutrition & Food Research*, 53, 1561-1572.

[12] Lescure, A., Rederstorff, M., Krol, A., Guicheney, P., & Allamand, V. (2009). Selenoprotein function and muscle disease. *Biochimica et Biophysica Acta*, 1790, 1569-1574.

[13] Lobinski, R., Edmonds, J. S., Sukuki, K. T., & Uden, P. C. (2000). Species-selective determination of selenium compounds in biological materials. *Pure and Applied Chemistry*, 72, 447-461.

[14] Bleys, J., Navas-Acien, A., Stranges, S., Menke, A., Miller, E. R., III, & Guallar, E. (2008). Serum selenium and serum lipids in US adults. *The American Journal of Clinical Nutrition*, 88, 416-423.

[15] Johnson, C. C., Fordyce, F. M., & Rayman, M. P. (2010). Symposium on "geographical and geological influences on nutrition": factors controlling the distribution of selenium in the environment and their impact on health and nutrition. *Proceedings of the Nutrition Society*, 69, 119-132.

[16] Navarro-Alarcon, M., & Cabrera-Vique, C. (2008). Selenium in food and the human body: a review. *Science of the Total Environment*, 400, 115-141.

[17] Schrauzer, G. N., & Surai, P. F. (2009). Selenium in human and animal nutrition: resolved and unresolved issues. A partial historical treatise in commemoration of the fiftieth anniversary of the discovery of the biologically essentiality of selenium, dedi-

cated to the memory of Klaus Schwarz (1914-1978) on the occasion of the thirtieth anniversary of his dead. *Critical Reviews in Biotechnology*, 29, 2-9.

[18] World Health Organization. (1987). Selenium. *Geneva; WHO.*

[19] Thompson, H. J. (2001). *In Selenium: Its Molecular Biology and Role in Human Health*, Dolph L. Hatfield, Ed.; Kluwer Academic. Publishers; Boston, MA, 283-297.

[20] Hatfield, D. L. (2001). *In Selenium: Its Molecular Biology and Role in Human Health*, Dolph L. Hatfield, Ed.; Kluwer Academic. Publishers; Boston, MA.

[21] Bols, M., Lopez, O., & Ortega-Caballero, F. (2007). In Kamerling, J. P., Ed., *Comprehensive Glycoscience: From Chemistry to Systems Biology*, Elsevier Science: Oxford, 1, 815-884.

[22] Soriano-Garcia, M. (2004). Organoselenium Compounds as Potential Therapeutic and Chemopreventive Agents: A review. *Current. Medicinal Chem*, 11, 1657-1169.

[23] Mugesh, G., du Mont, W.-W., & Sies, H. (2001). Chemistry of Biologically Important Synthetic Organoselenium Compounds. *Chemical Reviews*, 101, 2125-2180.

[24] Rayman, M. P. (2005). Selenium in cancer prevention: a review of the evidence and mechanism of action. *Proceedings of the Nutrition Society*, 64, 527-542.

[25] Papp, L. V., Holmgren, A., & Khanna, K. K. (2010). Selenium and Selenoproteins in Health and Disease. *Antioxidants & Redox Signaling*, 12, 793-795.

[26] Jablonska, E., Gromadzinska, J., Sobala, W., Reszka, E., & Wasowicz, W. (2008). Lung cancer risk associated with selenium status is modified in smoking individuals by Sep15 polymorphism. *European Journal of Nutrition*, 47, 47-54.

[27] Kryukov, G. V., Castellano, S., Novoselov, S. V., Lobanov, A. V., Zehtab, O., Guigo, R., et al. (2003). Characterization of mammalian selenoproteomes. *Science*, 300, 1439-1443.

[28] Ashton, K., Hooper, L., Harvey, L. J., Hurst, R., Casgrain, A., & Fairweather-Tait, S. J. (2009). Methods of assessment of selenium status in humans: a systematic review. *The American Journal of Clinical Nutrition*, 89, 2025S-2039S.

[29] Ip, C., Dong, Y., & Ganther, H. E. (2002). New concepts in selenium chemoprevention. *Cancer and Metastasis Reviews*, 21, 281-289.

[30] Schrauzer, G. N. (2009). Selenium and selenium-antagonistic elements in nutritional cancer prevention. *Critical Reviews in Biotechnology*, 29, 10-17.

[31] Evans, P., & Halliwell, B. (2001). Micronutrients: oxidant/antioxidant status. *British Journal of Nutrition*, 85, 67S-74S.

[32] Wolf, G. (2005). The discovery of the antioxidant function of vitamin E: the contribution of Henry A. Matill. *Journal of Nutrition*, 135, 363-366.

[33] Sies, H., & Murphy, M. E. (1991). Role of tocopherols in the protection of biological systems against oxidative damage. *Journal of Photochemistry and Photobiology*, 8, 211-218.

[34] Brigelius-Flohe´, R., & Traber, M. G. (1999). Vitamin E: function and metabolism. *The FASEB Journal*, 13, 1145-1155.

[35] Yap, S. P., Yuen, K. H., & Wong, J. W. (2001). Pharmacokinetics and bioavailability of alpha-, gamma-, and delta-tocotrienols under different food status. *Journal of Pharmacy and Pharmacology*, 53, 67-71.

[36] Yash, B. J., & Pratico, D. (2012). Vitamin E in aging, dementia and Alzheimer's disease. *Biofactors*, 38, 90-97.

[37] Hacquebard, M., & Carpentier, Y. A. (2005). Vitamin E: absorption, plasma transport and cell uptake. *Current Opinion in Clinical Nutrition & Metabolic Care*, 8, 133-138.

[38] O'Byrne, D., Grundy, S., Packer, L., Devaraj, S., Baldenius, K., Hoppe, P. P., Kraemer, K., Jialal, I., & Traber, M. G. (2000). Studies of LDL oxidation following alpha-, gamma-, or delta-tocotrienyl acetate supplementation of hypercholesterolemic humans. *Free Radical Biology & Medicine*, 29, 834-845.

[39] Horiguchi, M., Arita, M., Kaempf-Rotzoll, D. E., Tsujimoto, M., Inoue, K., & Arai, H. p. (2003). pH-dependent translocation of alpha-tocopherol transfer protein (alpha-TTP) between hepatic cytosol and late endosomes. *Genes Cells*, 8, 789-800.

[40] Uchida, T., Abe, C., Nomura, S., Ichikawa, T., & Ikeda, S. (2012). Tissue distribution of alfa- and gamma- tocotrienol and gama-tocopherol in rats and interference with their accumulation by alpha-tocopherol. *Lipids*, 47, 129-139.

[41] Kokarnig, S., Kuehnelt, D., Stiboller, M., Hartleb, U., & Francesconi, K. A. (2011). Quantitative determination of small selenium species in human serum by HPLC/ICPMS following a protein-removal, pre-concentration procedure. *Analytical & Bioanalytical Chemistry*, 400, 2323-2327.

[42] Suzuki, K. T., Kurasaki, K., Okazaki, N., & Ogra, Y. (2005). Selenosugar, trimethylselenonium among urinary Se metabolites: dose- and agerelated changes. *Toxicology and Applied Pharmacology*, 206, 1-8.

[43] Klotz, L. O., Kroncke, K. D., Buchczyk, D. P., & Sies, H. (2003). Role of copper, zinc, selenium, tellurium in the cellular defense against oxidative and nitrosative stress. *Journal of Nutrition*, 133, 1448S-1451S.

[44] Valko, M., Rhodes, C. J., Moncol, J., Izakovic, M., & Mazur, M. (2006). Free radicals, metals, antioxidants in oxidative stress-induced cancer, *Chemico-Biological Interactions*, 160, 1-40.

[45] Susan, J. F. T., Yongping, B., Martin, R. B., Rachel, C., Dianne, F., John, E. H., et al. (2011). Selenium in human health and disease. *Antioxidants & Redox Signaling*, 14, 1337-1383.

[46] Pasco, J. A., Jacka, F. N., Williams, L. J., Evans-Cleverdon, M., Sharon, L., Brennana, S. L., Kotowicza, M. A., Nicholsone, G. C., Ball, M. J., & Berk, M. (2012). Dietary selenium and major depression: a nested case-control study. *Complementary Therapies in Medicine*, 20, 119-123.

[47] Maes, M., Galecki, P., Chang, Y. S., & Berk, M. (2011). A review on the oxidative and nitrosative stress (O&NS) pathways in major depression and their possible contribution to the (neuro) degenerative processes in that illness. *Progress in Neuropsychopharmacology & Biological Psychiatry*, 35, 676-692.

[48] Ng, F., Berk, M., Dean, O., & Bush, A. I. (2008). Oxidative stress in psychiatric disorders: evidence base and therapeutic implications. *The International Journal of Neuropsycho- pharmacology*, 11, 851-876.

[49] Berk, M., Ng, F., Dean, O., Dodd, S., & Bush, A. I. (2008). Glutathione: a novel treatment target in psychiatry. *Trends in Pharmacological Sciences*, 29, 346-351.

[50] Herken, H., Gurel, A., Selek, S., Armutcu, F., Ozen, M. E., Bulut, M., et al. (2007). Adenosine deaminase, nitric oxide, superoxide dismutase, and xanthine oxidase in patients with major depression: impact of antidepressant treatment. *Archives of Medical Research*, 38, 247-252.

[51] Sarandol, A., Sarandol, E., Eker, S. S., Erdinc, S., Vatansever, E., & Kirli, S. (2007). Major depressive disorder is accompanied with oxidative stress: short-term antidepressant treatment does not alter oxidativeantioxidative systems. *Human Psychopharmacology*, 22, 67-73.

[52] Whanger, P. D. (2001). Selenium and the brain: a review. *Nutritional Neuroscience*, 4, 81-97.

[53] Machado, M. S., Rosa, R. M., Dantas, A. S., Reolon, G. K., Appelt, H. R., Braga, A. L., et al. (2006). An organic selenium compound attenuates apomorphine-induced stereotypy in mice. *Neuroscience Letters*, 410, 198-202.

[54] Malhi, G. S., & Berk, M. (2007). Does dopamine dysfunction drive depression? *Acta Psychiatrica Scandinavica Supplement*, 433, 116-124.

[55] Ishrat, T., Parveen, K., Khan, M. M., Khuwaja, G., Khan, M. B., Yousuf, S., et al. (2009). Selenium prevents cognitive decline and oxidative damage in rat model of streptozotocin-induced experimental dementia of Alzheimer's type. *Brain Research*, 1281, 117-127.

[56] Cardoso, B. R., Ong, T. P., Jacob-Filho, W., Jaluul, O., Freitas, M. I., & Cozzolino, S. M. (2010). Nutritional status of selenium in Alzheimer's disease patients. *British Journal of Nutrition*, 103, 803-806.

[57] Finley, J. W., & Penland, J. G. (1998). Adequacy or deprivation of dietary selenium in healthy men: clinical and psychological findings. *Journal of Trace Elements in Experimental Medicine*, 11, 1-27.

[58] Mokhber, N., Namjoo, M., Tara, F., Boskabadi, H., Rayman, M. P., Ghayour-Mobarhan, M., et al. (2011). Effect of supplementation with selenium on postpartum depression: a randomized doubleblind placebo-controlled trial. *Journal of Maternal-Fetal and Neonatal Medicine*, 24, 104-108.

[59] ISAAC Steering Committee. (1998). Worldwide variation in prevalence of symptoms of asthma, allergic rhinoconjunctivitis, and atopic eczema: ISAAC. *Lancet*, 351, 1225-1232.

[60] Allan, K., & Devereux, G. (2011). Diet and asthma: nutrition implication from prevention to treatment. *Journal of the American Dietetic Association*, 111, 258-268.

[61] Comhair, S., Bhathena, P., Farver, C., Thunnissen, F., & Srzurum, S. (2001). Extracellular glutathione peroxidase induction in asthmatic lungs: evidence for redox regulation of expression in human airway epithelial cells. *FASEB Journal*, 15, 70-78.

[62] Horváthovà, M., Jahnová, E., & Gazdik, F. (1999). Effect of selenium supplementation in asthmatic subjects on the expression of endothelial cell adhesion molecules in culture. *Biological Trace Element Research*, 69, 15-26.

[63] Hasselmark, L., Malmgren, R., Zetterstrom, O., & Unge, G. (1993). Selenium supplementation in intrinsic asthma. *Allergy*, 48, 30-36.

[64] Picado, C., Deulofeu, R., Lleonart, R., et al. (2001). Dietary micronutrients/antioxidants and their relationship with bronchial asthma severity. *Allergy*, 56, 43-49.

[65] Shaheen, S., Sterne, J., Thompson, R., Songhurst, C., Margetts, B., & Burney, P. (2001). Dietary antioxidants and asthma in adults. Population-based case-control study. *American Journal of Respiratory and Critical Care Medicine*, 164, 1823-1828.

[66] Burney, P., Potts, J., Makowska, J., et al. (2008). A case-control study of the relation between plasma selenium and asthma in European populations: a GAL2EN project. [see comment][erratum appears in Allergy. 2008; 63 1647.], *Allergy*, 63, 865-871.

[67] Thomson, C. D., Wickens, K., Miller, J., Ingham, T., Lampshire, P., Epton, Town. G. I., Pattemore, P., & Crane, J. (2012). Selenium status and allergic disease in a cohort of New Zealand children. *Clinical & Experimental Allergy*, 42, 560-567.

[68] Nogueira, C. W., Zeni, G., & Rocha, J. B. (2004). Organoselenium and organotellurium compounds: toxicology and pharmacology. *Chemical Reviews*, 104, 6255-6285.

[69] Silva Avila, D., Benedetto, A., Au, C., Manarin, F., Erikson, K., Antunes Soares, F., Teixeira Rocha, J. B., & Aschner, M. (2012). Organotellurium and organoselenium compounds attenuate Mn-induced toxicity in Caenorhabditis elegans by preventing oxidative stress. *Free Radical Biology & Medicine*, 52, 1903-1910.

[70] Fetoui, H., Mahjoubi-Samet, A., Jamoussi, K., Ellouze, F., Guermazi, F., & Zeghal, N. (2006). Energy restriction in pregnant and lactating rats lowers bone mass of their progeny. *Nutrition Research*, 26, 421-426.

[71] Moreno-Reyes, R., Egrise, D., Boelaert, M., Goldman, S., & Meuris, S. (2006). Iodine deficiency mitigates growth retardation and osteopenia in selenium-deficient rats. *Journal of Nutrition*, 136, 595-600.

[72] Ren, F. L., Guo, X., Zhang, R. L., Wang, Sh. J., Zuo, H., Zhang, Z. T., et al. (2007). Effects of selenium and iodine deficiency on bone, cartilage growth plate and chondrocyte differentiation in two generations of rats. *Osteoarthritis and Cartilage*, 15, 1171-1177.

[73] Pahuja, D. N., & De Luca, H. F. (1982). Thyroid hormone and vitamin D metabolism in the rat. *Archives of Biochemistry and Biophysics*, 213, 293-298.

[74] Amaraa, I. B., Troudia, A., Soudania, N., Guermazib, F., & Zeghala, N. (2012). Toxicity of methimazole on femoral bone in suckling rats: Alleviation by selenium. *Experimental and Toxicologic Pathology*, 64, 187-195.

[75] UNAIDS. (2008). Report on the global AIDS epidemic. *Geneva: UNAIDS*.

[76] Fawzi, W. W., Msamanga, G. I., Spiegelman, D., et al. (2004). A randomized trial of multivitamin supplements and HIV disease progression and mortality. *The New England Journal of Medicine*, 351, 23-32.

[77] Diamond, A. M., Hu, J. Y., & Mansur, D. B. (2001). Glutathione peroxidase and viral replication: Implications for viral evolution and chemoprevention. *Biofactors*, 14, 205-210.

[78] Pace, G. W., & Leaf, C. D. (1995). The role of oxidative stress in HIV disease. *Free Radical Biology & Medicine*, 19, 523-528.

[79] Stone, C. A., Kawai, K., Kupka, R., & Fawzi, W. W. (2010). Role of selenium in HIV infection. . Nutrition Reviews , 68, 671-681.

[80] St Germain, D. L., Galton, V. A., & Hernandez, A. (2009). Minireview: Defining the roles of the iodothyronine deiodinases: current concepts and challenges. *Endocrinology*, 150, 1097-1107.

[81] Zhang, S., Rocourt, C., & Cheng, W-H. (2010). Selenoproteins and the aging brain. *Mechanisms of Ageing and Development*, 13, 253-260.

[82] Schweizer, U., Brauer, A. U., Kohrle, J., Nitsch, R., & Savaskan, N. E. (2004). Selenium and brain function: a poorly recognized liaison. *Brain Research Reviews*, 45, 164-178.

[83] Hock, A., Demmel, U., Schicha, H., Kasperek, K., & Feinendegen, L. E. (1975). Trace element concentration in human brain. Activation analysis of cobalt, iron, rubidium, selenium, zinc, chromium, silver, cesium, antimony and scandium. *Brain*, 98, 49-64.

[84] Prohaska, J. R., & Ganther, H. E. (1976). Selenium and glutathione peroxidase in developing rat brain. *Journal of Neurochemistry*, 27, 1379-1387.

[85] Trapp, G. A., & Millam, J. (1975). The distribution of ^{75}Se in brains of selenium-deficient rats. *Journal of Neurochemistry*, 24, 593-595.

[86] Nakayama, A., Hill, K. E., Austin, L. M., Motley, A. K., & Burk, R. F. (2007). All regions of mouse brain are dependent on selenoprotein P for maintenance of selenium. *Journal of Nutrition*, 137, 690-693.

[87] Kyriakopoulos, A., Rothlein, D., Pfeifer, H., Bertelsmann, H., Kappler, S., & Behne, D. (2000). Detection of small selenium-containing proteins in tissues of the rat. *Journal of Trace Elements in Medicine and Biology*, 14, 179-183.

[88] Hawkes, W. C., & Hornbostel, L. (1996). Effects of dietary selenium on mood in healthy men living in a metabolic research unit. *Biological Psychiatry*, 39, 121-128.

[89] Ashrafi, M. R., Shabanian, R., Abbaskhanian, A., Nasirian, A., Ghofrani, M., Mohammadi, M., Zamani, G. R., Kayhanidoost, Z., Ebrahimi, S., & Pourpak, Z. (2007). Selenium and Intractable Epilepsy: Is There Any Correlation? *Pediatric Neurology*, 36, 25-29.

[90] Rayman, M. P. (2000). The importance of selenium to human health. *Lancet*, 356, 233-241.

[91] Chen, J., & Berry, M. J. (2003). Selenium and selenoproteins in the brain and brain diseases. *Journal of Neurochemistry*, 86, 1-12.

[92] Berry, M. J., Banu, L., & Larsen, P. R. (1991). Type I iodothyronine deiodinase is a selenocysteine-containing enzyme. *Nature*, 349, 438-440.

[93] Hill, K. E., McCollum, G. W., Boeglin, M. E., & Burk, R. F. (1997). Thioredoxin reductase activity is decreased in selenium deficiency. *Biochemical and Biophysical Research Communications*, 234, 293-295.

[94] Duntas, L. H. (2010). Selenium and the Thyroid: A Close-Knit Connection. *The Journal of Clinical Endocrinology & Metabolism*, 95, 5180-5188.

[95] Hatfield, D. L., Yoo, M.-H., Carlson, B. A., & Gladyshev, V. N. (2009). Selenoproteins that function in cancer prevention and promotion. *Biochimica et Biophysica Acta*, 1790, 1541-1545.

[96] Steinbrenner, H., & Sies, H. (2009). Protection against reactive oxygen species by selenoproteins. *Biochimica et Biophysica Acta*, 1790, 1478-1485.

[97] Ip, C. (1998). Lessons from basic research in selenium and cancer prevention. *Journal of Nutrition*, 128, 1845-1849.

[98] Shamberger, R. J. (1970). Relationship of selenium to cancer: inhibitory effect of selenium on carcinogenesis. *Journal of the National Cancer Institute*, 44, 931-936.

[99] Clark, L. C., Combs, G. F., & Turnbull, B. W. (1996). Effect of selenium supplementation for cancer prevention in patients with carcinoma of the skin: a randomized clinical trial. *Journal of the American Medical Association*, 279, 1975-1982.

[100] Bjelakovic, G., Nikolova, D., Simonetti, R. G., & Gluud, C. (2008). Systematic review: primary and secondary prevention of gastrointestinal cancers with antioxidant supplements. *Alimentary Pharmacology & Therapeutics*, 28, 689-703.

[101] Brinkman, M., Buntinx, F., Muls, E., & Zeegers, M. P. (2006). Use of selenium in chemo- prevention of bladder cancer. *The Lancet Oncology*, 7, 766-774.

[102] Lu, J., & Jiang, C. (2005). Selenium and cancer chemoprevention: hypotheses integrating the actions of selenoproteins and selenium metabolites in epithelial and non-epithelial target cells. *Antioxidants & Redox Signaling*, 7, 1715-1727.

[103] Waters, D. J., Shen, S., Glickman, L. T., Cooley, D. M., Bostwick, D. G., Qian, J., et al. (2005). Prostate cancer risk and DNA damage: translational significance of selenium supple- mentation in a canine model. *Carcinogenesis*, 26, 1256-1562.

[104] Battin, E. E., Perron, N. R., & Brumaghim, J. L. (2006). The central role of metal coordination in selenium antioxidant acitivity. *Inorganic Chemistry*, 45, 499-501.

[105] Rayman, M. P. (2009). Selenoproteins and human health: Insights from epidemiological data. *Biochimica et Biophysica Acta*, 1790, 1533-1540.

[106] Bleys, J., Navas-Acien, A., & Guallar, E. (2007). Serum selenium and diabetes in US adults. *Diabetes Care*, 30, 829-834.

[107] Akbaraly, T. N., Arnaud, J., Rayman, M. P., Hininger-Favier, I., Roussel, A. M., Berr, C., & Fontbonne, A. (2010). Plasma selenium and risk of dysglycemia in an elderly French population: results from the prospective Epidemiology of Vascular Ageing Study. *Nutrition & Metabolism*, 7, 21-27.

[108] Bleys, J., Navas-Acien, A., & Guallar, E. (2007). Selenium and diabetes: more bad news for supplements. *Annals of Internal Medicine*, 147, 271-272.

[109] Faure, P., Ramon, O., Favier, A., & Halimi, S. (2004). Selenium supplementation decreases nuclear factor-kappa B activity in peripheral blood mononuclear cells from type 2 diabetic patients. *European Journal of Clinical Investigation*, 34, 475-481.

[110] Stranges, S., Marshall, T. R., & Natarajan, R. (2007). Effects of long-term selenium supplementation on the incidence of type 2 diabetes: a randomized trial. *Annals of Internal Medicine*, 147, 217-223.

[111] Agarwal, A., & Prabhakaran, S. A. (2005). Mechanism, measurement and prevention of oxidative stress in male reproductive physiology. *Indian Journal of Experimental Biology*, 43, 963-974.

[112] Hansen, J. C., & Degachi, Y. (1996). Selenium and fertility in animals and men: a review, *Acta Veterinaria Scandinavica*, 37, 19-25.

[113] Wantanobe, T., & Endo, A. (1991). Effects of selenium deficiency on spermmorphology and spermatocyte chromosomes in mice. *Mutation Research*, 262, 93-96.

[114] Olson, G. E., Winfrey, V. P., Hill, K. E., & Burk, R. F. (2004). Sequential development of flagellar defects in spermatids and epididymal spermatozoa of selenium deficient rats. *Reproduction*, 127, 335-341.

[115] Lenzi, A., Gandini, L., Lombardo, F., Picardo, M., Maresca, V., Panfili, E., et al. (2002). Polyunsaturated fatty acids of germ cell membranes, glutathione and blutathione dependent enzyme-PHGPx: from basic to clinic. *Contraception*, 65, 301-305.

[116] Hawkes, W. C., Alkan, Z., & Wong, K. (2009). Selenium supplementation does not affect testicular selenium status and semen quality in North American men. *Journal of Andrology*, 30, 525-533.

[117] Seaton, A., Godden, D. J., & Brown, K. (1994). Increase in asthma: a more toxic environment or a more susceptible population. *Thorax*, 49, 171-174.

[118] University of Maryland. Asthma [Online]. (2011). cited on August 10, 2011]. Available from URL: http://www.umm.edu/altmed/articles/asthma-000015.htm Citation Date= July 4, 2012].

[119] Kloner, A. R., Przyklenk, K., & Whittaker, P. (1989). Deleterious effects of oxygen radicals in ischaemia-reperfusion: resolved and unresolved issue. *Circulation*, 80, 1115-1127.

[120] Prithviraj, T., & Misra, K. P. (2000). Reversal of atherosclerosis-fact or fiction? *Cardiology Today*, 4, 97-100.

[121] Sattler, W., Maiorino, M., & Stocker, R. (1994). Reduction of HDL and LDL associated cholestrylester and phospholipids hydroperoxides by phospholipids hydroperoxide-glutathione peroxidase and ebselen (Pz 51). *Archives of Biochemistry and Biophysics*, 309, 224.

[122] Ricetti, M. M., Guidi, G. C., Tecchio, C., et al. (1999). Effects of sodium selenite on in vitro interactions between platelets and endotelial cells. *International Journal of Clinical and Laboratory Research*, 29, 80-82.

[123] Kharb, S. (2003). Low blood glutathione levels in acute myocardial infarction. *Indian Journal of Medical Sciences*, 57, 335-337.

[124] Riaz, M., & Mehmood, K. T. (2012). Selenium in human health and disease: a review. *Journal of Postgraduate Medical Institute*, 26, 120-133.

[125] Suleyman, Ö., Mustafa, N., Mesut, Ç., Vedat, B., & Flores-Arce, M. F. (2011). Effects of Different Medical Treatments on Serum Copper, Selenium and Zinc Levels in Patients with Rheumatoid Arthritis. *Biological Trace Element Research*, 142, 447-455.

[126] Peretz, A., Siderova, V., & Neve, J. (2001). Selenium supplementation in rheumatoid arthritis investigated in a double blind, placebo- controlled trial. *Scandinavian Journal of Rheumatology*, 30, 208-212.

[127] Cunningham, F. G., & Lindheimer, M. D. (1992). Hypertension in Pregnancy: Current concepts. *The New England Journal of Medicine*, 326, 927-932.

[128] Sharma, J. B. (2001). Benefits of Selenium during pregnancy. *Obstetrics and Gynaecology*, 6, 459-462.

[129] Barrington, J. W., Lindsay, P., Names, D., Smith, S., & Robert, A. (1996). Selenium deficiency and miscarriage: a possible link? *British Journal of Obstetrics and Gynecology*, 2, 130-132.

[130] Tinggi, U. (2008). Selenium: its role as antioxidant in human health. *Environmental Health and Preventive Medicine*, 13, 102-108.

[131] Beck, M. A. (2001). Antioxidants and viral infections: host immune response and viral pathogenicity. *Journal of the American College of Nutrition*, 20, 384S-388S.

[132] Beck, M. (2006). Selenium, viral infections. Hatfield DL, Berry MJ, Gladyshev VN, editors, *Selenium: its molecular biology and role in human health*, New York: Springer, 287-298.

[133] Broome, C. S., McArdle, F., Kyle, J. A., Andrews, F., Lowe, N. M., Hart, C. A., et al. (2004). An increase in selenium intake improves immune function and poliovirus handling in adults with marginal selenium status. *The American Journal of Clinical Nutrition*, 80, 154-162.

[134] Baum, M. K., & Campa, A. (2006). Role of selenium in HIV/AIDS. In: Hatfield DL, Berry MJ, Gladyshev VN, editors. *Selenium- its molecular biology and role in human health*, New York: Springer, 299-310.

[135] McKenzie, R. C., Beckett, G. J., & Arthur, J. R. (2006). Effects of selenium on immunity and aging. In: Hatfield DL, Berry MJ, Gladyshev VN editors. *Selenium-its molecular biology and role in human health*, New York: Springer, 287-298.

[136] Sheweita, S. A., & Koshhal, K. (2007). Calcium metabolism and oxidative stress in bone fractures: role of antioxidants. *Current Drug Metabolism*, 8, 519-525.

[137] Finkel, T., & Holbrook, N. J. (2000). Oxidants, oxidative stress and biology of ageing. *Nature*, 408, 147-239.

[138] Cetinus, E., Kilinc¸, M., Uzel, M., Inanc¸, F., Kurutas, E. B., Bilgic, E., et al. (2005). Does long-term ischemia affect the oxidant status during fracture healing? *Archives of Orthopedic and Trauma Surgery*, 125, 376-380.

[139] Sandukji, A., Al-Sawaf, H., Mohamadin, A., Alrashidi, Y., & Sheweita, S. A. (2010). Oxidative stress and bone markers in plasma of patients with long-bone fixative surgery: Role of antioxidants. *Human and Experimental Toxicology*, 30, 435-442.

[140] Azzi, A., Aratri, E., Boscoboinik, D., Clement, S., Ozer, N. K., Ricciarelli, R., & Spycher, S. (1998). Molecular basis of α-tocopherol control of smooth muscle cell proliferation. *Biofactors*, 7, 3-14.

[141] Brigelius-Flohe′, R., & Galli, F. (2010). Vitamin E: A vitamin still awaiting the detection of its biological function. *Mol. Nutr. Food Res.*, 54, 583-587.

[142] Cicek, N., Eryilmaz, O. G., Sarikaya, E., Gulerman, C., & Genc, Y. (2012). Vitamin E effect on controlled ovarian stimulation of unexplained infertile women. *Journal of Assisted Reproduction and Genetics*, 29, 325-328.

[143] Khanna, S., Parinandi, N. L., Kotha, S. R., Roy, S., Rink, C., Bibus, D., & Sen, C. K. (2010). Nanomolar vitamin E alpha-tocotrienol inhibits glutamate-induced activation of phospholipase A2 and causes neuroprotection. *Journal of Neurochemistry*, 112, 1249-1260.

[144] Reiter, E., Jiang, Q., & Christen, S. (2007). Anti-inflammatory properties of alpha- and gamma-tocopherol. *Molecular Aspects of Medicine*, 28, 668-691.

[145] Naito, Y., Shimozawa, M., Kuroda, M., Nakabe, N., Manabe, H., Katada, K., Kokura, S., Ichikawa, H., Yoshida, N., Noguchi, N., & Yoshikawa, T. (2005). Tocotrienols reduce 25-hydroxycholesterol-induced monocyteendothelial cell interaction by inhibiting the surface expression of adhesion molecules. *Atherosclerosis*, 180, 19-25.

[146] Burlakova, E. B., Krashakov, S. A., & Khrapova, N. G. (1998). The role of tocopherols in biomembrane lipid peroxidation. *Membrane and Cell Biology*, 12, 173-211.

[147] Coquette, A., Vray, B., & Vanderpas, J. (1986). Role of vitamin E in the protection of the resident macrophage membrane against oxidative damage. Archives Internationales de Physiologie. de Biochimie et de Biophysique, 94 ., 29S-34S.

[148] Pratico, D., Tangirala, R. K., Rader, D. J., Rokach, J., & FitzGerald, G. A. (1998). Vitamin E suppresses isoprostane generation in vivo and reduces atherosclerosis in ApoE-deficient mice. *Nature Medicine*, 4, 1189-1192.

[149] Moriguchi, S., & Muraga, M. (2000). Vitamin E and immunity. *Vitamins & Hormones*, 59, 305-336.

[150] Wu, D., et al. (2000). In vitro supplementation with different tocopherol homologues can affect the function of immune cells in old mice. *Free Radical Biology & Medicine*, 28, 643-651.

[151] Palamanda, J. R., & Kehrer, J. P. (1993). Involvement of vitamin E and protein thiols in the inhibition of microsomal lipid peroxidation by glutathione. *Lipids*, 28, 427-431.

[152] Agarwal, A., & Sekhon, L. H. (2011). Oxidative stress and antioxidants for idiopathic oligoasthenoteratospermia: ıs it justified? *Indian Journal of Urology*, 27(1), 74-85.

[153] Bouayed, J., & Bohn, T. (2010). Exogenous antioxidants-double-edged swords in cellular redox state. Health beneficial effects at physiologic doses versus deleterious effects at high doses. *Oxidative Medicine and Cellular Longevity*, 3, 228-237.

[154] Chappell, L. C., Seed, P. T., Briley, A. L., et al. (1999). Effect of antioxidants on the occurrence of pre-eclampsia in women at increased risk: a randomized trial. *Lancet*, 354, 810-816.

[155] Fairfield, K. M., & Fletcher, R. H. (2002). Vitamins for chronic disease prevention in adults: scientific review. *Journal of the American Medical Association*, 287, 3116-3126.

[156] Forstrom, J. W., Zakowski, J. J., & Tappel, A. L. (1978). Identification of the catalytic site of rat liver glutathione peroxidase as selenocysteine. *Biochemistry*, 17, 2639-2644.

[157] Tamura, T., & Stadtman, T. C. (1996). A new selenoprotein from human lung adenocarcinoma cells: purification, properties, and thioredoxin reductase activity. *Proceedings of the National Academy of Sciences*, 93, 1006-1011.

[158] Kiremidjian-Schumacher, L., Roy, M., Wishe, H. I., Cohen, M. W., & Stotzky, G. (1994). Supple- mentation with selenium and human immune cell functions. II. Effect on cytotoxic lymphocytes and natural killer cells. *Biological Trace Element Research*, 41, 115-127.

[159] Baker, S. S., & Cohen, H. J. (1983). Altered oxidative metabolism in selenium-deficient rat granulocytes. *The Journal of Immunology*, 130, 2856-2860.

[160] Sakaguchi, S., et al. (2000). Roles of selenium in endotoxin-induced lipid peroxidation in the rats liver and in nitric oxide production in J774A. 1 cells. Toxicology Letters, 118, 69-77.

[161] Kim, S. H., Johnson, V. J., Shin, T. Y., & Sharma, R. P. (2004). Selenium attenuates lipopoly- saccharide-induced oxidative stress responses through modulation of p38 MAPK and NF-kappaB signaling pathways. *Experimental Biology and Medicine*, 229, 203-213.

[162] Maehira, F., Miyagi, I., & Eguchi, Y. (2003). Selenium regulates transcription factor NF-kappaB activation during the acute phase reaction. *Clinica Chimica Acta*, 334, 163-171.

[163] Beck, M. A., & Matthews, C. C. (2000). Micronutrients and host resistance to viral infection. *Proceedings of the Nutrition Society*, 59, 581-585.

[164] Ip, C., & Medina, D. (1987). Current concepts of selenium and mammary tumorigenesis. Medina D, Kidwell W, Heppner GH, Anderson E, editors, *Cellular and molecular biology of mammary cancer*, New York: Plenum press, 479.

[165] Ip, C., & Daniel, F. B. (1985). Effects of selenium on 7, 12 di-methyl benzanthracene induced mammary carcinogenesis and DNA adduct formation. *Cancer Research*, 45, 61-68.

[166] Lotan, Y., Goodman, P. J., Youssef, R. F., Svatek, R. S., Shariat, S. F., Tangen, C. M., Thompson, I. M. Jr., & Klein, E. A. (2012). Evaluation of Vitamin E and Selenium Supplementation for the Prevention of Bladder Cancer in SWOG Coordinated SELECT. *The Journal of Urology*, 187, 2005-2010.

[167] Jemal, A., Siegel, R., Xu, J., et al. (2010). Cancer statistics. *CA A Cancer Journal of Clinicians*, 60, 277-300.

[168] Samanic, C., Kogevinas, M., Dosemeci, M., Malats, N., Real, F. X., Garcia-Closas, M., Serra, C., Carrato, A., et al. (2006). Smoking and bladder cancer in Spain: effects of tobacco type, timing, environmental tobacco smoke, and gender. *Cancer Epidemiology, Biomarkers and Prevention*, 15, 1348-1354.

[169] Botteman, M. F., Pashos, C. L., Redaelli, A., et al. (2003). The health economics of bladder cancer: a Comprehensive review of the published literature. *Pharmacoeconomics*, 21, 1315-1330.

[170] Amaral, A. F. S., Cantor, K. P., Silverman, D. T., & Malats, N. (2010). Selenium and bladder cancer risk: a meta-analysis. *Cancer Epidemiol Biomarkers Prev*, 19, 2407-2415.

[171] Bardia, A., Tleyjeh, I. M., Cerhan, J. R., et al. (2008). Efficacy of antioxidant supplementation in reducing primary cancer incidence and mortality: systematic review and meta-analysis. *Mayo Clinic Proceedings*, 83, 23-34.

[172] Lippman, S. M., et al. (2009). Effect of selenium and vitamin E on risk prostate cancer and other cancers: the Selenium and Vitamin E Cancer Prevention Trial (SELECT). *Journal of American Medical Association*, 301, 39-51.

[173] Ledesma, M. C., Jung-Hynes, B., Schmit, T. L., Kumar, R., Mukhtar, H., & Ahmad, N. (2011). Selenium and vitamin E for prostate cancer: Post-SELECT (Selenium and Vitamin E Cancer Prevention Trial) status. *Molecular Medicine*, 17, 134-143.

[174] Beck, M. A. (2007). Selenium and vitamin E Status: Impact on viral pathogenicity. *The Journal of Nutrition*, 137, 1338-1340.

[175] Jiang, L., Yang, K-h., Tian, J-h., et al. (2010). Efficacy of Antioxidant Vitamins and Selenium Supplement in Prostate Cancer Prevention: A Meta-Analysis of Randomized Controlled Trials. *Nutrition and Cancer*, 62, 719-727.

[176] Jin, X., Hidiroglou, N., Lok, E., Taylor, M., Kapal, K., et al. (2012). Dietary Selenium (Se) and Vitamin E (VE) Supplementation Modulated Methylmercury-Mediated Changes in Markers of Cardiovascular Diseases. *Rats Cardiovascular Toxicology*, 12, 10-24.

[177] El-Desoky, G., Abdelreheem, M., AL-Othman, A., Alothman, Z., Mahmoud, M., & Yusuf, K. (2012). Potential hepatoprotective effects of vitamin E and selenium on hepatotoxicity induced by malathion in rats. *African Journal of Pharmacy and Pharmacology*, 6, 806-813.

Chapter 18

Antioxidant Role of Ascorbic Acid and His Protective Effects on Chronic Diseases

José Luis Silencio Barrita and
María del Socorro Santiago Sánchez

Additional information is available at the end of the chapter

http://dx.doi.org/10.5772/52181

1. Introduction

Ascorbic acid (AA), commonly known as vitamin C, plays an important role in the human body, although its function at the cellular level is not yet clear. It is necessary for the synthesis of collagen, a protein that has many connective functions in the body. Among the substances and structures that contain collagen are bone, cartilage and the surrounding material, as well as carrier substances and materials of union muscle, skin and other tissues. It also requires (AA) for the synthesis of hormones, neurotransmitters and in the metabolism of certain amino acids and vitamins. Participate in the liver for detoxification of toxic substances and blood level for immunity. As an antioxidant reacts with histamine and peroxide for reducing inflammatory symptoms.

Its antioxidant capacity is associated with reduced incidence of cancer. The requirement for vitamin C for adults is well defined but they have not been uniform across different cultures, so their need has been defined as culture-specific. They have also defined other roles in cellular processes and reactions. Some epidemiological data mentioned its usefulness in reducing cold with increasing consumption of foods rich in vitamin, so people sometimes ingest an overdose of it. In most reports mention that discrete increases in blood levels of this vitamin reduces the risk of death in all conditions. Although there are many functions of vitamin C, his role in health is discussed mostly in relation to its role as an antioxidant and its effects on cancer, blood pressure, immunity, drug metabolism and urinary excretion of hydroxyproline.

Antioxidants play important roles in cellular function and have been implicated in processes associated with aging, including vascular, inflammatory damage and cancer. In the case of

AA its antioxidant role is useful since it contributes to the maintenance of the vascular system and the reduction of atherogenesis through regulation in collagen synthesis, production of prostacyclin and nitric oxide. In addition to this antioxidant role, the AA has actions at the molecular level because it acts as a cofactor of enzymes such as dopamine hydroxylase (EC 1.14.17.1), influencing neurotransmitter concentration, improves lysosomal protein degradation and mediates consumer monosodium glutamate.

1.1. History

Since the nineteenth and early twentieth century research on these compounds led to the discovery of vitamins.[4] Since 1901, a publication of Wildiers first described the stimulating effect of small amounts of organic material in the growth of yeast; this effect was the subject of many publications and only after several years was universally accepted. Wilders gave the name "bios" to the substance or substances causing increased growth of yeast. In the years since it was shown that bios were multiple in nature, and changes fractionated as bios I, bios IIA, IIB and others.[5]

1747. - Lind cured scurvy in British sailors with oranges and lemons.

1907. - It is reproduced experimental scurvy in guinea pigs.

1928. - Eascott identified bios I as meso-inositol. Szent-Gyorgy and Glen published the isolation of vitamin C or hexuronic acid.

1933. - Allison, Hoover and Burk describe a compound that promotes the respiration and growth of Rhizobium, which designated "Coenzyme R". Then, it is defined molecular structure and synthesis of vitamin C.

1937. - the Nobel Prize in Chemistry was awarded to Walter Haworth for his work in determining the structure of ascorbic acid (shared with Paul Karrer for his work on vitamins) and the Nobel Prize for medicine was awarded to Albert von Szent-Györgyi Nagyrápolt for his studies on the biological functions of ascorbic acid.

2. Definition

Vitamin C is defined as hexuronic acid, cevitáminic acid or xiloascórbic acid. The term vitamin C is generally used to describe all these compounds, although the representative of which is ascorbic acid.

2.1. Structure, formula and chemical characteristics

Ascorbic acid is the enolic form of one α-ketolactone. Ascorbic acid solution is easily oxidized to the diketo form referred to as dehydroascorbic acid, which can easily be converted into oxalic acid , diketogulonic acid or threonic acid.

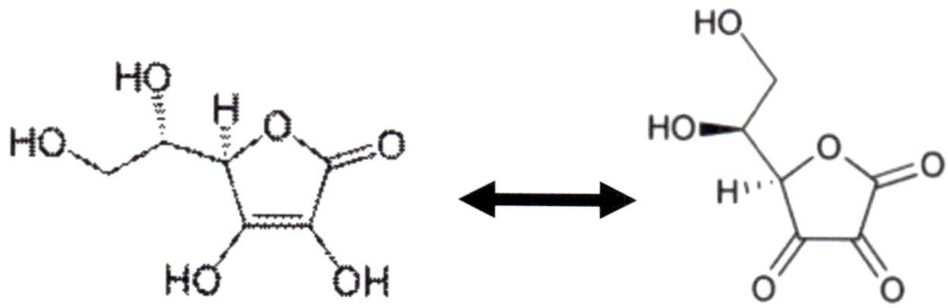

Figure 1. Structure of ascorbic acid and dehydroascorbic acid.

Name (IUPAC) systematic: (R) -3, 4-dihydroxy-5-((S) -1, 2-dihydroxyethyl) furan-2 (5H)-one; CAS Number 50-81-7: Formula: $C_6H_8O_6$; mol wt. 176.13 g / mol

2.2. Physical and chemical properties

Ascorbic acid contains several structural elements that contribute to their chemical behavior: the structure of the lactones and two enolic hydroxyl groups and a primary and secondary alcohol group. Enediol structure motivates their antioxidant properties, as can be oxidized easily enediols to diketones. Therefore, the carbonyl groups endioles neighbors are also called reductive.

Ascorbic acid forms two bonds intermolecular hydrogen bonds (shown in red in the figure) that contribute substantially to the stability and with it the chemical qualities of the structure endiol.

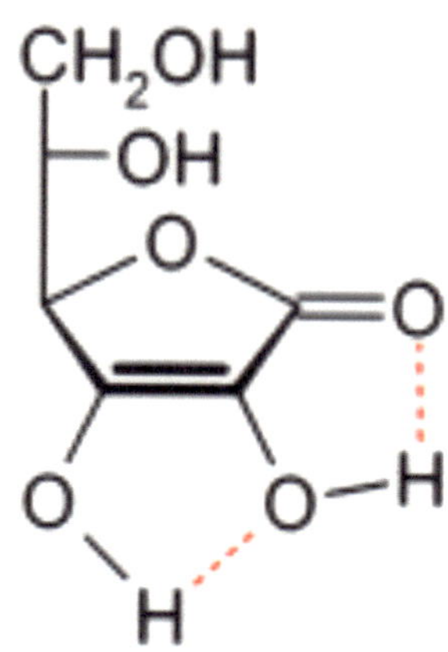

Figure 2. Hydrogen bridges formed by ascorbic acid

Ascorbic acid is rapidly interconvert in two unstable diketone tautomers by proton transfer, though it is most stable in the enol form. The proton of the enol is lost, and again acquired by the electrons from the double bond to produce a diketone. This reaction is an enol. There are two possible ways: 1, 2-diketone and 1, 3-diketone (figure 3).

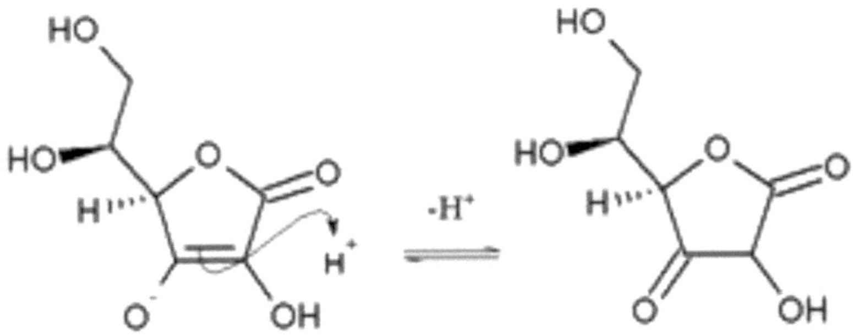

Figure 3. Nucleophilic attack on the proton ascorbic enol to give 1,3 diketone.

2.3. Vitamers or vitameric forms

The vitamer of a particular vitamin is any chemical compound which generally has the same molecular structure and each shows a different vitamin activity in a biological system which is deficient of the vitamin.

The vitamin activity of multiple vitamers is due to the ability (sometimes limited) of the body to convert one or many vitamers in another vitamer for the same enzymatic cofactor which is active in the body as the most important form of the vitamin. As part of the definition of the vitamin, the body can not completely synthesize an optimal amount of vitamin activity of foodstuffs simple, without a certain minimum amount of vitamer as base. Not all vitamers have the same vitamin power by mass or weight. This is due to differences in the absorption and the variable interconversion several vitamers in the vitamin.

For ascorbic acid per se, may be mentioned the following vitamers: dehydroascorbic acid, erythorbic acid (figure 4) and the following salts: sodium ascorbate, calcium ascorbate, and others. (Rogur, 2010)

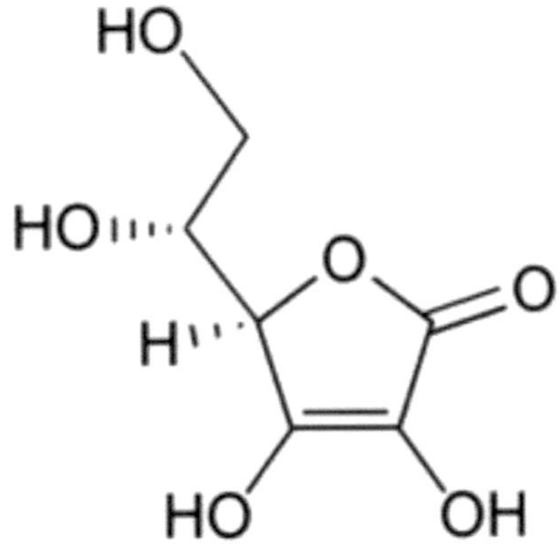

Figure 4. Structure of erythorbic acid, isoascorbic or Arabian-ascorbic

Chemical data: IUPAC Name: D-isoascorbic acid, CAS Number: [89-65-6] Molecular formula: $C_6H_8O_6$, molar mass: 176.13 g/mol; Melting point: 169-172 ° C

3. Relationship with other nutrients

3.1. Vitamin. A and E

A short-term supplementation with physiological doses of antioxidant vitamins, carotenoids and trace elements during alcohol rehabilitation clearly improves micronutrient status indicators. Heavy smokers in particular seem to respond to vitamin C supplementation (Gueguen, 2003)

In the liver accumulate micronutrients such as Vitamin A, E and iron, and therefore, in patients with hepatic impairment is a deficit of them due to reduced intake, as well as intestinal transport and liver stores. The alteration of fat soluble vitamins is especially important in patients with steatorrhea or cholestasis.

Moreover, in the alcoholic patient levels water-soluble vitamins are low due to the effect of ethanol on its metabolism, producing pyridoxine deficiency, retinol, cobalamin, folate and niacin. In fact, in chronic alcoholism may develop Wernicke encephalopathy. It has also shown a direct relationship between oxidative stress and disease severity liver, requiring the micronutrients with antioxidant activity, being increased the needs of vitamin E and C.

There are few data on vitamin needs and trace elements in burned patients. the large tissue loss, decreased gastrointestinal absorption, increased urinary losses, changes in distribution, and a high degree catabolism, that are made increased the needs of vitamins and trace elements. Clinical guidelines recommend giving also established daily requirements and additional doses of certain micronutrients.

Also, increase the intake of vitamin C (1,000 mg/day) as it promotes the healing process, and vitamin A (10,000 IU/day) for its effect immune and protective skin and mucous membranes. Also, is required additional vitamin D due to high risk of fractures in these group patients but have not yet been established Daily exact requirements.

Zinc supplement is suitable dose of 220 mg/day, as is involved in protein synthesis and tissue regeneration. Furthermore, Chan et al indicate that in the week post-injury, there are high losses exuding of copper, being necessary to increase their requirements (4.5 mg/day of copper sulfate).

The increased production of oxygen species reactive in this clinical situation requires administration of antioxidants (ascorbic acid, glutathione, carotenoids, vitamin A and E) been shown to reduce mortality, protecting the micro vascular circulation, reducing the peroxidation tissue lipid. According to some authors, surgical stress may necessitate supplementation of ascorbic acid, alpha tocopherol and trace elements, associating too low preoperative levels of vitamin A (<0.77 mol/L) with an increase of postoperative infection and mortality. At present, it is unknown whether supplementation micronutrient for a short period of time could restore plasma antioxidant levels after surgery. Some authors suggest that antioxidants could lead to improved metabolism and ventricular function after cardiac surgery. Also state that patient's major surgery may benefit from selenium, even before surgery, to action at the level of oxidative stress. The ESPEN recommended in these surgical patients treated with parenteral nutrition, supplement micronutrient recommended daily doses, the vitamin supplementation is unnecessary if the patient is on concomitant enteral nutrition, oral or parenteral.

3.2. Minerals such as selenium, iron and zinc

Recent studies have shown that in patients with inflammatory bowel disease there is a correlation between the level of some antioxidants such as selenium, vitamin C and E and clinical improvement and reduction in serum levels of TNF-α and decrease in steroid dose to 65%. Low serum zinc levels have been correlated with increased blood pressure, disease coronary type II diabetes mellitus and hyperlipidemia. Also, high intake of magnesium (> 500-1000 mg/day) can lower high blood pressure, and be effective in acute myocardial infarction and atherosclerosis. Houston recommended to prevent the emergence and development of hypertension, administration of additional vitamins and trace elements. Finally, oxidative stress plays an important role in the initiation and maintenance of the pathogenesis of cardiovascular disease and its complications. For some authors, antioxidants such as vitamins E and C, beta-carotene, selenium and zinc, may act by reducing the cardiovascular risk, although evidence is limited.

Malnutrition is a common feature of inflammatory intestinal diseases, being frequent the deficit of vitamins B_{12}, A, D, E and K, steatorrhea, ileal resection or extensive lesions in the intestine. GI bleeding contributing to iron losses, and through diarrhea and fistulas is loss of electrolytes and trace elements (copper, magnesium, selenium and zinc).

The role of antioxidant micronutrients in the clinical and functional improvement has been described by different authors. Thus, a low intake of selenium, beta-carotene and vitamins E and C, can reduce the body's natural defenses and increase inflammation of the airways, and therefore the selenium (100-200 ug/d) been associated with improved lung function, especially in smokers. Gazdik et al indicate that supplementation of 200 ug/d of selenium in asthmatic patients produced a statistically significant decrease in the use of corticosteroids. Loannidis and McClave et al indicate that antioxidants such as selenium, vitamin A, vitamin C and vitamin E reduce pancreatic inflammation and pain, and prevent the occurrence of exacerbations. Recently in a double blind study in patients with chronic pancreatitis were given daily supplements of 600 ug of selenium, 9,000 IU of beta-carotene, 540 mg of vitamin C, 270 IU of vitamin E and 2,000 mg methionine, the pain was reduced significantly, as well as stress markers oxidative normalized plasma concentrations of antioxidants.

For some authors, parenteral administration of ascorbic acid can lower the morbidity and mortality of these patients in a randomized, double-blind placebo-controlled; we observed that mortality at day 28 decreased in the group of patients who received ascorbate and vitamin E by intravenous infusion. Some authors recommend increasing the contribution of antioxidants such as vitamin C, retinol, vitamin E, beta-carotene and selenium. Also appear to require thiamine, niacin, vitamin A, E and C, B complex, zinc (15-20 mg/day and 10 mg/L intestinal leaks) and selenium (rise up to 120 ug/day) in patients with sepsis.

3.3. Polyunsaturated fatty acids

Consumption of 0.6 mg equivalents of alpha tocopherol/g linoleic acid is suitable for human adults. The minimum requirement of vitamin E related to the consumption of fatty acids with a high degree of unsaturation can be calculated with a specific formula that must take into account the peroxibility of polyunsaturated fatty acids is based on the results of animal experiments. But still no convincing evidence of how much vitamin E is required in relation to consumption of polyunsaturated fatty acids: EPA (20:5 n-3) and DHA (22:6 n-3).

Studies so far show that the effects of supplementation with EPA and DHA lipid peroxidizacion increase even when the amounts of vitamin E present are adequate in relation to the oxidative potential of these fatty acids. On the other hand the calculation of the requirement for vitamin E using current data from recent consumption, show that a reduction in total fat intake with a concomitant increase in consumption of polyunsaturated fatty acids, including EPA and DHA results in an increased need intake of vitamin E. In fact the methods used to investigate the requirements of vitamin E and polyunsaturated fatty acid intake (erythrocyte hemolysis) and the techniques used to assess lipid peroxidation (malondialdehyde analysis, MDA) may be inappropriate for measuring a quantitative relationship between the two loads.

Therefore, further studies are needed to establish the requirement of vitamin E when intake of unsaturated fatty acids of longer chain increases. For this purpose it is necessary to use functional techniques based on the measurement of in vivo lipid peroxidation. Until then or until the available data suggest using the index of 0.6 mg of alpha tocopherol per gram of ingested PUFA. However it is likely that higher levels are necessary for vitamin fats are rich in fatty acids containing more than two double bonds (Valk, 2000).

The diet of our ancestors was less dense in calories, being higher in fiber, rich in fruits, vegetables, lean meat and fish. As a result, the diet was significantly lower in total fat and saturated fat, but containing equal amounts of n-6 essential fatty acids and n-3. Linoleic acid (LA) is the major n-6 fatty acid and alpha-linolenic acid (ALA) is the major n-3 fatty acid.

In the body, LA is metabolized to arachidonic acid (ARA), and ALA is metabolized to eicosapentaenoic acid (EPA) and docosahexaenoic acid (DHA). The essential fatty acid ratio n-6: n-3 is 1 to 2:1, with higher levels of long chain polyunsaturated fatty acids (PUFA), such as EPA, DHA and ARA, today's diet. Today this ratio is about 10 to 1 or 20 and 25 to 1, indicating that Western diets are deficient in n-3 fatty acids compared with the diet that humans evolved and established patterns genetic.

The n-3 and n-6 are not interconvertible in the human body and are important components of practically all cell membranes. The n-6 fatty acids and n-3 influence eicosanoid metabolism, gene expression, and intercellular communication cell to cell. The polyunsaturated fatty acid composition of cell membranes is largely dependent on food ingestion. Therefore, appropriate amounts of n-6 fatty acids and n-3 in the diet should be considered in making dietary recommendations.

These two classes of polyunsaturated fatty acids must be distinguished because they are metabolically and functionally distinct and have opposing physiological functions, balance is important for maintaining homeostasis and normal development. Studies with nonhuman primates and human newborns indicate that DHA is essential for normal functional development of the retina and the brain, especially in premature babies. A balanced n-6/n-3 ratio in the diet is essential for normal growth and development and should lead to reduced cardiovascular disease and other chronic diseases and improve mental health. Although a recommended dietary allowance for essential fatty acids does not exist, an adequate intake (AI) has been estimated for essential fatty acids n-6 and n-3 by an international scientific working group. The final recommendations are for Western societies, reduce the consumption of n-6 fatty acids and increased intake of n-3 fatty acids. (Simopoulus, 2000)

4. Vitamin C sources

4.1. Food sources in Mexico

The main sources of ascorbic acid are presented in Table 1. The results of vitamin C are shown as the mean and correspond to the official tables of composition of Mexican foods.

FOOD	CONCENTRATION (mg/100g EP)
poblan chili	364
"trompito" chili	320
yucca flower	273
guava	199
Marañon	167
cauliflower	127
red bell pepper	160
guajillo chili	100
garlic	99
Nanche fruit	86
orange	76
manila mango	76
serran chili	65
pumpkin	58
watercress	51
beet	20
cucumber	13

EP = edible portion

Table 1. Main sources of ascorbic acid in México

4.2. Fruits and vegetables

Vitamin C is a major constituent of fruits and vegetables, which also contain citric acid, oxalates and substances such as anthocyanins, coloring agents and carotenoids that are difficult to quantify when using colorimetric methods.

Currently there is great interest in relation to consumption of natural foods and mainly on the content of nutrients in fruits, vegetables and vitamin C. This interest is due in part to vitamin C is probably one of the most widely used nutrients in the food and pharmaceutical industry. Used as a supplement, additive, preservative, as an antioxidant in processed foods. Table 2 shows the main foods that are good sources of vitamin C.

FOOD	CONCETRATION (mg/100g)
Acerola	1743
Coriander leaves dry	567
red peppers, spicy, ripe, raw	369
Orange juice, dehydrated	359
Grape juice, dehydrated	350
common Guavas, whole, raw	242
Dried Tomato juice	239
red pepper, spicy, immature, raw	235
Lemon juice	230
Orange juice, canned	229
Col common, dehydrated	211
Peppers, sweet, ripe, red, raw	204
Currants, black, European, raw	200
parsley, dry	172
Orange juice	144
green radish, raw	139
Grape juice	138
Orange peel, raw	136
Pokeroot fruit	136
Mustard seeds	130
leaves of kale, boiled, drained	93
Broccoli cooked	90
Brussels sprouts cooked	87
Lamb crude	80
Leaves and stems of cress, raw	79
Cauliflower, raw	78
Brussels sprouts, cooked in water	76
Cabbage, red, raw	61
Strawberries Raw	59
Papaya	56

Table 2. Acid content of ascorbic acid in different foods and different presentations

In Table 3 shows the analysis of vitamin C in different parts of the food such as edible portion thereof, the seed or plant center and the shell and stalks that are normally discarded.

Food	Edible portion	Shell	Seed	Stem
Apple	458.1±17.8 (84.5)	848.9±11.2 (81.8)	--	--
chayote	7.2±0.9 (94.2)	19.9±0.5 (94.2)	266.6±6.7 (90.5)	--
onion	17.0±0.7 (87.6)	--	--	456.6±1.8 (93.0)
lime	256.2±17.2 (87.1)	1916.5±186.7 (69.5)	--	--
pomarrosa	531.7±36.4 (83.6)	623.0±35.1 (82.5)	1044.5±50.9 (65.7)	--
guanábana	140.7±3.6 (80.0)	82.1±4.9 (81.8)		
beat	70.8±6.8 (85.3)	115.8±1.2 (85.8)		
Potato	74.6±6.7 (80.3)	4.2±1.0 (85.8)		

Table 3. Distribution of ascorbic acid (mg/100g) in some fruits and vegetables produced in México

Below the edible portion the moisture (%) is indicated in parenthesis. The data shown in Tables 3 through 6 are original and have not been published yet. EP = Edible portion

Table 4 shows the concentration of ascorbic acid in the main fruits and vegetables consumed in Mexico. Data are presented as mean and standard deviation. It shows that even the concentration of vitamin C is lower in the edible portion in the shells reported in Table 5.

Food	Concentration (mg / 100 g EP)	Moisture (%)
squash	7.2±0.9	94.2
spinach	8.5±0.2	91.6
potatoes	74.6±6.7	79.7
cucumber	93.0±7.1	96.0
green tomato	222.8±10.7	90.9
poblano chile	191.0±7.8	91.9
green pepper	195.5±9.5	94.3
nopales	268.9±30.1	95.2
cambray onion	17.0±0.7	93.0
carrot	50.4±5.6	87.2
white cabbage	184.7±17.2	93.2
grapefruit	261.3±10.7	87.8
mango	319.6±5.3	84.6
watermelon	56.2±8.9	92.5
banana	333.7±6.3	76.5
orange	279.8±39.4	84.0
mamey	31.6±1.2	66.8
plum	331.1±17.9	88.1
grape	66.1±13.6	84.3
apple	458.1±17.8	84.5
beet	70.8±6.8	85.3
lemon	39.4±2.5	88.3
avocado	256.2±61.9	84.3
sweet lime	306.8±23.4	89.8

Table 4. Content of ascorbic acid in Mexican fruits and vegetables

The shell or skin of many fruits and vegetables is usually discarded, but these same wastes have significant amounts of ascorbic acid, which is shown in Table 5, where the lime peel, rose apple and apple are the best examples.

Shell of	concentration (mg/100g)	moisture (%)
CARROT	129.1±10.1	88.5
APPLE	848.9±11.2	81.8
BITTER LEMON	39.4±2.5	71.7
AVOCADO	187.1±27.2	82.0
BEET	115.8±1.2	85.8
SMALL POTATO	4.2±1.0	85.5
BIG POTATO	2.1±0.1	80.3
POMARROSA	622.9±35.1	82.5
LIME	1916.4±186.7	69.4
CHAYOTE	19.9±0.5	94.2

Table 5. Content of ascorbic acid in the shell of some fruits and vegetables

Table 6 shows the values of ascorbic acid in some plant species used as flavoring for Mexican dishes. In most cases the amount used for the preparation of food is very low and sometimes do not amount to more than 2% by weight of the end plate. However their presence in cooked food gives organoleptic properties suitable for the acceptance of it and especially the potential of the flavors of food.

spices	concentration (mg /100g EP)	moisture (%)
coriander	56.0±7.3	85.3
yerbabuena	226.9±21.9	86.8
epazote leaves	18.3±1.1	86.3
chard leaves	8.5±0.9	92.7
chard stem	15.7±0.9	95.0
parsley leaves	243.9±17.2	84.5
parsley stem	182.1±11.0	90.2
celery leaves	5.6±0.2	88.4
celery stem	7.8±0.9	90.3
corn grain	35.3±6.0	70.7
corn "hair"	55.9±1.2	85.4

Table 6. Content of ascorbic acid in some food used as spices

http://dx.doi.org/10.5772/52181

Food-industrialized

Vitamin C (ascorbic acid) is water soluble and sensitive to oxygen. For this reason, it may partially destroyed in foods during processing, if exposed to air during storage or if treated with water. Manufacturers can protect them from oxidation by adding vitamin C. The addition of ascorbic acid as an antioxidant should be appropriately marked in the list of ingredients on the label of the final product

Ascorbic acid and its salts are practically insoluble in lipids (fats), for this reason that is often used in the food industry as an antioxidant and preservative greasy foods, in order to avoid rancid. Their salts are usually used with a solubilizing agent (usually a monoglyceride) to improve its implementation. Also this often used in the processing industry of cosmetics products. Sodium ascorbate is a sodium salt of ascorbic acid (vitamin C) and formula $C_6H_7NaO_6$. This form is used in the food industry for their functions antiseptic, antioxidants, and preservatives. Ascorbyl palmitate is an ester formed by ascorbic acid (vitamin C) and palmitic acid creating a liposoluble form of vitamin C. It is used in the food industry as an antioxidant (code E 304). It is wrong to think that is a natural antioxidant

Use as a preservative

Is usually used as a food preservative and as antioxidant in the food industry, a typical case is found as a bread improver additive. In industry collecting fruit prevents the color oxidative change known browning. Is often added to foods treated with nitrite in order to reduce the generation of nitrosamines (a carcinogen), so commonly found in sausages and cold cuts. In the same way is generally used in the food industry as an acidity regulator.

Ascorbic acid and its sodium, potassium and calcium salts are used widely as antioxidants and additives. These compounds are soluble in water, so that fats do not protect against oxidation. For this purpose may be used ascorbic acid esters with fat soluble long chain fatty acids (palmitate and ascorbyl stearate).

5. Deficiencies

5.1. Primary deficiency – scurvy. Signs and symptoms

A frank deficiency of vitamin C causes scurvy, a disease characterized by multiple hemorrhages. Scurvy in adults is manifested by lassitude, weakness, irritability, vague muscle pain, joint pain and weight loss. Early signs objectives are as bleeding gums, tooth loss and gingivitis.

The diagnosis of scurvy, is achieved by testing plasma ascorbic acid, low concentration indicates low levels in tissues. It is generally accepted that ascorbic acid concentration in the layer of coagulated lymph (20-53 ug/10^8 leukocytes) is the most reliable indicator of nutritional status regarding vitamin C and its concentration in tissues. In the most extreme cases scurvy shows: bleeding gums and skin, perifollicular bleeding and ocular petechiae, salivary and lachrymal glands "drier", functional neuropathy, lower limb edema, psychological disturbances, anemia and poor healing of wounds. The consumption of snuff lowers blood levels of vitamin C.

Clinical manifestations of Vitamin C deficiency

Clinical manifestations of deficiency have been described at several levels: a) mesenchymal. - By the presence of petechiae, ecchymosis, curly hair, peri-follicular hemorrhages, bleeding gums, swollen, hyperkeratosis, Sjôgren syndrome, dyspnea, arthralgia, edema and poor healing, b) systemics, characterized by fatigue, weakness and lassitude c) psychological and neurological, by the presence of depression, hysteria, hypochondriasis and vasomotor instability.

5.2. Deficiencies secondary and association with other diseases

Severe deficiency of vitamin C leads to Scurvy. This rarely occurs but can be observed deficiencies in those who consume a diet without vegetables and fruits, alcoholism, in older people with limited diets, severely ill patients with chronic stress and in infants fed cow's milk.

Symptoms of scurvy are follicular hyperkeratosis, gingival swelling and inflammation (in gums), bleeding gums, loose teeth, dry mouth and eyes, hair loss and dry skin, among other symptoms that can lead to death.

By deficiency of collagen, the wounds do not heal scars and wounds of previous rupture and may lead to secondary infections. Neurotic disorders are common, consisting of hysteria and depression, followed by decreased psychomotor activity. It is not safe indiscriminate administration of ascorbic acid, since as the body becomes saturated, decreased absorption, and giving large doses, abruptly deleted. So if you continue with diet low in vitamin, may appear "rebound scurvy". In addition to "rebound scurvy," gastric intolerance and kidney, its use decreases the cobalamin (vitamin B_{12}), a substance synthesized by the body.

Eating a balanced and varied diet high in fruits and vegetables, the minimum dose of vitamin C, this completely covered. The daily requirement in an adult male is 90 mg/d and a woman of 75 mg/d (mg/day), although there are always situations where it is necessary to increase the dose of vitamin A through supplementation. Such circumstances or situations are:• Pregnancy and Lactation, • Alcoholics and smokers, • diabetics, • Allergic and asthmatic, • People who take daily medications or medications such as oral contraceptives, cortisone, antibiotics, etc.

Anemia by Vitamin C Deficiency

Anemia of vitamin C Deficiency is a rare type of anemia that is caused by a severe and very prolonged lack of vitamin C. In this type of anemia, the bone marrow produces small red blood cells (microcytosis). This deficiency is diagnosed by measuring the values of vitamin C in white blood cells. One tablet of vitamin C per day corrects the deficiency and anemia cure.

Associated symptoms

Associated deficiency or lack of vitamin C (ascorbic acid) can produce or be reflected by: Swollen and bleeding gums, dry, rough skin, spontaneous bruising, Impaired wound healing, bleeding nose, joint pain and swelling, anemia and dental enamel weakened. Very small amounts of vitamin C may be associated with signs and symptoms of deficiency, including: • Anemia, • Bleeding gums, • Decreased, ability to fight infection. • Decrease the rate of wound healing. • Dry and separated strands in the hair. • Tendency to bruising. • Gingivitis (gum inflammation). • nosebleeds. • Possible weight gain because of slowed metabolism. • Rough, dry and scaly. • Pain and swelling of the joints. • Weakened tooth enamel.

6. Laboratory methods for its measurement in foods and biological fluid

In the analysis of vitamin C, for the methods commonly used consume many time and therefore overestimates the concentration, due to other oxydizable species different of vitamin C; the determination by liquid chromatography of high resolution (HPLC), with electrochemical detection, for example, they require equipment not always available in smalls laboratories and also, is very expensive. However this method quantifies all the forms of the vitamin C present in the sample, and even it detects an epimer of ascorbic acid, the eritorbic acid or isoascorbic acid. The samples of vitamin C saturation are used to establish the deficiency of ascorbate in tissue and are useful to confirm the diagnostic of scurvy when the patient has a normal absorption (Engelfried, 1944). It has been described 3 types of tests to determine the tissue saturation, the first 2 are easy to make but they don't cover the problem on totally, the third test is complicated and it's only useful in research work listed below:

a. Measurement of blood levels with and without a test sample:

    - The vitamin C in the plasma is not found doing a metabolic function; it is rather in a transit

    - from one tissue to another. Its lost or decrease does not indicates the intracellular status of this vitamin. A well-nourish adult with a free acid ascorbic diet decreases his serum levels of acid to cero in about 6 weeks; however only after many weeks of more deprivation the scurvy symptoms appears. So that for this reason the scurvy patients have low levels of ascorbic acid in the plasma. The vitamin C determination in plasma after a charge dose generally reflects the vitamin proportion, which has been stored by the tissues, however is tough to do a completely quantitative technique, because in high doses of this vitamin, the plasmatic concentration exceeds the kidney threshold which causes the lost of this vitamin in urine. For a specific measure most be given multiple small doses, to avoid an excess in the blood levels above the renal threshold.

b. Measurement of kidney excretion with and without sample dose: The most important problems in this measurement are those concerning to the collection of urine, more than the vitamin measurement. The excretion of the vitamin has been correlated with the creatinine excretion. This is because the creatinine is used as a real and simple indicator of glomerular filtration.

c. Tissue Measurement This is the one of the 3 techniques which gives a real representation of desaturation of the vitamin. This measurement is difficult because of problems in the sampling of tissue. In the case of ascorbic acid is recommended two methods to measure tissue levels. In the first method is measured its concentration in the buffy coat and platelets which correlates good with the first signs of scurvy, making the most recommended technique. The second method determines the tissue saturation grade for an intradermal test, using dichlorophenol indophenol, which depends of skin reductor substances, which made this nonspecific.

6.1. Spectrophotometric method

Another proposed method is highly sensitive colorimetric determination of ascorbic acid with 2,4-dinitrophenylhydrazine.

The method can be applied in virtually any laboratory is simple to perform and requires little complicated equipment compared to HPLC. Samples can be from serum to food. In foods the determination may be affected if the food contains natural dyes interfering reading the wavelength of detection. However, it is an easy to implement, since it is inexpensive and sensitive.

The use of UV-Visible spectroscopy (UV-Vis), for the determination of AA, is widely used in research with food, since this acid has strong electronic transitions in the UV region, facilitating their identification and quantification by this technique.

Vitamin C is easily affected by such factors as moisture, light, air, heat, metal ions such as iron and copper, oxygen and the alkaline medium. After decomposition under these conditions is easily transformed into various compounds such as: oxalic acid, L-threonic acid, L-xylonic acid and L-dehydroascorbic acid, and in turn the latter are irreversibly transformed into acid-diketo 2.3 L gluconic acid, which is its main degradation product.

Within the official methods described for the analysis of vitamin C in tablet, one of the most widely used is the direct titration with 2,6-dichlorophenol indophenol by simple and rapid result. The method is valid if it is known that the composition of the sample no interfering substances and the concentration of dehydroascorbic acid is negligible, therefore, can be applied to a freshly prepared sample, but not useful in stability studies of vitamin C.

6.2. Chromatographic method (HPLC)

However, high performance liquid chromatography (HPLC) ensures detection and quantitation limits lower, which also facilitate the elimination of the effects caused by the matrix (interference to other methods of analysis); this technique used as an essential tool in detailed kinetic studies.

Quantification of Ascorbic Acid by HPLC.

Using a standard solution of 50 mg AA/L phosphoric acid, 0.05 N and taking into account the various reports, on the conditions for the quantification of AA in samples, it optimizes the mobile phase and the wavelength of the detector (scanning from 200 to 320 nm) to obtain the greatest sensitivity and resolution in the chromatographic signal. These tests are performed with the following mobile phases:

2% KH_2PO_4, pH = 2.3

Acetonitrile-Water (70:30)

1% NaH_2PO_4, pH = 2.7

Methanol-buffer solution: 0.03 M KH_2PO_4, pH = 2.7, (99:1)

Water-methanol-acetonitrile (74.4: 25.0: 0.6)

The test is performed with the standard addition curve and calibration curve AA patterns between 1.0 to 25.0 mg/L phosphoric acid, 0.05 N, to determine the effect matrix in the method of quantification. The proposal is a column Hypersil ODS C_{18}, 5μm x 4.0 mm x 250 mm.

6.3. Other methods reported

Iodometric titration

Ascorbic acid or Vitamin C ($C_6H_8O_6$) can be determined by means of an iodometric titration. Vitamin C is a mild reducing agent that reacts rapidly with tri-iodide ion, this reaction is generated in a known excess of tri-iodide ion (I_3-) by reacting iodate iodide, is allowed to react and then the excess is titrated by I_3-back with a solution of thiosulphate. -The method is based on the following reactions:

$$8I^- + IO_3^- + 6H^+ \; ------\rightarrow 3H_2O + 3I_3$$

$$C_6H_8O_6^- + H_2O \;+\; I_3^- ------\rightarrow C_6H_8O_6 \;+\; 2H^+ + 3I$$

Ascorbic acid dehydroascorbic acid

$$I^{3-} + 2(S_2O_3)^{-2} \; ------\rightarrow 3\,I^- + (S_4O_6)^{-2}$$

thiosulfate tetrathionate

7. Requirements and recommendations in Mexico

The recommended daily ingestion (intake) is of 60 to 100 mg to avoid the appearance of disease symptoms that are produced by deficiencies of this vitamin. The infants require a little more of 100mg/day, although there is controversy over the minimum amount of this vitamin. We must take into account that this vitamin is very labile at heat and oxygen presence. The ascorbic acid is specific in the treatment of scurvy; the required dose could be better measured by the urinary excretion after a saturation dose. Depending of the required saturation velocity is the daily dose recommended which varies between 0.2 and 2.0 g/d. In the vitamin C deficiency, the tissue saturation is obtained with 3 daily doses of 700 mg each one for 3 days. Harris and cols. defined the tissue saturation as a sufficient storage of ascorbic acid where occurs a excretion of 50 mg or even more in a period of 4 to 5 hours after 1 dose of 700 mg/d.

The decreased levels in smokers are basically explained because they consume fewer sources of the vitamin. In this kind of population will be required a 50% more of the recommended dose of the vitamin. Because of the daily recommendation (RDA) it is defined as the daily ingestion average of food which is enough to cover the nutriments required for healthy people in a group of the population, it's necessary to continually assess these recommendations for vitamin C. The totalities of the reviewed information suggest that a consumption of 90-100 mg of this vitamin is enough for the optimum reduction of chronic disease risk in non-smoking men and women. Although some reports are suggested amounts up to 120 mg/day.

8. Toxicity and hypersensitivity

High doses of the vitamin (5-15 g/day), may cause osmotic diarrhea because it is ingested more vitamin of which can be absorbed. Also ascorbic acid can provoke intestinal cramps and acidification of the urine, leading to the formation of oxalate stones in the kidney of urinary tract. An exaggerated complementation during pregnancy may high the fetal requirement and result in the presence of scurvy in the newborn. It is also credited with the destruction of vitamin B_{12} of food during the ingestion.

Since the oxalic acid is a metabolite of the catabolism of the ascorbic acid is likely to be the formation of oxalate crystals in kidney in patient's susceptible to nucleation and therefore the formation of crystals or "kidney stones" when it is consumed excess of the vitamin. This relation however does not extend to subjects which are not susceptible to the formation of these kidney stones.

9. Biochemical functions

9.1. Paper as an antioxidant

Vitamin C is a soluble antioxidant important in biological fluids. An antioxidant is defined as "any substance which, when present in lower concentrations compared with the oxidable substrates (for example, proteins, lipids and carbohydrates and even nucleic acids) avoids or prevent significantly the oxidation of this substratum". The definition also given by the Food Nutrition Board is "A dietary antioxidant is a substance present in food which decreases significantly the adverse effects of the reactive species of oxygen (ROS), reactive species of nitrogen (RNS), or both for the normal physiology function in humans.

http://dx.doi.org/10.5772/52181

9.2. Interaction with ROS

Vitamin C quickly debug the reactive species of oxygen and nitrogen just as superoxide, hidroperoxile radicals, aqueous peroxyl radicals, singlet oxygen, ozone, peroxynitrite, nitrogen dioxide, nitroxide radicals and hypochlorous acid thereby protecting in fact other substrate of the oxidative damage[16].

Although the AA (ascorbic acid) reacts quickly with the hydroxyl radicals (constant speed > 10^9 $Lmol^{-1}s^{-1}$) is clumsy to debug this radical preferentially over other substrates. This is because hydroxyl radicals are very reactive and they will combine immediately with nearest substratum in their environment at a limited speed because of its diffusion. Vitamin C can also act as a co-antioxidant when regenerate the α-tocopherol (vitamin E) from the α-tocopheroxil radical produced when this is debugged from the lipid-soluble radicals just made. This is a function potentially important because in the *in vitro* experiments have shown that α-tocopherol can act as a pro-oxidant in absentia of co-oxidants just as vitamin C. However the relevance *in vivo* of the interaction of both vitamins it's not that clear yet.

AA can regenerate urates, glutathione and β-carotene *in vitro* from their respective oxidation products with one unpaired electron (urate radicals, glutathionil radicals and cations of β-carotene radicals). Two important properties of vitamin C make it an ideal antioxidant. The first one is its low potential reduction of ascorbate (282 mV) and its oxidation product with an electron, the ascorbile radical (2174 mV), which is derivates from its functional group en-diol in the molecule. This low potential of reduction of the ascorbate and the ascorbile radical makes them potentially appropriate for oxidation-reduction reactions and that why the vitamin acts as a soluble antioxidant terminal molecule. The second property which makes the vitamin an ideal antioxidant is the stability and the low reactivity of the just made ascorbile radical when the ascorbate debug the reactive species of oxygen and nitrogen. (equation 1).

The ascorbyl radical disproportionate rapidly to form ascorbate and dehydroascorbic acid (equations 1 and 2), or it is retro-reduced to ascorbate by an enzyme semi-dehydroascorbate reductase dependent of NADH. The oxidation product of two AA electrons, dehydroascorbic acid, can be reduced by itself to ascorbate because of the glutathione, by enzymes dependents of glutathione: glutaredoxin (dehydroascorbate oxidoreductase [glutathione dehydrogenase (ascorbate)]) or by an enzyme dependent of selenium (seleno-enzyme): the tioredoxine reductase. Alternatively, the dehydroascorbic acid gets quickly and irreversibly hydrolyzed to 2,3-dicetogulonic acid (ADCG) (equation 3):

$$\mathbf{AH_2 \leftrightarrow A\bullet_2 \leftrightarrow A} \tag{1}$$

$$\mathbf{A\bullet_2 + A\bullet_2 \rightarrow AH_2 + A} \tag{2}$$

$$\mathbf{A \rightarrow ADCG} \rightarrow \text{oxalate, threonate, etc.} \tag{3}$$

Where in equation 1 shows the reversible oxidation of 2 ascorbate electrons (AH_2) to the ascorbile radical ($A\bullet_2$) and dehydroascorbic acid (A) respectively; the equation 2 shows the dismutation of the ascorbile radical to transform ascorbate and dehydroascorbic acid; and equation 3 show the hydrolysis of the dehydroascorbic acid to ADCG, which decomposes to oxalate, treonate y many other products. Vitamin C has been recognized and accepted by the FDA as one of 4 dietary antioxidants, the other 3 are vitamin E, β-carotene precursor of vitamin A and selenium as an essential component of antioxidant enzymes glutathione peroxidase (GPx) and thioredoxin reductase.

Although there is substantial scientific evidence of the role of antioxidant vitamin C and its effects on human health are needed more research that guarantee the role of vitamin both *in vivo* and *in vitro*, particularly because the AA is a redox active compound, which may act not only as an antioxidant but also as pro-oxidant in the presence of ion redox active transition metal.

The reduction of metallic ions like iron and copper for the vitamin C *in vitro* (equation 4) results in the formation of hydroxyl radicals highly reactive way to the reaction of this ions with hydrogen peroxide, a process known as the Fenton chemistry (equation 5), The lipid hydroperoxides can also "break" because of reduced metallic ions, forming lipid alkoxy radicals (equation 6) which can begin and spread chain reactions of the lipidic peroxidation. However the shown mechanism in the equation 5 requires the availability of free ions, redox active metallic ions and a low index vitamin C/metallic ion, conditions unlikely under normal conditions *in vivo*. Although has been shown that in biological fluids like plasma, the vitamin C acts like an antioxidant towards the lipids even in presence of free active ions.

$$\mathbf{AH_2 + M(\mathit{n}+1) \rightarrow A\bullet_2 + M\mathit{n} + H+} \tag{4}$$

$$\mathbf{H_2O_2 + M\mathit{n} \rightarrow \bullet OH + OH_2 + M(\mathit{n}+1)} \tag{5}$$

$$\mathbf{LOOH + M\mathit{n} \rightarrow LO\bullet + OH_2 + M(\mathit{n}+1)} \tag{6}$$

Where in equation 4 shows the reduction of metallic active redox ions M(*n*+1)] because of the ascorbate to form de ascorbile radical and the reduced metal (M*n*), the equation 5 shows the productions of hydroxyl radicals highly reactive (•OH) of the reaction of the hydrogen peroxide (H_2O_2) with the reduced metallic ions and the equation 6 shows the reaction of the lipid hydroperoxides (LOOH) with the reduced metallic ions to form alkoxy radicals (LO•). Although there no convincing evidence of a prooxidant effect of vitamin C on humans, exist a substantial evidence of its antioxidant activity. Interestingly its antioxidant activity does not correlate directly in its anti-curvy effect.

Because of this, the experts considerate that if the antioxidant activity of Vitamin C is accepted *in vivo* and that if this is relevant for human health, then scurvy should not be considerate as the only criterion for the nutritional fitness or for determine the ideal quantity or required of the vitamin.

http://dx.doi.org/10.5772/52181

9.3. Molecular mechanisms intracellular

9.3.1. As an enzyme cofactor

The molecular mechanisms of the anti-scurvy effect of vitamin C are very broad and so low studied. Also Vitamin C is a cofactor of many involved enzymes in the collagen biosynthesis, carnitine and neurotransmitters. The pro-collagen dioxygenase (proline hydroxylase) and the procolagene-lisine-5-dioxigenese are two enzymes involved in the synthesis of the collagen which needs ascorbic acids for maximum activity. The posttranslational hydroxylation of the lysine and proline residues of these enzymes are indispensable for the synthesis and formation of the stable helix which forms the collagen. So that the difference of the vitamin leads to the formation of weak structures causing lost of teeth, pain in joints, disorders of the connective tissue, and poor healing, characteristic signs of scurvy.

Two dioxigenases involved in the carnitine synthesis also require vitamin C for its activity. Carnitine is essential for the transport of long chain fatty acids to the mitochondria so one deficiency of vitamin C will bring consequences just as fatigue and lethargy which are late symptoms of scurvy. Besides that the vitamin C is a cofactor for the synthesis of catecholamines, in particular for the conversion of dopamine to norepinephrine catalyzed by the enzyme dopamine-β-monooxygenase. Depression, hypochondriasis, and behavioral changes are common in scurvy as a result of deficient dopamine hydroxylation.

Other kind of enzymes where vitamin C acts as a cofactor are the ones involved in the peptides amidations and in the tyrosine metabolism (this are also of the mono and dioxygenases kind). It is also implicated in the cholesterol metabolism to bile acids, way of 7-α-monooxygenase and in the adrenal steroids metabolism. The hydroxylation of aromatic drugs and carcinogens by cythocrome P-450, gets better also by reducing agents like vitamin C.

The role of vitamin C, due to its redox potential is to reduce metal ions present in the active sites of enzymes mono and dioxygenases. Ascorbate for instance acts as a co substrate in these reactions, not as a coenzyme. The reduction of iron, involved by the presence of vitamin improves the intestinal absorption of dietary non heme iron. Other proposals include the maintenance of the thiol groups of proteins, keeping in its reduced form of glutathione addition, a cellular antioxidant and enzyme cofactor, and tetrahydrofolate as a cofactor required for the synthesis of catecholamine.

10. Ascorbic acid in cancer

The recommended daily allowance (RDA) for ascorbic acid varies from 100 to 120 mg/day for adults. Have been attributed many benefits just like its antioxidant power, antiatherogenic, anticarcinogenic, immunomodulatory and anti-cold. However these benefits have been subject of debate and controversies because of the danger in the use of mega doses often used and its prooxidant effects and antioxidants. Discussed even if ascorbic acid cause cancer or promote or interfere with cancer therapy, the experts panels of dietary antioxidants and related compounds have been concluded that the data *in vivo* does not shows clearly a direct relation between the excess ingestion and the formation of kidney stones, the prooxidant effects and the excess absorption of iron.

The epidemiological and clinic study does not shows conclusive benefic effects in many kinds of cancer, with the exception of stomach cancer. Recently it has tested several derivatives of ascorbic acid on cancer cells as ascorbic acid spheres. The ascorbyl stearate is a compound which inhibits the human carcinogenic cell proliferation, by interfering with the progression of the cellular cycle and inducing apoptosis by modulation of signal transduction pathways. The cancer is a global public health problem with increasing levels of mortality. Although exists a great variety and types of cancer, we can remark the role of vitamin C and its effects in this suffering. AA is effective protecting against the oxidative damage in tissue and also suppressing the carcinogens formation like nitrosamines.

Although vitamin C is a cytotoxic agent for tumor cells and non toxic for normal cells, in modern medicine and conventional favors more the use of powerful toxic chemotherapeutic agents. A great amount of studies have shown that the consumption of vitamin C is inversely related with cancer with protective effects in cancer of lung, pancreas, stomach, cervix, rectum and oral cavity.

The guanine oxidation, a DNA purine, gets reduced significantly after the vitamin C supplementation, but the adenine oxidation, another purine, it's up high which suggest the antioxidant role of the vitamin. Other extensive studies both *in vivo* and *in vitro* have shown its ability to prevent, reduce or increase the adverse effects of chemotherapy. The combination of vitamin C and vitamin K already given in the chemotherapy increases the survival and the effects of various chemotherapeutic agents in a tumor-ascitic-murine model. The vitamin C has shown be safe even with the radiotherapy. The co administration of vitamin A, β-carotene, E and C can reduce the incidence and delay the progression of several cancers, such as skin, colon, stomach, esophagus, mammary gland and matrix.

Epidemiologic studies have revealed an inverse relation between the consumption of vitamin A, β-carotene, E and C and the incidence of several human cancers. There are a decrease in the risk and incidence of cancer in populations with high content of vitamins in plasma. The carcinogenesis is related with the cell differentiation, progression, metabolism and synthesis of collagen. The basic mechanism for the carcinogenesis is the cell differentiation because the cancer develops when a lost in this differentiation exists. And here is where the mentioned vitamins have a wide influence over de cell growth and its differentiation. Vitamin C is a strong antioxidant that acts synergistically with vitamin E in the purification of free radicals which are carcinogenic. The AA as sodium ascorbate exerts marked cytotoxic effects over many human cell lines when they are cultured. These effects are dose dependent.

Lupulescu reported that vitamin C (up to 200 ug/mL) did not cause any morphological change in mouse melanoma, neuroblastoma, and mouse and rat gliomas but is lethal for neuroblastoma cells. Cytotoxic effects are dependent cell also because they are stronger in human melanoma cells compared to mouse melanoma. It has even been suggested that the cytotoxicity induced by vitamin C is mediated primarily by the formation of hydrogen peroxide in the cell surface. The cytotoxic activity may also be mediated by the presence of cupric ions (Cu^{2+}) in malignant melanoma cells that react with vitamin C to form free radicals in solution. Vitamin C also invests into cells, transforming them chemically to a normal phenotype fine.

Studies of cell surface and ultrastructure suggest that cancer cells after administration of vitamin C had cytolysis, cell membrane damage, mitochondrial changes, nuclear and nucleolar reduction and an increase in the formation of phagolysosomes. Changes in cell surface as cytolysis showed predominantly increased synthesis of collagen and disruption of the cell membrane with increased phagocytic activity and apoptotic.

The quantitative estimation of cellular organelles shown that vitamin C affects the intracellular distribution of the organelles, event that plays an important role in the citodifferentiation of the carcinogenic cell and this is the shared effect that not only vitamin C has, but also vitamin A and E. Changes in the Golgi complex and apoptotic activity and autophagic addition to changes in cell surface and in some cases even the reversal of transformed cells to their normal cell types are needed in the possible reduction in incidence of various cancers.

Have also been associated changes in the protein synthesis, DNA and ARA with the differentiation and proliferation cell. But these mechanisms are not clear yet. It have been mentioned that many of this metabolic effects are mediated by the transcription and translation at genomic level. This vitamins modulate the DNA synthesis and the genetic expression in a similarly to hormones and steroids.Their effects can affect the chemical mutagenicity and the cell status. The vitamins can control the cell replication affecting the DNA, RNA and proteins in specific places which are target of electrophyles, promoting the rearrangement of codons in the altered cells and the translocation of specific genes or carcinogens. In this way, vitamins A, E and C affects directly the DNA, RNA and the protein synthesis in the carcinogen cells.

The vitamin C administration decrease the DNA synthesis in the core, the RNA synthesis in the nucleolus and the protein in the cytoplasm of these cells. This inhibition is accompanied by ultrastructural changes mentioned which decreases the cancer progression.

Mechanism of action:

Have been proposed many mechanisms of the vitamin C activity in the prevention and treatment of cancer:

1. Would improve the immune system by increasing the lymphocytes production.
2. Stimulation in the collagen synthesis.
3. Inhibition of the hyaluronidase, keeping the substances around the tumor intact avoiding metastasis.
4. Inhibition of carcinogen virus
5. Correction of a likely ascorbate deficiency, seen in patients with cancer

6. Adequate healing after the surgery.
7. Improvement in the effect of some chemotherapeutic agents, just like tamoxifen, cisplatina, DTIC and others.
8. Reduction of the toxicity of other chemotherapeutic agents like adriamicine.
9. Prevention of the cell damage by free radicals.
10. Neutralization of carcinogenic substances.

Patients with cancer tend to immune-undertake, showing low levels of ascorbate in their lymphocytes. The survival of immune system is important both for inhibit the carcinogen cell growth phase and to prevent its proliferation. The supplementation with ascorbate increases the number and the effectiveness of the lymphocytes and upgrades the phagocythosis

The characteristics of the neoplastic cell and its behavior (invasiveness, selective nutrition and possibly accelerated growth) are caused by microenvironmental depolymerization. This destabilization of the matrix is favored by constant exposure to lysosomal glycosidases continually released by the neoplastic cell. The AA is then involved in the control and restriction of this degradative enzyme activity.

The synthesis of collagen is a major factor for the encapsulation of tumors or metastases decreased via the development of a nearly impermeable barrier. AA is necessary for synthesis of collagen and its stabilization. A loss of ascorbate significantly reduces the hydroxylation of proline and hydroxyproline and hydroxylysine to lysine respectively, affecting the cross linking of collagen. This disrupts the structure of collagen triple helix, which increases its catabolism s. *In vitro*, vitamin C also increases the synthesis of collagen in fibroblasts.

11. Ascorbic acid in diabetes mellitus

It has been demonstrated *in vitro* competition between glucose and ascorbic acid by the cell membrane transporter, granulocytes and fibroblasts, and under conditions of substantial and significant changes in chemotaxis of PMN leukocytes and mononuclear cells. There are significant changes to various chemoattractants, and a significant correlation with the decrease in AA.

These results are consistent with the hypothesis that the chronic hyperglycemia associated with leukocyte AA deficiency, an acute inflammatory response damaged and altered susceptibility to infections, and failure to repair bleeding in these patients, further changes are observed sustained hyperglycemia.

The concentrations of ascorbic acid (AA) are decreased in tissues and plasma in diabetes. These values can be normalized with extra supplements of 20-40 mg/d or corresponding to its maximum synthetic rate. Treating diabetic rats with this scheme prevents the decreased activity of granulation tissue of proline hydroxylase (Prolasa) an AA-dependent enzyme, required to maintain the normal properties of collagen. The decrease in AA concentration in plasma and Prolasa activity in diabetes can be normalized by the inhibitor of aldose reductase. We conclude that in diabetic animals there is a deficiency of AA, which may be responsible for the observed changes in collagen in diabetes.

The decrease in plasma ascorbic acid in diabetes plays an important role in the abnormalities of collagen and proteoglycans. These are the 2 major constituents of the extracellular matrix and its abnormalities are associated with the pathogenesis and complications of diabetes. The structural similarity between glucose metabolism and AA and can interact at the level of the membrane and transporters.

Ascorbic acid enhances the collagen and proteoglycans in fibroblast culture media. This stimulatory action is inhibited by high concentrations of glucose (25 mM). This effect however, is not mediated AA consumption of fibroblasts. Insulin removes the inhibitory effect of glucose on the production of collagen, but the mechanism is not yet known. Thus high concentrations of glucose in diabetes damage the action of ascorbic acid at the cellular level.

12. Ascorbic acid in essential hypertension (HAS)

High blood pressure is a powerful indicator of heart disease and stroke. And in many cases is "asymptomatic" or people who have it doing not give importance. However there have been great efforts to use its measurement in the detection of primary or secondary essential hypertension for decades.

Virtually the observed declines in blood pressure and its control in recent years due to better control among individuals diagnosed as hypertensive. In this regard dietary factor has been the best for control. Obesity, dietary sodium and alcohol consumption are strongly associated with low or high blood pressure values.

A high intake of polyunsaturated fatty acids and magnesium are associated with for instance with low pressure. It has also shown an inverse association between plasma vitamin C and blood pressure. Subjects with serum levels of vitamin C equal to 0.5 mg/dL have a systolic pressure 122 mm Hg average compared to subjects with average pressure is 113 mm Hg and whose values of serum vitamin C are 0.9 mg/dL, showing a relative difference of 7%.

These subjects have a similar difference in diastolic pressure ranging from 78 to 73 mm Hg, a difference of 6%. The prevalence of hypertension was 7.5% in the group of subjects with low serum vitamin C and only 1% in subjects with high values of the vitamin in the serum. These results were consistent in several studies regardless of quintiles being compared.

Such relationships have also been identified in Chinese-American population; both men and women aged 60-96 years without antihypertensive treatment. This study revealed a statistically significant difference between the values of systolic and diastolic pressure in upper and lower quintiles of 14% (21 mm Hg) and 9% (8mmHg) respectively. It appears that vitamin C has a lowering effect on systolic rather than diastolic pressure. Supplementation with vitamin C (1g/day) does not influence the diastolic pressure. Subjects with low vitamin C levels in serum have a high risk of developing stroke compared with those with high values in plasma of the vitamin. Hypertensive subjects, usually overweight, and low levels of serum vitamin C have the same risk.

The increase in the consumption of vitamin C during periods of fat restriction occurs on the one hand a reduction in blood pressure. Low levels of AA in plasma are also associated with low concentrations of 6-keto-prostaglandin F, a prostacyclin. Thus dietary antioxidants enhance the production of prostacyclin for the purification of free radicals and peroxides that inhibit prostacyclin synthase. Vitamin C and blood pressure then are related, because it has a lowering effect on blood pressure especially when fat intake is low.

13. Ascorbic acid and cardiovascular disease

Vitamin C acts as a regulator of the catabolism of cholesterol into bile acids in the guinea pig and is an important factor in the regulation of lipid in several animal species (rabbit, horse, and rat).

Correlation studies in humans have shown an inverse relationship between vitamin C intake and mortality from cardiovascular disease.

Experimental and observational studies in humans have been inconsistent but indicate that individuals with high cholesterol consumption, greater than or equal to 5.20mM/L (200mg/dL) and lower in tissue saturation, increase concentration of vitamin C, which may have a beneficial effect on total cholesterol. This effect is explained by the promotion or inhibition of degradation of prostacyclin and its implications for thrombosis and atherogenesis, in addition to its protective effect on lipid peroxidation. In patients with high cardiovascular risk, supplementation with antioxidant vitamins shows no reduction in overall mortality or incidence of any vascular disease, cancer or other adverse events.

Recent findings indicate a relationship between the nutritional status of vitamin C (as measured by the concentration of ascorbate in serum), biological markers of infection and haemostatic factors and support the hypothesis that vitamin C may protect against cardiovascular events through effects on the haemostatic factors in response to infection.

This relationship is surprising given the uncertainty and potential error in the estimation of consumption of vitamin and vitamin C status assessment (determined mostly by food intake records of 24 h blood samples isolated). Add to this the wide variation between subjects is greater than within the same subject.

Lower socioeconomic status and smoking are associated with low concentrations of ascorbate and high concentrations of homeostatic factors that may be confounding factors in cross-sectional studies.

As expected smokers have lower concentrations of AA in serum than non-smokers. The relationship between concentrations of ascorbate and homeostatic factors is very consistent, but when cigarette smokers were excluded from the analysis, we obtain an association where smoking becomes a confounding variable. The relationship between fibrinogen, Factor VIIc and ascorbate concentrations were consistent in subjects taking supplements of vitamin C during 1 year, which indicates that the homeostatic factors relate the variability in the status of vitamin C within a range usual dietary.

The inverse association between homeostatic factors and serum concentrations of ascorbate is strong and consistent, however only some markers of infection (e.g. C-reactive protein and α1-antichymotrypsin) are related inversely and significantly with serum ascorbate. It is possible that this low concentration of ascorbate may be the result rather than the cause, of a biological response to infection. The strong relationship between serum ascorbate and dietary intake suggest however that their serum concentrations reflect the nutritional status of the vitamin.

The various studies reported in the literature indicate that vitamin does not prevent respiratory infection but may modulate the biological response, leading to less severe disease, so it has a protective function in lung function.

13.1. Effect of antioxidants in cardiovascular disease

It has been suggested a protective effect of antioxidants such as vitamin C, A (β-carotene) and E plus selenium in cardiovascular disease. Prospective studies so far have documented an inverse relationship between vitamin C intake and cardiovascular disease, and a strong protective effect of vitamin E supplementation on coronary patients.

Finnish and Swiss studies showed that blood levels of ascorbate and therefore a diminished nutritional status of vitamin predicts myocardial infarction. Low levels of vitamin C increased to 2.7 times the risk of myocardial infarction and this is independent of other risk factors. Mediterranean studies showed a 70% reduction in mortality and risk of myocardial infarction independent of the effect on blood pressure and lipids. Any protective effect on heart disease of these antioxidants is mediated by the oxidation of LDL cholesterol, but there may be other mechanisms of homeostasis and inflammation.

13.2. Infection, homeostasis, and cardiovascular disease

Fibrinogen and factor VII are recognized risk factors of myocardial infarction and stroke, in the same way that acute and chronic infection and increased white blood cell count are risk factors for cardiovascular disease. The infection may contribute to the inflammatory process observed in atherosclerosis.

C-reactive protein and alpha-1 antichymotrypsin are acute phase proteins are synthesized in hepatocytes in large numbers in inflammatory processes. This synthesis is mediated primarily by IL-6 produced by monocytes and macrophages. Elevated fibrinogen favors these mechanisms and therefore an increased cardiovascular risk. In this way a reduction in dietary intake in winter for instance, would lead to lower serum ascorbate levels, an increase in susceptibility to infection and the factors haemostatic factors and therefore to an increase in cardiovascular mortality.

According to the seasonal variations of vitamin C intake may be relatively low (<80 mg/day on average), corresponding to serum ascorbate concentrations of 50 umol/L. within this range may be variations in infection and homeostatic markers. Increased intake of vitamin C to 90-100 mg/day can increase in these subjects more than 60 umol/L, which has a significant effect on all risk factors.

14. Ascorbic acid and immunity

In stress situations the adrenal glands react liberating a large number of active and ready hormones. It has been suggested that 200 mg of vitamin C per day can reduce stress levels caused by these hormones. The stress suppresses the immune response. Megadoses of vitamin C increases the body levels of antibodies in animal models (rats stressed and unstressed) having the highest values stressed rats. Stressed animals may need more vitamin C for proper immune system function.

Healing is characterized by synthesis of connective tissue, whose main component is collagen. This molecule AA required for cross linking of the fibers in hydroxylated residues of prolyne and lysine. Ascorbic acid supplementation is necessary for healing since this is oxidized during the synthesis of collagen.

There is an undeniable evidence of the interaction of vitamin C and phagocytes. The collected cells from the blood, peritoneal or alveolar fluid usually contain high concentrations of vitamin C (1-2 ug/mg protein). Guinea pig neutrophils produced H_2O_2 and destroy staphylococci in the same way they do control cells. Both ascorbate as dehydroascorbic acid are used for phagocytic process. Neutrophils can avoid self-poisoning absorb extra amounts of ascorbic acid, which can neutralize the antioxidants. However glucolytic activity does not increase much in the neutrophils of guinea pigs supplemented and the stimulation of NADPH oxidase activity is depressed. The addition of ascorbate to the culture media of normal macrophages increases the concentration of cyclic GMP (cGMP) in addition to the route of pentose or hexose monophosphate.

Although the addition of large amounts of ascorbate can inhibit myeloperoxidase activity is not altered its bactericidal capacity. It has been an increase in the bactericidal activity in mouse peritoneal macrophages by the addition of ascorbate to the medium. Besides ascorbate increase the motility and chemotactic activity of these cells. The motor functions of cells as the random motion and chemotactic migration of neutrophils and macrophages is damaged in the absence of vitamin C. Ascorbic acid can also influence the ability of certain cell lines to produce interferon. The addition of AA to cultures of skin embryonic or fibroblasts leads to the production of interferon.

Vitamin C is also necessary for thymic function and operation of certain cells involved in the production of thymic humoral factor. Thymic content of dehydroascorbate diminishes in direct proportion to vitamin C intake. The hormonal activity of thymic extracts correlates with thymic ascorbate and inversely with dehydroascorbate.

15. Ascorbic acid and gallbladder

The gallbladder disease is highly prevalent in the U.S. and in Mexico, remains a serious public health problem. It has been estimated that only about in U.S. 20 million Americans have gallstones partially or entirely composed of cholesterol. Gallstones form when bile supersaturated with cholesterol is destabilized. AA affects a limiting step in the catabolism of cholesterol into bile acids in experimental animals, as described AA-deficient guinea pig common development of cholesterol gallstones.

Because of this it has been hypothesized that the deficiency in humans may be a risk factor for this disease in humans. The main findings in humans have shown an inverse relationship between serum levels of AA and biliary disease prevalence among women. It was also observed a low prevalence of clinical biliary disease between women taking ascorbic acid supplements. However, this prevalence has not been observed in males. Almost all the findings from different countries agree to respect.

In another study, Simon showed that the use of ascorbic acid supplementation correlates with biliary disease among postmenopausal women with coronary disease. Among women who consumed alcohol, the use of ascorbic acid supplementation was associated independently with a 50% reduction in the prevalence of gallstones and 62% for cholecystectomies.

Within the NHANES III study was not observed association, linear or nonlinear, between serum ascorbic acid and prevalence of biliary disease in men. Reflecting the low prevalence of the disease in men and reduced statistical power to detect such an association.

It has been hypothesized that the inverse relationship between AA and biliary disease, demonstrated in animals, affects the activity of the enzyme cholesterol-7-α-hydroxylase, which is the limiting step that regulates the metabolism of cholesterol into bile acids. Supplementation with ascorbic acid increases the activity of the enzyme up to 15 times compared with the vitamin-deficient animals that develop the formation of cholesterol gallstones. Additionally there is a hypersecretion of mucin, a glycoprotein that is secreted by the epithelium of the gallbladder, which precedes cholesterol destabilization and gallstone formation. Because oxygen and hydroxyl radicals stimulate mucin hypersecretion, inhibition of the oxidative changes in the vesicle due to AA can decrease the production of this glycoprotein.

16. Ascorbic acid in other conditions

16.1. Sjögren's syndrome

16.1.1. Vitamin C and Sjögren syndrome

Primary Sjögren's syndrome (SSP) is a chronic disorder of unknown cause, characterized by dry eyes and mouth. It usually occurs in middle-aged women with a prevalence of 1:5000. Patients may have swelling of joints, muscles, nerves, thyroid, kidneys or other body areas. These symptoms result from lymphocytic infiltration and destruction of these tissues.

The diagnosis is based on clinical examination of the eyes and mouth, blood tests specific (auto antibodies) and biopsy of minor salivary gland (taken from inside the inner lip). Sjögren's syndrome is not fatal, but must be addressed quickly to prevent complications due to dry mouth (caries, abscesses, gingivitis) and eyes (corneal erosion, infection). However, there is no therapy available that "removes" these symptoms because all therapies are directed at eliminating the symptoms and prevent complications.

On the other hand there are several studies that reported a direct relationship between the clinical manifestations of Sjögren's syndrome and those caused by deficiency of ascorbic acid (vitamin C): scurvy. A frank deficiency of vitamin C causes scurvy, a disease characterized by multiple hemorrhages. Scurvy in adults is manifested by latitude, weakness, irritability, vague muscle pain, joint pain and weight loss. Early signs are objective as bleeding gums, tooth loss and gingivitis.

The diagnosis of scurvy, is achieved by testing plasma ascorbic acid, low concentration indicates low levels in tissues. It is generally accepted that ascorbic acid concentration in the layer of coagulated lymph (20-53 ug/10^8 leukocytes) is the most reliable indicator of nutritional status regarding vitamin C and its concentration in tissues and serum.

16.2. Pharmacological data

Ascorbic acid is specific in the treatment of scurvy; the dose required can best be measured by determining urinary excretion after a dose of saturation, depending on the speed at which the saturation is required is the recommended daily dose ranging from 0.2 and 2.0 g/day. In the vitamin deficiency C tissue saturation is achieved with three daily doses of 700 mg c/u, for three days. Harris defined as saturation of tissues, a sufficient store where an ascorbic acid excretion 50 mg or more occurs in a period of 4 to 5 hours after a dose of 700 mg/day.

16.3. Previous experience in animals

Kessler in his study reported that rats which have been induced vitamin C deficiency, develop various manifestations of primary Sjögren's syndrome (SSP), such as infiltration of mononuclear cells in salivary and lachrymal glands and that these were more severe is female rats than in male rats, concluding that these pathological changes are similar to those that characterize the syndrome in humans.

In previous studies Hood reported a direct relationship between the manifestations of primary Sjögren's syndrome (xerostomia and xerophthalmia) and clinical signs of scurvy such as gingivitis, periodontal bleeding and loss of teeth. In their study, Hood study 5 subjects men whose diets did not contain ascorbic acid, for 84 to 97 days. During the deficiency, in the demonstrations that make Sjögren syndrome, observed that prostaglandins, particularly PGE_1, is important in the immune defects associated with the syndrome.

In 1992, Gomez et al, from the National Institute of Medical Sciences and Nutrition "Salvador Zubiran", observed values less than 0.2 mg AA/dl in plasma of patients with SSP (reference values were from 0.4 to 2.0 mg/dL), representing a frank deficiency of vitamin in 100% of cases with SSP.

16.4. Role of vitamin C in other body disorders

It is reported that the diabetic individual has low levels of vitamin C in plasma and leukocytes, which is our immune defense. However, more clinical studies, in a large scale, are needed to determine whether the supplementation with large doses of the vitamin are beneficial or not. Some studies have shown that supplementation with 2 g/d, decreased glucose levels in diabetics and reduce capillary fragility.

Megadoses of vitamin C can still be toxic in diabetics with kidney disorders. It was mentioned that vitamin C also helps to reduce body glycosylation, which shows abnormalities in the binding of sugars and proteins. In addition vitamin C reduces the accumulation of sugar sorbitol which damages eyes and kidneys. Vitamin C lowers blood pressure and plasma cholesterol helping to keep the blood flowing and protected from oxidation in a synergistic action with vitamin E. In doses of 1g/day protects the body against LDL lipoproteins.

Atherosclerosis is the best contributor to heart disease. Vitamin C prevents the formation of atheromatous plaque by inhibiting the oxidative modification of LDL's, which contributes to atherosclerotic process for their cytotoxic effects, inhibition of receptor radical scavengers and their influence on the motility of monocytes and macrophages.

Vitamin C also helps to prevent atherosclerosis through the synthesis of collagen in the arterial wall and prevent undesirable adhesion of leukocytes to the damaged artery.

Supplementation with 2 g/day reduces the adhesion of monocytes to blood vessels, effectively reverses the vasomotor dysfunction observed in patients with atherosclerosis. In addition these doses increase HDL, being highly protective against heart attacks and stroke. Risk is reduced by up to 62% in subjects consuming 700 mg/day compared with those consuming 60mg/day or less. Only Joel study has shown that low levels of ascorbic acid in serum (AAs) are marginally associated with an increased risk and fatal cardiovascular disease was significantly associated with an increased risk of mortality for all causes. Low levels of AAs are also a risk factor for cancer death in men, but unexpectedly it was associated with a decreased risk of cancer death in women.

Vitamin C has an effect antihistaminic. Subjects with low plasma vitamin C levels have elevated blood histamine and vitamin supplementation, reduces these levels. Table 7 shows the relationship of vitamin C reported in different conditions.

Low concentrations of AA	High concentration of AA
Rheumatoid arthritis (Lunec)	Cancer in women (Joel)
Cancer in Men (Joel)	Uric Acid Excretion (Stein)
Asthma (Ruskin)	Back pain and spinal discs.
Bronchospasm	Antioxidant (Kahn)
Cataract (Jackes)	Allergic process (Ruskin)
Aging (Jackes)	Blood pressure (Ringsdorff)
Retinopathy (Crary)	Constipation (Sindair)
(Macular Degeneration)	Probable association with menopause. (Smith)

Figures in parentheses indicate the reference.

Table 7. Association of serum vitamin C (ascorbic acid) in serum or plasma in different symptoms and diseases.

Author details

José Luis Silencio Barrita[1] and María del Socorro Santiago Sánchez[2]

1 Association of Chemists, National Institute for Medical Sciences and Nutrition "Salvador Zubiran", Mexico

2 Department of Nutrition and Dietetics, General Hospital No, 30 "Iztacalco", Mexico

References

[1] Akhilender Naidu K Vitamin C in human health and disease is still a mystery? An overview. Nutr J 2003, 2:1475-2891-2-7

[2] Anitra C Carr and Balz Frei.Toward a new recommended dietary allowance for vitamin C based on antioxidant and health effects in humans Am J Clin Nutr 1999;69:1086–107

[3] Ascencio C., Gomez E., Ramirez A., Pasquetti A., Bourges H. Dietary habits and nutrient intake in patients with primary Sjögren's syndrome.7th International Congress of Mucosal Immunology. August 16-20, 1992, Prague, Czechoslovaquia.

[4] Bergman F, Curstedt T, Eriksson H, van der Linden W, Sjo°vall J. Gallstoneformation in guinea pigs under different dietary conditions: effect of vitamin C on bile acid pattern. Med Biol. 1981; 59:92-98.

[5] Blomhoff R. Dietary antioxidants and cardiovascular disease Curr Opin Lipidol 16:47–54. # 2005

[6] Bowes and Church.- Food values of portions commonly used. Lippincot Co.,1975.

[7] Bourges H. Madrigal H. Chavéz A., Tablas del valor nutritivo de los alimentosmexicanos. Publicación L-12 INN, 1983, 13 ed. México.

[8] Bradley A.V. Tables of food values, Chas A. Bennet Co. 1976.

[9] Cameron E, Pauling L. Supplemental ascorbate in the supportive treatment of cancer: Prolongation of survival times in terminal human cancer. Proc Natl Acad Sci 1976; 73:3685-3689.

[10] Chen Qi, Espey M G, Krishna MC, Mitchell JB, Corpe CP, Buettner GR, Shacter E, Levine M. Pharmacologic ascorbic acid concentrations selectively kill cancer cells: Action as a pro-drug to deliver hydrogen peroxide to tissues. PNAS 2005, 102(38): 13604-13609

[11] Crary, E.J. and M.F. McCarty, 1984. Potential clinical applications for high-dose nutritional antioxidants. Med. Hypoth., 13: 77-98.

[12] Drake IM, Davies MJ, Mapstone NP, Dixon MF, Schorah CJ, White KLM, Chalmers DM, Axon ATR. Ascorbic acid may protect against human gastric cancer by scavenging mucosal oxygen radicals. Caranogenesis. 1996, 17 (3):.559-562,

[13] Engelfried J.J. The ascorbic acid saturation test. J.Lab. Med. 1944, 29:234

[14] F.J. de Abajo y M. Madurga. Vitamina C: aplicaciones terapéuticas en la actualidad. Med Clin (Barc) 1993; 101: 653-656

[15] Fisher E, McLennan SV, Tada H, Heffernan S, Yue DK, Turtle JR. (Interaction of ascorbic acid and glucose on production of collagen and proteoglycan by fibroblasts. Diabetes, 1991, 40(3):371-6)

[16] Gueguen S, Pirollet P, Leroy P, Guilland JC, Arnaud J, Paille F, Siest G, Visvikis S, Hercberg S, Herbeth B. Changes in Serum Retinol, α-Tocopherol, Vitamin C, Carotenoids, Zinc and Selenium after Micronutrient Supplementation during Alcohol Rehabilitation. J Am Coll Nutr, 2003, 22:303-310

[17] Gomez E, Silencio JL, Bourges H. Vitamin B2, B6 and C status in patients with primary sjögren's syndrome. 7th International congress of mucosal immunology, august 16-20, 1992, Prague Csechoslovaquia.

[18] Harris J.L. Vitamin C saturation test. Standarization meassurements at graded levels of intake. Lancet 1943, 1:515,

[19] Head KA, Ascorbic Acid in the Prevention and Treatment of Cancer. Altern Med Rev 1998; 3(3):174-186)

[20] Herrera, V. Más evidencia en contra del uso de vitaminas y antioxidantes en la prevención de enfermedades crónicas Evidencia Actualización en la Práctica Ambulatoria- 2002, 5(6): Nov-Dic

[21] Hoffman-La Roche F. Compendio de vitaminas, 1970, Basilea, Suiza.

[22] Hood J., Burns Ch.A., Hodges R.E. Sjögren's syndrome in scurvy. N. Engl. J. Med. 1970, 282:1120

[23] Horrobin D.F., Manku M.S. The regulation of prostaglandin E1 formation: A candidate for one of the fundamental mechanism involved in the actions of vitamin C. Med. Hypotheses. 1979, 5:849.

[24] Ip C. Interaction of vitamin C and selenium supplementation in the modification of mammary carcinogenesis in rats. J Natl Cancer Inst 1986;77:299-303.

[25] Iqbal K, Khan A Ali Khan Khattak MM. Biological Significance of Ascorbic Acid (Vitamin C) in Human Health: A Review Pakistan J Nutr 2004, 3(1):5-13,

[26] Jacob RA, Pianalto FS, Agee RE. Cellular ascorbate depletion in healthy men. J Nutr. 1992; 122:1111-1118.

[27] Jacques, P.F., L.T. Chylack, R.B. McGandy and S.C Hartz, 1988. Antioxidant status in persons with and without senile cataract. Arch Ophthalmol, 106: 337-340.

[28] Jenkins SA. Biliary lipids, bile acids and gallstone formation in hypovitaminotic C guinea-pigs. Br J Nutr. 1978;40:317-322

[29] Jenkins SA. Hypovitaminosis C and cholelithiasis in guinea pigs. Biochem Biophys Res Commun. 1977; 77:1030-1035.

[30] Kahn, H.A., H.M. Leibowitz and J.P.Ganley, 1977. The Framingham Eye Study: Outline and major prevalence findings. Am. J. Epidemiol., 106: 17-32.

[31] Kathleen A. Head, N.D. Ascorbic Acid in the Prevention and Treatment of Cancer Altern Med Rev 1998; 3(3):174-186

[32] Kessler S. A laboratory model for Sjögren syndrome. Lancet 1968, 52:671

[33] Khaw KT, Woodhouse P. Interrelation of vitamin C, infection, haemostatic factors, and cardiovascular disease BMJ 1995; 310:1559-1563

[34] Lunec, J.B., 1985. The determination of dehydroascorbic acid and ascorbic acid in the serum and sinovial fluid of patients with rheumatoid arthritis. Free Radical Research Communications, 1: 31-39.

[35] Lupulescu A The Role of Vitamins A, B Carotene, E and C in Cancer Cell Biology. Intern. Vit. Nutr. Res. 1993; 63:3-14.

[36] Marks J. A guide to the vitamins, their role in health and disease. MTP Medical and Technical Pub. 1975, England.

[37] McLennan S, Yue DK, Fisher E, Capogreco C, Heffernan S, Ross GR, Turtle JR. Deficiency of ascorbic acid in experimental diabetes. Relationship with collagen and polyol pathway abnormalities. Diabetes, 1988 ,37(3):359-361,

[38] MRC/BHF Heart Protection Study of antioxidant vitamin supplementation in 20.536 high-risk individuals: a randomised-placebo controlled trial. Heart Protection Study Collaborative Group. The Lancet 2002;360:23-33

[39] Mullan BA, Young IS, Fee Hd, McCance DR. Ascorbic Acid Reduces Blood Pressure and Arterial Stiffness in Type 2 Diabetes Hypertension. 2002; 40:804-809

[40] Omaye, T.,Scala J.H., Jacob R.A. Plasma ascorbic acid in adult males: effect of depletion and supplementation. Am. J. Clin. Nutr. 1986, 44: 257,

[41] Paul A.A., Southgate D.A. The composition of foods, 14th edition Elsevier/North Holland Biomedical Press, 1985.

[42] Pecoraro RE, Chen MS. Ascorbic acid metabolism in diabetes mellitus. Ann New York Acad Sci, 1987, 498(1): 248-258,

[43] Ringsdorf, W.M. Jr. and E. Cheraskin, 1981. Ascorbic acid and glaucoma: A Rev. J. Holistic. Med., 3: 167-172.

[44] Roe J.H., Kuether C.A. determination of ascorbic acid in wholeblood and urine through the 2,4-dinitotrophenylhydrazine derivative of dehydroascorbic acid. J. Biol. Chem, 1942, 147:399 .

[45] Rogur L, Lundblad and Fiona M . MacDonald. Handbook of Biochemistry and Molecular Biology, Fourth Edition, Edited by CRC Press 2010 Pages 243–250

[46] Rose C. R., Nohrwld L.D. Quantitative analysis of ascorbic acid and dehydroascorbic acid by High Performance Liquid Chromatography Anal. Biochem. 114:140, 1981.

[47] Rune Blomhoff. Dietary antioxidants and cardiovascular disease. Curr Opin Lipidol 16:47–54. # 2005

[48] Ruskin, S.L., 1947. Sodium ascorbate in the treatment of allergic disturbances. The role of adrenal cortical hormone-sodium-vitamin C. Am. J. Dig. Dis.,14: 302-306

[49] Sahyoun, R.N.,1996. Carotenoids, vitamins C and E and mortality in an elderly population. Am. J. epidemio.,144: 501-511.

[50] Sánchez-Moreno C, Cano MP, de Ancos B, Plaza L, Olmedilla B, Granado F, Martín A.Effect of orange juice intake on vitamin C concentrationsand biomarkers of antioxidant status in humansAm J Clin Nutr 2003; 78:454–60.

[51] Silencio Barrita JL Vitaminas: conceptos generales, Nutr clin 2006;9(3):36-44

[52] Simon JA., Esther SH, Jeffrey AT. Relation of serum ascorbic acid to mortality among US adults. J. Am. College of Nutr, 2001.20(3);456

[53] Simon JA, Hudes ES. Serum Ascorbic Acid and Gallbladder Disease Prevalence Among US Adults. The Third National Health and Nutrition Examination Survey (NHANES III) Arch Intern Med. 2000; 160:931-936

[54] Simon JA. Ascorbic acid and cholesterol gallstones. Med Hypotheses. 1993; 40:81-84.

[55] Simon J.A Vitamin C and cardiovascular disease: a review J Ame Collage of Nutr, 1992, 11 (2):107-125

[56] Simon JA, Hudes ES, Tice JA. Relation of serum ascorbic acid to mortality among US adults. J Am Coll Nutr 2001; 20:255–63.

[57] Simopoulos AP. Human requirement for N-3 polyunsaturated fatty acids. Poult Sci. 2000 Jul;79(7):961-70.

[58] Spittle C.R. Artherosclerosis and vitamin C. Lancet, december 11:1280, 1971.

[59] Stein, H.B., 1976. Ascorbic acid-induced uricosuria: a consequence of megavitamin therapy. Ann. Intern.Med., 84: 385-388.

[60] Sinclair AJ, Girling AJ, Gray L. an investigation of the relationshipbetween free radical activity and vitamin C metabolism in ederly diabetic subjects with retinopathy. Gerontology 1992, 38:266-274

[61] Smith CJ. Non-hormonal control of vaso-motor flushing in menopausal patients . Chicago med, 1984.

[62] Valk EE, Hornstra G Relationship between vitamin E requirement and polyunsaturated fatty acid intake in man: a review. Int J Vitam Nutr Res. 2000 Mar;70(2):31-42.

[63] Walingo KM. Role of vitamin C (ascorbic acid) on human health- A review. African J Food Agric Nutr Dev (AJFAND): 2005, 5(1):1-25

[64] Watt B.K. A.L. Composition of foods. Agriculture Handbook num. 9, United States, Departament of Agriculture, USA, 1975.

[65] Yi Li and Herb E. Schellhorn*New Developments and Novel Therapeutic Perspectives for Vitamin C J. Nutr. 2007, 137: 2171–2184,

Chapter 19

Protective Effect of Silymarin on Liver Damage by Xenobiotics

José A. Morales-González, Evila Gayosso-Islas, Cecilia Sánchez-Moreno, Carmen Valadez-Vega, Ángel Morales-González, Jaime Esquivel-Soto, Cesar Esquivel-Chirino, Manuel García-Luna y González-Rubio and Eduardo Madrigal-Santillán

Additional information is available at the end of the chapter

http://dx.doi.org/10.5772/51502

1. Introduction

The liver is the vertebrates' largest internal organ. It weighs nearly 1.5 kg, is dark red in color, and is situated in the upper right quadrant of the abdominal cavity. Among the functions that it performs are the following: the metabolism of lipids and carbohydrates, and the synthesis of proteins, coagulation factors, and biliary salts. Eighty percent of the hepatic parenchyma is made up of hepatocytes, which are the cells mainly responsible for maintaining every function that the liver in its entirety requires to sustain the body's normal physiological functions in general. In addition to hepatocytes, the liver possesses other cells, such as the so-called Kupffer cells (hepatic macrophages), Ito cells, endothelial cells. The hepatocytes are disposed in the liver in groups denominated lobules, which have a central orifice comprised of the bile duct and by means of which the biliary salts are excreted. The anatomical loss of the structure of the hepatic lobule is considered a symptom of severe damage to the liver; it can be accompanied by partial or total loss of some physiological function, as in the case of alcohol-related hepatic cirrhosis. [23].

2. Hepatic regeneration

Liver regeneration is a fundamental response of the liver on encountering tissue damage. The complex interaction of factors that determine this response involves a stimulus (experimentally, a hepatectomy), gene expression, and the interaction of other factors that modulate the response. This proliferation depends on the hepatocytes, epithelial bile cells, Kupffer cells, and Ito cells. [24].

The mechanisms of hepatic growth have been studied in detail in experimental models. In the latter, regeneration is induced whether by tissue resection (partial hepatectomy) or by death of the hepatocytes (toxic damage). The principles that govern the growth of this organ in these systems also apply to clinical situations, such as, for example, fulminating liver failure, acute and chronic hepatitis, partial hepatectomy for treating liver cancer, or even in liver transplant donors. Evidence that there is a humoral growth factor of the hepatocyte has been observed in animal models and in patients with liver disease from the 1980s. [1, 13, 34, 10].

3. Ethanol

On being ingested, alcohol (also called ethanol) produces a series of biochemical reactions that lead to the affectation of numerous organs involving economy, having as the endpoint the development of hepatic diseases such as alcoholic hepatitis and cirrhosis. Despite that much is known about the physiopathological mechanisms that trigger ethanol within the organism, it has been observed that a sole mechanism of damage cannot fully explain all of the adverse effects that ethanol produces in the organism or in one organ in particular. [37, 30].

A factor that is referred as playing a central role in the many adverse effects that ethanol exerts on the organism and that has been the focus of attention of many researchers is the excessive generation of molecules called free radicals, which can produce a condition known as oxidative stress, which triggers diverse alterations in the cell's biochemical processes that can finally activate the mechanism of programmed cell death, also known as apoptosis. [28, 26, 17, 19, 25].

Of particular importance for the objective of this chapter is the focus on a particular class of free radicals that are oxygen derivatives, because these are the main chemical entities that are produced within the organism and that affect it in general.

4. Ethanol metabolism

Ethanol is absorbed rapidly in the gastrointestinal tract; the surface of greatest adsorption is the first portion of the small intestine with 70%; 20% is absorbed in the stomach, and the remainder, in the colon. Diverse factors can cause the increase in absorption speed, such as gastric emptying, ingestion without food, ethanol dilution (maximum absorption occurs at a 20% concentration), and carbonation. Under optimal conditions, 80-90% of the ingested dose is completely absorbed within 60 minutes. Similarly, there are factors that can delay ethanol absorption (from 2-6 hours), including high concentrations of the latter, the presence of food, the co-existence of gastrointestinal diseases, the administration of drugs, and individual variations [14, 37].

Once ethanol is absorbed, it is distributed to all of the tissues, being concentrated in greatest proportion in brain, blood, eye, and cerebrospinal fluid, crossing the feto-placentary and hematocephalic barrier [44]. Gender difference is a factor that modifies the distributed ethanol volume; this is due to its hydrosolubility and to that it is not distributed in body fats, which explains why in females this parameter is found diminished compared with males.

Ethanol is eliminated mainly (> 90%) by the liver through the enzymatic oxidation pathway; 5-10% is excreted without changes by the kidneys, lungs, and in sweat [14, 30]. The liver is the primary site of ethanol metabolism through the following three different enzymatic systems: Alcohol dehydrogenases (ADH); Microsomal ethanol oxidation system (MEOS); Catalase system.

5. Liver regeneration and ethanol

Ethanol is a well known hepatotoxic xenobiotic because hepatotoxicity has been well documented in humans as well as in animals. Although aspects concerning the pathogenesis of liver damage have been widely studied, it is known that liver regeneration restores the functional hepatic mass after hepatic damage caused by toxins. Suppression of the regenerating capacity of the liver by ethanol is the major factor of liver damage. [45]. Although the effects of acute or chronic administration of ethanol on the proliferative capacity of the liver to regenerate itself has been studied, the precise mechanism by which ethanol affects hepatocellular function and the regenerative process are poorly explained. [31, 29, 38].

Liver regeneration induced by partial hepatectomy in rats represents an ideal model of controlled hepatocellular growth. This surgical procedure has been sufficiently employed to study the factors than can be implicated in the growth of the liver. Endogenous signals have been described to control hepatic regeneration. The first marker of DNA synthesis in partially hepatectomized rats (70%) occurring normally 24-28 hours postsurgery comprises an enormous action of growth factors and cytokines affecting expression of the gene of the hepatocytes, associated with initiation of the cell cycle. [2]. It has indicated that the hepatocytes enter into a state denominated "priming" to thus begin replication and response to growth factors, that is, which range from the quiescent to the G 1 phase of the cell cycle. The progression of hepatic cells requires the activation of cyclin-dependent kinases that are regulated by cyclins and cyclin-dependent kinase inhibitors. [39]. It has also been demonstrated that a dose of ethanol importantly diminishes the specific activity of two enzymes related with the metabolism of DNA synthesis, which are thymidine kinase and thymidylate synthetase. [47, 11, 32].

6. Free radicals

Free radicals are the result of the organism's own physiological processes, such as the metabolism of food, respiration, exercise, or even those generated by environmental factors, such as contamination, tobacco, or by drugs, chemical additives, etc. Free radicals (FR) are atoms or groups of atoms that in their atomic structure present one or more unpaired electrons (odd in number) in the outer orbit. This spatial configuration generates in the molecule distinct physical and chemical properties such as heightened reactivity and diminished lifetime, respectively. [5, 27].

This instability confers on these physical avidity for the uptake of an electron of any other molecule in its ambit (stable molecules), causing the affected structure to remain unstable with the purpose of reaching its electrochemical stability. Once the free radical has achieved trapping the electron that it requires for pairing with its free electron, the stable molecule that cedes the latter to it in turn becomes a free radical, due to its remaining with an unpaired electron, this initiating a true chain reaction that destroys our cells. [4, 7].

In aerobic cells, there are diverse pathways that lead to the production of Oxygen-derived free radicals (OFR). The main sources are enzymes associated with the metabolism of arachidonic acid, such as cycloxygenase, lipoxygenase, and cytochrome P-450. The presence and ubiquity of enzymes (superoxide dismutase, catalase, and peroxidase) that eliminate secondary products in a univalent pathway in aerobic cells suggest that the superoxide anions and hydrogen peroxide are important secondary products of oxidative metabolism. [40, 7]. Reactive oxygen species (ROS) can damage macromolecules such as DNA, carbohydrates, and proteins. These cytotoxic oxygen species can be classified as two types:

1. the free radicals, such as the superoxide radical (O_2) and the hydroxyl radical ($^{\cdot}OH$), and
2. non-radical oxygen species, such as hydrogen peroxide (H_2O_2), the oxygen singlette (O_1), which is a very toxic species, peroxynitrite (ONOO), and Hypochlorous acid (HOCL).

The instable radicals attack cell components, causing damage to the lipids, proteins, and the DNA, which can trigger a chain of events that result in cellular damage. [7, 21]. These reductive processes are accelerated by the presence of trace metals such as iron (Fe) and copper (Cu) and of specific enzymes such as monoxygenases and certain oxidases. [7, 21]

7. Oxidative stress

In 1954, an Argentine researcher, Rebeca Gerschman, suggested for the first time that FR were toxic agents and generators of disease. [12].

Due to the atomic instability of FR, the latter collide with a biomolecule and subtract an electron, oxidating it, losing in this manner its specific function in the cell. If lipids are involved (polyunsaturated fatty acids), the structures rich in these are damaged, such as the cell membranes and the lipoproteins. In the former, the permeability is altered, leading to edema and cell death, and in the latter, to oxygenation of the Low-density lipoproteins (LDL) and genesis of the atheromatous plaque. The characteristics of lipid oxygenation by FR involve a chain reaction in which the fatty acid, on being oxygenated, becomes a fatty acid radical with the capacity of oxidizing another, neighboring molecule. This process is known as lipid peroxidation and it generates numerous subproducts, many of these, such as Malondialdehyde (MDA), whose determination in tissues, plasma, or urine is one of the methods for evaluating oxidative stress. In the case of proteins, these preferentially oxidize the amino acids (phenylalanine, tyrosine, triptophan, histidine, and methionine), and consequently form peptide chain overlapping, protein fragmentation, and the formation of carbonyl groups, and these impede the normal development of their functions (ionic membrane transporters, receptors, and cellular messengers), enzymes that regulate the cell's metabolism, etc.). [7, 33].

8. Liver regeneration, ethanol, and free radicals

While ROS and FR are generated during ethanol metabolism, causing oxidative stress and lipoperoxidation in the liver, they can also form a significant pathway of damage to the regenerative process of the hepatocyte. In this process, ethanol-induced FR and the generation of ROS involve the mitochondria, the microsomal cytochrome P450 2E1, the iron (FE) ion, and less frequently, peroxisomes, cytosolic xanthines, and aldehyde oxidases, to regulate cellular proliferation, acting as direct or indirect factors. [37, 2].

In general, ROS-derived FR intervene in the persistent bombardment of molecules by reactive oxygen radicals, thus maintaining redox homeostasis, in such a manner that during liver regeneration, these can modify the metabolic response necessary for carrying out cellular mitosis in the hepatocyte. While FR are generated and utilized by the cells as neutrophils, monocytes, macrophages, eosinophils, and fibroblasts for eliminating foreign organisms or toxic substances, the increase of FR due to exposure to ethanol leads to cellular deterioration that in turn produces hepatic alterations, with an unfavorable influence on cell proliferative action. [25].

9. Antioxidants

Halliwell defines an antioxidant as all substances that on being found present at low concentrations with respect to those of an oxidizable substrate (biomolecule), delays or prevents the oxidation of this substrate. The antioxidant, on colliding with FR, is ceded to an electron, in turn oxidizing itself and transforming itself into a non-toxic, weak FR and, in some cases such as with vitamin E, it can regenerate itself into its primitive state due to the action of other antioxidants. Not all antioxidants act in this way: the so-called enzymatic antioxidants catalyze or accelerate chemical reactions that utilize substrates that in turn react with FR. Of the numerous classifications of antioxidants, it is recommended to adopt that which divides these into the following: exogenes or antioxidants that enter through the alimentary chain, and endogenes that are synthesized by the cell. Each antioxidant possesses an affinity for a determined FR or for several. Vitamin E, beta-carotene, and lycopene act within the liposoluble medium of the cell and their absorption and transport are found to be very much linked with that of the lipids. Vitamin E is considered the most important protector of lipid molecules. [27].

Life in the presence of molecular oxygen requires the possession of a multiple battery of defenses against the diverse oxygen FR, which on the one hand tend to impede their formation and on the other, neutralize them once they are formed. These defenses exert an effect at five levels [7, 21, 33]:

9.1. First level

This consists of editing univalent oxygen reduction through enzymatic systems capable of effecting consecutive tetravalent reduction without releasing the partially reduced intermediaries; this is achieved with great effectiveness by the cytochrome-oxidase system of the mitochondrial respiratory chain, which is responsible for more than 90% of oxygen reduction in the human organism.

9.2. Second level

This is constituted of enzymes specialized in the uptake of the superoxide anion radical (O 2 –). These are Superoxide dismutase (SOD), the methaloenzyme that catalyzes the dismutation of the superoxide anion radical to provide molecular oxygen and hydrogen peroxide, with such great effectiveness that it approaches the theoretical limit of diffusion. In the cells of the eukaryotic organisms, there are two of these: one is cytoplasmatic, and the other is mitochondrial. SOD was described by Fridovich in 1975.

9.3. Third level

This is conferred by a group of specialized enzymes on neutralizing hydrogen peroxide. Among these is catalase, which is found in the peroxisomes and which catalyzes the dismutation reaction.

Also in mammals, glutathione peroxidase (a cytoplasmic enzyme that contains selenium) is the most important.

9.4. Fourth level

Here the hydroxyl radical produced in the Haber-Weiss cycle can neutralized by vitamin E or alpha-tocopherol, which is an effective antioxidant and that due to its hydrophobicity is found in biological membranes in which its protection is particularly important. In addition, vitamin C or ascorbic acid is a reducer agent or electron donor and reacts rapidly with the OH– radical and with the superoxide anion.

9.5. Fifth level

Once the molecular damage is produced, there is a fifth level of defense that consists of repair. It has been demonstrated that FR were capable of causing breaks in the DNA chain and even of inducing mutagenesis, but there are enzymatic repair mechanisms that permit reestablishment of genetic information.

10. Antioxidants and their role in hepatoprotection

The term antioxidant was originally utilized to refer specifically to a chemical product that prevented the consumption of oxygen [6]; thus, antioxidants are defined as molecules whose function is to delay or prevent the oxygenation of other molecules. The importance of antioxidants lies in their mission to end oxidation reactions that are found in the process and to impede their generating new oxidation reactions on acting in a type of sacrifice on oxidating themselves. There are endogenous and exogenous antioxidants in nature. Some of the best-known exogenous antioxidant substances are the following: β-carotene (pro-vitamin A); retinol (vitamin (A); ascorbic acid (vitamin C); α-tocopherol (vitamin E); oligoelements such as selenium; amino acids such as glycine, and flavonoids such as *silymarin*, among other organic compounds [46, 36].

Historically, it is known that the first investigations on the role that antioxidants play in Biology were centered on their intervention in preventing the oxidation of unsaturated fats, which is the main cause of rancidity in food. However, it was the identification of vitamins A, C, and E as antioxidant substances that revolutionized the study area of antioxidants and that led to elucidating the importance of these substances in the defense system of live organisms. [36].

Due to their solubilizing nature, antioxidant compounds have been divided into hydrophilics (phenolic compounds and vitamin C) and lipophilics (carotenoids and vitamin E). The antioxidant capacity of phenolic compounds is due principally to their redox properties, which allow them to act as reducing agents, hydrogen and electron donors, and individual oxygen inhibitors, while vitamin C's antioxidant action is due to its possessing two free electrons that can be taken up by Free radicals (FR), as well as by other Reactive oxygen species (ROS), which lack an electron in their molecular structure. Carotenoids are deactivators of electronically excited sensitizing molecules, which are involved in the generation of radicals and individual oxygen, and the antioxidant activity of vitamin A is characterized by hydrogen donation, avoiding chain reactions. [7, 21, 33].

The antioxidant defense system is composed of a group of substances that, on being present at low concentrations with respect to the oxidizable substrate, delay or significantly prevent oxygenation of the latter. Given that FR such as ROS are inevitably produced constantly during metabolic processes, in general it may be considered as an oxidizable substrate to nearly all organic or inorganic molecules that are found in living cells, such as proteins, lipids, carbohydrates, and DNA molecules. Antioxidants impede other molecules from binding to oxygen on reacting or interacting more rapidly with FR and ROS than with the remainder of molecules that are present in the microenvironment in which they are found (plasma membrane, cytosol, the nucleus, or Extracellular fluid [ECF]). Antioxidant action is one of the sacrifices of its own molecular integrity in order to avoid alterations in the remainder of vitally functioning or more important molecules. In the case of the exogenic antioxidants, replacement through consumption in the diet is of highest importance, because these act as suicide molecules on encountering FR, as previously mentioned. [7, 21, 33].

This is the reason that, for several years, diverse researchers have been carrying out experimental studies that demonstrate the importance of the role of antioxidants in protection and/or hepatic regeneration in animals. Thus, in this chapter, the principal antioxidants will be described that play an important role in the regeneration of hepatic cells and in the prevention of damage deriving from alcohol.

11. Flavonoids

Flavonoids are compounds that make up part of the polyphenols and are also considered essentials nutrients. Their basic chemical structure consists of two benzene rings bound by means of a three-atom heterocyclic carbon chain. Oxidation of the structure gives rise to several families of flavonoids (flavons, flavonols, flavanons, anthocyanins, flavanols, and isoflavons), and the chemical modifications that each family can undergo give rise to >5,000 compounds identified by their particular properties. [16].

Flavonoid digestion, absorption, and metabolism have common pathways with small differences, such as, for example, unconjugated/non-conjugated flavonoids can be absorbed at the stomach level, while conjugated flavonoids are digested and absorbed at the intestinal level by extracellular enzymes on the enterocyte brush border. After absorption, flavonoids are conjugated by methylation, sukfonation, ands glucoronidation reactions due to their biological activity, such as facilitating their excretion by biliary or urinary route. The conjugation type the site where this occurs determine that metabolite's biological action, together with the protein binding for its circulation and interaction with cellular membranes and lipoproteins. Flavonoid metabolites (conjugated or not) penetrate the tissues in which they possess some function (mainly antioxidant), or are metabolized. [27].

On the other hand, the flavonoids possess implications in health; in recent years, the properties of these compounds have been studied in relation to diverse pathologies. In diabetes, these compounds present regulation of glycemia through diverse mechanisms that include the inhibition of some enzymes such as α-glucosidase, glucose 6 phosphatase, and phosphorylated glycogen. The flavonoids possess other characteristics such as the trapping of molecules of glioxal and methyl-glioxal molecules, which propitiate the formation of advanced final products of glycosylation that are found to be directly related with micro- and macrovascular complications. They also regulate the rise or fall of transporter proteins; the structure of some flavonoids appears to have important participation with regard to the studied benefits. [16].

More research is needed because great majority of the former has been conducted in animals, to determine effects and dosage. Flavonoids in the menopause result in controversial effects due to the population type studied, that is, Asiatic, absorption, metabolism, the binding of isoflavones to estrogen receptors, etc.; however, they appear to possess a beneficial effect in terms of the prevention of certain types of cancer and osteoporosis. [16].

The flavonoids absorb Ultraviolet light (UV) from the sun and possess direct and indirect antioxidant effects (through the induction of cytoprotector proteins). Topical application (on the human skin) of the polyphenolic fraction of green tea protects against immunosuppression and inhibits the erythema and the formation of pyrimidine dimers in DNA caused by UV. On representing one of the most important lifestyle factors, alimentation can importantly affect the incidence and initiation of cardiovascular or neurodegenerative diseases. The cardioprotector effect of flavonoids is based on reducing oxidation and blood concentrations of the binding of cholesterol to Low-density lipoproteins (LDL); flavonoids reduce endothelial dysfunction and blood pressure and increase the HDL-bound cholesterol concentration. Flavonoids possess a neuroprotector effect because they protect the neurons from oxidative stress by means of induction of antioxidant defenses, modulation of signaling cascades, mitochondrial interactions, apoptotic processes, or by synthesis/degradation of the β-amyloid peptide. The potential effect of flavonoids as neuroprotectors is due to three main factors: they prevent neurodegeneration; inhibit neuroinflammation, and reduce the diminution of age-related cognitive functions. [16].

In cancer, the flavonoids have been classified as chemopreventive, as blockers as well as inhibitors, given their functions in carcinogenesis, in which they modulate transduction signaling in cellular proliferation and angiogenesis, modulate enzymes for the metabolic activation of procarcinogens and the detoxification of carcinogens, and modulate enzymes in the biosynthesis of anti-oxidant-pro-oxidant estrogen activity estrogen (promoting oxidative homeostasis, rendering its antioxidative capacity as a contribution to antineoplastic as well as preventive as well as therapeutic activity due to inhibiting the activation of mitogenic kinases and transduction factors, while pro-oxidative activity increases the cell damage that promotes detention of the cell cycle and apoptosis). In obesity, the flavonoids have been identified as reducer factors of fat mass and as inhibitors of fat mass deposition and catabolic activity. [16].

The procyanidins and proanthocyanidins have demonstrated, in human population, to diminish visceral fatty mass (depending on the dose) with an associated increase of adiponectin. This diminution is linked with the malabsorption of carbohydrates and lipids due to enzyme inhibition. It has been observed that the procyanidins increase β-oxidation and inhibit the expression of genes that promote the synthesis of fatty acids. Epigallocatechin gallate can increase energy expenditure and lipid oxidation in humans; it is thought that this is possible because of the increase of thermogenesis and the inhibition of the activity of the lipase, as well as, according to studies *in vitro,* the inhibition of lipogenesis and apoptosis of the adipocytes. Catechins that alter the deposition of adipose tissue related with diminution of the respiratory co-efficient and greater oxygen consumption, and thermogenesis induced by the sympathetic nervous system. Phytoestrogens can improve obesity and its alterations on diminishing insulin resistance, thus lipogenesis, as well as inhibition of the mechanisms for cell differentiation and proliferation. The study of flavonoids and their effects on the prevention and treatment of obesity is a widespread, yet incomplete research field. [16].

The metabolism of phytoestrogens and their maximum concentration in serum presents great variability, depending on genetic differences and estrogen exposure in early life stages. [16].

12. Silimarina (*silybum marianum*)

Silymarin is a compound of natural origin extracted from the *Silybum marianum* plant, popularly known as St. Mary's thistle, whose active ingredients are flavonoids such as silybin, silydianin, and silycristin. This compound has attracted attention because of its possessing antifibrogenic properties, which have permitted it to be studied for its very promising actions in experimental hepatic damage. In general, it possesses functions such as its antioxidant one, and it can diminish hepatic damage because of its cytoprotection as well as due to its inhibition of Kupffer cell function. [41].

Silymarin, derived from the milk thistle plant named *Silybum marianum*, has been used since time past as a natural remedy for combating liver diseases. *Silymarin* and its active constituents (silybinin, silycristin, and silydianine, among others), have been classified as uptakers of free radicals and inhibitors of lipoperoxidation; some studies also suggest that that they increase the synthesis of hepatocytes, diminish the activity of tumor promoters, stabilize mastocyte cells, and act as iron chelates. [8].

Silybum marianum belongs to the Aster family (Asteraceae or Compositae), which includes daisies and thistles. The milk thistle is distributed widely throughout Europe, was the first plant that appeared in North America to the European colonizers, and is at present established in the South of the U.S., California, and South America. [22].

The name milk thistle is derived from the characteristics of its thorny leaves with white veins, which, according to the legend, were carried by the Virgin Mary. Its name *Cardo lechoso* derives from the same tradition. The mature plant has large flowers, of a brilliant purple color, and abundant thorns of significant appearance. The milk thistle grows in places where exposure to the sun is abundant. [15].

Extracts of the milk thistle have been used as medical remedies from ancestral Greece, when Dioscorides, a Greek herbalist, wrote that the seeds of the milk thistle could cure the bite of a poisonous snake. Pliny noted that the mixture of the juice of the plant and its honey were excellent for bile tract disorders. [9]. In 1596, Gerard mentioned *Silybum marianum* as a major remedy against melancholy or black bile. The milk thistle was sold for treating liver diseases. In the 1960s, observed that milk thistle was an excellent remedy for cleaning obstructions of the liver and spleen, notwithstanding that infusions of the fresh roots and seeds were effective for counteracting jaundice.

The main active agent of the milk thistle is *silymarin*, a mixture of flavonolignans, silydianine, silycristin, and silybinin, the latter the most biologically active extract; the flavonoids appear to be activated as trappers of free radicals and as plasmatic membrane stabilizers. Concentrations of *silymarin* are localized in the fruit of the plant, as well as in the seeds and leaves, from which *silymarin* is extracted with 95%-proof ethanol, achieving a brilliant yellow liquid. The term flavonoid is derived from *flavus*, which denotes yellow. [20, 16].

The standardized extract of *silymarin* contains 70% *silymarin*. Pharmacokinetic studies have shown that there is rapid absorption of silybinin into the bloodstream after an oral dose. Steady-state plasma concentrations are reached after 2 hours and the elimination half-life is 6 hours. [Lorenz et al., 1984, 3]. From 3-8% of an oral dose is excreted in the urine and from 20-40% is recovered in the bile as glucuronide and sulfate. [42].

Silybinin works as an antioxidant, reacting rapidly with oxygen free radicals as demonstrated *in vitro* with hydroxyl anions and hypochlorous acid. Reported activities include the inhibition of hepatocyte lipoperoxidation, the microsomal membrane in rats, and protection against genomic damage through the suppression of hydrogen peroxide, superoxide anions, and lipoxygenase. It is thought that silybinin also increases the synthesis of the proteins of the hepatocyte through stimulation of the activity of the ribosomal RNA (rRNA) polymerase. In addition, silybinin diminishes hepatic and mitochondrial oxidation induced by an iron overcharge and acts as an iron chelate. [16].

13. Antioxidant and hepatoprotector action

Silymarin is an active principle that possesses hepatoprotector and regenerative action; its mechanism of action derives from its capacity to counterarrest the action of FR, which are formed due to the action of toxins that damage the cell membranes (lipid peroxidation), competitive inhibition through external cell membrane modification of hepatocytes; it forms a complex that impedes the entrance of toxins into the interior of liver cells and, on the other hand, metabolically stimulates hepatic cells, in addition to activating RNA biosythesis of the ribosomes, stimulating protein formation. In a study published by [41], the authors observed that *silymarin*'s protector effect on hepatic cells in rats when they employed this as a comparison factor on measuring liver weight/animal weight % (hepatomegaly), their values always being less that those of other groups administered with other possibly antioxidant substances; no significant difference was observed between the *silymarin* group and the *silymarin*-alcohol group, thus demonstrating the protection of *silymarin*. On the other hand, *silymarin* diminishes Kupffer cell activity and the production of glutathione, also inhibiting its oxidation. Participation has also been shown in the increase of protein synthesis in the hepatocyte on stimulating polymerase I RNA activity. *Silymarin* reduces collagen accumulation by 30% in biliary fibrosis induced in rat. An assay in humans reported a slight increase in the survival of persons with cirrhotic alcoholism compared with untreated controls [2].

Silymarin is a flavonoid derived from the *Silybum marianum* plant that has been employed for some 2,000 years for the treatment of liver diseases. At present, its use as an alternative drug has extended throughout Europe and the U.S. *Silymarin* acts as a hepatoprotector due to its antioxidant effect, which has been observed to inhibit liver damage due to the releasing of the substances of free radicals, such as ethanol, acetaminophen, and Carbon tetrachloride (CCL_4), in addition to increasing the activity of SOD and glutathione. As a uptaker of free radicals, *silymarin* can inhibit the lipid peroxidation cascade in the cell membranes. The hepatoprotector effect of this flavonoid also can be explained by an anti-inflammatory effect, in which it has been observed that *silymarin* acts on the functions of the Kupffer cells. Inhibition also has been reported in the activation of the Nuclear kappa-Beta [NK-B) transcription factor. [2, 16, 7, 21, 33].

14. *Silymarin* and Exercise

During physical activity, oxygen consumption increases, which produces oxidative stress that leads to the generation of free radicals, which are highly toxic for the cell, because these interact with organic molecules susceptible to being oxidized, such as unsaturated fatty acids, which causes lipoperoxidation. To avoid this damage by FR, there are the following antioxidant systems: Superoxide-dismutase (SOD), and Catalase (CAT), in addition to other protector substances such as vitamins A, C, and E and the flavonoids, which trap free radicals. (unpublished data)

In experiments conducted by our research group on groups of rats that were submitted to daily aerobic exercise in a physical-activity cage for 20 minutes during 4 weeks (5 days/week) and on another group of rats submitted to physical activity plus administration of *silymarin* (200 mg/kg of weight) prior to exercise, with daily quantification of physical performance and at the at the end of the experiment, quantification of DNA in serum and of SOD and CAT activity in liver. We found that in the group with physical activity, MDA increased 134% (in serum) and 123% (in liver) vs. control rats. In the group with exercise plus *silymarin*, MDA returned to normality (in serum and in liver). Catalase activity increases during exercise (118%) and with exercise plus *silymarin* (137%). SOD activity exhibited no modifications in any treatment. Finally, we found an increase of physical activity in the group administered *silymarin* (27%) in comparison with the group in which no *silymarin* was administered. (unpublished data)

A protector effect was found of *silymarin* during exercise, because it diminishes MDA levels in serum as well as in liver, which translates into diminution of the production of free radicals, causing as a consequence less cellular damage, which in turn leads to an increase in physical performance.

15. Conclusions

The process of the induction of oxidative stress generated in the liver due to the presence of ethanol implies the conjugation of various factors. The role that these factors play in the de-

http://dx.doi.org/10.5772/51502

velopment of oxidative stress depends in part on whether acute or chronic intoxication is involved. The factors that contribute to the development of oxidative stress imply disequilibrium among pro- and antioxidant factors. It can occur that oxidative stress develops if the xenobiotic increases the pro-oxidant factors (the generation of Oxygen-generated free radicals [OFR]) or decreases intracellular antioxidant factors. In whichever of the two cases, the general result is important damage to the hepatocyte that can lead to general damage to the DNA that, in turn, can comprise a determining factor in the induction of the apoptotic system (programmed cell death) of the cell, thus accelerating its death and destruction.

The study of the factors that determine the increase in the generation of OFR in the liver, originating due to acute or chronic intoxication with a xenobiotic, is of great importance because it will allow diminishing the damage that these reactive species produce within the hepatocyte. On the other hand, despite that at present much is known concerning the physiopathological mechanisms of ethanol ingestion-related liver damage and the role that the production of oxygen-generated free radicals plays in these processes, the exact extent of this damage, as well as how to prevent it, remains unknown with precision. There is evidence obtained from laboratory models that the ingestion of natural antioxidants, such as vitamins A, C, and E, oligoelements (selenium), amino acids (glycine), and principally flavonoids, such as *silymarin*, can in the future be a potential treatment for all persons who present hepatic alterations. However, beyond the remedy, the cooperation of the patient is required to regulate his/her ethanol consumption; as long as this does not take place, taking antioxidant vitamins can be considered within the regular therapy of a patient with alcoholism, taking care above all that this supplement does not reach toxic concentrations, in particular in the case of the vitamin that possess the tendency to accumulate in the liver.

The use of novel experimental procedures that determine the degree of damage caused by xenobiotics, and in particular by free radicals, is of great importance in the management of diseases caused by this type of substance, especially if they damage the liver, because this organ comprises a vital part of our organism on having in its charge the metabolic support of the latter.

Author details

José A. Morales-González[1], Evila Gayosso-Islas[1], Cecilia Sánchez-Moreno[1], Carmen Valadez-Vega[1], Ángel Morales-González[2], Jaime Esquivel-Soto[3], Cesar Esquivel-Chirino[3], Manuel García-Luna y González-Rubio[3] and Eduardo Madrigal-Santillán[1]

1 Instituto de Ciencias de la Salud, UAEH, México

2 Escuela Superior de Computo, IPN, México

3 Facultad de Odontología, UNAM, México

References

[1] Arakaki, N., Kawatani, S., Nakamura, O., & Ohnishi, T. (1995). Evidence for the presence of an inactive precursor of human hepatocyte growth factor in plasma and sera of patients with liver diseases. *Hepatology*. Vol. 22, pp. 1728-1734.

[2] Baptista, P. (2012). *Liver Regeneration*. Ed. Intech, Croatia, 252 pp.

[3] Barzaghi, N., Crema, F., Gatti, G., et al. (1990). Pharmacokinetic studies in IdB1016, a silybin-phosphatidylcholine complex, in healthy human subjects. *Em J Drug Metab Pharmacokinet*. Vol. 15, pp.333-338.

[4] Bergendi, LL., Benes, Z., & Durackiova y, M. F. (1999). Chemistry, physiology and pathology of free radicals. *Life Sciences*. Vol.64, pp. 1865-1874.

[5] Brunk, U., & Cadenas, E. (1988). The potential intermediate role of lysosomes in oxygen free radical pathology. *Review article*. Vol. 96, pp. 3-13.

[6] Burneo-Palacios, ZL. (2009). Determinación del contenido de compuestos fenólicos totales y actividad antioxidante de los extractos totales de doce especies vegetales nativas del sur del Ecuador (Tesis) Loja, Ecuador: Universidad Técnica Particular de Loja. Disponible en: http://es.scribd.com/doc/43393190/TESIS-ANTIOXIDANTES

[7] Camacho-Luis, A., Mendoza-Pérez, JA. (2009). La naturaleza efímera de los radicales libres. Química y bioquímica de los radicales libres. In *Los antioxidantes y las enfermedades crónico degenerativas*. Morales-González, JA., Fernández-Sánchez, AM., Bautista-Ávila, M., Vargas-Mendoza, N., Madrigal-Santillán, EO (ed). Ed. UAEH, Pachuca, Hidalgo, México, pp. 27-76.

[8] Flora, K., Hahn, M., Rosen, H., & Benner, K. (1998). Milk thistle (Silybum marianum) for the therapy of liver disease. *American Journal of Gastroenterology*. Vol 93, pp. 139-143.

[9] Foster, S. (1991). Milk thistle: Silybum marianum. Austin, TX: AmericanBotanical Council, No. 305.

[10] Fujiwara, K., Nagoshi, S., Ohno, A., Hirata, K., Ohta, Y., & Mochida, S. (1993). Stimulation of liver growth factor by exogenous human hepatocyte growth factor in normal and partially hepatectomized rats. *Hepatology*. Vol. 18, pp. 1443-1449.

[11] George, D., Liatsos, MD., et al. (2003). Effect of Acute Ethanol Exposure on Hepatic Stimulator Substance (HSS) leves During Liver Regeneration. *Digestive Diseases and Sciences*. Vol 48, pp. 1929-1938.

[12] Gerschman, R. (1954). Oxigen poisoning and X-Irradiation. A mechanism in common. *Science*. Vol. 119, pp. 623-626.

[13] Ghoda, E., Tsubouchi, H., Nakayama, H., & Hirono, S. (1988). Purification and partial characterization of hepatocyte growth factor from plasma of a patient with fulminant hepatitis failure. *J Clin Invest*. Vol. 81, pp. 414-419.

[14] Goldfrank, L., Flomenbaum, N., & Lewin, N. (2002). Goldrank´s Toxicology Emergencies. 7th. Ed. McGraw-Hill, USA, pp. 952-962.

[15] Greive M. (1981). A modern herbal, vol. 2. New York: Dover Publications.

[16] Guillén-López, S., Álvarez-Salas, E., & Ochoa-Ortiz, E. (2009). Antioxidantes en el tratamiento de las enfermedades: flavonoides. In *Los antioxidantes y las enfermedades crónico degenerativas*. Morales-González, JA., Fernández-Sánchez, AM., Bautista-Ávila, M., Vargas-Mendoza, N., Madrigal-Santillán, EO (ed). Ed. UAEH, Pachuca, Hidalgo, México, pp. 593-615.

[17] Gutiérrez Salinas, J., & Morales-González (2004). Producción de radicales libres derivados del oxígeno y el daño al hepatocito. *Revista de Medicina Interna de México. Vol.* 20, pp. 287-295.

[18] Gutierrez-Salinas, J. (2007). Daño al hígado por radicales libres derivados del oxígeno. In *Alcohol, alcoholismo y cirrosis. Un enfoque multidisciplinario.* Morales-González, JA (ed). Ed. UAEH, Pachuca, Hidalgo, México, pp. 97-109.

[19] Gutiérrez-Salinas, J., & Morales-González, JA. (2006). La ingesta de fluoruro de sodio produce estrés oxidativo en la mucosa bucal de la rata. *Revista Mexicana de Ciencias Farmacéuticas. Vol.* 37, pp. 11-22.

[20] Harnisch, G., & Stolze, H. (1983). Silybum marianum: Mariendistel. In: BewaehrteP-flanzendrogen in Wissenschaft und Medizin. Notamed Verlag, pp. 203-215.

[21] Hernández-Ceruelos, MCA., Sánchez Gutiérrez, M., Fragoso Antonio, S., Salas Guzmán, D., Morales-González, JA., Madrigal Santillán, EO. (2009). Quimioprevención de fitoquímicos. In *Los antioxidantes y las enfermedades crónico degenerativas*. Morales-González, JA., Fernández-Sánchez, AM., Bautista-Ávila, M., Vargas-Mendoza, N., Madrigal-Santillán, EO (ed). Ed. UAEH, Pachuca, Hidalgo, México, pp. 77-89.

[22] Hobbs, C. (1992). Milk thistle: The liver herb. Capitola, CA: Botanical Press.

[23] Koolman, J. Rohm. (2005). Bioquímica, Texto y Atlas. 3ra edición, panamericana. pp. 306.

[24] Michalopoulos, GK., & DeFrances, MC. (1997). Liver regeneration. *Science* Vol. 276, pp. 60-66.

[25] Morales-González, JA., Barajas-Esparza, L., Valadez-Vega, C., Madrigal-Santillán, E., Esquivel-Soto, J., Esquivel-Chirino, C., Téllez-López, AM., López-Orozco, M., & Zúñiga-Pérez, C. (2012). The Protective Effect of Antioxidants in Alcohol Liver Damage In: *Liver Regeneration*. Baptista, P. (ed). Ed. Intech, Croacia, pp. 89-120.

[26] Morales-González, JA., Bueno-Cardoso, A., Marichi-Rodríguez, F., & Gutiérrez-Salinas, J. (2004a). Programmed cell death (apoptosis): the regulating mechanisms of cellular proliferation. *Arch Neurocien*. Vol. 9, pp. 124-132.

[27] Morales-González, JA., Fernández-Sánchez, Bautista-Ávila, M., Vargas-Mendoza, N., & Madrigal-Santillán, EO. (2009). *Los antioxidantes y las enfermedades crónico degenerativas*. Ed. UAEH, Pachuca, Hidalgo, México, 751 pp.

[28] Morales-González, JA., Gutiérrez-Salinas, J., & Hernández-Muñoz, R. (1998). Pharmacokinetics of the ethanol bioavalability in the regenerating rat liver induced by partial hepatectomy. *Alcoholism Clinical and Expimental Research*. Vol. 22, pp. 1557-1563.

[29] Morales-González, JA., Gutierrez-Salinas, J., & Piña, E. (2004b). Release of Mitochondrial Rather than Cytosolic Enzymes during Liver Regeneration in Ethanol-Intoxicated Rats. *Archives of Medical Research*. Vol. 35, pp. 263-270.

[30] Morales-González, JA., Gutiérrez-Salinas, J., Arellano-Piña, G., Rojas-López, M., & Romero-Pérez, L. (1998). El metabolismo hepático del etanol y su contribución a la enfermedad hepática por etanol. *Revista de Medicina Interna de México*. Vol. 14, pp. 180-185.

[31] Morales-González, JA., Gutiérrez-Salinas, J., Yánez, L., Villagómez, C., Badillo, J. & Hernández, R. (1999). Morphological and biochemical effects of a low ethanol dose on rat liver regeneration. Role of route and timing of administration. *Digestive Diseases and Sciences*. Vol. 44, No. 10 (October), pp. 1963-1974.

[32] Morales-González, JA., Jiménez, L., Gutiérrez-Salinas, J., Sepúlveda, J., Leija, A. & Hernández, R. (2001). Effects of Etanol Administration on Hepatocellular Ultraestructure of Regenerating Liver Induced by Partial Hepatectomy. *Digestive Diseases and Sciences*. Vol. 46, No. 2 (February), pp. 360–369.

[33] Muñoz Sánchez, JL. (2009). Defensas antioxidantes endógenas. In *Los antioxidantes y las enfermedades crónico degenerativas*. Morales-González, JA., Fernández-Sánchez, AM., Bautista-Ávila, M., Vargas-Mendoza, N., Madrigal-Santillán, EO (ed). Ed. UAEH, Pachuca, Hidalgo, México, pp. 93-118.

[34] Nakamura, T., Nawa, K., Ichihara, A. (1984). Partial purification and characterization of hepatocyte growth factor from serum of hepatectomized rats. *Biochem Biophys Res Commun*. Vol. 122, pp. 1450-1459.

[35] Orr, WC., Sohal, RJ. (1994). Extension of life-span by overexpression of superoxide dismutase and catalase in drosophila melanogester. *Science*. Vol. 263, pp. 1128-1130.

[36] Parra-Vizuet, J., Camacho-Luis, A., Madrigal-Santillán, E., Bautista, M., Esquivel-Soto, J., Esquivel-Chirino, C., García-Luna, M., Mendoza-Pérez, JA., Chanona-Pérez, J., & Morales-González, JA. (2009). Hepatoprotective effects of glycine and vitamin E during the early phase of liver regeneration in the rat. *African Journal of Pharmacy and Pharmacology*. Vol.3, No. 8, pp. 384-390 (August, 2009).

[37] Piña-Garza, E., Gutiérrez-Salinas, J., Morales-González, JA., & Zentella de Piña, M. (2003). ¿Es tóxico el alcohol?" In: *Temas Bioquímicos de vanguardia*. Riveros Rosas, H.,

Flores-Herrera, O., Sosa-Peinado, A., Vázquez-Contreras, E. (ed). Ed. Facultad de Medicina UNAM, pp. 121-146.

[38] Ramírez-Farías, C., Madrigal-Santillán, E., Gutiérrez-Salinas, J., Rodríguez-Sánchez, N., Martínez-Cruz, M., Valle-Jones, I., Gramlich-Martínez, I., Hernández-Ceruelos, A., & Morales-González JA. (2009). Protective effect of some vitamins against the toxic action of ethanol on liver regeration induced by partial hepatectomy in rats. *World Journal of Gastroenterology* Vol. 14, pp. 899-907.

[39] Riehle, KJ., Dan, YY., Campbell, JS., & Fausto, N. (2011). New concepts in liver regeneration. *Journal of Gastroenterology and Hepatology*. Vol. 26, Suppl. 1, pp. 203–212.

[40] Rybczynska, M. (1994). Biochemical aspects of free radical mediated tissue injury. *Postepy Hig Med Dows*. Vol. 48, pp. 419-441.

[41] Sandoval, M., Lazarte, K., & Arnao, I. (2008). Hepatoprotección antioxidante de la cáscara y semilla de Vitis vinífera L. (uva) (2008). En: *Anales de la Facultad de Medicina* (citado el 3 de septiembre de 2011). Disponible en: http://www.scielo.org.pe/scielo.php?pid=S1025-55832008000400006&script=sci_arttext

[42] Schandalik, R., Gatti, G., & Perucca E. (1992). Pharmacokinetics of silybin in bile following administration of silipide and silymarin in cholecystectomy patients. *Arzneimittel-Forschung*. Vol. 42, pp. 964-968.

[43] Sohal, RS., Sohal, BH., & Orr, WC. (1995). Mitochondrial superoxide and hydrogen peroxide generation, protein oxidative damag, and longevity in different species of flies. *Free Radic Biol Med*. Vol. 19, pp. 499-504.

[44] Téllez, J., & Cote, M. (2006). Alcohol etílico: un tóxico de alto riesgo para la salud humana socialmente aceptado. *Revista Facultad de Medicina de la Universidad Nacional de Colombia*. Vol 54, pp. 32-47.

[45] Tzu-Chen, Y., Kwam –Liang, K., & Hish-Chen, L. (1994). Age dependent increase of mitochondrial DNA deletions together with lipid peroxide and superoxide dismutase in human liver mitocondria. *Free Radic Biol Med*. Vol. 16, pp. 207-214.

[46] Venereo Gutiérrez, JR. (2002) Daño oxidativo, radicales libres y antioxidantes. En: Revista Cubana de Medicina Militar, Febrero 2002, Disponible en: http://bvs.sld.cu/revistas/mil/vol31_2_02/MIL09202.pdf

[47] Yoshida, Y., Komatsu, M., Ozeki, A., Nango, R., & Tsukamoto, I. (1997). Ethanol represses thymidylate synthase and thymidine kinase at mRNA level in regenerating rat liver after partial hepatectomy. *Biochim Biophys Acta*. Vol. 1336, pp. 180-186.